Southeast Asian Conference on Geophysics (SEACG 2016)

IOP Conference Series: Earth and Environmental Science
Volume 62

Bali, Indonesia
31 August – 3 September 2016

Editors:

R.M. Rachmat Sule
Andri Dian Nugraha
Rexha Verdhora Ry

ISBN: 978-1-5108-3958-8
ISSN: 1755-1307

Printed from e-media with permission by:

Curran Associates, Inc.
57 Morehouse Lane
Red Hook, NY 12571

CURRAN ASSOCIATES INC.
proceedings.com

Some format issues inherent in the e-media version may also appear in this print version.

Printed by Curran Associates, Inc. (2017)

Institute of Physics
Dirac House, Temple Back
Bristol BS1 6BE UK

Phone: 44 1 17 929 7481
Fax: 44 1 17 920 0979

techtracking@iop.org

Additional copies of this publication are available from:

Curran Associates, Inc.
57 Morehouse Lane
Red Hook, NY 12571 USA
Phone: 845-758-0400
Fax: 845-758-2633
Email: curran@proceedings.com
Web: www.proceedings.com

Table of contents

Volume 62

Southeast Asian Conference on Geophysics

31 August to 3 September 2016, Bali, Indonesia

Accepted papers received: 6 April 2017
Published online: 24 April 2017

Preface

Rock- and Paleomagnetism

Near-Surface Geophysics

Microseismic for New and Renewable Energy Exploration

Developments and New Approaches

Geophysical Approaches in Geothermal Exploration

Volcanology

Environmental Geophysics

Current Developments on Potential Method

Exploration Seismology

Earthquake Seismology

IOP Conf. Series: Earth and Environmental Science **62** (2017) 011001 doi:10.1088/1755-1315/62/1/011001

Preface

R. M. Rachmat Sule, Andri Dian Nugraha, Rexha Verdhora Ry

Department of Geophysical Engineering, Faculty of Mining and Petroleum Engineering, Institut Teknologi Bandung, Indonesia

email: seacg2016@gmail.com

This issue of Journal *IOP Conference Series*: *Earth and Environmental Science* contains the outstanding papers presented at *Southeast Asian Conference on Geophysics* (SEACG) *2016*. The SEACG is the first event which was organized by Department of Geophysical Engineering, Faculty of Mining and Petroleum Engineering, Institut Teknologi Bandung. It was held on August 31st until September 3rd, 2016 in Bali – a small beautiful island and a part of Indonesia archipelago.

This conference provided a venue for collaboration, sharing and developing new ideas and technologies. The SEACG created an opportunity for academics, professionals, and students to promote and discuss the scientific results related to recent development in geophysical methods and related science. Topics to be during the conference including: (1) rock- and paleomagnetism; (2) near-surface geophysics; (3) geophysical approaches in hydrocarbon exploration; (4) geophysical approaches in geothermal exploration; (5) microseismic for new and renewable energy exploration; (6) environmental geophysics; (7) volcanology; (8) earthquake seismology; (9) exploration seismology; (10) current developments on potential method; (11) development and new approaches.

The main objective of this conference is to lead to develop new ideas and technologies in geophysical methods and related science. This year, our theme was "*Development of Geophysical Methods for Sustainable Energy and Environments*", presented by the invited speakers. The invited speakers are Prof. Jeannot Trampert (Univ. Utrecht), Prof. James M. Russell (Brown Univ.), Prof. Toshifumi Matsuoka (Kyoto University), Prof. Sri Widiyantoro (ITB). Prof. Djoko Santoso (ITB), Dr. Satria Bijaksana (ITB), and Dr. Haposan Napitupulu (Government).

The organizing committee would like to thank the sponsors of this conference: Institute of Research and Community Services-ITB, Faculty of Mining and Petroleum Engineering ITB, The SATREPS Program of Gundih CCS Pilot Project. We would like to express our sincere thanks to all authors and presenters for their valuable research. We hope to see you again during the next event SEACG 2018 which will be held in Labuan Bajo, Indonesia.

IOP Conf. Series: Earth and Environmental Science **62** (2017) 011001 doi:10.1088/1755-1315/62/1/011001

1. Photographs

IOP Conf. Series: Earth and Environmental Science **62** (2017) 011001 doi:10.1088/1755-1315/62/1/011001

2. List of Committee

Chairman	Dr. rer. nat. M. Rachmat Sule	
Steering Committees	Dr. Andri Dian Nugraha	
	Prof. Sri Widiyantoro, Ph.D.	
General Affair Manager	Billy Sugiartono Prabowo, M.T.	
Secretariat Officer	Rexha Verdhora Ry, M.T.	
Treasurer and Sponsorship	Katrine Siska, S.T.	
	Widya	
Public Relation and Information	Ir. Indra Gunawan, M.Sc.	
	Fernando Lawrens, M.T.	
General Secretary	Anisa Agustina, M.T.	
	Shindy Rosalia, M.T.	
Logistic and Field Trip	Kadek H. Palgunadi, S.T.	
	Irfan Aufa, S.T.	
Design	Fadhilaz Diqdaya, S.T.	
Official Team	Dadi Abdurahman, M.T.	
	Dr. Zulfakriza	
Team of Reviewer	Dr. Tedi Yudistira	Institut Teknologi Bandung
	Dr. Hendra Grandis	Institut Teknologi Bandung
	Prof. Dr. Satria Bijaksana	Institut Teknologi Bandung
	Prof. Sri Widiyantoro, Ph.D.	Institut Teknologi Bandung
	Prof. Dr. Ir. Djoko Santoso	Institut Teknologi Bandung
	Prof. Dr. Awali Priyono	Institut Teknologi Bandung
	Prof. Dr. Nanang T Puspito	Institut Teknologi Bandung
	Prof. Phil Cummins, Ph.D.	Australian National University
Editors	Dr. rer. nat. M. Rachmat Sule	Institut Teknologi Bandung
	Dr. Andri Dian Nugraha	Institut Teknologi Bandung
	Rexha Verdhora Ry, M.T.	Institut Teknologi Bandung

Peer review statement

All papers published in this volume of *IOP Conference Series: Earth and Environmental Science* have been peer reviewed through processes administered by the proceedings Editors. Reviews were conducted by expert referees to the professional and scientific standards expected of a proceedings journal published by IOP Publishing.

IOP Conf. Series: Earth and Environmental Science **62** (2017) 012001 doi:10.1088/1755-1315/62/1/012001

Correlation Between Magnetic Susceptibility And Gold (Au) Content In Bombana Gold Mine Area (BGMA), Southeast Sulawesi, Indonesia

Ngkoimani L O[1], Jahidin[1], Haryati[1], Razak S[1], Makkawaru A[2], Asfar S[2], Muin M R[3]
[1] Faculty of Earth Sciences and Technology, Universitas Halu Oleo
[2] Mineral and Energy Services Agency, Southeast Sulawesi Province
[3] Faculty of Mathematic and Natural Sciences, Tadulako University

Email: laodem@yahoo.com

Abstract. The measurement of magnetic susceptibility and the content of gold (Au) on sediments in the Bombana Gold Mining Area (BGMA) has been done. The samples analyzed totaled 121 samples. The magnetic susceptibility values was measured using Low Field MS2B Bartington Susceptibility meter, while the contents of gold (Au) was measured using X-Ray Nitton. Generally, the lower the value of magnetic susceptibility, contents of gold (Au) on the sediment is higher. Based on this result, the magnetic susceptibility can be used as one of the parameters and/or the alternative methods on the exploration of gold in BGMA.

1. Introduction

Environmental magnetic methods based on the study of variations in magnetic susceptibility and rockmagnetic parameters have been successfully used over several decades to characterize and quantify the degree of pollution of air, water vegetation and land systems [1]. Geological conditions of the Wumbubangka area is reflected in the morphological, structural geology, various rock types making up to stratigrafinya types. Wumbubangka area composed by rock formations that have Langkowala flat morphology. The geology of lowland Langkowala composed by Langkowala Formation which is one of the rock formations are part of clastic sediments Sulawesi molasa. Langkowala formation contained of shale sandstones and conglomerates. These formations are widespread in areas that are Langkowala mainland tertiary sedimentary basin [2]. These formations are widespread in areas that are Langkowala mainland tertiary sedimentary basin. Stratigraphy study area is composed by lithologies of the young to the old form of schist, conglomere sandstones, limestones and alluvial. Schists consists of schist mica, schist chlorite and schist amphibolite with foliasi structure. In the metamorphic lithologies there are chunks of rocks and quartz cockade structure, vuggy and fracture-filling foliasi schist rock fractures [3]. Langkowala formation composed of shale sandstones and conglomerates. These formations are widespread in areas that are Langkowala mainland tertiary sedimentary basin [2].

IOP Conf. Series: Earth and Environmental Science **62** (2017) 012001 doi:10.1088/1755-1315/62/1/012001

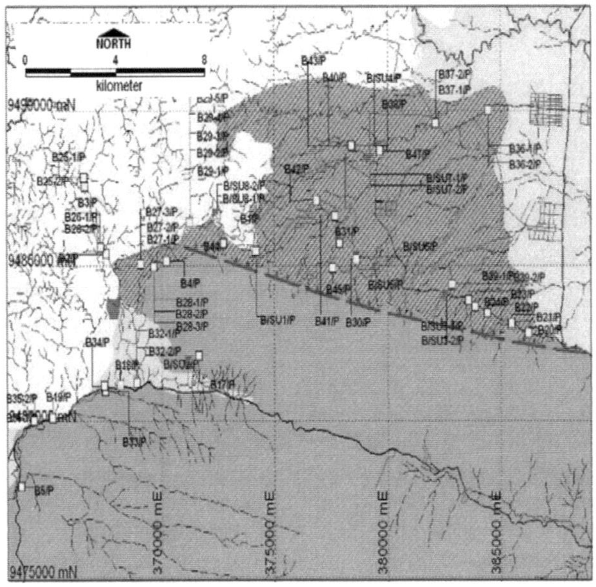

Figure 1. Regional Geology Map Bombana [4]
Note. Yellow squares are sampling point

The method most commonly used is the measurement of magnetic susceptibility. Value of magnetic susceptibility of a material can be determined either on the sample in the laboratory and in the field carried on the surface of the soil or outcrop. Pricing experimental magnetic susceptibility can be performed using a tool called Suseptibility Meter. The principle of this tool is an electromagnetic circuit that serves to detect changes in inductance when a sample is placed inside the coil [5]. Suseptibility Meter in general can work at two different frequencies, namely low frequency ordenya hundreds of hertz (465 Hz) and high-thousand hertz frequency (4650 Hz). XRF (X-Ray fluorescence) elemental analysis is a technique that forms a material with a basic X-ray interaction with the analyte material. This technique is widely used in the analysis of rock as it requires a relatively small number of samples (approximately 1 gram). This technique can be used to measure many elements contained in rocks or minerals. Mineral concerned can be ground, liquids, powders, filter material or any other form. In addition, XRF (X-Ray fluorescence) can also be used to determine the depth and composition of the coating. Detailed are the results of magnetic susceptibility measurements to correlating Between Magnetic Susceptibility And Gold (Au) Content In situ soil at Bombana Gold Mine Area (BGMA), Southeast Sulawesi, Indonesia.

2. Methods

The study, entitled "The correlation between the magnetic susceptibility and mineral content of gold in the ground at the site of gold Wumbubangka Rural District of North Rarowatu Bombana" was held in September 2015, and located in the village of Wumbubangka District of North Rarowatu Bombana Figure 1.

This research was conducted by collecting data or samples directly in the field, then measuring the magnetic susebtibilitas value and mineral content of gold at the Geophysical Laboratory. And as the supporting data or supporting data obtained from the literature. The stages of the sampling process is as follows; Soil sampling in outcrop carried around gold mining, Determining the position of the coordinates of the acquisition of land by using GPS, sampling is done vertically aims to facilitate calculate the depth sampled in outcrop, Measuring high outcrop using the meter, taking samples at any depth, samples were taken at every interval of 20 cm, Noting the point of measurement and sampling point every point and depth in the sample bag. Detailed

The stages are performed on sample preparation is as follows; The first preparatory do is dry the sample is wet or damp, Once dry samples until finely by using a tool called mortal, samples were sieved using a sieve of 100 mesh to obtain delicate parts, samples that have been delicately inserted into the container tubular (holder) with a size of 2.5 cm diameter and 2.5 cm height also, labeling holder includes code samples and depth. Example: A(0), the sample mass measurement begins by weighing the mass of an empty holder, then the next stage is to consider holder containing the sample, run the software Multisus. By choosing MS2B later measurements with measurement devices are set to high frequency (χHF) and low frequency (χLF). Result and Discussion

The sample used in this study were taken at 5 soil profile to a depth of sampling different and positioned vertically straight down. The geographical position and the maximum depth of sampling sites for all profiles can be seen in Table 1.

Table 1. Description of Sample researchs

Profile	Position	ID	depth of sampling (cm)
1	S : 04^{0}37’46.5” E : 121^{0}54’38,1”	St.1	700
2	S : 04^{0}39’37.3” E : 121^{0}53’26,7”	St.2	400
3	S : 04^{0}37’38.3” E : 121^{0}54’44,5”	St.3	420
4	S : 04^{0}37’39.2” E : 121^{0}53’32,0”	St.4	500
5	S : 04^{0}37’38.1” E : 121^{0}54’46,9”	St.5	500

Figure 2, 3, 4, 5 and 6 is a graph of the value of susceptibility to elements of metallic minerals. The third profile is the exploration of mining land is still in the processing because the amount of gold that is still in large enough quantities. Type of material contained in the area is a kind of alluvial material that is composed of sandstone, mudstone, and mica schist.

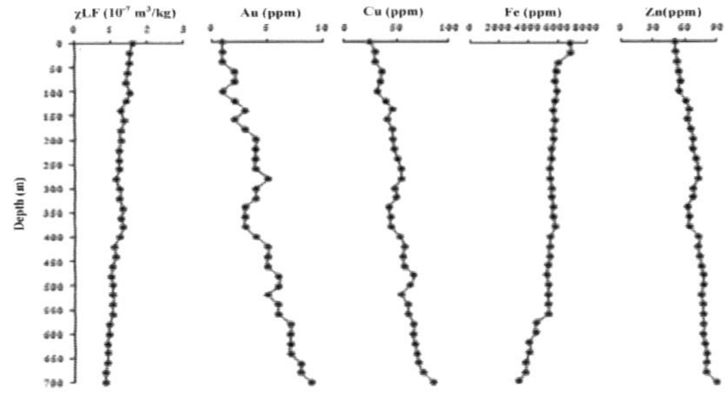

Figure 2. Relationships χLF with mineral elements Au, Cu, Fe and Zn to Profile 1,

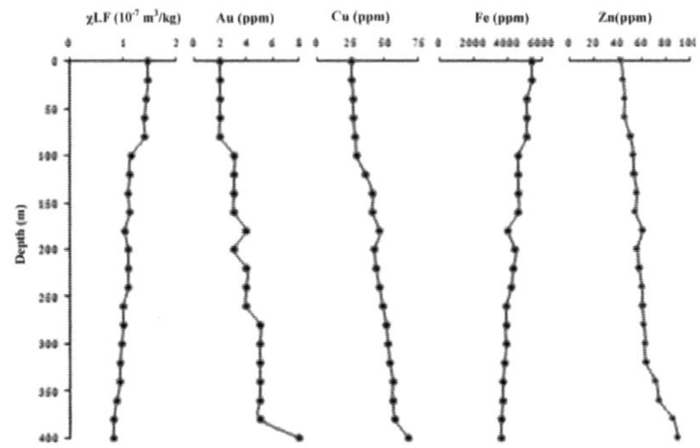

Figure 3. Relationships χLF with mineral elements Au, Cu, Fe and Zn to Profile 2

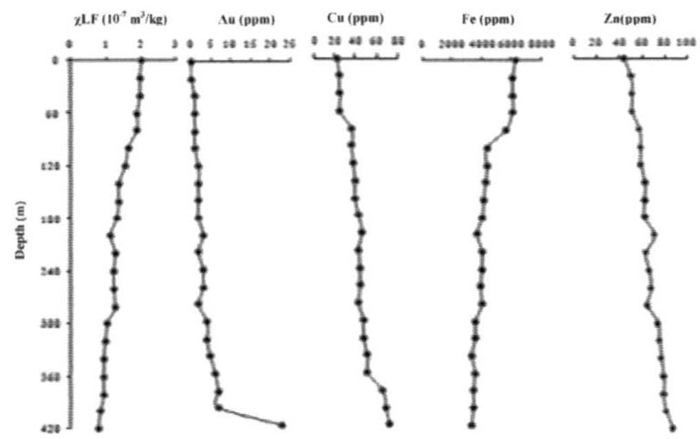

Figure 4. Relationships χLF with mineral elements Au, Cu, Fe and Zn to Profile 3

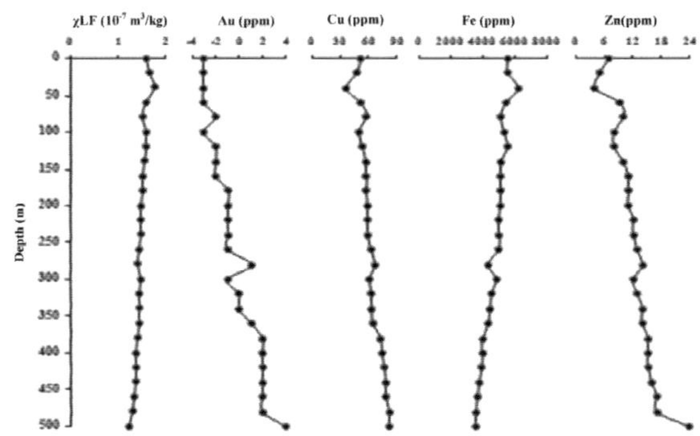

Figure 5. Relationships χLF with mineral elements Au, Cu, Fe and Zn to Profile 4

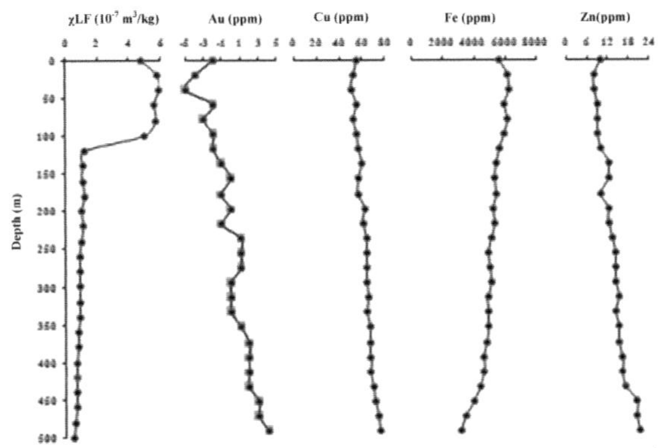

Figure 6. Relationships χLF with mineral elements Au, Cu, Fe and Zn to Profile 5

Value susceptibility maximum at profile 4 obtained at a depth of 40 cm with a value of 1.76 x $10^{-7}m^3$ / kg (susceptibility χLF) whereas the profile 5 found at a depth of 40 cm with a susceptibility value of 5.89 x 10^{-7}m3 / kg, minimum susceptibility values obtained at a depth of 500 cm with a value of 1.2 x $10^{-7}m^3$ / kg (susceptibility χLF) whereas the profile 5 found at a depth of 500 cm with a susceptibility value of 0.5 x $10^{-7}m^3$ / kg. A depth of 80 to 180 cm in 4 profiles obtained uniform susceptibility value which is about 1.5 x $10^{-7}m^3$ / kg, at a depth of 200 to 360 cm its susceptibility values also tend to be the same, namely 1.4 x $10^{-7}m^3$ / kg, and on into 380 to 460 cm susceptibility value obtained is 1.3 x $10^{-7}m^3$ / kg. While the profile 5 susceptibility values were uniformly found at depths of 120 to 220 cm with a value of approximately 1.2 x $10^{-7}m^3$ susceptibility / kg, at a depth of 240 to 340 cm with a value of approximately 0.8 x $10^{-7}m^3$ / kg, and on 360 to 440 cm depth with susceptibility value of approximately 0.7 x $10^{-7}m^3$ / kg. Similarly, in the image (1), (2) and (3) factors affecting the high, uniform, and the low value of susceptibility is influenced by the type of material, the temperature factor / temperature and grain size.

The relationship between the value of susceptibility is inversely related to the mineral content of some elements of metallic minerals such as (Au, Cu, Zn), can be seen in the above image the greater the value its susceptibility of the mineral elements Au, Cu, and Zn is getting smaller, while the mineral elements Fe if the value of susceptibility obtained the greater mineral content of Fe.

3. Conclusion

Based on the results of research and discussion, we can conclude; Type of material contained in the profile 1, 2, and 3 is a kind of alluvial material composed of sandstone, mudstone, and mica schist, while the 4 and 5 are profile-Sandstone Unit consisting of metamorphic sandstone, sandstone and claystone.

The relationship between the value of the susceptibility profile 1,2,3,4, and 5 is inversely proportional to the mineral content in the metal element (Au, Cu, and Zn) and proportional to the mineral content of Fe. Value susceptibility in five (5) that profile is influenced by several factors, including the type of material, temperature and grain size.

4. Acknowledgements

This research was supported by UNIVERSITAS HALU OLEO (UHO). We thank to The Indonesian Natural Magnetic Society for constructive comments of the manuscript.

References

[1] Petrovsky E, Ellwood B B 1999 Magnetic monitoring of pollution of air, land and waters. In: Maher, B.A., Thompson, R. (Eds.), *Quaternary Climates, Environments and Magnetism, Cambridge University Press, Cambridge*, pp. 279-322.

[2] Surono 2013 Geologi Sulawesi *LIPI Press*, Jakarta.

[3] Simandjuntak T O 1993 Neogene tectonics and orogenesis of Indonesia**,** *Bulletin of the Geological Society of Malaysia* **33** 43-64.

[4] Kisman 2011 Gold mineralization occurrences associated with cinnabar In District Rarowatu Bombana Southeast Sulawesi Province *Geological Resources Bulletin* **6** 23

[5] Dearing J A 1999 Environmental Magnetic Susceptibility *British Library Cataloguing in Publication Da* p.54.

[6] Bombana Map 2016 (https://www.google.co.jp/search?q)

Magnetic Susceptibility Analysis of Soil Affected by Hydrocarbon in Wonocolo Traditional Oil Field, Indonesia

Melianna Ulfah, Bambang Wijatmoko and Dini Fitriani

Geophysics Study Program, Faculty of Mathematics and Natural Sciences, Universitas Padjadjaran, Jl. Raya Bandung - Sumedang KM. 21, Jatinangor 45363, Indonesia

E-mail: melomeliana@gmail.com

Abstract. Magnetic susceptibility of soil affected by hydrocarbon was studied through cored soil samples in two zones (Zone One and Zone Two) of an oil field in Wonocolo Village, East Java. We also collected soil samples as the background from a residential area near the oil field (Zone Three). The Zone One, consisted two cores near producing well; the Zone Two consisted two cores obtained from near a dry hole well and a discontinued well; and the Zone Three consisted two cores to validate the initial soil magnetic susceptibility value in this area. The hydrocarbon content measurement was also done for the upper part of each cores using distillation method to identify the correlation between magnetic susceptibility and hydrocarbon content. From magnetic susceptibility measurement in dual frequency, samples from the Zone One and Zone Two have magnetic susceptibility range from 6,1 x 10^{-8} m^3kg^{-1} - 160 x 10^{-8} m^3kg^{-1} and 15,7 x 10^{-8} m^3kg^{-1} - 417,9 x 10^{-8} m^3kg^{-1}, respectively. Whereas background samples from Zone Three have magnetic susceptibility range from 4,8 × 10^{-8} m^3kg^{-1} to 81,1 × 10^{-8} m^3kg^{-1}. We found low χ_{fd} (%) in samples with high magnetic susceptibility values, shown that there was no indication of superparamagnetic minerals in the samples. The hydrocarbon content measurement shows the value range of 8% - 14% only exists in the upper part of all cores in Zone One and one core in Zone Two. From this analysis, we assume that other than the volume of the hydrocarbon content in soil, the period of petroleum hydrocarbon deposition in soil and the fossil fuel combustion generated in the study site could differently increase the soil magnetic susceptibility value in this area. Positive correlation between the two parameters hopefully could contribute to develop environmental magnetic methods for detecting oil spills in soil, especially to remediate former hydrocarbon exploration and production area.

1. Introduction

Magnetic methods have been used as one of hydrocarbon exploration surveys as several studies show the association of magnetic anomaly with the presence of hydrocarbon in subsurface [1]. In further research, many researchers use magnetic analysis on a smaller scale, such as on samples of sediment or even soil in certain size, known as rock magnetic method, to identify the magnetic characteristics of each sample. The study of rock magnetism for sediment affected by hydrocarbon confirms that there is a positive correlation between magnetic susceptibility anomaly with the presence of hydrocarbon from samples of sediments from core data in Barinas-Apure Basin, one of hydrocarbon exploitation area in Venezuela [2]. The magnetic susceptibility profile showed the positive correlation, that there were high magnetic susceptibility values measured in samples from the depth corresponding to the presence of reservoir from an oil-producing well.

Rock magnetic methods also contribute to detect and monitor contamination in soil or sediment from various contaminants, known as environmental magnetism [3]. The study of environmental magnetism in unconsolidated sediment affected by hydrocarbon was done in a former military site contaminated by petroleum hydrocarbon due to leak in oil storage tanks [4]. The results of this study show that there is also a positive correlation between high magnetic susceptibility value and the presence of hydrocarbon. Magnetic analysis of the samples indicates that hydrocarbon accumulation triggers geomicrobiological activity that forms magnetite in sediment with hydrocarbon fluctuation [4].

The purpose of this research is to determine the magnetic characteristic of soil in hydrocarbon exploitation area with petroleum hydrocarbon content due to oil exploitation and production activity. The magnetic characterization mainly focuses on magnetic susceptibility analysis. The result of the analysis is used to determine the type and grain size domination of magnetic minerals in samples and then used to analyze the cause of the presence of magnetic susceptibility anomaly value that measured on samples. For surface samples, the magnetic susceptibility values were then correlated with petroleum hydrocarbon concentration. The correlation found in this study would determine whether the use of magnetic parameters to detect hydrocarbon content in unconsolidated sediment is applicable for an early stage of soil remediation.

2. Research Methodology

2.1. Study Area

The study area is a traditional petroleum hydrocarbon exploitation area in Wonocolo Oil Field, Bojonegoro, Indonesia. The geological map shown in Figure 1 was mapped from field observation and drawn using ArcGIS with basemap from ESRI Topographical Map. Petroleum hydrocarbon contaminated the site due to the exploration and exploitation activity since late 1800 where the locals exploit petroleum hydrocarbon mechanically with traditional equipments to lift up fossil fuels from the subsurface. The area consists of an upper layer of Quaternary sediments namely Wonocolo Formation, while the reservoir is composed of Miocene and Pliocene sandstones deposited in Ledok Formation with 18% maximum porosity [5].

We obtained cored soil samples from three zones, namely Zone One, Zone Two and Zone Three. Zone One and Zone Two are located at the oil exploitation activity area. We also collected cored soil samples from the residential area or Zone Three near the oil field to be considered as background samples. The location of the residential area is about one kilometer from the oil field, and there is apparently no oil spills in the soil of this area.

2.2. Field Sampling

Soil samples were collected by manual core using hand drilling with one-inch sample diameter. In Zone One we obtained two cores consists of Core W1A and W1B. Core W1A was taken near an active oil well and oil distillation point. The location of core W1B was surrounded by oil wells and far from oil distillation point. Zone Two consists of core W2A and W2B, where the location of Core W2A was near a dry hole well and distillation point and Core W2B was taken near an inactive oil well that has been discontinued since June 2015. Zone Three consists of core W3A and W3B. Core W3A and W3B is apparently not influenced by petroleum hydrocarbon and unaffected by contamination since there is no oil exploitation activity nearby. The location of each core is also shown in Figure 1. The depths of each core are in the range of 45 to 65 cm. All cores were divided into five centimeters interval for magnetic susceptibility measurement. The samples are then put into sample holders to prevent any change in magnetic properties. Totally we have 63 single samples. For samples from 0-5 cm depth, we also measured hydrocarbon content to identify the correlation between magnetic susceptibility value and the oil content of soil on the upper part of the cores.

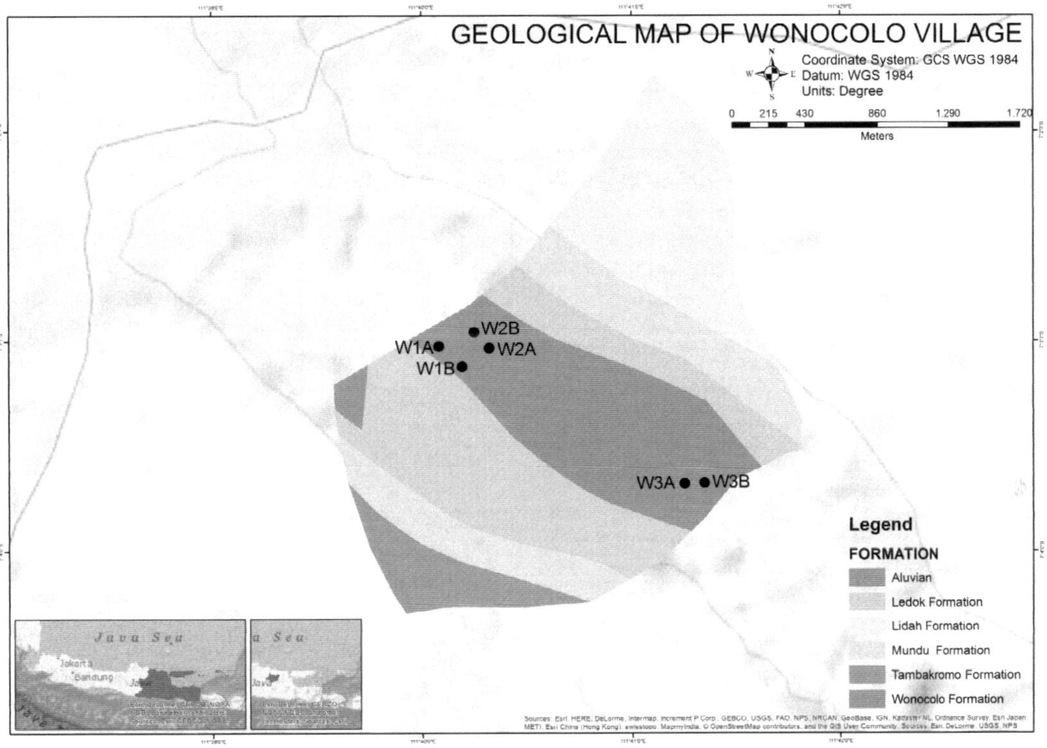

Figure 1. Geological Map of Wonocolo Village with sampling points.

2.3. Sample Characterization Methods

2.3.1. Magnetic Susceptibility. Magnetic susceptibility were measured using Bartington Magnetic Susceptibility meter with sensor B (MS2B) that operates on two frequencies, 470 Hz and 4700 Hz. This instrument measured the bulk magnetic susceptibility and the value was normalized with the mass to obtain mass-specific magnetic susceptibility (χ). Magnetic susceptibility measured at 470 Hz is known as χ_{lf}. The relative differences of the magnetic susceptibility value in two frequencies were calculated to obtain χ_{fd} (%) to identify the existence of superparamagnetic minerals. This value, along with χ_{lf} was then used to analyze the domination of grain size and domain of the magnetic minerals in samples.

2.3.2. Hydrocarbon Content. Homogenized samples were put into a ten mL steel container for hydrocarbon content measurement. Retort Kit measured the oil and water content with distillation principle.

3. Results and Discussions

3.1. Magnetic Susceptibility

From magnetic susceptibility measurement, the samples have χ_{lf} in the range of 4,8 × 10^{-8} m^3kg^{-1} - 417,9 × 10^{-8} m^3kg^{-1}. There are 45 samples that have χ_{lf} greater than 10 × 10^{-8} m^3kg^{-1}. Based on [6], samples with χ_{lf} values greater than 10 × 10^{-8} m^3kg^{-1} were dominated by ferrimagnetic minerals. The χ_{fd} (%) values are in the range of 0-11,1 %.

Samples in Zone One have χ_{lf} between 8,4 × 10^{-8} m^3kg^{-1} and 160 × 10^{-8} m^3kg^{-1} with χ_{fd} (%) ranges of 0% to 11,9%. Highest magnetic susceptibility value in this zone was found in upper part of core

W1A. The fossil fuel combustion is thought to cause the high magnetic susceptibility value in this upper part of the core W1A. This process is defined as the contact between hydrocarbon and oxygen that could form multi-domain or pseudo single domain magnetic minerals [6]. The χ_{lf} values in W1A decreases over depth. Zone Two has magnetic susceptibility range of 15,7 × 10^{-8} m^3kg^{-1} to 417,9 × 10^{-8} m^3kg^{-1} and χ_{fd} (%) of 0% to 5,05%. Zone Three samples have range of magnetic susceptibility values from 4,8 × 10^{-8} m^3kg^{-1} to 81,1 × 10^{-8} m^3kg^{-1} with χ_{fd} (%) range from 1,59% to 10,13%. The magnetic susceptibility value differences of Zone Three and two other zones are thought to indicate that there are magnetic mineral differences between samples from those zones. Among all samples, the highest magnetic susceptibility value was found in core W2B at the depth of 25 to 30 cm.

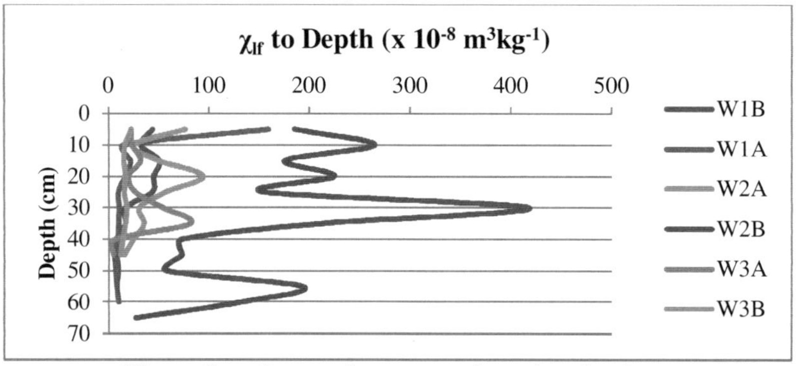

Figure 2. χ_{lf} from each core sample against depth

We also measured the magnetic susceptibility of the petroleum hydrocarbon acquired from the study site. The magnetic susceptibility value of the petroleum hydrocarbon is -0,4 × 10^{-8} m^3kg^{-1}, indicates that it is diamagnetic. According to [7], the accumulation of the contaminant in soil triggers magnetic formation in soil, and the period of contaminant deposition could also affect the value of soil magnetic susceptibility. We assume that the difference of soil magnetic susceptibility values in W1A, W1B and W2B is because each site has different duration length of petroleum hydrocarbon deposition in soil.

The considerably high values of χ_{fd} (%) on samples were dominantly found in samples with low magnetic susceptibility value. According to [6] that condition can be described as weak samples because measurements of one sample in two frequencies show different value due to unstable state of magnetism in the sample. This finding indicates that all samples contained no magnetic minerals with the domination of superparamagnetic domain. It concludes that samples with high magnetic susceptibility values have the domination of either multi-domain or single domain magnetic grain size.

3.2. Hydrocarbon Content

The hydrocarbon content of six surface samples shows a wide variation. Core W1A contains the highest amount of hydrocarbon, reaching 14% of ten mL measured samples. Core W1B and W2B have hydrocarbon concentrations of 12% and 8%, respectively. Meanwhile core W2A has the hydrocarbon content of 0%, as well as core W3A and W3B.

There is a correlation between magnetic susceptibility and hydrocarbon content that concluded from values measured from surface core samples. Figure 3 shows the trend of χ_{lf} values and hydrocarbon concentration. Surface samples with hydrocarbon content such as from core W1A, W1B and W2B have higher χ_{lf} values than the χ_{lf} values of the background soils. Core W2A has high χ_{lf} ; however the hydrocarbon content is 0%. We assume that this anomaly was caused by fossil fuel combustion since the core was obtained near the oil distillation point.

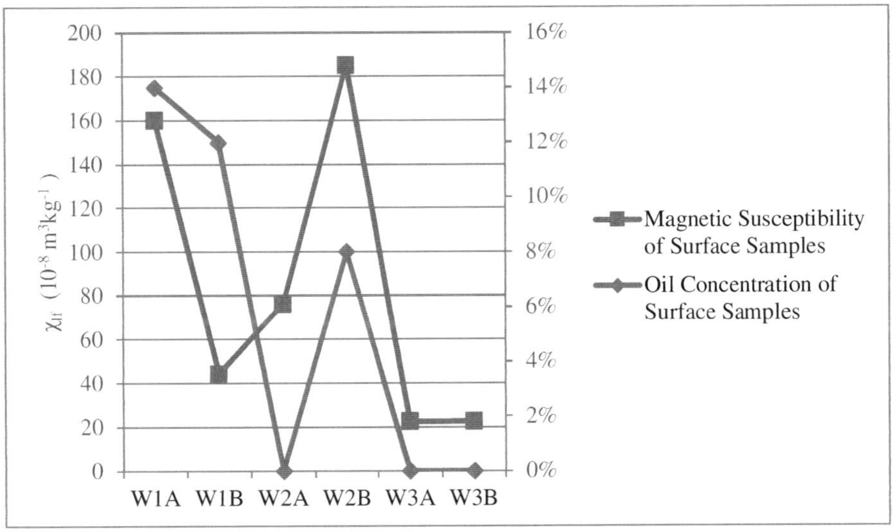

Figure 3. Measured χ_{lf} and oil concentration from surface samples

4. Conclusions

From the result of magnetic susceptibility measurement with dual frequency, the χ_{lf} values of 45 samples out of 63 samples are categorized into ferrimagnetic because they have χ_{lf} values higher than 10×10^{-8} m^{3}kg^{-1}. The samples acquired from Zone One and Zone Two, mainly samples from the upper parts of each core have higher magnetic susceptibility values than samples acquired from Zone Three that apparently show the initial soil magnetic susceptibility value of the study area. We found low χ_{fd} (%) in samples with high magnetic susceptibility values, shown that there was no indication of superparamagnetic minerals in the samples. We then conclude that samples with high magnetic susceptibility values have the domination of either multi-domain or single domain magnetic grain size. The hydrocarbon content measurement shows the value range of 8% - 14% only exists in the upper part of all cores in Zone One and one core in Zone Two. From this analysis, we assume that other than the volume of the hydrocarbon content in soil, the period of petroleum hydrocarbon deposition in soil and the fossil fuel combustion generated in the study site could differently increase the soil magnetic susceptibility value in this area. This correlation could be used to determine which treatment could be best to be applied for soil remediation in some former exploitation well site in this oil field.

References

[1] Piskarev A L and Tchernyshev M Y 1997 Magnetic and Gravity Anomaly Patterns related to Hydrocarbon Fields in Northern West Siberia *Geophysics* **62** 831-841

[2] D'Onofrio L, Perez-Perez A, Bosch M and Zapata E 2011 Association between Magnetic Susceptibilities and Hydrocarbon Deposits in the Barinas-Apure Basin, Venezuela *Geophysics* **76** 35-41

[3] Fitriani D, Huliselan E K and Bijaksana S 2010 Magnetic Properties as a Proxy Indicator of Environmental Pollution *Jurnal JTM* **16** 231-236

[4] Rijal M L, Appel E, Petrovsky E and Blaha U 2010 Change of Magnetic Properties due to Fluctuations of Hydrocarbon Contaminated Groundwater in Unconsolidated Sediments *Environmental Pollution* **158** 1756-1762

[5] Riestyastuti W 2012 Evaluasi Pompa Electric Submersibel (ESP) Sumur KWG WK di Lapangan Kawengan Area Cepu PT. Pertamina EP Region Jawa *Jurnal Ilmiah MTG* **5**

[6] Dearing J 1999 *Environmental Magnetic Susceptibility Using the Bartington MS2 System*

[7] Kirana K, Aufa N, Huliselan E and Bijaksana S 2011 Magnetic and Electrical Properties of Leachate *ITB Journal of Sciences* **43** 165-178

Neogene Tectonics And Paleomagnetism Of The Western And Eastern Of Palu Bay, Central Sulawesi, Indonesia

Muin M R, [1,4*],Pramumijoyo S[2], Warmada I W[2], Suryanto W[3]

[1]PhD Candidate, Geological Engineering, Gadjah Mada University
[2]Geological Engineering, Gadjah Mada University
[3]Geophysics, Gadjah Mada University
[4]Geophysics, Tadulako University

*Email : rusli8998@yahoo.com

Abstract. Geologic observations and paleomagnetic studies of the western and eastern Palu Bay was undertaken, in order to study the extent of rotated areas and to develop constraints on tectonic models concerning the formation of this bay. Characteristic of remanent magentization (ChRM) results from Palu bay, the western show that; yield a pole at -39.116^{0}N, 119.895^{0}E with R=13.99, α_{95} =0,42, D_m/I_m=61.76/-26.7 and the Eastern Palu bay show that; yield a pole at -81.855^{0}N, 52.913°E, R=8.99, α_{95}=0.29, D_m/I_m =67.1/-5.42. The granitic rocks in these areas are dominated by granite and Monzonit-Quartz in composition, with high-K calc-alkaline photasik (KAP) and show metaluminous-peraluminous affinity. This suggests that the rotational motion on both sides of the bay are similar during the Neogen.

1. Introduction

Sulawesi Island, central part of the Indonesian Archipelago, is located at the junction between the converging Pacific-Philippine, Indo-Australian Plates and the Sundaland, i.e. the south-eastern part of the Eurasian Plate. Sulawesi tectonic evolution results from successive collisions of continental slivers, island arcs, and oceanic domains with the Sundaland [1]. The Palu Koro Fault System (PKFS), one of the major structures in SE Asia, cuts across Sulawesi Island from NW to SE,connecting the North Sulawesi subduction zone to the Banda Sea deformation zones [1].

The multi-armed shape of the island, with different lithologic sequences suggests that it is a complex assemblage of tectonic terranes, leading to the wide range of interpretation for the evolution of the island [2]. One of the most problematic features in Sulawesi island is the occurrence of various contrasting igneous rocks in space and time. For example, whereas HK (high-K or shoshonitic series) and CAK (high-K calc-alkaline series) igneous rocks are only restricted in the younger (less than 14 Ma) rocks from the southern part of the Western Sulawesi Province, the CA (calc-alkaline) to low-K series igneous rocks are widely distributed in the central and northwestern part of the province which yielded a relatively older magmatic series (more than 16 Ma). Despite intensive scientific research, there is an ongoing controversy about the source of these magmatisms and the processes responsible for their petrogenesis as well as the geodynamic setting [2]. One of main components of the magmatic products in this island is the Neogen granitoids which extend from the Western to the Northern part of the island. Granitic rocks are the most abundant plutonic rocks in the upper crust and plate tectonic processes including crustal evolution. Many granitic rocks are associated with mineralization [3]. Another important consideration in studying granitic rocks is Paleomagnetism.

The paleomagnetism of the Sulawesi Islands are important for the study of tectonic evolution. Paleomagnetism is the study of the record of the Earth's magnetic field in rocks. The Earth's magnetic field has its origin in the convection of the iron rich material of the outer core of the Earth. At any point of the earth surface, the magnetic field (F) can be defined by two angles [4]. Declination (Dec) is the angle between the horizontal component of F and the geographic north, ranging from 0 to 360 ° and defined as positive clockwise; inclination (Inc) is the vertical angle or the angle between F and the

In this paper, the paleomagnetic data were analysed Characteristik Of Remanent Magnetization (ChRM) of the Granitic rocks to assess the relationships between Granitism and tectonic evolution in the Palu Koro Fault Zone (PKFZ). The work focused to measurements rock magnetism on hand and core sampling in the granitic rocks.

2. Regional Geology and Tectonic Setting Tectonic evolution Sulawesi Island

Sulawesi can be divided into four tectonic provinces namely (1) the Western and North Sulawesi Pluto-Volcanic Arc, (2) the Central Sulawesi Metamorphic Belt, (3) the East Sulawesi Ophiolite Belt and (4) the Banggai-Sula and Tukang Besi continental fragments [2]. The tectonic history of Sulawesi began in the Cretaceous when several Gondwana fragments accreted to the eastern Sundaland margin, now forming the present day basement of a large part of Western Sulawesi. Western Sulawesi subsequently separated from east Borneo by rifting that began around the Middle Eocene and led to the formation of the Makassar Strait. Calc-alkaline volcanism took place along the length of Western Sulawesi, but appears to have been localized in both space and time. It ceased by the end of the Eocene in SW Sulawesi but continued into the Oligocene further north [5].

In Neogen, convergence between the Australian plate and the Eurasian plate was absorbed in several ways in the Sulawesi region: it involved counter-clockwise rotation of Western Sulawesi (together with Borneo and Java), and contraction, uplift and erosion in East and SE Sulawesi [6]. Hall has advanced an alternative model that envisages the vertical movements to be the result of lithospheric flow in an overall extensional environment [6].

3. Location and Analytical Methods

Location Sampling The Neogen Granitic rock [7] for paleomagnetic analyses were performed mainly along two area across the Palu bay, one in the western part of the bay and the other in the eastern part of the bay. 40 oriented Granitic block samples were collected from 8 sites; Kabonga (0.76 S, 119.77 E), Watusampu (0,82 S, 119.78 E), Salena (0.79 S, 119.80 E), Matantimali (0.90 S, 119.79 E), Layana (0.82 S, 119.94 E), Taipa (0.77 S, 119.94 E) and Rano (0.75 S, 119.92 E). of the exposed rocks of Palu Koro Fault Zone (PKFZ) (Figure 1). All the samples were drilled and cut into standard specimens (2.2 cm length and 2.5 cm diameter) [9] . As the earth's magnetic field is a vector, samples need to be oriented in situ to get a geographic unambiguous position and orientation in order to provide a reference system to which refer the data obtained during the laboratory analyses. Therefore, samples are oriented in the field by a magnetic compass coupled to an orienting device with hand waterpass [8,9,10].

The main laboratory analysis consists in progressive demagnetization of the samples and the subsequent measurement of the NRM in order to identify and isolate the different paleomagnetic components recorded during the rock-geological history. Progressive demagnetization of the samples can be achieved by alternating field demagnetization (AFD) using a Molspin AF demagnetizer. The Minispin is a low cost, high sensitivity portable fluxgate spinner magnetometer, manufactured by MOLSPIN, Ltd. [8,11]. Vector directions are described in terms of declination (D) and inclination (I). After the complete demagnetization of the samples, these directions are displayed with a projection that depicts threedimension information on a two-dimension environment [9,11]. Although projections on a sphere are commonly used, the most common representation of results of progressive demagnetization are vector end point or orthogonal projection diagrams, also known as Zijderveld diagrams [12]. Magnetic components are then extracted from the Zijderveld diagrams), and the most stable and

IOP Conf. Series: Earth and Environmental Science **62** (2017) 012003 doi:10.1088/1755-1315/62/1/012003

consistent component that can be isolated is referred to as the characteristic remanent magnetisation (ChRM) [3,12].

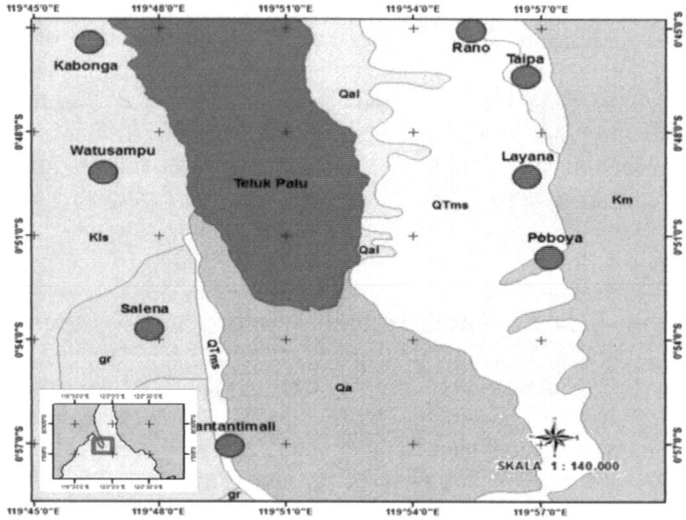

Figure 1. Map Showing Sampling Sites (Red Solid), With the name of the sample as follows; Kabonga (KAB), Watusampu (WTS), Salena (SLN), Matantimali (MTL), Poboya (PBY), Layana (LYN, Taipa (TPA), Rano (RNO). Note. Qa, Alluvium; Km, Amphibiolit; Qtms, Limestone; Qal, Clay; Kls, Sandstone, Conglomerate, gr, Granitic rocks

4. Results and Discussion

Paleomagnetik studies using samples of granite has been done by researchers [13,14,15]. Some rock specimens used in this study provide good results in stability testing magnetism. Natural Remanent Magnetization intensity rock sample from Palu granitic showed between 0,348 (*m*A /m). With the average value was 40.083 (*m*A/m). Result of the stability properties of intensity and direction of the magnetic field measurement Alternating Field/AF Demagnetization can be seen in Figure 2. The result CharacteristicRemanent Magnetization (ChRM) as presented in Table 1.

Based on test stability with decay curve, Zijderveld plots and equal-angle stereoplots have been done obtained the movement magnitude of paleomagnetic on the site research, they are Matantimali by 54,01^{0}S, for Rano indicate -21,43^0 S, for Salena sites indicate a movement is 19,17^0 S, for the Taipa -31,01^{0}S to 61, 76^{0}S, for the Watusampu sites indicate a movement is 42,8^0 S, for Layana -44,16^{0}S, for Poboya -45,17^0 LS, and Kabonga sites shows a movement is 67,06^{0}S.

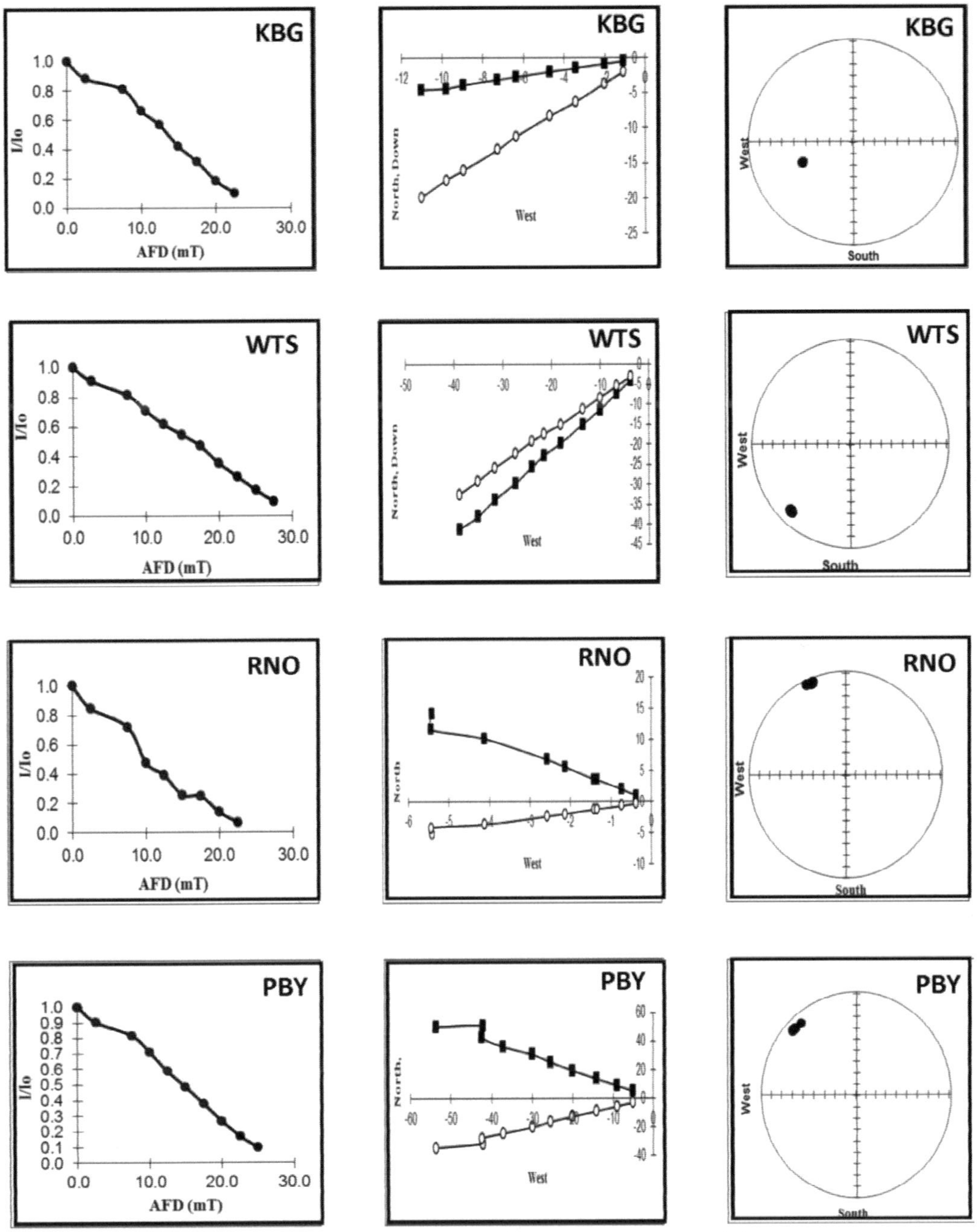

(a) (b) (c)

Figure 2. (a) Decay curve, (b) Zijderveld plots and (c) equal-angle stereoplots of Alternating Field demagnetisation behavior of selected samples

Table 1. Chractaeristic Remanent Magnetic Results (ChRM) for the Palu granites

Specimen	R	N	K	α_{95}	$D_c(^0)$	$I_c(^0)$	$\lambda_p(^o)$	$\phi_p(^o)$
KBG	8.999	9	31487.58	0.29	247.1	58.8	-37.26	111.27
WTS	10.978	11	15708.41	0.36	222.8	29.1	14.24	119.66
SLN	7.998	8	40455.4	0.27	19.2	35.5	33.37	180.69
MTL	5.890	6	45.64918	10	234.1	9,4	-29.69	122.90
PBY	9.490	10	1766.173	1.15	314.8	25.1	52.55	106.54
LYN	9.044	10	9410.033	0.49	135.8	41.7	-51.55	120.18
TPA	13.867	14	9767.061	0.40	61.8	-26.8	-32.09	119.96
RNO	8.997	9	3656.871	0.85	338.6	18.2	-4.29	264.36

Note. R, resultante; N, total number of samples from the site; K (kappa), concentration parameter; α_{95}, radius of the confidence circle; D_c, mean declination; I_c, mean inclination; λ_p, Pole Latitude; ϕ_p, Pole Longitude.

5. Conclusion

Based on remanent magnetic analyses for specimen samples from 8 sites within the Palu Granites, located in Palu Koro Fault Zone (PKFZ), we conclude the following:

- The samples acquired a stable magnetization in the presence of an apparently normal and reversed geomagnetic field. This field direction has been accurately recorded within the granitic rocks.
- Determination of tectonic processes seen on the site movement. This process can be determined based on the parameters paleomagnetic that is based on the value of ChRM.
- The paleomagnetic data are complex and some may have rotated, but motion direction paleolatitude site research is predominantly directed research area from southwest to northeast.

Acknowledgements

The authors would like to thanks to all my student in Tadulako University for grate help on the sampling process. Thanks also to SEACG 2016 committee for immeasurable help of publishing this paper.

References

[1] Bellier O, Sebrier M, Seward D, Beaudouin T, Villeneuve M, Putranto E 2006 *Tectonophysics* **413** 201–220

[2] Maulana A, Watanabe K, Imai A and Yonezu, K 2011 Geochemical variation of granitic rocks in Sulawesi Island, Indonesia *Proceeding of International Symposium on Earth Science and Technology 2011* 379–387.

[3] Beamud E, Garcés M, Cabrera L, Muñoz, J A, and Almar Y 2003 *Earth and Planetary Science Letters* **216**(4) 501-514.

[4] Dubey A K 2014 Understanding an Orogenic Belt *Springer Geology, Springer International Publishing Switzerland* 7-34

[5] Maulana A, Watanabe K, Imai A, and Yonezua K 2013 Origin of magnetite- and ilmenite-series granitic rocks in Sulawesi, Indonesia: magma genesis and regional metallogenic constraint *Procedia Earth and Planetary Science* 50-57

[6] Hall R 2011 Australia – SE Asia collision: plate tectonic and crustal flow. In: The SE Asian gateway: history and tectonic of Australian – Asia collision (eds) Hall, R., Cottam, M.A. & Wilson, M.E.J. *Geological Society (London) Special Publication* 75-105

[7] Priadi B 1996 Kompleks granitoid Neogen di Sulawesi Tengah: Tinjauan geokimia *Tesis ITB*
[8] Bijaksana S 2002 Analisa Mineral Magnetik dalam Masalah Lingkungan *Jurnal Geofisika* **1** 19 – 27.
[9] Tauxe L 1998 Paleomagnetic Principles and Practice *Kluwer Academic Publishers*
[10] Butler R F 1992 Paleomagnetism *Blackwell, Oxford* **319**
[11] Molspin Ltd. Minispin Operators Manual. Bartington Instrument. *Newcastle, England*
[12] Zijderveld J D A 1967 AC demagnetization of rocks: Analysis of results, In: Methods in Paleomagnetism, ed. D W Collinson, K M Creer and S K Runcorn, *Elsevier, Amsterdam,* 254-286
[13] Lung S C 1991 Paleomagnetism of Late Mesozoic Granitic Intrusions in Hong Kong Implications for Upper Cretaceous Reference Pole of South China *Journal Of Geophysical Research* **96** B1 327-335
[14] Kenichi W, Hideo T, Nobutatsu M, Yuhji Y, and Yutaka T 2006 Paleomagnetism of the middle Cretaceous Iritono granite in the Abukuma region, northeast Japan *Tectonophysics* **421** 161 171
[15] Shatsillo A V, Fedyukin I V, and Powerman I 2014 Paleomagnetism of the Late Paleozoic granites of the Angara–Vitim batholith and the host rocks of the Baikal–Patom folded area: tectonic implications *Russian Geology and Geophysics* **55** 864–880

IOP Conf. Series: Earth and Environmental Science **62** (2017) 012004 doi:10.1088/1755-1315/62/1/012004

Interpretation of Self-Potential anomalies for investigating fault using the Levenberg-Marquardt method: a study case in Pinggirsari, West Java, Indonesia

Fajriani, Wahyu Srigutomo, and Prihandhanu M Pratomo

Earth Physics and Complex System Research Group, Faculty of Mathematics and Natural Sciences, Institut Teknologi Bandung, Ganesha 10 St., Bandung 40132, Indonesia

Email: fajriani.nst@gmail.com

Abstract. Self-Potential (SP) method is frequently used to identify subsurface structures based on electrical properties. For fixed geometry problems, SP method is related to simple geometrical shapes of causative bodies such as a sphere, cylinder, and sheet. This approach is implemented to determine the value of parameters such as shape, depth, polarization angle, and electric dipole moment. In this study, the technique was applied for investigation of fault, where the fault is considered as resembling the shape of a sheet representing dike or fault. The investigated fault is located at Pinggirsari village, Bandung regency, West Java, Indonesia. The observed SP anomalies that were measured allegedly above the fault were inverted to estimate all the fault parameters through inverse modeling scheme using the Levenberg-Marquardt method. The inversion scheme was first tested on a synthetic model, where a close agreement between the test parameters and the calculated parameters was achieved. Finally, the schema was carried out to invert the real observed SP anomalies. The results show that the presence of the fault was detected beneath the surface having electric dipole moment K = 41.5 mV, half-fault dimension a = 34 m, depth of the sheet's center h = 14.6 m, the location of the fault's center x_o = 478.25 m, and the polarization angle to the horizontal plane θ = 334.52° in a clockwise direction.

1. Introduction

The self-potential method is a geophysical method which is used for measuring naturally occurring potential differences expressed in mV between two points at the surface. The measured data are potential anomalies caused by natural subsurface electrical phenomena. SP anomalies can be generated through several mechanisms such as diffusion and membrane potentials, bioelectric potential, streaming potential, and mineral potential. SP Method has been widely applied in the exploration of sulfides, hydrogeology, volcanic activity and geothermal systems, hydrocarbon exploration, archeology, and environmental geophysics.

Quantitative interpretation of SP anomaly can be classified into two categories: continuous modeling and fixed-body modeling. The first category uses continuous subsurface model including 2-D and 3-D. This category has electric current density, resistivity, and estimation of structure depth as its parameters. The second one is developed using fixed simple geometrical model. This type deals with determining depth and shape of buried structure whose shape is assumed to be fixed (e.g., sheet-type structure, spheres, horizontal cylinders, and vertical cylinders). The advantages of this category are that it does not require current density, resistivity and depth approximation but only SP data, and the interpretation are fast and accurate.

Several methods have been developed to estimate the model parameters (shape, depth, angle of polarization, and the electric dipole moment) of buried structure from based on SP anomaly profiles. Those include, for example, using the characteristic points, distances, curves and nomograms method [1,2], least-squares method [3,4], derivatives analysis method [5], spectral analysis approach using Fourier transform and wave number domain [6], and non-linear inversion method [7,8].

In this study, the author proposes an implementation of non-linear inversion technique using Levenberg-Marquardt (LM) method, applied in fixed geometry model of the sheet structure. In the field, sheet structure can be considered as the subsurface condition with fault structures. Because the fault can frequently be associated with the presence of mineral or fluid's flow path, an inclined conductive plane will be formed between the layers.

2. Theoretical Formulation

2.1. Forward modeling

SP anomaly (V) due to an inclined sheet structure (Figure 1) at a point P(x) along a profile perpendicular to the structure's strike is given by the following formula:

$$V(r_1, r_2, k) = K \log \frac{r_1^2}{r_2^2} \tag{1}$$

Where r_1^2 and r_2^2 are the distances from the top and bottom edges of the sheet to the measurement point (m), respectively. An inclined sheet structure was modeled by a set of five model parameters, namely electric moment dipole K $(K = I\rho / 2\pi)$ where I is the current density and ρ is the resistivity of the surrounding medium, abscissa of the center of the x_o, depth of the center h, the half-width of the sheet a, and inclination angle θ. The general equation of SP anomaly $V(x)$ at any point P on a profile perpendicular to the strike of inclined sheet is written as follows [9,10]:

$$V(x) = K \ln \left[\frac{\{(x - x_0) + a\cos\theta\}^2 + (h - a\sin\theta)^2}{\{(x - x_0) - a\cos\theta\}^2 + (h + a\sin\theta)^2} \right] \tag{2}$$

Equation 2 is used to calculate the forward response where K, a, h, x_o, and θ

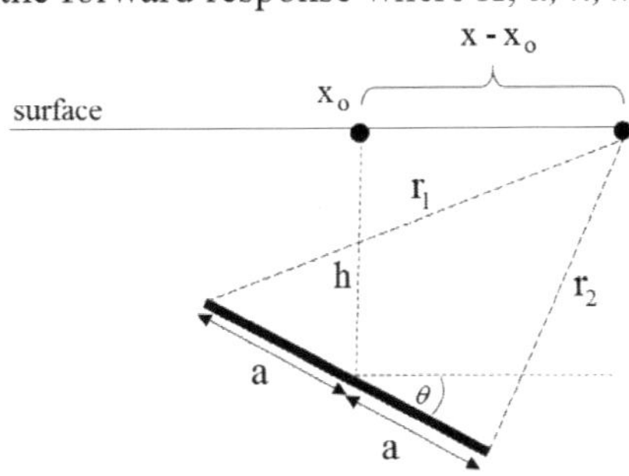

Figure 1. Geometry of sheet which is clock-wisely inclined by θ

2.1.1. Levenberg-Marquardt Inversion

In this study, we propose an implementation of an iterative inversion that estimates all model parameters using Levenberg-Marquardt method. This algorithm guarantees good convergence and non-singularity of the solution. It takes the advantages of Gauss-Newton method which is fast but having uncertain convergence and Steepest-Descent method which is very slow but having certain convergence [11].

In this problem, the model parameters and data are related a non-linear system of equations:

$$\mathbf{d} = \mathbf{G}(\mathbf{m}) \tag{3}$$

Where **d** is a vector containing data of M observations, **m** is a vector of n model parameters which will be estimated (K, a, h, x_o, and θ), and **G** is mathematical model or kernel matrix that relates the data **d** to the parameters **m.**

Assuming measurement errors are normally distributed thus error can be minimized using their data standard deviation:

$$f(\mathbf{m}) = \sum_{i=1}^{M} \left(\frac{\mathbf{G}(\mathbf{m})_i - \mathbf{d}_i}{\sigma_i} \right)^2, \quad i = 1,2,3,\ldots,M \tag{4}$$

$$\text{or } f(\mathbf{m}) = \sum_{i=1}^{M} (f_i(\mathbf{m}))^2 \tag{5}$$

In the LM method, equations for successive iterations are expressed as:

$$\left(\mathbf{J}(\mathbf{m}^k)^T \mathbf{J}(\mathbf{m}^k) + \lambda \mathbf{I} \right) \Delta \mathbf{m} = -\mathbf{J}(\mathbf{m}^k)^T \mathbf{F}(\mathbf{m}^k) \tag{6}$$

where k is a number of iteration, and **J(m)** is Jacobian matrix which can be written as:

$$\mathbf{J}(\mathbf{m}) = \begin{bmatrix} \frac{\partial f_i(\mathbf{m})}{\partial m_1} & \cdots & \frac{\partial f_i(\mathbf{m})}{\partial m_n} \\ \vdots & \ddots & \vdots \\ \frac{\partial f_m(\mathbf{m})}{\partial m_1} & \cdots & \frac{\partial f_m(\mathbf{m})}{\partial m_n} \end{bmatrix} \tag{7}$$

I is identity matrix, and λ is a damping parameter to maintain the inversion convergence stability. **F(m)** is a vector containing differences between observed data and calculated data:

$$\mathbf{F}(\mathbf{m}) = \begin{bmatrix} f_1(\mathbf{m}) \\ \vdots \\ f_m(\mathbf{m}) \end{bmatrix} \tag{8}$$

Therefore, updated parameters equation is given by:

$$\mathbf{m}^{k+1} = \mathbf{m}^k + \left(\mathbf{J}(\mathbf{m}^k)^T \mathbf{J}(\mathbf{m}^k) + \lambda \mathbf{I} \right)^{-1} \mathbf{J}(\mathbf{m}^k)^T \mathbf{F}(\mathbf{m}^k) \tag{9}$$

Before application of the inversion scheme to the field data was applied first to synthetic data generated from the forward calculation of test parameters. The aim is to validate the accuracy of the inversion scheme in reconstructing the model parameters. The parameters used to create synthetic data are depth of sheet = 25 m, half-length of the sheet = 15 m, inclination angle = 30°, electric dipole moment = 100 mV, and a midpoint of the sheet = 37 m.

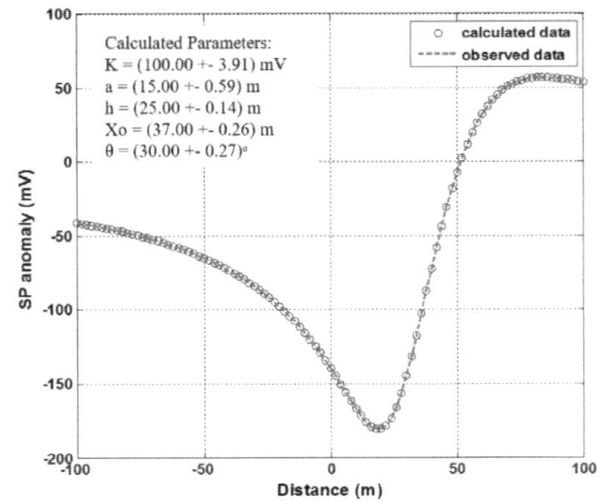

Figure 2. Results of LM inversion using synthetic data: profile of observed vs. calculated data and obtained model parameters (K, a, h, x_o, and θ).

In Figure 2, it is shown that synthetic data were inverted using LM inversion with the same standard deviations for all data, which was assumed to be one mV. The estimated parameters agreed well with observed parameters, and the synthetic data were successfully recovered by the calculated data. Based on the fitness between the synthetic data and the calculated data and similarity between the test parameters and the inverted parameters, the inversion scheme was considered ready to be applied to field data.

3. Observed Data

Fault structures can be characterized based on lithological differences which are visible on the surface, and on the difference in depth of subsurface structure with the same lithology seen from the geological cross section. Information of the fault presence was obtained from the geological map of Garut and Pameungpeuk quadrangle, Java (Figure 3). The field data acquisition was undertaken on 24 May 2016 in Pinggirsari village, West Java, Indonesia. The measurement line was set by crossing fault structures with South-North direction regarding the geological cross section. The length of acquisition line was more than 1 km, with a distance between electrodes of 25 m.

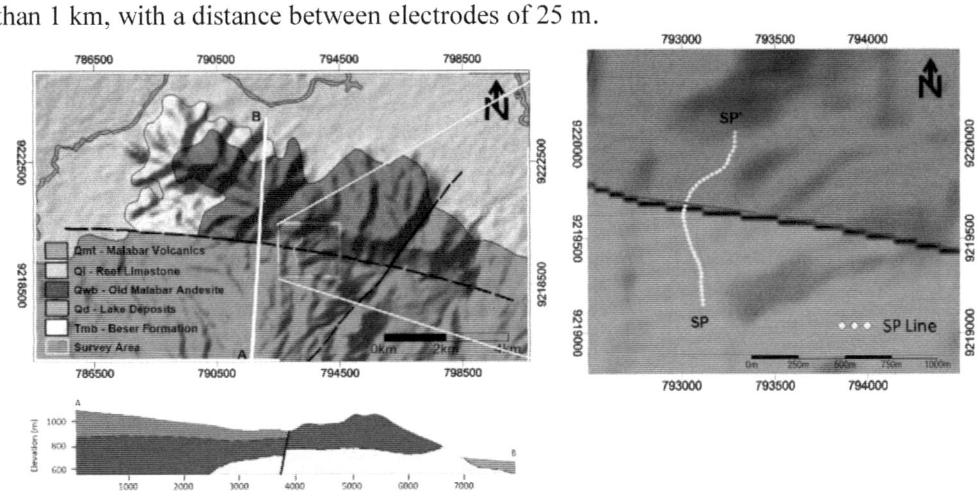

Figure 3. Location of the study of self-potential methods (the presence of fault based on the geological map) [12]

From the field data acquisition, SP anomalies were obtained as shown in Figure (3). It is shown that anomalies values are fluctuating, this is probably due to differences in soil condition along the acquisition line, either contrast porosity or dissimilarity on fluid content of rocks. From the measured anomalies, it can be seen that there is a sharpening anomaly which is estimated to be attributable to the presence of a fault.

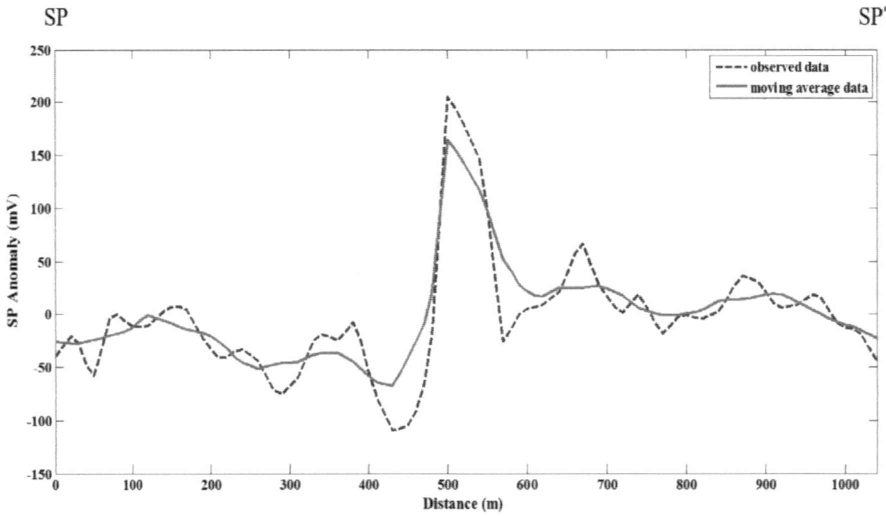

Figure 4. SP anomaly profile from measurement in SP-SP′ line

Furthermore, data smoothing of observed SP anomaly data was performed using moving average. Its purpose was to sharpen anomaly effect and eliminate soil heterogeneity effect. Thus it obtained smoothed data as shown by the red curve in Figure 4. In the inversion process, the observed data to use was data resulted from the smoothing process.

4. Result and Discussion

From the results of field data acquisition using SP method (Figure 4), the observed SP anomaly showed the presence of anomalous potential difference located between 400 m - 600 m with maximum values (V_{max}) of 162.70 mV and minimum values (V_{min}) of -69.88 mV. This anomaly is estimated to be caused by their fault structure below the earth's surface.

In the inversion process, interpreted electric dipole moment K = 41.5 mV, half-fault dimension a = 34 m, depth of the sheet's center h = 14.63 m, the location of the fault's center x_o = 478.25 m, and the polarization angle to the horizontal plane $\theta = 334.52^{\circ}$ in a clockwise direction. Inversion results in Figure 4 show the data calculation curve was successfully following the pattern of the field data.

The inversion codes were constructed in MATLAB and executed using a CPU with 1.70 GHz processor and 8.00 GB RAM. The time needed for running program was 0.0725 seconds. We used standard error (SE) as statistical criteria for comparing the observed data to the inverted data using the above approaches,

$$SE = \sqrt{\frac{\sum_{i=1}^{m}[D(x_i) - V(x_i)]^2}{m - n}} \tag{10}$$

Where $D(x_i)$ is the observed SP anomaly at a discrete points x_i , m is the number of points and $n = 5$ (i.e., K, a, h, x_o, θ), $V(xi)$ denotes the SP anomaly data calculated using the interpreted parameters. The obtained value of standard error for this method (*SE*) was 5.43. Standard error represent incompatibility the observed and calculated data about 5.43%. This mismatch data is rated low so that the parameters of the calculation can be accepted, and its inversion results can be seen in Figure 5.

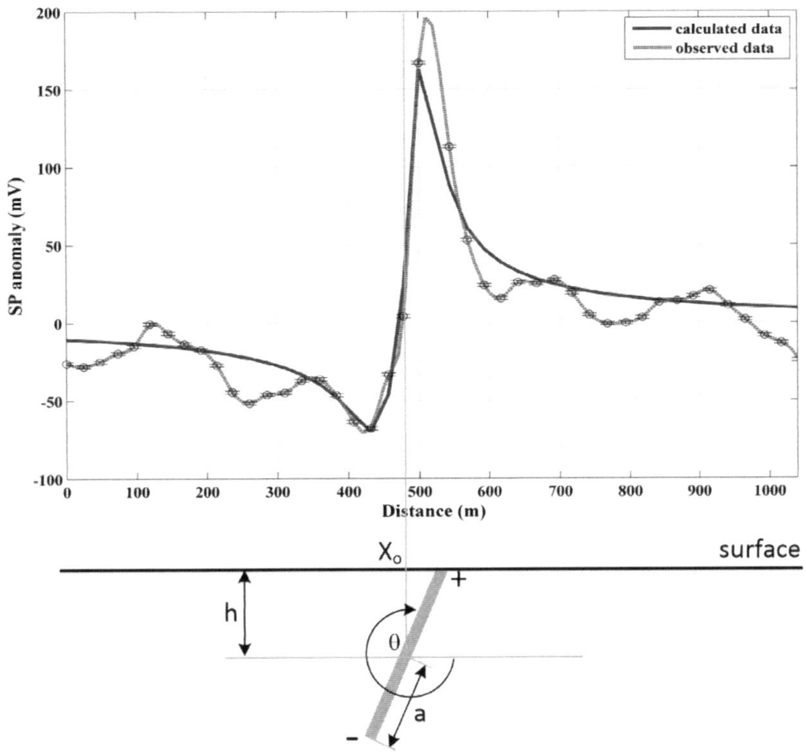

Figure 5. The results of observed data inversion anomaly SP using LM

5. Conclusion

The fault in Pinggirsari village was identified using SP method. The Levenberg-Marquardt method was used for quantitative interpretation of self-potential anomalies due to the fault structure to determine model parameters (electric dipole moment, depth, half-length, location midpoint, and inclination angle). This method was applied to synthetic data where a good agreement between the observed and the calculated data was achieved. To further evaluate the performance of this method, interpretation of SP field anomalies using the method were then carried out. The result shows that calculated data followed the pattern of anomalies, and the recovered model parameters were realistic and physically logical. Based on the measured SP anomaly which has only minimum to maximum value of about 230 mV, the reconstructed depth of the fault is considerably shallow having only length of $2a$ or 68 m.

References

[1] Murthy B V S and P Haricharan 1985 Nomogram for the Complete Interpretation of Spontaneous Potential Profiles Over Sheet-like and Cylindrical 2-D Source *Geophysics* **50** 1127-1135

[2] Rao D A and Babu H V R 1988 Rapid Graphical Method For The Interpretation of the Self-Potential Anomaly Over A Two-Dimensional Inclied Sheet Of Finite Depth Extent *Geophysics* **53** 1126 1128

[3] Abdelrahman E M and Sharafeldin S M 1997 A Least-Square Approach to Depth Determination From Residual Self-Potential Anomalies Caused by Horizontal Cylinders and Sphere *Geophysics* **62** 44-48
[4] Abdelrahman E M, Saber H S, Essa K S and Fouda M A 2004 A Least-Square Approach to Depth Determination From Numerical Horizontal Self-Potential Gradients *Pure and Applied Geophysics* **161** 399-411
[5] Abdelrahman E M, Ammar A A, Hassanein H I and Hafez M A 1998 Derivative Analysis of SP Anomalies *Geophysics* **63** 890-897
[6] Rao D A, Babu H V R and Sinha G D J S 1982 A Fourier Transform Method for The Interpretation of Self-Potential Anomalies Due to Two-Dimensional Inclined Sheets Of Finite Depth Extent *Pageoph* **120** 355-374
[7] Srigutomo W, Agustine E and Zen M H 2006 Quantitative Analysis of Self-Potential Anomaly: Derivative Analysis, Least-Square Method and Non-Linear Inversion *Indonesian Journal of Physics* **17** 49-55
[8] Sharma S P and Biswas A 2013 Interpretation of Self-Potential Anomaly Over a 2d Inclined Structure Using Very Fast Simulated-Annealing Global Optimization – An Insight About Ambiguity *Geophysics* **78** wb3-wb15
[9] Asfahani J and Tlas M 2005 A Constrained Nonlinear Inversion Approach to Quantitative Interpretation of Self-Potential Anomalies Caused by Cylinders Spheres and Sheet-like Structures *Pure Appl. Geophys.* **162** 609-624
[10] Sundararajan N, Rao P S and Sunitha V 1998 An Analytical Method to Interpret Self-Potential Anomalies Caused by 2-D Inclined Sheets *Geophysics* Vol. **63** 1551-1555
[11] Marquardt D W 1963 An Algorithm for Least-Square Estimation of Nonlinear Parameters *Journal of the Society for Industrial and Applied Mathematics* **11**(2) 431-441
[12] Alzwar M, Akbar N and Bachri S 1992 Peta Geologi Lembar Garut dan Pameungpeuk, Jawa *Pusat Penelitian dan Pengembangan Geologi*

Acknowledgement
The authors gratefully thank the Earth Physics Lab., Physics Dept., Institut Teknologi Bandung for providing tools and financial support to set up the field measurement and we are also indebted to the assistant team of the laboratory during the measurement.

Application of 2-D Resistivity Imaging and Seismic Refraction Method in Identifying the Structural Geological Contact of Sedimentary Lithologies

M. K. A. Nur Amalina[1*], M.M. Nordiana[2], Andy Anderson Bery[2], Mohammad Noor Akmal bin Anuar[1], Umi Maslinda[1], Nabila Sulaiman[1], Muhamad Afiq Saharudin[1], Hazrul Hisham[1], A. N. Nordiana[1], Z. M. Taqiuddin[1]

[1]Geophysics Postgraduate Student, Geophysics Section, School of Physics, Universiti Sains Malaysia
[2]Geophysics Lecturer, Geophysics Section, School of Physics, Universiti Sains Malaysia

*E-mail: nuramalinamka93@gmail.com

Abstract. The interest of this study was to prove the existence of geological contact to field models with presences of outcrops as references. The physical relief of the outcrops can be determined by geological events of faulting, fracture, and folding. Geological contact plays important roles in environmental studies. 2-D resistivity imaging is the best method used for identifying the geological structures of study area located in Guar Jentik, Perlis and Bukit Kukus, Kedah. Besides that, seismic refraction method also applied at the study area. Results from both methods were integrated to get data correlation. There is good correlation produced which have successfully proved the existence of the faults and contact zones in study areas. Resistivity result shows that first study area has two main zones, red mudstone with resistivity value of 1 Ωm – 100 Ωm, sandstone with resistivity value of 2000 Ωm – 9000 Ωm, and Seismic refraction has provided the result on velocity of each zone, mudstone zone is 200 m/s - 1800 m/s and sandstone zone is >2000 m/s. The geological contact of fault is determined between the red mudstone zone and sandstone zone. In the second study area, the fracture was found within chert zone and contact zone is located between the chert zone and mudstone zone. In addition, the fold is found to form in the chert zone. Mudstone zone has resistivity value of 1 Ωm – 1500 m/s and chert zone has resistivity value of 2600 Ωm – 35000 Ωm. The first layer of the seismic section is consisting of mudstone with a velocity of <800 m/s and the velocity obtained for the second layer was generally >1200 m/s is interpreted as chert zones.

1. Introduction

Structural geology features are central to any study of past and present about the rock or sediment which has deformed relative to their original orientations or positions. The deformations may occur due to deformational conditions either stress and strain [1]. Fault and fold are very common structural features. Plate tectonic is the large-scale process that is the ultimate cause of the most deformation of the rocks and could lead to a mountain formation which can change the structure of the subsurface. As a conclusion, the type and direction of the plate tectonic of geologic structure could be determined by the observation of faulting, folding and layering present today. 2-D resistivity and seismic refraction are the best utilized to study subsurface characterization. 2-D resistivity methods were successfully used for detection in the shallow subsurface for maintaining the geo-environment and could get detail with deeper penetration for shallow subsurface study [2]. Geophysical exploration should not use as a stand-alone solution to problems. It will become a more effective component of site investigation if it was combined with another method such as seismic refraction. Near-surface seismic waves refraction can provide useful information about the near-surface for investigation of engineering and environmental purpose [3]. Besides that, this method will measure both depth and quality of hard rock regarding seismic velocity which could give more information about each of layer existence inside the subsurface. Thus, 2-D resistivity imaging and seismic refraction surveying were carried out to relocate the structural geology features such as fault and fold. This study was about the application of both methods to field models at selected area in Peninsular Malaysia with sedimentary outcrops.

1.1. 2-D Resistivity Imaging

Electrical resistivity is a bulk property of a material to inhibit the flow of electrical injection current. The current between a pair of surface electrodes (A and B) was inserted into the ground, and next pair of an electrode (M and N) used to measure the resulting potential on other electrodes by using current (I). The potential value was measured in unit voltage (V). An equation was produced from this relationship as in equation 1.

$$\rho a = k\frac{v}{I} \tag{1}$$

Thus, resistivity was interpreted as the ability of rock, soil or ground water to resist the flow of an electrical current. The increase in electrode spacing will cause the penetration deeper. Resistivity contrast could occur in the subsurface between different layers such as dry and water-bearing sediments or different lithology and different weathering layers. Electrical resistivity used to map subsurface variability in electrical properties caused by the change in lithology, structure, alteration, and contamination due to human activities. The electrical imaging system is being carried out by using multi-electrode resistivity meter system. The resistivity meter used was an ABEM SAS4000 with a pole-dipole array, and the instrument uses an automatic data longer which had allowed the data to be recorded as the survey progresses for later downloading to a computer for processing and presentation.

1.2. Seismic refraction method.

The seismic wave will travel across the boundaries by reflected their energy to the surface when there are two different rock types, and the remainder continues on its way at a different angle or being refracted [4]. Seismic refraction is the best tools in the study of earth structure to identify the characteristic and lithology. This method is depending on the ray path produce from discontinuities of wave velocity. Elastic moduli and densities of rock are considered to determine the seismic wave velocity.

2. General Geology

Perlis and Kedah were occupied with sedimentary lithology which was bounded in the Western Belt of Peninsular Malaysia.

2.1. Guar Jentik, Perlis

Hill B is an outcrop section that cut in a south-north direction. The stratigraphic of Hill B is well exposed where it does show transitional sequence from carbonate to clastic deposition [5]. There are two types
of formation in this area which is known as Jentik Formation and Kubang Pasu Formation. Hill B is divided into three informal units including unit 4 which is well bedded of dark limestone, unit 5 with the present of black mudstone interbedded with chert with slump structures, and unit 6 is proposed to

have thick beds of brownish-red mudstone interbedded with sandstone. From the excavation in this area, several types of marine fossil have been found to prove the existences of deep marine [6]. The electrical techniques used in this study area are 2-D Resistivity to study the structure of fault in the sedimentary outcrop and result obtained is supervised with Seismic Refraction method.

2.2. Bukit Kukus, Kedah

This small hill is proposed to have stratigraphic of chert unit which comprised black laminated mudstone, interbedded sandstone, mudstone, interbedded tuffaceous sandstone, tuff with a paraconglomerate bed, interbedded tuffaceous sandstone, siliceous shale, chert, bedded cert, tuffaceous mudstone and interbedded chert and siliceous mudstone [7]. Chert found at the site is observed and had been proposed to contain radiolarians which strongly indicated of late Early Permian to Middle Triassic of age by [8]. Bukit Kukus is mapped in Semanggol formation in Kuala Ketil Kedah and has being divided into the Siliceous and clastic rock into three members of the unit.

3. Study Area

The study was carried out at two different locations with the features of fracture, contact, and fold which is mainly focused on shallow subsurface investigations. The study area is located at Hill B, Guar Jentik in Perlis and Bukit Kukus in Kedah. Both study area is located in Peninsular Malaysia.

3.1. Guar Jentik, Perlis

The hilly ridge of the study area chooses is known as Hill B are located in the vicinity of Guar Sanai, Kampung Guar Jentik, Beseri District in Northwest Perlis, Malaysia such as shown in figure 1.

3.2. Bukit Kukus, Kedah

Bukit Kukus located in the small town of Kuala Ketil, Kedah is mainly exposed to sedimentary rocks from Semanggol Formation. 2-D Resistivity method and Seismic Refraction is adopted to map and characterize the structure of fault in this shallow subsurface of sedimentary outcrop such as shown in figure 2.

Figure 1. 2-D resistivity imaging and seismic refraction survey line at Guar Jentik, Perlis.

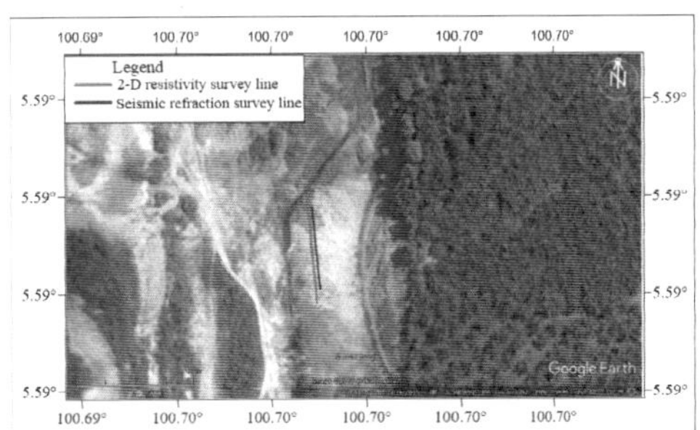

Figure 2. 2-D resistivity imaging and seismic refraction survey line at Kuala Ketil, Kedah.

4. Methodology

The 2-D resistivity imaging used the pole-dipole array with different electrode spacing on each area depends on accessibility on site. The space between electrodes is 1 m at Guar Jentik, Perlis and 1.5 m at Bukit Kukus, Perlis. Hence, the total length of the survey line at Guar Jentik was about 40 m and the total length of survey line at Bukit Kukus was about 60 m. This survey was conducted using ABEM SAS4000 Terrameter with smart cables and resistivity data was modeled using the Res2Dinv software. While for Seismic refraction method, the seismic wave produced by sledgehammer refracted from different reflector boundary and detected by geophones (receiver). The total distance of survey line for seismic refraction is 46 m for both locations with geophone spacing of 2 m.

5. Result and discussion

The result from 2-D resistivity method is supervised by using the result from seismic refraction method. Resistivity pseudosection is produced from inverted data processing has displayed the resistivity data of earth materials. While seismic cross-sections can provide the earth profile used to interpret depth and dip direction of geological contact in each layer of the subsurface.

5.1. Guar Jentik, Perlis

Figure 3 shows the contouring model resistivity at Hill B which had successfully classified the structural geology of contact zone and the type of soils. The study area was divided into two main zones, red mudstone with resistivity value of 1 – 100 Ωm and sandstone with resistivity value of 2000 Ωm – 9000 Ωm. Figure 4 shows the countering model from the seismic refraction survey. Even though the depth of penetration does not enough as compared to resistivity model, the result from seismic refraction is able to characterize the geological structure at the study area and match with the resistivity survey.

Figure 5 shows the result of resistivity survey. There is contact between sandstone and red mudstone. The contrast value of the resistivity contouring was suggested to be the present of the contact zone between sandstone and red mudstone. The result from inversion model was chosen to overlap with the outcrop to supervise this study. There is high contrast at distance of 23 m which indicated the location of contact zone at Hill B. Figure 6 shows the results of the seismic cross sections of the Hill B which consists of two layers.

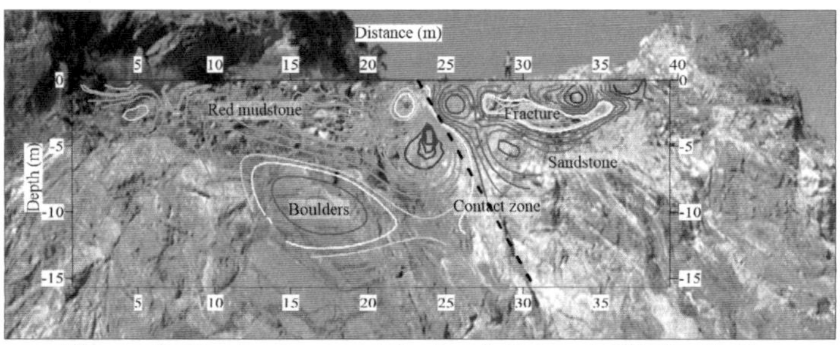

Figure 3. Contouring model of resistivity at Hill B, Perlis.

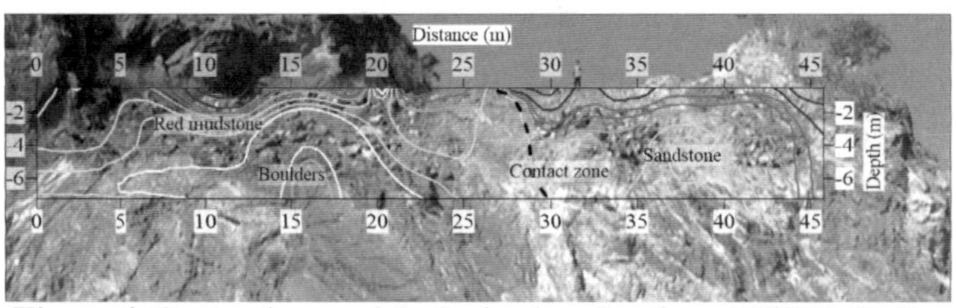

Figure 4. Contouring model of seismic refraction at Hill B, Perlis.

The distance scale of digitized value on these graphs are relative to the resistivity line respectively. Based on seismic refraction profile in figure 6, the seismic velocity of the mudstone zone is 200 m/s to 1800 m/s and the velocity of the sandstone zone is more than 2000 m/s. The result shows the first and second zone was abrupt by the velocity change as an outlined which was suggested the location of geological contact that presents between sandstone zone and red mudstone zone.

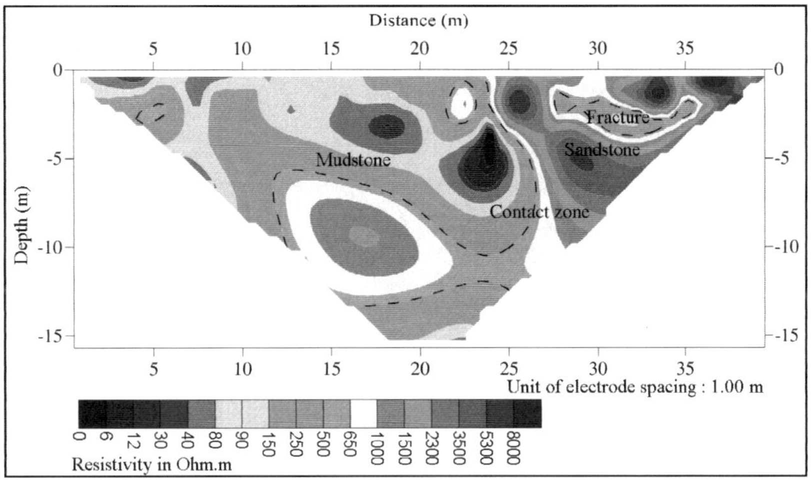

Figure 5. Inversion model resistivity at Hill B, Perlis

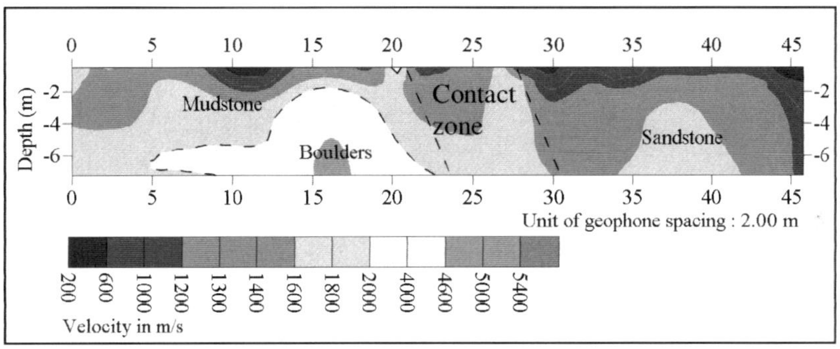

Figure 6. Seismic section at Hill B, Perlis

5.2. Bukit Kukus, Kedah

Figure 7-8 shows the contouring model resistivity and seismic refraction at Bukit Kukus. Based on figure 9, the resistivity results generally show the study area divided into two main zones, mudstone with resistivity value of 1 – 1500 Ωm and chert zone with resistivity value of 2600 – 35000 Ωm.

The contrast between low and high resistivity value indicates fault zone. Another contrast value shows the contact zone between mudstone and chert. Chert has higher resistivity due to its chemical properties which are rich in silica. This rock is mainly composed of microcrystalline silicon oxide (SiO_2).

The physical properties of chert are very hard and compact. Hence, electrical current could not easily penetrate within the rock and cause the resistivity be higher. Fold zone cause the resistivity value become too high because of folding of the rock has makes the rock to be more compact and electric current is flowing without having any impediment.

Resistivity result was strongly supported by the seismic cross section as shown in figure 10 conducted on the same line. The result shows the study area generally consisted of two main layers. The first layer is predominantly consisting of mudstone with a velocity of <800 m/s while the second layer with a velocity of >1200 m/s is interpreted as chert zones. The dashed lines indicate the fracture/fault and contact zone at the study area.

IOP Conf. Series: Earth and Environmental Science **62** (2017) 012005 doi:10.1088/1755-1315/62/1/012005

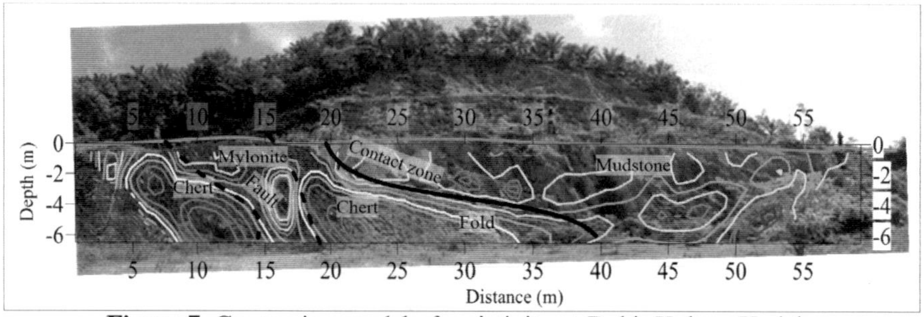

Figure 7. Contouring model of resistivity at Bukit Kukus, Kedah.

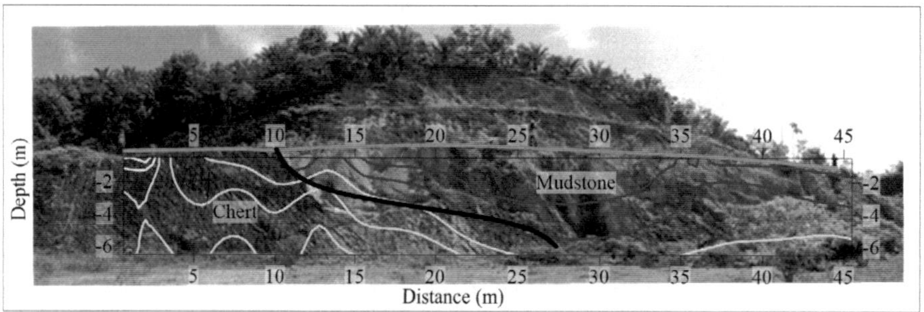

Figure 8. Contouring model of seismic refraction at Bukit Kukus, Kedah.

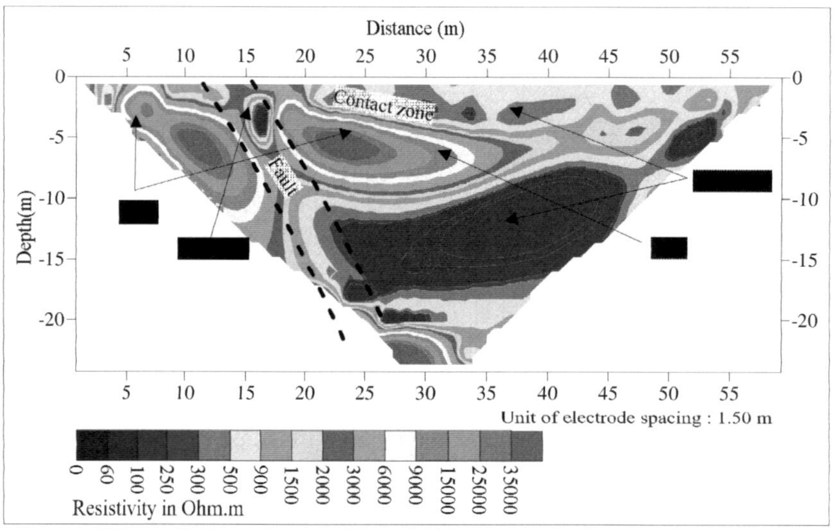

Figure 9. Inversion model resistivity at Bukit Kukus, Kedah.

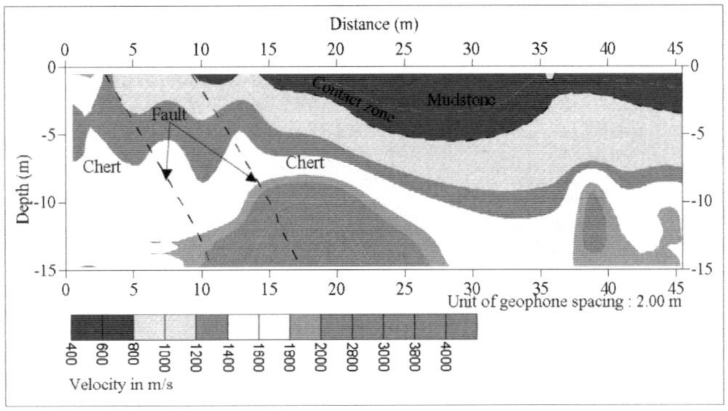

Figure 10. The seismic cross section at Bukit Kukus, Kedah.

6. Conclusion

In conclusion, the 2-D resistivity imaging and seismic refraction method have successfully proved the interpretation of shallow subsurface structural geology characterization of the study area in sedimentary lithology. The contrast between low and high resistivity value indicates the location of contact zone and fracture zone. The result obtains from both method in all study area might be different in their depth of penetration due to the different spacing number used for each survey line. The smaller the spacing, only small depth of penetration was obtaining but the better the resolution. Some of the survey lines have a different spacing was due to the limitation access at the study area which is making hard to plant the electrode/geophone. Both methods are good in identifying the geological structures of the subsurface.

Acknowledgements

The authors express their sincere thanks to all postgraduate student and technical staffs School of Physics, Universiti Sains Malaysia, and financial support to this research was given by the IPS Conference Fund, Universiti Sains Malaysia.

References

[1] Stein S, Geoffrey C P and Rundle J B 1988 The Growth of Geological Structures by Repeated Earthquakes 2. Field Examples of Continental Dip-Slip Faults **93** p 88

[2] Nordiana M M, Rosli S, Nawawi M N M, Azwin I N, and Mohamad E T 2013 Case Study: Shallow Subsurface Geology Mapping Using 2-D Resistivity Imaging with EHR Technique *APCBEE Procedia* **5** 134–140.

[3] Nisa A, Rosli S and Nordiana M. M. 2013 Integration of Seismic Refraction and 2D Electrical Resistivity in Locating Geological Contact 7–12.

[4] Griffths D H and King R F 1981 Applied Geophysics for Engineers and Geologist *Pergamon, London* 223.

[5] Hassan M H and Peng L C 2002 Stratigraphy of the Jentik Formation, the transitional sequence from the Setul Limestone to the Kubang Pasu Formation at Guar Sanai, Kampung Guar Jentik, Beseri, Perlis - a preliminary study *Bull. Geol. Soc. Malaysia, Annu. Geol. Conf. 2002 Issue* **45** 171–178.

[6] Aye K A, Hassan M H and Ng T F Discovery of Late Devonian (Frasnian) conodonts from the ' Sanai limestone ', Guar Jentik, Perlis, Malaysia *Geol. Soc. Malaysia* **59** 93–99.

[7] Harun Z, Jasin B, Mohsin N and Azami A 2009 Thrust in the semanggol formation, kuala ketil, kedah *Bull. Geol. Soc. Malaysia* **55** 61–66.

[8] Jasin B and Harun Z 2007 Stratigraphy and sedimentology of the chert unit of the Semanggol Formation *Geol. Soc. Malaysia* **53** 103–109

Correlation between Resistivity and Ground Penetrating Radar (GPR) Methods in Understanding the Signatures in Detecting Cavities

Muhamad Afiq Saharudin [1], Umi Maslinda[1], Hazrul Hisham[1], Z.M. Taqiuddin[1], M.K.A Nur Amalina[1], Nordiana Ahmad Nawawi[1], Nabila Sulaiman[1], M.M Nordiana[1], I.N, Azwin[1]

[1] Geophysics Department, School of Physics, 11800 Universiti Sains Malaysia, Penang, Malaysia.

muhamad.afiq@yahoo.com.my

Abstract. The research was conducted using Resistivity and Ground Penetrating Radar (GPR) methods in detecting in-filled cavities and air-filled cavities. The importance of this study is to see the difference in conductivity value of the in-filled and air-filled cavity. The first study location in which the known target is air-cavity located at School of Language, Literacies, and Translation (SoLLAT). The next study location is at Desasiswa Bakti Permai, which the known target is a bunker with both were located at Universiti Sains Malaysia, Penang and the last location is at Gua Musang, Kelantan with suspected in-filled cavity. The result from Gua Musang is compared with both of the results that have been done at Universiti Sains Malaysia. The resistivity value of the first location that indicates the possible tunnel is about 500 Ωm to 800 Ωm and the conductivity value is about 0.0017 S/m. The resistivity value for the second location located at Desasiswa Bakti Permai that indicates the bunker is about 50 Ωm to 250 Ωm and the conductivity value is about 0.1104 S/m. The resistivity value from Gua Musang is about 50 Ωm to 100 Ωm and the conductivity value is about 0.0101 S/m. The velocity of the in-filled cavities is much lower compared with the velocity of the air-filled cavities. Based on the characteristics, Gua Musang area was dominated with in-filled cavities.

1. Introduction

Universiti Sains Malaysia largely consists of granite which is an igneous type of rock while Gua Musang, Kelantan mostly made up from limestone. Detecting cavities or voids by applying geophysical methods have gained wide interest in the past few decades. Geophysical methods play a fundamental role in a process to identify the evidence of cavities at the three different places. The resistivity method used in this research in order to study the electrical properties and electrical conductivity of the subsurface area. Electrical methods utilize direct currents or low frequency alternating currents to investigate the electrical properties of the subsurface [1]. The resistivity method is a method using artificially-generated electric currents is injected into the ground and measured potential differences are at the surface. Ground Penetrating Radar (GPR) is a non-destructive and cost-effective geophysical survey method that can be used in solving many environmental problems such as assessing construction sites and environmental objects [2]. The short bursts of electromagnetic (EM) wave being reflected with a range of frequencies and transmitted into the ground and the reflected pulses were recorded as functions of time and the position of the antenna pair along with a survey line.

The application of GPR on dry soils will give a better resolution of the buried structures since the electromagnetic (EM) wave can penetrates deeper [3]. There are many GPR antennas such as 500 MHz, 250 MHz, and 100 MHz antennas that can be used with each one of them will give different depth penetration. The most important parameter in GPR is the frequency used in the studies because it constrains the values of many other parameters to obtain a good GPR profile with better resolution depends on the suitable frequency. GPR signals can be collected rapidly and sufficiently with all the data are recorded digitally. The initial interpretations of GPR data can be made with minimal data processing required, making the GPR is suitable for shallow geophysical studies [4].

2. Study Area

The study for this research was conducted at Universiti Sains Malaysia (USM), Penang, Malaysia from Figure 1 and Batu Papan at Gua Musang, Kelantan from Figure 2. The study area covered at different locations as shown in Figure 1. The first study location was at School of Languages, Literacies and Translation labelled (A) at Universiti Sains Malaysia, starting from N 5°21'23.24", E 100°18'25.72" to N 5°21'23.24", E 100°18'26.20", second location was at Bakti Desasiswa hostel labelled (B) also at Universiti Sains Malaysia starting from N 5°21'21.8", E 100°18'00.5" to N 5°21'22.8", E 100°18'00.9", the final location is at Gua Musang, Kelantan labelled (C) start with N 04°51'04.5540", E 101°57'03.4596" and the end point is at N 4°51'05.1588", E 101°57'04.3668".

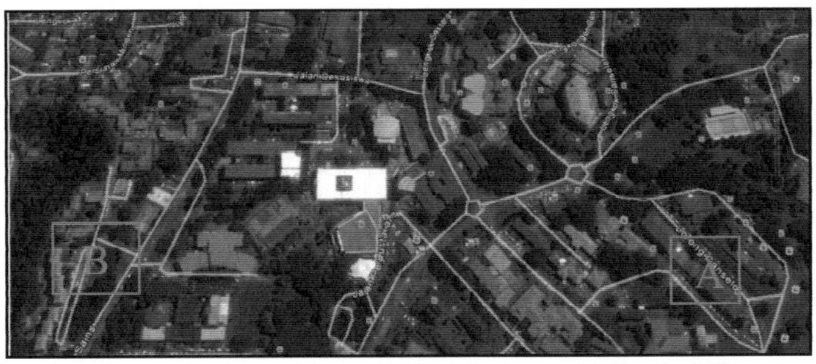

Figure 1. Survey area at USM, Penang, Malaysia [5].

Figure 2. Survey area at Gua Musang, Kelantan, Malaysia [5].

3. Methodology

The minimum electrode spacing at the air-cavity area, (A) is 0.5 m spacing, at the bunker area, (B) is 1.5 m spacing and the minimum electrode spacing for the survey area at Gua Musang, (C) is 2.5 m spacing. The type of array used in the resistivity survey at the air-cavity study area, and the bunker is Pole-Dipole while at the Gua Musang is Wenner-Schlumberger. The length of the resistivity survey line at the tunnel is 20 m, at the bunker is 60 m and at the Gua Musang is 100 m. The distance for Ground Penetrating Radar (GPR) survey lines lies on the same line as for the resistivity survey lines for all the study areas. The total length of the GPR survey line at all study areas are 30 m.

4. Results and Discussions

Based on the inversion profile of the resistivity data, the possible location of the suspected cavity is about 1.50 m in depth from Figure 3. Water or clay-filled and air-filled features produce low and high resistivity anomalies respectively [6]. High resistivity value between 500 Ωm until 800 Ωm indicates the suspected cavity that being filled with air at a distance from 11 m until 13 m. The conductivity value obtained from the resistivity data is 0.0016 S/m and the calculated conductivity value is 0.0017 S/m. The data at Location A shows the hyperbolic curve which indicated as the target of suspected cavity (Figure 4). From Figure 4 the suspected cavity was located at depth about 1.57 m

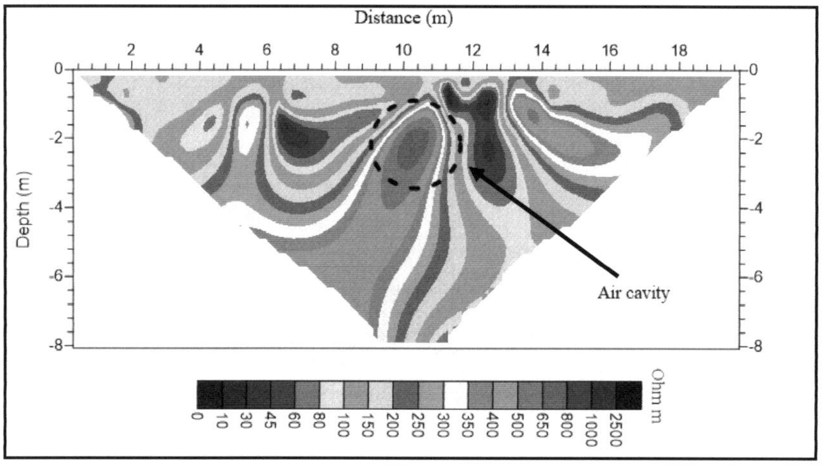

Figure 3. Inversion profile of the resistivity data at Location A.

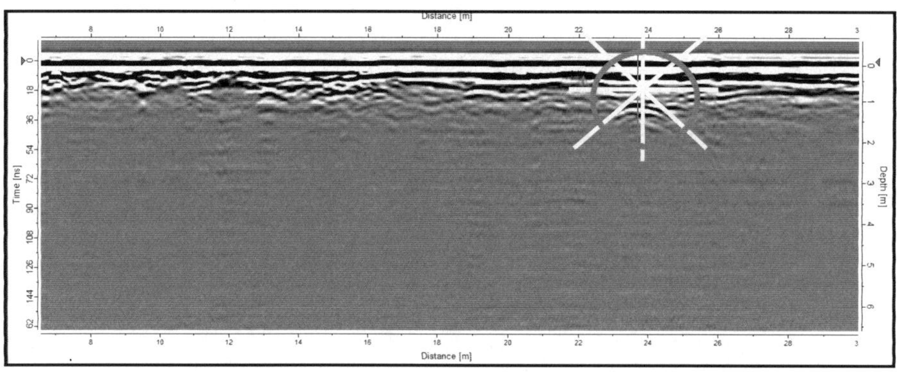

Figure 4. GPR profile at Location A with the target of suspected cavity.

From the inversion profile of the resistivity data (Figure 5), the depth of the bunker is about 1.0 m from the subsurface area. The resistivity value that indicates the bunker is between 50 Ωm until 250 Ωm at a distance starting from 23 m until 55 m. The conductivity obtained from the resistivity data is 0.0909 S/m while the calculated conductivity value is 0.1104 S/m. The velocity at Location B is lower compared with the Location A due to the sedimentation and concrete structure. Location B acts as the model for the in-filled cavities. The presence of soil or air-filled void can be represent by a bigger scale of the hyperbolic curves taking an inverted chevron shape, while a homogeneous materials such as clay-rich soils, bedrock or groundwater will have a reflection free pattern because all the three materials are considered as highly conductive dissolved minerals that generate the attenuation process of the EM wave produced by the GPR antenna [7]. The data at location B shows the suspected target at 0.87 m depth which represents by a hyperbolic pattern at the bunker located at Desasiswa Bakti Permai (Figure 6).

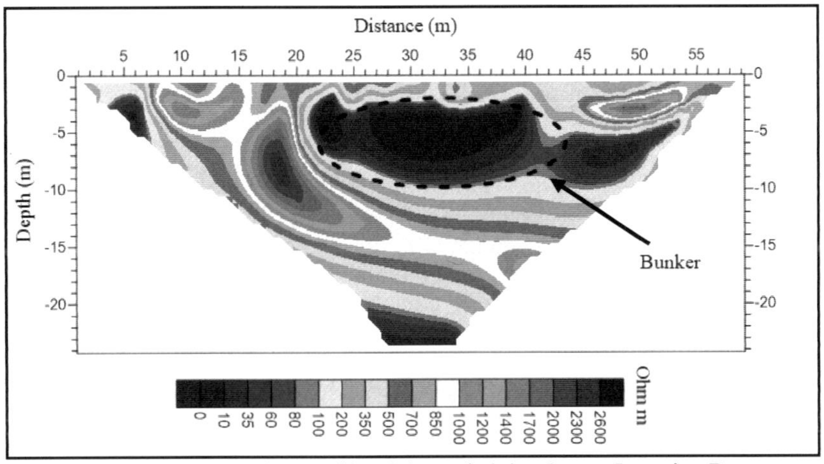

Figure 5. Inversion profile of the resistivity data at Location B.

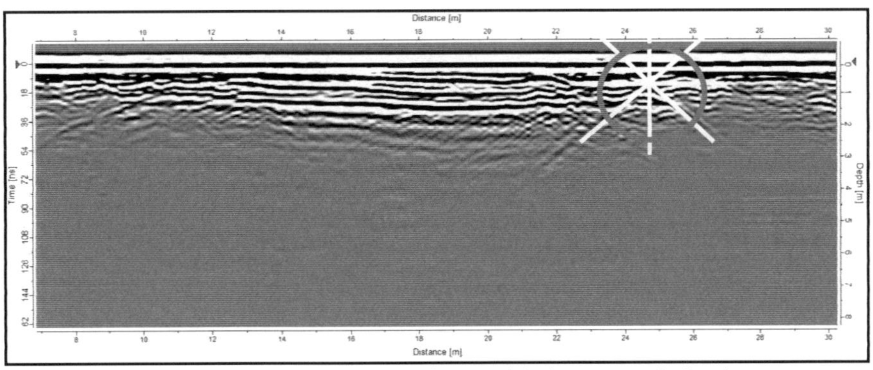

Figure 6. GPR profile at Location B with the target of a bunker.

The inversion profile of the resistivity data from Figure 7 indicates the possible cavities at a certain depth between 0 m until 10 m. The resistivity value of 50 Ωm until 100 Ωm at a distance starting from 10 m until 90 m represents the possible cavities at the subsurface area. The conductivity value obtained from the pseudosection is 0.009764 S/m and the calculated conductivity value is 0.0101 S/m. The velocity of Location C is lower compared with Location A and Location B. The resistivity value at Location C is lower, and that indicates the in-filled cavities. Based on the GPR data (Figure 8), there are two types of pattern that clearly can be seen at Location C. The attenuation pattern is because of the reclaimed area between 0 m until 17 m in length. The hyperbolic curve can be seen on the starting from 18 m until 30 m in length of the survey line. The possible hyperbolic curve that represents the cavity is at depth 1.70 m from Figure 8.

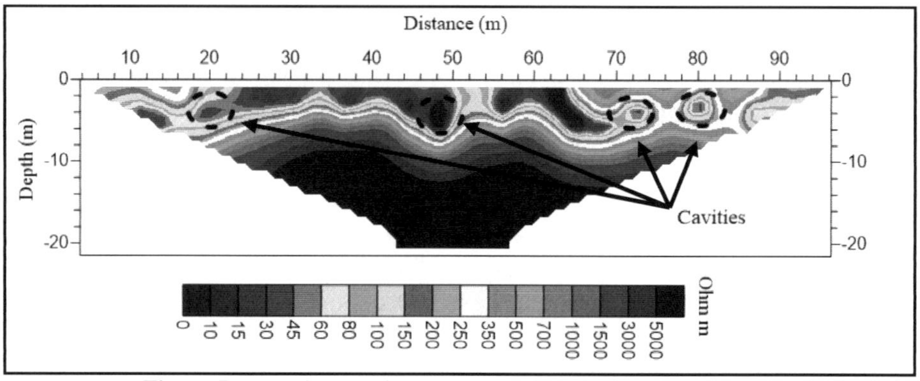

Figure 7. Inversion profile of the resistivity data at Location C.

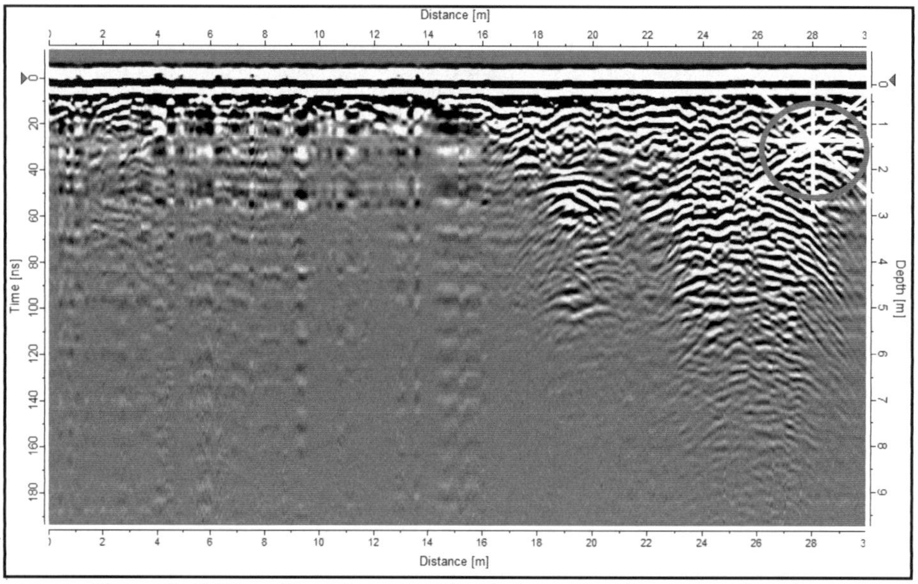

Figure 8. GPR profile at Location C with the target of a suspected cavity.

6. Conclusion

The higher resistivity value which in between 500 Ωm until 800 Ωm represents the air-filled cavity while the lower resistivity value between 10 Ωm until 250 Ωm represents the in-filled cavity. The resistivity value of 50 Ωm until 250 Ωm represents the in-filled cavity associated with the concrete structure while the resistivity value of 50 Ωm until 100 Ωm represents the in-filled cavity associated with the sediment filled cavity. The lowest conductivity value is represents the air-filled cavity with 0.0017 S/m. The conductivity value for the in-filled cavity associated with the concrete structure is 0.1104 S/m and the in-filled cavity associated with the sediment filled cavity conductivity value is 0.0101 S/m. Resistivity and Ground Penetrating Radar (GPR) are two geophysical methods that are effectively can be used to study the cavities and able to distinguished between air-filled cavity and in-filled cavity.

References

[1] Kearey P, Brooks M, and Hill I 2002 An Introduction to Exploration Geophysics, Third Edition, 183 - 4.

[2] Timo S, and Pekka M 2011 The Use of GPR in Road Rehabilitation Projects Roadscanners Oy, Finland

[3] María V B, Matías D L V, and Néstor B 2011 Contribution of the resistivity method to characterize mud walls in very dry region and comparison with GPR *Journal of Archaeological Science.*

[4] Steve C 2002 Subsurface Investigation Using Ground Penetrating Radar *Department of Geology and Geophysics, University of Missouri-Rolla, Rolla, MO*

[5] Google Earth, 2015.

[6] R S Wadhwa, N Ghosh, M S Chaudhari, V Chandrashekhar and Rajib K S 2008 Delineation of cavities in a canal bed by Geophysical Survey in Navargaon Project Area Maharashtra, J. Ind. *Geophys. Union* **12** 55 - 62.

[7] Bristow C S, and Jol H M 2003 An introduction to ground penetrating radar (GPR) in sediments *Geological Society, London* **211** 1 - 7.

Acknowledgement

The authors would like to appreciate all the technical staffs of the Geophysics laboratory, Geophysics lecturers, postgraduate students and final year students from School of Physics, Universiti Sains Malaysia for their assistance during this research. The authors also would thanked the Institute of Postgraduate Studies Universiti Sains Malaysia (IPS USM) for providing the Conference Fund for this conference and USM for also providing the Research University Grant (RUI) with the title "Geophysical Application and Approaches in Engineering and Environmental Problems" (1001/PFIZIK/811323).

IOP Conf. Series: Earth and Environmental Science **62** (2017) 012007 doi:10.1088/1755-1315/62/1/012007

Integration of Ground Penetrating Radar (GPR) and 2-D Resistivity Imaging methods for soil investigation

Nabila Sulaiman[1], M.M. Nordiana[1], I.N. Azwin[1], Z.M. Taqiuddin[1], Umi Maslinda[1], Hazrul Hisham[1], M.K.A. Nur Amalina[1], Muhamad Afiq Saharudin[1] and A.N. Nordiana[1]

[1]School of Physics, Universiti Sains Malaysia, 11800 USM, Penang, Malaysia

E-mail: nabilasulaiman93@gmail.com

Abstract. Rock lithology influences the electrical properties representing soils or rocks. Electrical conductivity value can be measured using geophysical methods like Ground Penetrating Radar (GPR) and 2-D resistivity imaging. The objective of this survey is to integrate GPR, 2-D resistivity imaging and borehole log based on the conductivity value with soil description and N-value from borehole. Borehole is conducted in the middle of the survey line at a distance of 20 m. GPR method used 250 MHz frequency antenna. The result was filtered using Band Pass, Time Varying Gain and DC removal. 2-D resistivity imaging used two arrays; Wenner-Schlumberger and pole-dipole with total distance of 40 m and 1 m minimum electrode spacing using ABEM SAS4000. The results of both arrays are represented in the form of inversion models. Electrical conductivity values for GPR are calculated based on the conductivity values obtained by 2-D resistivity imaging. The conductivity values calculated from GPR are in good agreement with the values from 2-D resistivity imaging method. Electrical conductivity for the top soil is 0.7 – 3.0 mS/m with no soil description and N-value due to imprecise sample of the loose soil condition. The results showed that soil composed of loose silty gravel with some sand at the depth of 1.81 – 2.99 m has higher value of conductivity (0.4 – 3.0 mS/m) while soil dominated by very stiff sandy silt with some rock fragment (gravel) at the depth of 3 – 3.5 m has lower conductivity values of 0.4 mS/m to 0.7 mS/m. Soil having low electrical conductivity is probably due to the stiff condition (minimum water content) confirmed by greater N-value. Integration of geophysical methods and geotechnical method is a success and the geophysical parameters can be used in understanding soil condition.

1. Introduction

Electrical conductivity value is vital in understanding the subsurface condition as each type of soils or rocks has specific electrical properties. Both GPR and 2-D resistivity imaging method implement the conductivity value to acquire the data. Borehole log will provide more details of the ground in terms of soil type, grain size, colour consistency, texture, plasticity and N-value. N-value is the blow count as the borehole sampler penetrates into the ground by automatic trip hammer. N-value indicates the relative density of overburden at that particular depth [1]. The soil mineralogy especially the cohesive soils like clayey silt and sandy silt will affect the electrical conductivity [2]. The integration between GPR, 2-D resistivity imaging and borehole log will produce more reliable information of the study area.

GPR works by emitting electromagnetic (EM) wave in tiny pulse of energy using a transmitter into the ground and receive the reflected wave using a receiver. Materials underground that have contrast of electrical conductivity and dielectric permittivity from the surrounding ground will reflect the signal and produce hyperbola signatures on the radargram based on the time taken and amplitude. Strong reflected signal with high amplitude value is produced when the difference of the conductivity and dielectric permittivity between two materials is pronounced. The conductivity value can be calculated from GPR radargram by calculating the dielectric permittivity [3].

2-D resistivity imaging method measures the electrical potential distribution by injecting the current using two electrodes into the ground [4]. The inversion of the data will produce the value of resistivity and conductivity of the datum point. The values can be used to plot contour map of the study area for better interpretation.

GPR has the ability to detect near surface anomaly on any type of surfaces such as concrete, soil, asphalt and even water while resistivity method is suitable to image near surface and deeper conditions of the subsurface depending on the electrode spacing and type of array used [4]. Since electromagnetic wave from the GPR tend to attenuate when encounter presence of water within the subsurface [5], the 2-D resistivity imaging is used to facilitate in identifying the lost data from GPR.

2. Study area

The study area is located in Universiti Sains Malaysia (USM), Penang with coordinate of N5.362310° E100.306370° and N5.362350° E100.306670°. The survey line was at a flat field with a small slope from the 0 – 18 m of the survey line and the ground during the survey was dry due to the hot weather. In general, Penang is dominated by granitic rocks. The area has geology of medium to coarse grained biotite granite [6].

3. Methodology

One survey line was conducted in the study area with a total length of 40 m for both GPR and 2-D resistivity imaging methods. Borehole was conducted at the centre of the survey line at 20 m distance. The geophysical methods were conducted on the same line to correlate the results of the area.

The GPR method used shielded antenna of 250 MHz.. The radargram is filtered using Band Pass, DC removal and Time Varying Gain to improve the result for better interpretation.

The arrays used for 2-D resistivity imaging method are Wenner-Schlumberger and Pole-dipole due to different maximum depth obtained and the sensitivity in terms of lateral and vertical resistivity variations of each array. Wenner-Schlumberger is adequately sensitive for lateral and vertical structures with lower maximum depth obtain while pole-dipole is good for horizontal coverage with deeper penetration [4]. The survey used ABEM SAS4000 with minimum electrode spacing of 1 m. The data was processed using RES2DINV software to plot the inversion model of the area. Surfer 10 software was used to plot the contour map and the point for every datum point.

Electrical conductivity value for 2-D resistivity imaging method was obtained from the RES2DINV software. There are several calculations that needed to be done before the conductivity value for GPR method is obtained. Velocity, v of the EM wave is calculated using the value of depth, d and time, t from the radargram substituted into equation (1).

$$v = \frac{2d}{t} \quad (1)$$

The relative dielectric permittivity value, ε_r is calculated using equation (2) where c is the speed of light. Radar energy attenuation, α is calculated using equation (3) where σ is electrical conductivity value acquired from 2-D resistivity imaging method. α is in the unit of dB/m [7]. From Maxwell's equation, α can also be calculated in the form of equation (4) and can be derived to form equation (5).

Equation (5) is used to calculate σ that represents electrical conductivity for GPR method where ω is the angular frequency, ε is the dielectric permittivity and μ_0 is the magnetic permeability constant.

$$\varepsilon_r = \left(\frac{c}{v}\right)^2 \tag{2}$$

$$\alpha = 1.69\frac{\sigma}{\sqrt{\varepsilon_r}} \tag{3}$$

$$\alpha = \omega\sqrt{\varepsilon\mu_0}\left\{\frac{1}{2}\left[\sqrt{1+\left(\frac{\sigma}{\omega\varepsilon}\right)^2}-1\right]\right\}^{\frac{1}{2}} \tag{4}$$

$$\sigma = \omega\varepsilon\left[\left(\frac{2\alpha^2}{\omega^2\varepsilon\mu_0}+1\right)^2-1\right]^{\frac{1}{2}} \tag{5}$$

4. Results and discussions

GPR result is shown in figure 1 in the form of radargram. The maximum depth of penetration is approximately 8 m. The 20 m position is chosen for calculation of electrical conductivity of GPR because the value obtained for that position is used to compare with the 2-D resistivity imaging method and borehole log. There are several shapes that have hyperbolic features on the radargram but the shapes are not distinct. The chaotic signal from 27 m to 37 m distance is due to tree roots. The maximum depth for conductivity calculation is 3.5 m. Figure 2 shows brief soil description that is obtained from the borehole log and is used to integrate with the results from GPR and 2-D resistivity imaging method. The position of borehole log is denoted on the radargram as shown in figure 3.

The results of 2-D resistivity imaging method for Wenner-Schlumberger and pole-dipole arrays are shown in figure 4 and 5. The position of the borehole which is at 20 m distance is shown on the inversion results. The inversion of pole-dipole array showed better resistivity variations laterally than the inversion result from Wenner-Schlumberger array. Both inversions showed that the area at 20 m distance possibly composed of loose soil (150 – 1500 Ωm) with mixed grain sizes which can affect the resistivity value due to the ability to hold water. There is a boulder (1000 – 5000 Ωm) that can be seen in both inversions that situated near to the 20 m distance position. Saturated zone can also be seen in the pole-dipole array inversion with values ranging from 0 Ωm to 200 Ωm. The conductivity values at 20 m distance for certain depth are obtained from 2-D resistivity imaging data.

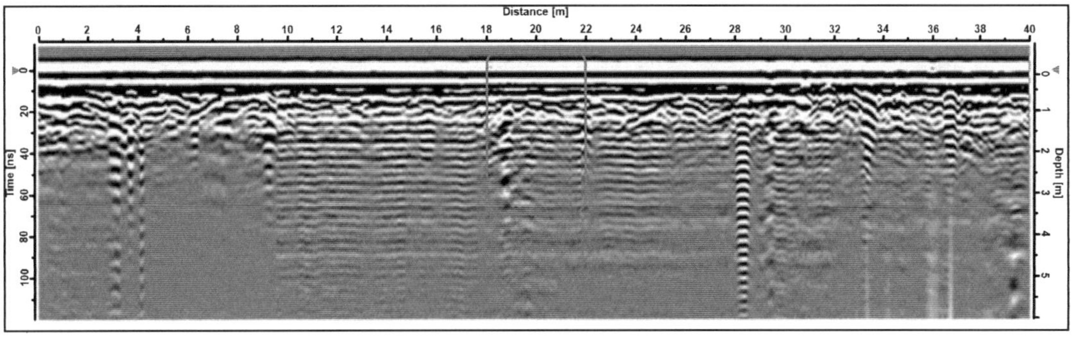

Figure 1. The radargram for the survey line.

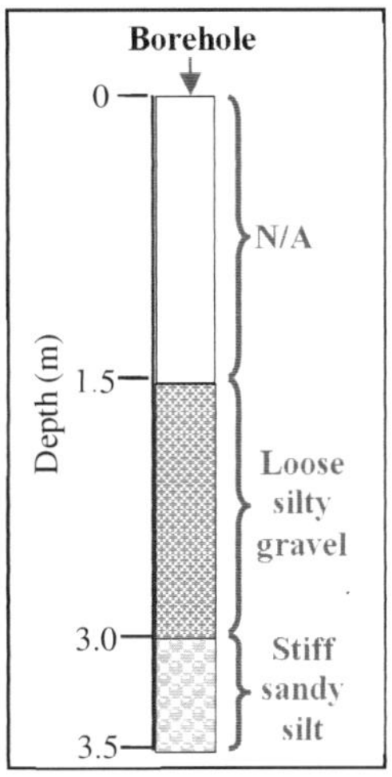

Figure 2. Brief soil description from the borehole log.

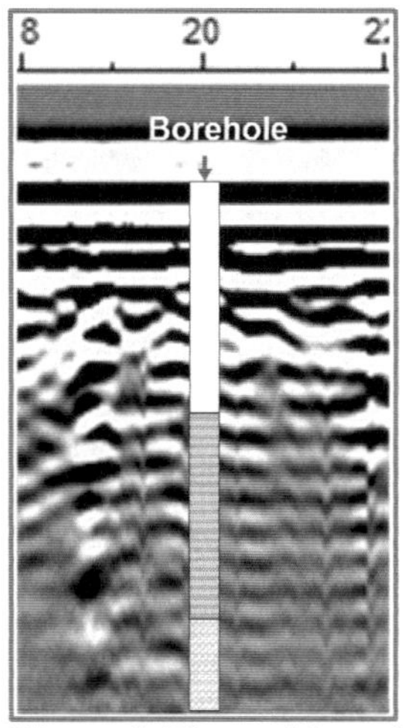

Figure 3. The position of the borehole is shown on the radargram.

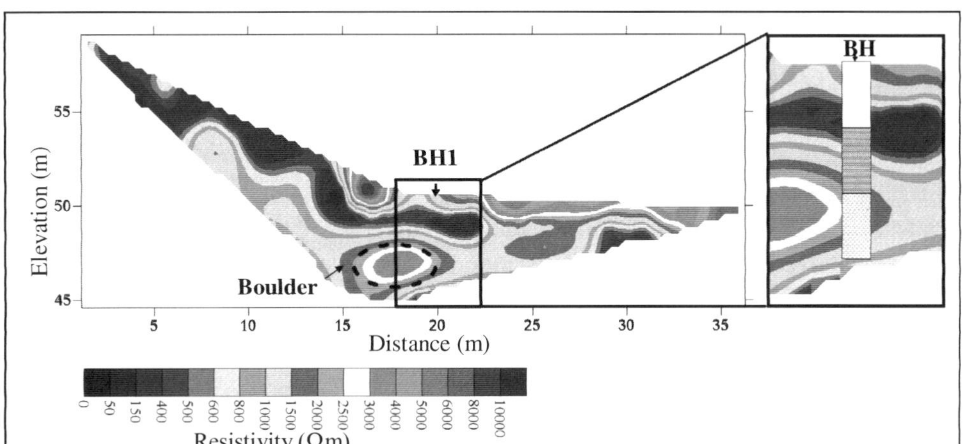

Figure 4. Inversion model of 2-D resistivity imaging method using Wenner-Schlumberger array.

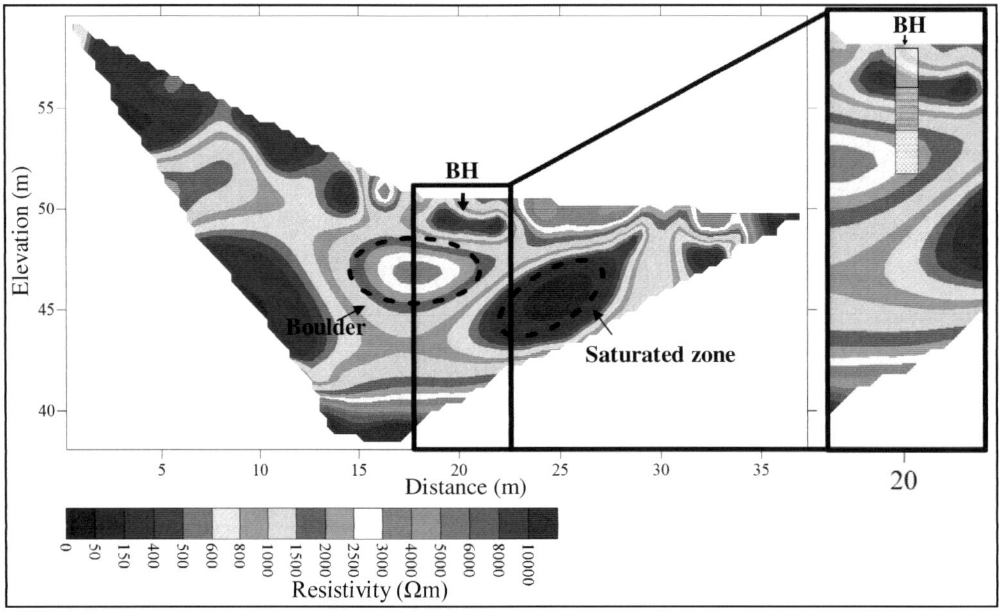

Figure 5. Inversion model of 2-D resistivity imaging using pole-dipole array.

Table 1 shows the electrical conductivity values from GPR and 2-D resistivity imaging (Wenner-Schlumberger and pole-dipole array) with soil descriptions and N-value from borehole log. From Table 1, near the surface (depth of 0.29 – 1.3 m) shows that the value of conductivity keeps on increasing with 3.0 mS/m as the highest value at depth of 1.3 m. The soil description for near surface depth is not available from the borehole log due to the loose top soil condition. The recovery for loose soil sample is not available due to the imprecise information of the soil. Hence, there is no recovery and N-value of the top soil. The conductivity value decreases from 2.0 mS/m to 0.4 mS/m which indicate loose silty gravel (depth of 1.81 – 2.99 m). Lower conductivity value may be due to the presents of gravel. The N-value representing depth of 1.81 – 2.99 m is 6. Sandy silt has lower conductivity from 0.4 – 0.7 mS/m due to the stiff condition of the soil that also contain some rock fragment (gravel). Since electrical conductivity is mainly affected by the water content [5], very stiff sandy silt with low conductivity values shows that presence of water is low. This is proven by the borehole log which indicated the N-value of 19 at depth of 3 – 3.5 m. The existence of silt in the area restricts the penetration depth of EM wave from GPR due to higher electrical conductivity [5]. The table conclude that soil having low electrical conductivity value (sandy silt) exhibits stiff condition that signifies low water content with greater amount of N-value.

Table 1. The integration of conductivity value, soil description and N-value.

Depth	Conductivity, mS/m		Borehole Log	
	2-D resistivity	GPR	Description of soil	N-value
0.29	0.74 - 0.97	0.70 - 1.00	N/A	N/A
0.6	1.08 - 1.30	1.05 - 1.35		
0.96	2.37 - 2.57	2.30 - 2.70		
1.3	2.76 - 2.92	2.70 - 3.00		
1.81	1.67 - 1.97	1.60 - 2.0	Loose, light brown reddish silty GRAVEL of low palsticity with some sand	6
1.89	1.61 - 2.94	1.60 - 3.00		
2.32	0.81 - 0.93	0.80 - 0.96		
2.99	0.43 - 0.45	0.40 - 0.46		
3	0.47 - 0.66	0.45-0.70	Very stiff, yellowish brown reddish sandy SILT of high plasticity with a little gravel, (rock fragment)	19
3.5	0.41 - 0.51	0.40-0.53		

5. Conclusions

The conductivity values calculated from GPR are in range with the values from 2-D resistivity imaging method. Electrical conductivity for the top soil is 0.7 – 3.0 mS/m. The soil description and N-value is not available due to imprecise sample of the loose soil condition. The integration of both methods with borehole showed the value of conductivity for loose silty gravel soil (N-value = 6) are betwee 0.4 – 3.0 mS/m. Presence of gravel may have produced the lower value of conductivity (0.4 mS/m). Very stiff sandy silt with rock fragment (N-value = 19) has conductivity value of 0.4 – 0.7 mS/m. Lower values of electrical conductivity consist lower amount of water content resulting stiff soil condition which is confirmed by the higher N-value. Integration of geophysical methods and geotechnical method is a success and the geophysical methods parameters can be used in understanding soil condition.

References

[1] Price D G 2009 *Engineering Geology*, ed de Freitas M H (UK: Springer) p 160
[2] Glendinning S, Jones C J F P and Lamont-Black J 2005 *Ground Improvement Case Histories* vol 3, ed I Buddhima and J Chu (UK: Elsevier) p 997
[3] Davis J L and Annan A P 1989 *Geophysical Prospecting* **37** 531-51
[4] Loke M H 1999 Electrical Imaging Surveys for Environmental and Engineering Studies **6574525** 4
[5] McNeill J D 1980 *Electrical Conductivity of Soils and Rock* (Ontario: Geonics Limited) pp 10-4
[6] Ahmad F, Yahaya A S and Farooqi M A 2006 *J. of Environmental Sc.* **2** 121-28
[7] Takahashi K, Igel J, Preetz H, and Kuroda S 2012 *Problems, Perspectives and Challenges of Agricultural Water Management*, ed M Kumar (Croatia: InTech) chapter 8 pp 155-80

Acknowledgements

The authors would like to extend sincere gratitude to all Geophysics Department Staff, Postgraduate students of School of Physics and Institute of Postgraduate Studies, Universiti Sains Malaysia for their assistance in acquiring the geophysical data and making the research a success.

Evaluation of Semanggol Formation (Permian Facies) Using Electrical Resistivity Tomography and Seismic Refraction Tomography Parameter

Hazrul Hisham[1], Dr. M.M. Nordiana[2], Dr. Teoh Ying Jia[2]

[1]Postgraduate student of Geophysics Section, School of Physics, Universiti Sains Malaysia, 11800 Penang, Malaysia
[2]Senior Lecturer of Geophysics Section, School of Physics, Universiti Sains Malaysia, 11800 Penang, Malaysia

E-mail: hazrul_hisham@outlook.com

Abstract. Outcrop studies are a fascinating part of geology as it evidently shows the aftermath of how the earth forming processes billion years ago. Outcrops do not cover majority of the Earth's land surface as it is covered by soils or vegetation thus cannot be seen clearly. In Kedah, Malaysia, there are many outcrops exposed in the state. The aim of this research studies was to correlate the parameters of geophysical survey with the properties of the Permian facies of Semanggol Formation in Kedah. The Permian facies consists of bedded chert and claystone. Two geophysical technique, electrical resistivity tomography (ERT) and seismic refraction tomography (SRT) were applied at the same line on top of the outcrop at Bukit Kukus beside Kulim – Baling (Kedah) road. The arrays used for ERT are Pole–dipole and Wenner–Schlumberger. The spacing between electrodes for ERT is 1.5 m while the geophone spacing for SRT is 2 m. Both ERT and SRT line is 60 m and 46 m respectively. Based on the results of both geophysical techniques, relating the porosity and permeability (poroperm) with geophysical parameters, it can be concluded that the bedded chert of low poroperm having seismic velocity and resistivity values at range of 1500 m/s – 2500 m/s and 1400 Ωm – 45000 Ωm. Whereas for claystone, it is very soft and laminated, concluding having high poroperm with seismic velocity between 600 m/s – 1200 m/s and resistivity values between 400 Ωm – 1000 Ωm.

1. Introduction

Outcrop can be defined as a visible exposure of bedrock or ancient superficial deposits on the surface of the earth. Due to frequent tectonic uplift or erosion processes, some locations of outcrops may be exposed such as on steep hillside, mountain ridges and riverbanks.

In Kedah, Malaysia, there are many outcrops and geological formations which having a certain number of rock strata that have a comparable lithology or other similar properties. The main focus of the outcrop studies is the Permian facies of Semanggol Formation located in Kedah. To understand the petrophysics of the Permian facies, thin section of the chert sample has been made by the petrography department at Mineralogy and Geosciences Department, Ipoh (JMG). From the thin section, the texture characteristics of the chert can be observed under the polarized microscope. Two types of geophysical survey are conducted on top of the outcrop; Electrical Resistivity Tomography (ERT) and

Seismic Refraction Tomography (SRT). In supporting the results of ERT and SRT, the petrophysics of chert specifically the porosity and permeability (poroperm) are analyzed. These methods are applied on top of the outcrops simultaneously. Both ERT and SRT gave parameters that can be related to the porosity and permeability of rocks.

2. Geological Setting of the Study Area

The study area is specified at the chert unit of Kulim – Baling area. In Kuala Ketil area as shown in Figure 1, isolated hills locally called Bukit Kukus has been excavated and exposed a part of the Semanggol Formation. The orientation of the section is almost north – south direction [1]. The northern part of the section is cut and leveled into one terrace contrarily to the southern part of the section which consists of two terraces. Generally, both sections exposed moderate to highly weathered mudstone, sandstone, tuffaceous sandstone, tuff, paraconglomerate, siliceous shale and chert that illustrate the chert unit of the Semanggol Formation [2].

There are four lithofacies recognized in the Permian part of the Semanggol Formation at Bukit Kukus. In an ascending sequence, these facies are laminated black mudstone, interbedded mudstone and sandstone, volcanogenic sediments and thinly bedded chert. The bedded chert is a pelagic facies which formed in the deep marine environment. [3].

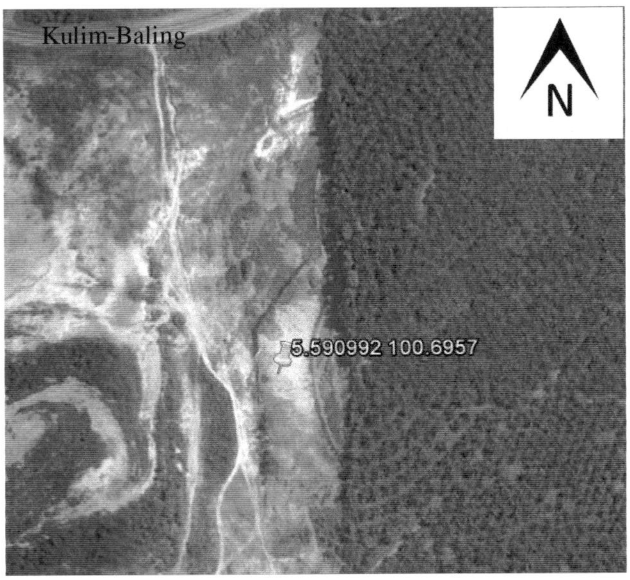

Figure 1. Google Earth map of the study area in Kuala Ketil.

3. Methodology

The layout of the survey line is conducted exactly on top of the 6.5 m height Permian facies outcrop of the Semanggol Formation as show in Figure 2. The electrode spacing for ERT is 1.5 m whereas for SRT, the geophone spacing is 2 m. Both ERT and SRT line is 60 m and 46 m respectively. For ERT, two arrays are applied on the field, Pole – dipole and Wenner – Schlumberger array. The Pole – dipole array has good horizontal coverage. Different from other common array, this array is an asymmetrical array thus over symmetrical subsurface structures the apparent resistivity anomalies in the pseudosection are asymmetrical. The Wenner – Schlumberger array is a combination form of the classical Schlumberger array and is averagely sensitive to both vertical and horizontal structures [4].

While for SRT, the shot points are between geophone 1 and 2, 6 and 7, 12 and 13, 18 and 19, 23 and 24. The offset is at -10 m and 30 m. Hammer is used as the source. The equipment used for ERT and SRT are ABEM Terrameter SAS4000 and ABEM Terraloc MK8 respectively.

After data acquisitions, ERT data is processed using Res2Dinv software while for SRT, it is processed using SeisOpt2D software. The tomography produced for both is then correlated with the outcrops. In correlating the geophysical tomography with the rock properties, rock sample is taken from the site for laboratory test. The rock sample of chert unit is send to the laboratory for porosity test measured by water immersion under vacuum. To get the intrinsic permeability, the rock sample is tested using nitrogen gas permeability test. From the result of porosity permeability, it is then correlated with the results of the geophysical tomography.

Figure 2. Permian facies outcrop of Semanggol Formation at Bukit Kukus.

4. Results and Discussion

Generally, the results of both ERT and SRT shows a good characterization of the outcrops. Since there are two types of facies presents at the study area; chert and claystone, therefore two distinct ranges of resistivity values and seismic velocities can be categorized. As shown in Figure 3, the Pole – dipole array clearly shows the similarity between the outcrops and the tomography result. There are two distinguishable low and high resistivity values. The resistivity values for claystone is between 400 Ωm to 1000 Ωm, whereas for chert, it has high resistivity values which is between 1400 Ωm to 45000 Ωm. The layout of the survey line is conducted exactly on top of the 6.5 m height chert unit outcrop of the Semanggol Formation as show in Figure 2.

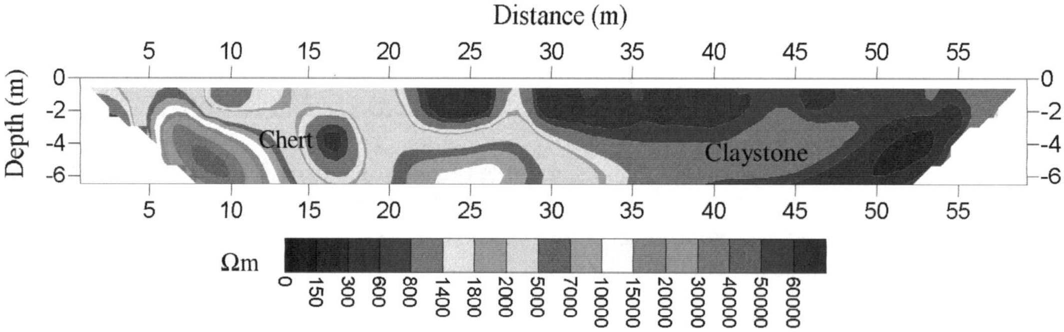

Figure 3. Inversion model of 2-D resistivity of Pole-dipole array at Bukit Kukus.

Figure 4 shows the inversion model for Wenner – Schlumberger array with a result of shallower depth compared to pole – dipole array, the values of the resistivity is the same as Pole – dipole array. The claystone resistivity is between 400 Ωm to 1000 Ωm. The resistivity of chert is from 1400 Ωm to 45000 Ωm. Therefore, tomography results also show good correlation with the outcrop.

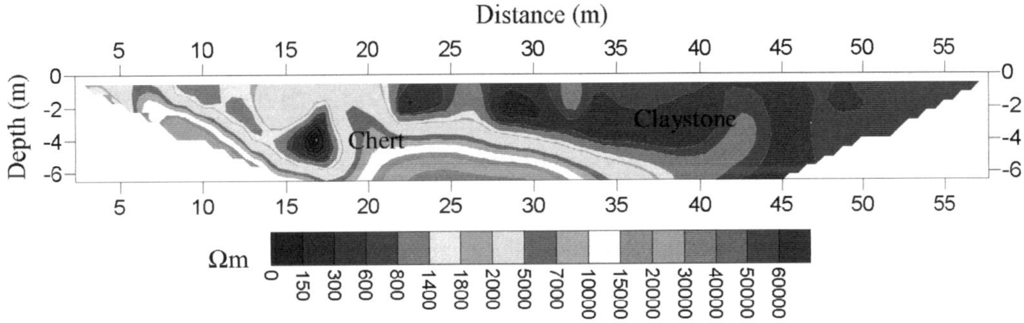

Figure 4. Inversion model of 2-D resistivity of Wenner-Schlumberger array at Bukit Kukus.

Based on the thin section of chert, the reason why the resistivity value is high is because of the radiolarian and sponge spicules is filled with quartz minerals as shown in Figure 5. Quartz minerals have high resistivity value from 4 x 10^{10} Ωm up to 2 x 10^{14} Ωm [5]. The thin section of claystone could not be made because of its soft characteristic. It is believed that there are many spaces between the laminated claystone that caused its soft properties thus lowering the resistivity value.

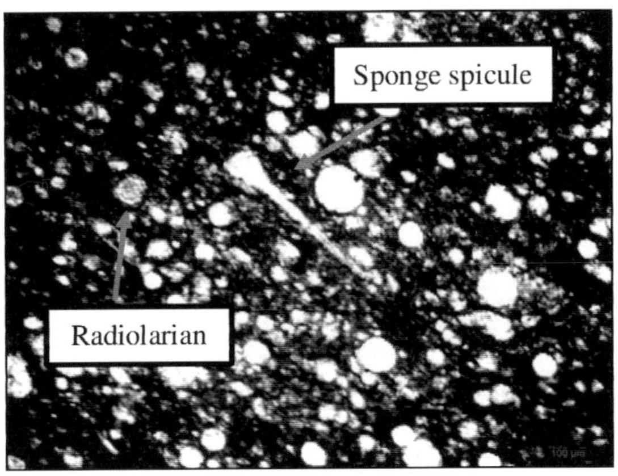

Figure 5. Thin section of chert.

Figure 6 shows an example of overlaying the inversion model of pole-dipole array on the outcrops to show that ERT results have good correlation to characterize the outcrop.

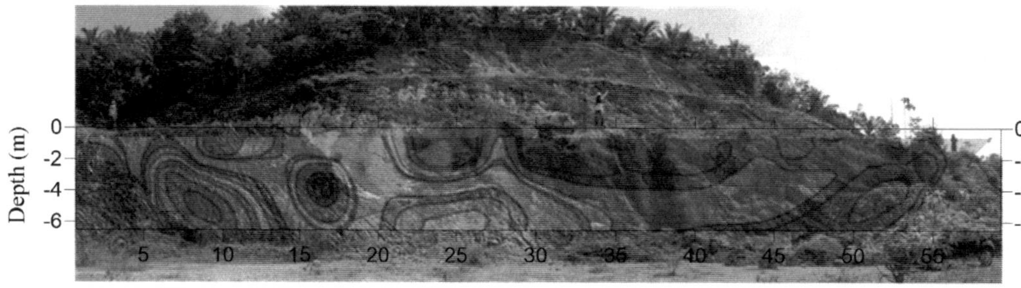

Figure 6. Overlaying inversion model of 2-D resistivity on the outcrop.

The tomography of seismic refraction survey shows a good correlation with the ERT results as shown in Figure 7. Seismic refraction results illustrate that chert has higher velocity values compared to the claystone due to its hard and brittle properties. The seismic velocity of claystone is between 600 m/s to 1200 m/s, whereby for chert the velocity is >1500 m/s. The in-filled radiolarian and sponge spicules with quartz has promotes the efficiency of sound waves to travel through the chert, compared to the laminated claystone having many air spaces between the layers.

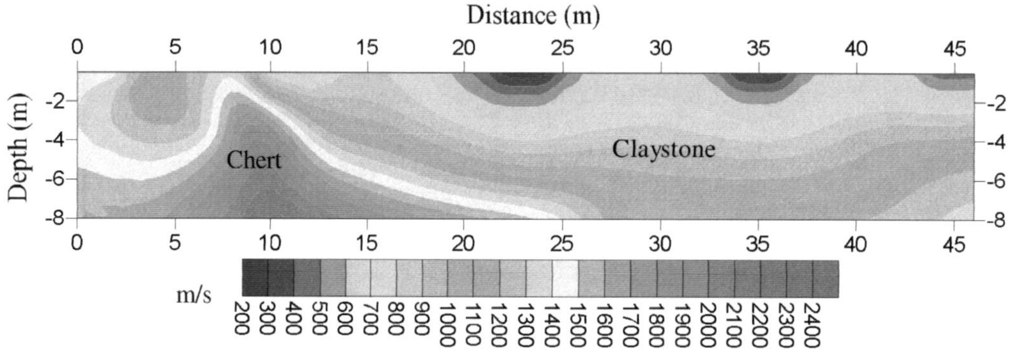

Figure 7. Seismic refraction tomography of chert unit at Bukit Kukus.

To correlate the results of both ERT and SRT with the properties of rocks specifically in terms of porosity and permeability, two types of separated test is performed. Firstly, by conducting nitrogen permeability gas test to get the value of intrinsic permeability. Secondly, the porosity of rock samples is measured by water immersion under vacuum. In this study, only chert is tested due to its dense, formed by diagenetic concentration of silica within the purest biogenic calcareous – diatomaceous sediments that accumulated with only little dilution by terrigenous material [6] thus can be handle by the instruments. Claystone could not be done due to its soft properties thus consider having high porosity and permeability. The porosity result of the chert is 3% whereas permeability value is 1.98 x 10^{-4} μd.

5. Conclusion

In this geophysical survey, both ERT and SRT gave good correlation with the outcrop physically. The characterization of the outcrop referring to the results of both geophysical tomography is nearly the same. This study contributes the application of geophysical method (ERT and SRT) in describing the properties of rock specifically in terms of porosity and permeability. Table 1 summarise the Permian facies of Semanggol Formation (SF) with its geophysical parameters and rock properties.

Table 1. Table of Permian facies of SF with its geophysical parameters and rock properties.

Permian Facies	ρ (Ωm)	v (m/s)	Porosity (%)	Permeability (μd)
Claystone	400 – 1000	600 – 1200	High	High
Chert	1400 – 45000	>1500	3	1.98×10^{-4}

References
[1] Harun Z, Jasin B, Mohsin N and Azami A 2009 Thrust in the Semanggol Formation, Kuala Ketil, Kedah. *Bulletin of the Geological Society of Malaysia* **55** pp. 61 – 66
[2] Jasin B and Harun Z 2007 Stratigraphy and sedimentology of the chert unit of the Semanggol Formation, *Geological Society of Malaysia, Bulletin* **53** pp. 103 – 109
[3] Baioumy H M and Ulfa Y 2016 Facies analysis of the Semanggol Formation, South Kedah, Malaysia: A possible Permian-Triassic boundary section. *Arabian Journal of Geosciences*, Vol. **9** No 8
[4] Loke M H 1999 *A practical guide to 2-D and 3-D resistivity surveys*
[5] Telford W M, Geldart L P and Sheriff R E 1990 *Applied Geophysics Second Edition*
[6] Behl R J and Garrison R E 1994 The Origin of Chert in the Monterey Formation of California (USA) Proc. *29th International Geological Congress* Part C pp. 101 – 132

Acknowledgement
The authors would like to thank all postgraduate students, laboratory staffs and lecturers of Geophysics Section, School of Physics, Universiti Sains Malaysia for their involvement in this research studies. The authors also would like to thank the Ministry of Higher Education (MOHE) for sponsoring the school fees under the MYBRAIN15 program and also to the Institue of Postgraduate Studies (IPS), Universiti Sains Malaysia for providing funds for the conference trip under the conference funds program.

Two-Dimensional Boundary Element Method Application for Surface Deformation Modeling around Lembang and Cimandiri Fault, West Java

M J Mahya[1], T A Sanny[2]

[1]Undergraduate Program of Geophysical Engineering, Institut Teknologi Bandung, Ganesha 10, Bandung 40132, Indonesia
[2]Exploration and Engineering Geophysics Research Group, Institut Teknologi Bandung, Ganesha 10, Bandung 40132, Indonesia

E-mail: milamuthia09@gmail.com

Abstract. Lembang and Cimandiri fault are active faults in West Java that thread people near the faults with earthquake and surface deformation risk. To determine the deformation, GPS measurements around Lembang and Cimandiri fault was conducted then the data was processed to get the horizontal velocity at each GPS stations by Graduate Research of Earthquake and Active Tectonics (GREAT) Department of Geodesy and Geomatics Engineering Study Program, ITB. The purpose of this study is to model the displacement distribution as deformation parameter in the area along Lembang and Cimandiri fault using 2-dimensional boundary element method (BEM) using the horizontal velocity that has been corrected by the effect of Sunda plate horizontal movement as the input. The assumptions that used at the modeling stage are the deformation occurs in homogeneous and isotropic medium, and the stresses that acted on faults are in elastostatic condition. The results of modeling show that Lembang fault had left-lateral slip component and divided into two segments. A lineament oriented in southwest-northeast direction is observed near Tangkuban Perahu Mountain separating the eastern and the western segments of Lembang fault. The displacement pattern of Cimandiri fault shows that Cimandiri fault is divided into the eastern segment with right-lateral slip component and the western segment with left-lateral slip component separated by a northwest-southeast oriented lineament at the western part of Gede Pangrango Mountain. The displacement value between Lembang and Cimandiri fault is nearly zero indicating that Lembang and Cimandiri fault are not connected each other and this area is relatively safe for infrastructure development.

1. Introduction

Lembang and Cimandiri faults are active faults in West Java that yield surface deformation as the effect of fault creeping process. Surface deformation must be analyzed to examine the disaster mitigation and infrastructure development planning. To determine the surface deformation pattern, GPS measurement was conducted in the area along Lembang and Cimandiri fault then the horizontal velocity values of each GPS station were corrected by the effect of Sunda plate horizontal movement. The horizontal velocity value then used as the input data of BEM calculation besides the data of shear modulus value, Poisson's ratio, and fault orientation to determine the displacement value at each point of observation along Lembang and Cimandiri fault that bounded by coordinate -7.4° S, 106.34°E and -6.6° S, 107.9° E.

IOP Conf. Series: Earth and Environmental Science **62** (2017) 012009 doi:10.1088/1755-1315/62/1/012009

BEM is a method to solve the partial differential equation besides Finite Element Method (FEM), and Finite Difference Method (FDM). The use of BEM instead of FEM and FDM in this research suits the distribution of GPS stations that only obtained at the particular point along the fault, as the observation boundary, agree with BEM characteristic. Moreover, BEM can reduce the problem complexity [1]. BEM in this research is computed numerically and generate the displacement value at each point of observation along Lembang and Cimandiri fault.

Lembang and Cimandiri fault are located on Sunda block that resulted from the subduction of Indo-Australia plate beneath Eurasia plate with velocity 67 mm / year [2]. The subduction with northwest direction at Oligocene dan Miocene generated the southwest-northeast oriented structure like Cimandiri fault, and the current subduction generate the east-west oriented structure like Lembang fault [3]. Lembang fault is an active fault and segmented into the older eastern segment and the younger western segment [4]. Cimandiri fault that formed at Miocene era has three major lineament orientations: east-west, northwest-southeast, and southwest-northeast [5].

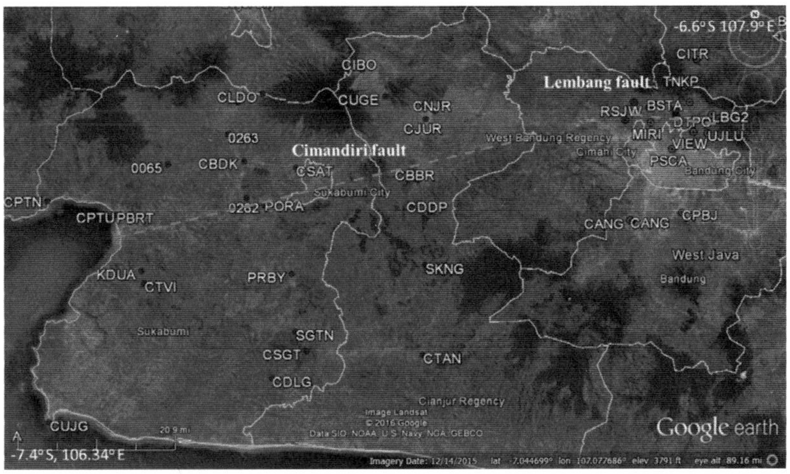

Figure 1. GPS stations distribution along Lembang and Cimandiri fault (Google Earth. 2015)

2. Method

To investigate the surface displacement distribution around strike-slip faults in Lembang and Cimandiri fault region, two-dimensional boundary element method (BEM) is applied. The BEM employs the differential equations of Laplace equation, similar to the finite-element method, but only requires the discretization of observation area boundaries. Model boundaries are defined as S and observation area is defined as D. Laplace equation is defined by equation (1) [6].

$$\nabla^2\varphi(\mathbf{p}) = \frac{\partial^2\varphi(\mathbf{p})}{\partial x^2} + \frac{\partial^2\varphi(\mathbf{p})}{\partial y^2} = 0 \quad (\mathbf{p} \in D) \tag{1}$$

Where : $\varphi(\mathbf{p})$ = the displacement value at observation point **p**

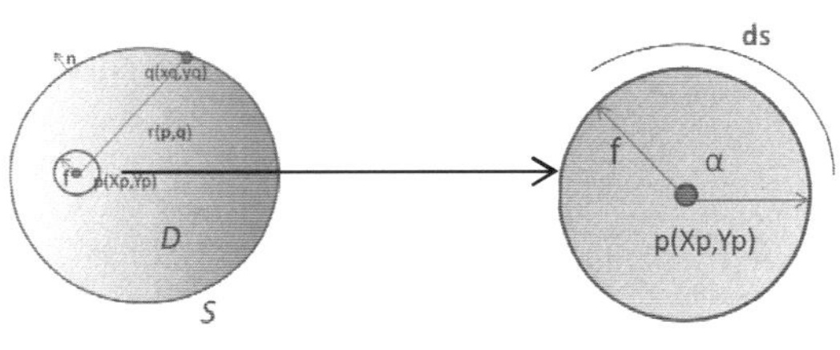

Figure 2. Observation area and observation boundary [6]

Green's function that defined at equation (2) is the fundamental solution of 2-D Laplace equation [6]. Green's function is a function that correlates the value of **p** point to **q** point where the displacement is known and has the distance *r* (**p**,**q**) from **p**.

$$G(\mathbf{p},\mathbf{q}) = \frac{1}{2\pi}\ln(|r(\mathbf{p},\mathbf{q})|) \tag{2}$$

$$\frac{\partial G(\mathbf{p},\mathbf{q})}{\partial \boldsymbol{n}} = \frac{\partial G(\mathbf{p},\mathbf{q})}{\partial f}\frac{\partial f}{\partial \boldsymbol{n}} = \frac{1}{2\pi f} \tag{3}$$

Because **p** is sorrounded by the area with radius *f*, so the boundary became $S + Sf$ and the observation area became $D - Df$. By applying Green's second identity in equation (4), the value of $\varphi(\mathbf{p})$ at any observation point can be determined [6].

$$\int_{D-Df}\left[\varphi\nabla^2 G(\mathbf{p},\mathbf{q}) - G(\mathbf{p},\mathbf{q})\nabla^2\varphi\right]dD = \int_{S+Sf}\left[\varphi\frac{\partial G(\mathbf{p},\mathbf{q})}{\partial \boldsymbol{n}} - G(\mathbf{p},\mathbf{q})\frac{\partial\varphi}{\partial \boldsymbol{n}}\right]dS \tag{4}$$

The value of $\nabla^2\varphi = 0$ and the value of $\nabla^2 G$(p,q) = 0 in each point of observation, so the left segment of equation (4) is equal to zero [6].

$$-\int_S\left[\varphi(\mathbf{q})\frac{\partial G(\mathbf{p},\mathbf{q})}{\partial \boldsymbol{n}} - G(\mathbf{p},\mathbf{q})\frac{\partial\varphi(\mathbf{q})}{\partial \boldsymbol{n}}\right]dS = \int_{S_f}\left[\varphi(\mathbf{p})\frac{\partial G(\mathbf{p},\mathbf{q})}{\partial \boldsymbol{n}} - G(\mathbf{p},\mathbf{q})\frac{\partial\varphi(\mathbf{p})}{\partial \boldsymbol{n}}\right]dS \tag{5}$$

$$-\int_{S_f}\left[\varphi(\mathbf{q})\frac{\partial G(\mathbf{p},\mathbf{q})}{\partial \boldsymbol{n}} - G(\mathbf{p},\mathbf{q})\frac{\partial\varphi(\mathbf{q})}{\partial \boldsymbol{n}}\right]dS = \frac{1}{2\pi}\int_0^{2c(\mathbf{p})\pi}\left[\varphi(\mathbf{p})\frac{1}{f} - \ln\frac{1}{f}\frac{\partial\varphi(\mathbf{p})}{\partial \boldsymbol{n}}\right]f d\alpha = c(\mathbf{p})\varphi(\mathbf{p}) \tag{6}$$

Where : $c(\mathbf{p}) = \begin{cases} 1, & \text{if } \mathbf{p} \text{ is on observation area} \\ \frac{1}{2}, & \text{if } \mathbf{p} \text{ is on observation boundary} \\ 0, & \text{if } \mathbf{p} \text{ is outside observation area} \end{cases}$

By substituting equation (6) to equation (5), and applied to any i-**p** and j-**q**, and for any point i=j, equation (7) is obtained, and the value of $\frac{\partial\varphi(\mathbf{q}_j)}{\partial \boldsymbol{n}}$ can be determined. By substituting the value of $\frac{\partial\varphi(\mathbf{q}_j)}{\partial \boldsymbol{n}}$ to equation (5), the value of $\varphi(\mathbf{p}_i)$ in each point of observation can be determined.

$$\left[c(\mathbf{p}_i) + \int_{S_j}\frac{\partial G(\mathbf{p}_i,\mathbf{q}_j)}{\partial \boldsymbol{n}}dS\mathbf{q}_j\right]\varphi(\mathbf{q}_j) = \int_{S_j} G(\mathbf{p}_i,\mathbf{q}_i)\,dS\mathbf{q}_j\sum_{j=1}^{N}\frac{\partial\varphi(\mathbf{q}_j)}{\partial \boldsymbol{n}} \tag{7}$$

In this research the model boundaries are Lembang and Cimandiri fault lines and the observation points are distributed along the faults. BEM models are therefore useful for determining the geometry of prescribed fault. The application of the boundary element method in this research requires a mesh of the displacement value at the boundary of the observation area. The displacement value that used in this research are the horizontal slip rate values of GPS stations around Lembang and Cimandiri fault that have been corrected by the effect of Sunda block movement to separate the displacement that resulted by fault creeping process from the displacement that resulted by Indo-Australia plate subduction beneath Eurasia plate. GPS data is obtained from 2006 until 2011 for Lembang fault and from 2008 until 2013 for Cimandiri fault. The assumptions that used at the modeling stage are the deformation occurs in homogeneous and isotropic medium, and the stresses that acted on faults are on elastostatic condition. As the observed medium is homogeneous and isotropic, the value of shear modulus and Poisson's ratio is defined as the elastic value of volcanic rock as observed at observation area even the lithological units are varied along the observation area. BEM in this research is calculated numerically in MATLAB R2012a and requires the discretization of observation boundary and observation area. Observation area in this research is discretized using 0.05° grid because it yields the same result with 0.01° grid but requires less computational memory. The Lembang fault line as observation boundary was discretized into 26 elements and the Cimandiri fault line was discretized into 34 elements. The computational solution then results the approximate solution in displacement parameter at 544 selected points in the observation area as present at figure 3.

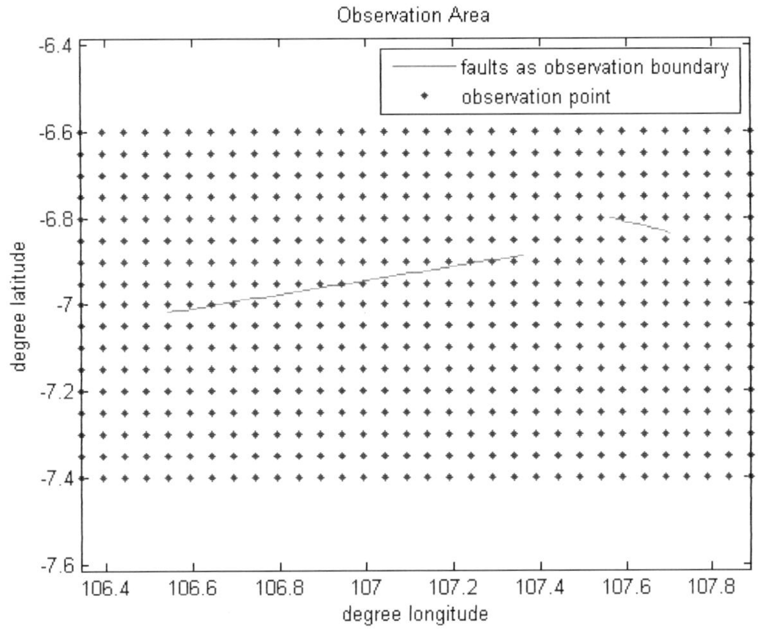

Figure 3. Lembang and Cimandiri fault as observation area boundary and distribution of observation points

3. Result

Research results are presented with displacement distribution map that overlayed with topography map. Displacement distribution is divided in the x-direction (UX) and the y-direction (UY). Positive value in the x-direction (UX) displacement distribution map indicates the displacement is toward the north, and the positive value in the y-direction (UY) displacement distribution map indicates the displacement direction is toward the east. The approximated Lembang fault line is shown with a black dashed line, the approximated Cimandiri fault line is shown with a red dashed line and the interpreted fault's segment boundary are shown with a white dashed line.

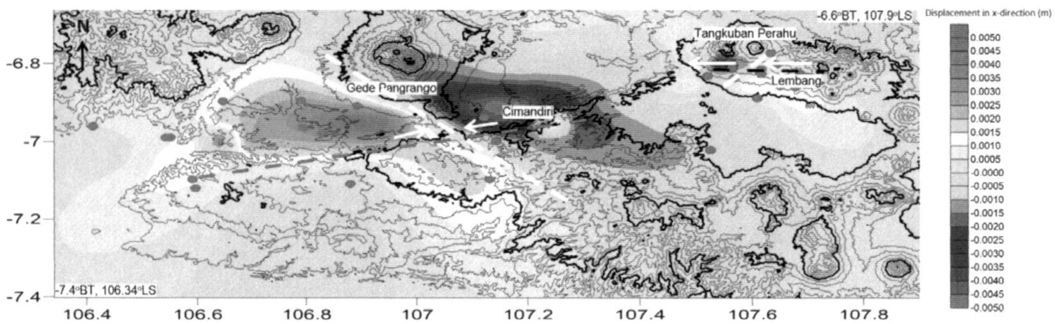

Figure 4. Displacement distribution map in x-direction (UX)

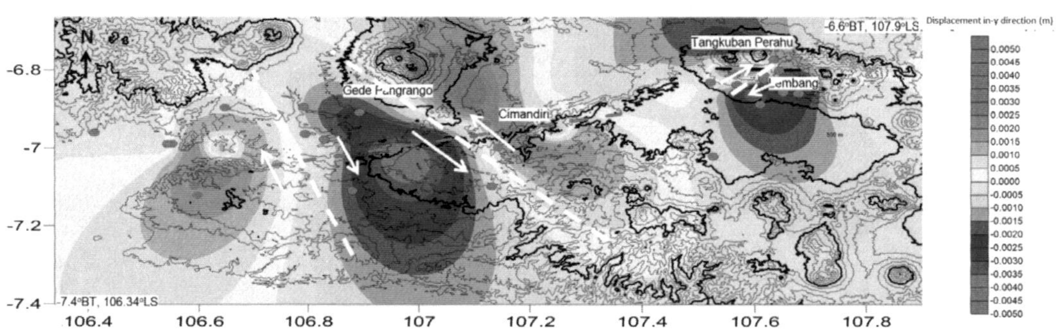

Figure 5. Displacement distribution map in y-direction (UY)

The displacement in the x-direction at the eastern part of Lembang fault has positive value while the western part is dominated with the negative value. Lembang fault has left-lateral slip component as the displacement value at the northern block is less than the displacement value at the southern block, so the northern and the southern blocks are moving to the left direction relatively to each other. Displacement distribution map the in y-direction also shows the displacement direction difference between the eastern and the western part of Lembang fault. The eastern part tends to displace with the negative value and the western part tends to displace with the positive value. The direction difference boundary is located near Tangkuban Perahu Mountain, the same place with the direction difference boundary shown at the displacement distribution map in the x-direction. From the similarities of the lineament shown in displacement distribution map in the x-direction and the y-direction, Lembang fault is interpreted to be divided into two segments, the eastern and the western segment, that separated by a lineament with northeast-southwest orientation with right-lateral slip component as shown in the displacement distribution map in the y-direction.

According to the displacement distribution map in the x-direction, Cimandiri fault is segmented into the eastern segment and the western segment. The eastern segment has left-lateral slip component that shown with the greater negative displacement value at the northern block compared with the southern block of Cimandiri fault. The western segment has right-lateral slip component that shown with the greater positive displacement value at the northern block compared with the southern block of Cimandiri fault. The boundary between the eastern and the northern segment of Cimandiri fault forms a lineament oriented at northwest-southeast direction at the western part of Gede Pangrango Mountain.

The displacement distribution map in the y-direction indicates that the western part of Cimandiri fault is divided into the eastern part of the western segment that tends to displace in the negative value and the western part of the western segment that tends to displace in the positive value. The eastern segment of Cimandiri fault has positive displacement value. The boundary between the eastern segment and the eastern part of the western segment of Cimandiri fault shows a lineament in the same place and orientation with the lineament that shown at the displacement distribution map in the x-direction of Cimandiri fault. This lineament has left-lateral slip component pursuant to the displacement distribution map in the y-direction. The boundary between the eastern part of the western segment and the western part of the western segment also shows a lineament that oriented at northwest-southeast direction with right-lateral slip component pursuant to the displacement distribution map in the y-direction. The displacement value between Lembang and Cimandiri fault is nearly zero as shown at the displacement distribution map in the x-direction and the displacement distribution map in the y-direction and indicates that the Lembang and Cimandiri fault are not connected each other.

4. Conclussion

The models that results from BEM calculation can show the fault segmentation with the input of displacement value near the fault. Lembang fault has left-lateral slip component and is interpreted to be divided into two segments, the eastern and the western segment, that separated by a lineament with northeast-southwest orientation with right-lateral slip component. Cimandiri fault is segmented into the eastern segment and the western segment. The eastern segment has left-lateral slip component and the western segment has right-lateral slip component. The western part of Cimandiri fault is divided into the eastern part of the western segment that tends to displace to south direction and the western part of the western segment that tends to displace to north direction. The displacement value between Lembang and Cimandiri fault is nearly zero indicates that the Lembang and Cimandiri fault are not connected each other, so the area between Lembang and Cimandiri fault is relatively safe for infrastructure development.

References

[1] Maerten F, Maerten L and Cooke M 2010 *Comput Geosci* **14**: 551. doi:10.1007/s10596-009-9170-x

[2] DeMets C, Gordon R G, Argus D F and Stein S 1994 Effect of recent revisions to the geomagnetic reversal time scale on estimates of current plate motions *Geophys. Res. Lett* **21**: 2191–94. doi:10.1029/94GL02118

[3] Simons W J F., et al. 2007 A decade of GPS in Southeast Asia: Resolving Sundaland motion and boundaries *J. Geophys. Res* **112**, B06420, doi:10.1029/2005JB003868

[4] Dam M 1997 The late quaternary evolution of the Bandung basin, West-Java, Indonesia *Misc, Geological Research Centre, Ministry of Mines* 240

[5] Malod J A, Karta K, Beslier M O and Zen M T 1995 From normal to oblique subduction: Tectonic relationships between Java and Sumatra *Journal of Southeast Asian Earth Sciences* **12**(1-2): 85-93

[6] Becker A 1992 The boundary element method in engineering *Mcgraw-Hill*, ISBN-13: 978-0077074395

Horizontal-vertical Spectral Ratio Method in Microtremor to Estimate Engineering Bedrock Thickness at Sedati Mud Volcano

Alif Prabawa Arwananda[1*], Bagas Aryaseta[1], Hafidz Dezulfakar[1], Yosar Fatahillah[1], Juan Pandu Gya Nur Rochman[1]

[1]Geophysical Engineering Department, Institut Teknologi Sepuluh Nopember Surabaya, Geomatics Building Second floor, Kampus ITS Sukolilo, Surabaya, 60111

[1*]Corresponding author: alifp.arwananda94@gmail.com

Abstract. Based on field study, Sedati Mud Volcano located in a line with Gunung Anyar Mud Volcano and occurred by increased pressure in the compression area and rapid loss of gas. The combination of both fast-growing constructions of infrastructures and the presence of the mud volcanoes brings new challenges in Sidoarjo city. The purpose of this scientific research is to determine the sedimentary thickness around Sedati mud volcano. Only a few data show real amplitude spectrum, which represent high contrast impedance. At some point, there are several peaks indicating the presence of contrast impedance between layers. Based on 20 processed data, Sedati Mud Volcano has a 30 – 70m engineering bedrock thickness and natural frequency between 0.5 until 14.4 Hz. The enhancement of natural frequency tends to occur along decrement of layer thickness in the upper basement layer. The result shows the natural frequency parameter and its amplification is slightly variated around Sedati Mud Volcano, as caused by sedimentary lateral depth variation and/or the presence of variation on existing rock. Further analysis indicates a fault inside the area of mud volcano as possible reason behind the occurring mudflow.

1. Introduction

Mud Volcano has become a popular discussion lately, especially after the tragedy of Lapindo Mud Volcano in Sidoarjo. Although, it is still debated for the cause of its eruption.

Mud volcanoes phenomenon indicates a strong active tectonic setting. It refers to the expression of the natural topography of the volcano cone-shaped formations created by geological factors and the released sedimentary liquid, clay fragment, liquid, and gas. Most of the released gases are methane, with some carbon dioxide and nitrogen [1][2].

Many cases of earthquakes show that the damage depends on topography and the surface geology. The earthquake that should not have a destructive character can be dangerous due to the amplification of the earthquake. One method that is used for mapping prone areas to earthquake damage (seismic microzonation) is microtremor method. This passive geophysical method shows how the influence of the effect in the earthquake zone and the soft soil of the building in an area [3].

The purpose of this scientific research is to determine the sedimentary thickness by using frequency of spectral ratio from microtremor data. Thus, microzonation may explain the value of earthquake vulnerability in each region.

The thickness of engineering bedrock is one aspect that needs to be considered in determining the location of construction project. As it will affect the durability and safety of the building infrastructure. Sidoarjo is a city that is growing rapidly nowadays. Hence, it takes geophysical information relating to the scope of the geotechnical planning for urban infrastructure such as roads, bridges, housing, etc. This research evaluate the dynamic characteristics of soil from the wave propagation. Some physical parameters can be seen from the propagation of the wave is the wave velocity, amplitude variations, wave period and the natural frequencies [4].

2. Methodology

2.1. HVSR Method

Amplitude Horizontal to Vertical Fourier Spectral Ratio Method or HVSR also known as Quasi-Transfer Spectra or QTS was first introduced by Yutaka Nakamura in 1989 [4]. We used this method to estimate the natural frequency and amplification of the regional geology from microtremor data. Throughout the development, this method arose its capability to estimate the vulnerability index for land and building vulnerability index [5][6].

The main parameter from HVSR is the natural frequency or f0 and amplification. Measured HVSR in the soil applied in order to characterize the region, natural frequency and amplification to which corresponds to subsurface physical parameters [7].

In connection with HVSR to local geological characterization, one should understand the factors effecting the subsurface parameters that contribute to change in the natural frequency (f0) and amplification. This serves three purposes. First, the correctly interpretation of microzonation with using HVSR. Second, determining the cause of the damage will be determined with certainty. Third, as a consideration in extracting the subsurface parameter from HVSR curve. [8] introduced the application of HVSR equation for vibration measurement at the surface.

HVSR analysis method as developed by [4], calculates the ratio of Fourier spectrum of the microtremor signal from horizontal component to the vertical component. The result of HVSR analysis will show a peak in the predominant frequency spectrum [4]. Dominant frequency (f0) and the amplification factor (A) describes the dynamic characteristics of the land resulting from the analysis of HVSR [9].

One of the main aspect of using microtremor in the study area is the rarity of occurring earthquake. Hence, the research only requires ambient noise from around the study area. Furthermore, this method does not require any use of costly and time-consuming methods such as borehole or seismic methods.

2.2. Engineering Bedrock Thickness Calculation

[10] Gives an equation used to determine the thickness value of engineering bedrock layer. This research utilizes Equation (1) to calculate thickness of layer of soft sedimentary bedrock (m). The process starts from determining the value of Vs30 from USGS with respect to the research location to which corresponds with Vo, fr or dominant natural frequency from microtremor data and X as exponential factor.

$$m = \left[\frac{Vo(1\text{-}X)}{4fr} + 1 \right]^{1/1-X} - 1 \tag{1}$$

2.3. Data Acquisition

In this research, we acquired 20 data from the Sedati Mud Volcano Area. Each data was taken through 40-minute and 60-minute recording time. This difference in recording time provided simpler time window picking process and obtaining better f0 value.

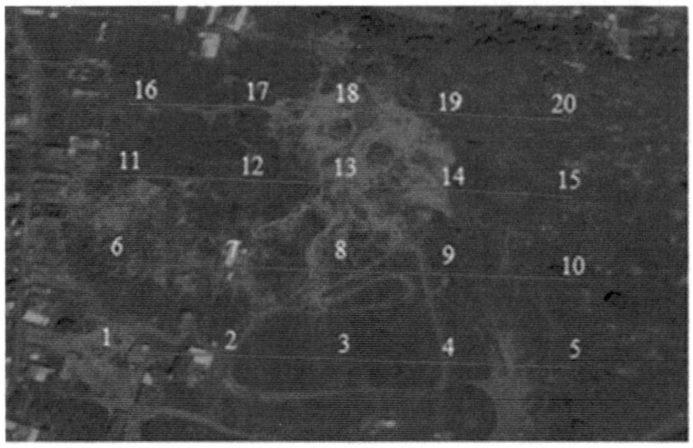

Figure 1. Line acquisitions shown in four red line, moving from West to East.

3. Result and Analysis

In this study, HVSR curve is evaluated based on [11] criteria. This evaluation determines whether the H/V curve and peak is reliable or unusable. The process will determine the accuracy and reliability of natural frequency obtained.

We processed the data by using 20 – 50 second window length. Based on [11] suggestion, data should have minimal value of $lw = 10/f0$, with lw by means of window length and f0 as resonance frequency. Thus, each window provided the 10 minimum significant cycle.

The result of processing 20 data yields predominant frequency around 0.5 Hz – 14.4 Hz. HVSR curves and f0 value are divided into two categories, the first category can be seen in figure 2 which have a wide H/V value at the beginning of the data and narrowed at the end. Peak frequency of first category has a range values between 0.5 Hz – 0.8 Hz. The location of the measurement also has the same characteristic, at point 1, 2, 3, and 11. According to [11], H/V curve of first category is the result of measurement on natural soil.

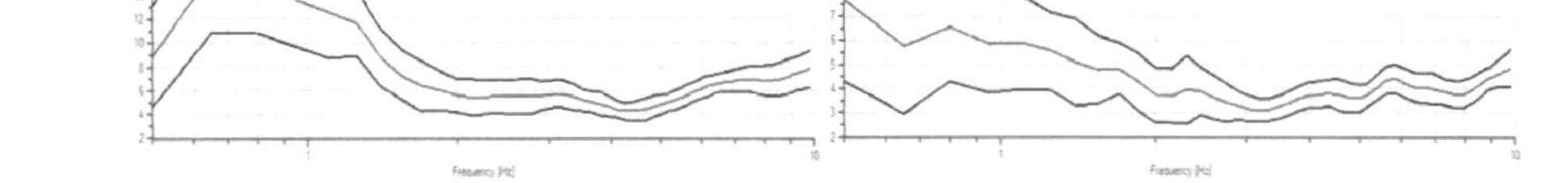

Figure 2. H/V curve for first category has a wide H/V value at the beginning and narrows at the end of the curve. This type of H/V curve occurred at first measurement (a), second (b), third (c), and the eleventh (d). This type of curve is listed on [11] as natural soil H/V curve (e).

The second category is H/V curve with low value at the beginning of the data and getting higher at the end of the data. This HVSR curve has a peak frequency between 7 Hz – 16 Hz. Based on geological condition of measurement locations, point 6-10 and 15-19 are located at the edge of Sedati Mud Volcano crater. Second category of HVSR curve indicated that measurements was conducted on a thin surface layer [12], research conducted by [13] also proved that microtremor data with more than

10 Hz indicate a thin surface layer. Nevertheless, data with f0 value of more than 10 Hz can't be used as the microtremor's maximum frequency is 10 Hz.

The remaining data is not included in the two categories because it has its own characteristics on its curve shape and predominance frequency. Point 4 has similarities as in category 2, the difference is the amount of environmental noise occurred at point 4 where the data obtained does not have a clear peak and constant form. Point 5 is the only point with a clear peak; we assumed that the measurement point is not in geological area within the mud volcano. Point 12 has similar characteristics to category 2; the difference lies in the different value of frequencies (0.781 Hz). The differences occur on the location of cracked and split ground, so an optimal vertical wave was not recorded. Furthermore, data measured near the mud volcano crater has curve similar to point 12, with H/V range of values between 4 – 8 with 0.6 Hz and 0.8 Hz frequency, respectively. Low frequencies were caused by the depth of layer in the Sedati Mud Volcano crater.

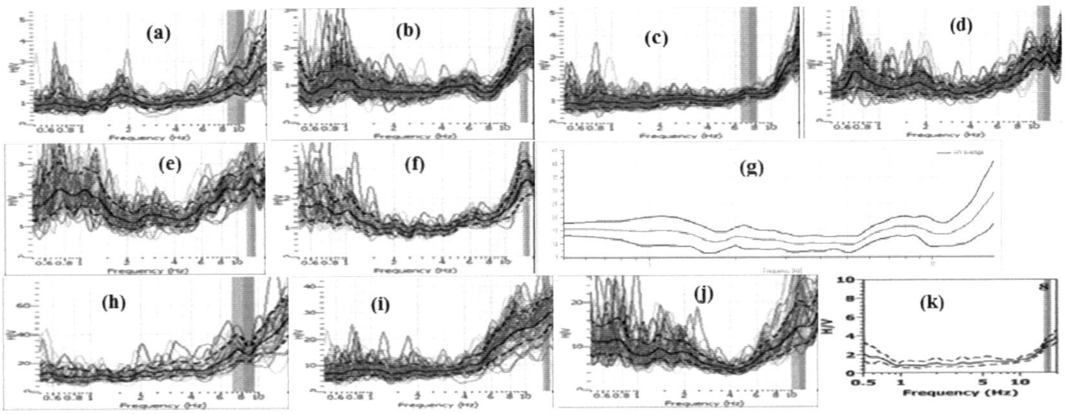

Figure 3. H/V curve for second category has a narrow H/V value at the beginning and peaks at the end of the curve. This type of H/V curve occurred at sixth measurement (a), seventh (b), eight (c), ninth (d), tenth (e), fifteenth (f), sixteenth (g), seventeenth (h), eighteenth (i), and nineteenth (j). This type of curve is listed on [12][13] as thin surface layer H/V curve (k).

As valid microtremor data has a natural frequency value less than 10 (f0 < 10 Hz) and with amplification factor more than 2 (A > 2), only several f0 and A0 data can be used. In these measurements, only 9 out of 20 data meet these requirements. Other 7 data is unused due to natural frequencies exceeding 10 Hz and 4 other data due to amplification value below 2. Unused data due to invalid natural frequency and amplification, can be seen In table 2 as labelled with yellow and pink colour, respectively.

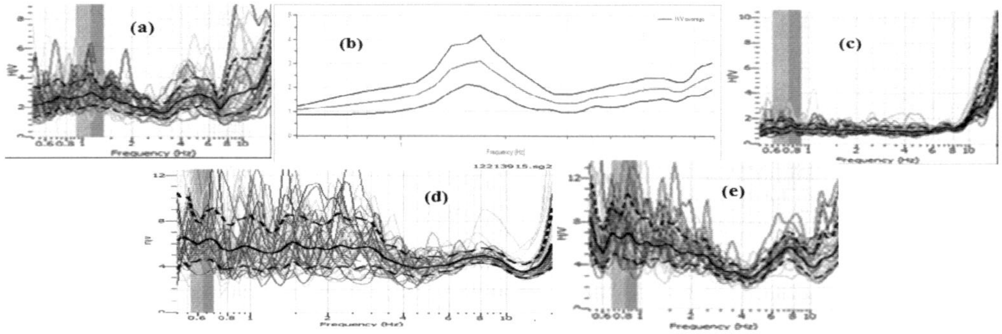

Figure 4. Uncategorized HVSR curve because of its own characteristic, fourth measurement (a), fifth (b), twelve (c), thirteenth (d), and fourteenth (e).

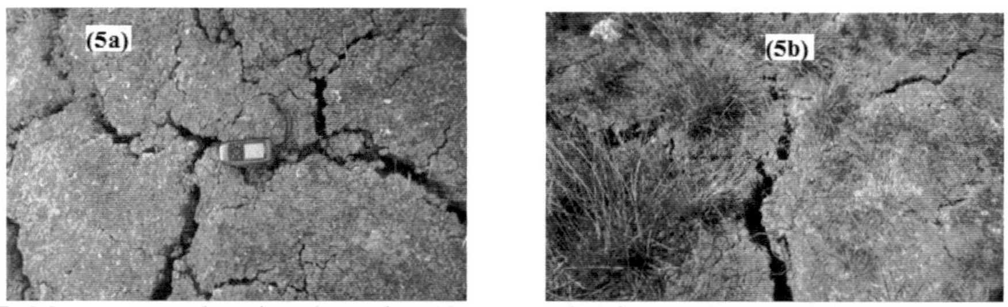

Figure 5. Measurement location for point 12 (5a) and point 9 (5b), many other areas have same geological surface condition which affecting data quality and resulting the second category or uncategorized HVSR curve.

From 9 reliable data, engineering bedrock thickness maps is created using Vs30 and coordinate point of measurements. As can be seen from figure 6 that soft sedimentary layer or engineering bedrock has a ±70m thickness on the east side of measurements (point 1, 6, 11, and 16) and diminish towards the west side. Figure 7 is slicing map from figure 6 to show the conditions below the surface. Correlation with [10] data lies in the presence of southwest – northeast regional trend. Correlation between the two data took place in locations where the regional trend is directing towards southwest - northeast (figure 6) which indicates the possibility of continuity of Watukosek fault that crosses this area. In figure 6, contour line for 70 – 78m depth symbolizes the fault trending southwest – northeast. From the interpretation of previous research [4][10] it can be seen that Watukosek fault have continuity heading towards Madura Island, and the location of the centre point of the mud volcano bursts Gunung Sedati located on the eastern side of Watukosek fault line.

Point	Name	A0	F0	x	y	vs30	h
1	21121228	2.16159	0.729616	697353	9181547	210	71.95566
2	21130631	16.5	0.8	697433	9181542	210	65.625
3	21135716	16	0.8	697480	9181543	210	65.625
4	21144358	2.99244	1.14041	697554	9181542	210	46.03607
5	21153451	3	1.7	697589	9181535	210	30.88235
6	21162539	1.97914	9.59679	697366	9181602	210	5.470579
7	21184444	2.05223	14.0039	697432	9181610	210	3.748956
8	21193153	1.30373	7.29059	697478	9181603	210	7.201063
9	21201904	2.32894	12.2059	697540	9181605	210	4.301199
10	22101336	2.54615	12.6325	697569	9181615	210	4.155947
11	21171245	1.15402	4.99618	697363	9181656	210	10.50803
12	21211518	1.40906	0.781512	697420	9181646	210	67.17747
13	23103808	7.37603	0.681167	697474	9181679	210	77.07361
14	23112610	6.95482	0.837098	697536	9181663	210	62.71667
15	22111302	2.59993	13.531	697565	9181652	210	3.879979
16	23122255	13.2134	0.734559	697362	9181708	210	71.47145
17	23130855	29.0464	7.54541	697411	9181727	210	6.957872
18	23135623	30.0473	14.4934	697484	9181716	210	3.622338
19	23145144	14.4091	11.3954	697543	9181715	210	4.607122
20	23154243	25.1199	0.808827	697599	9181714	210	64.90881

Table 1. Used data for measuring engineering bedrock thickness labelled with blue colour, while yellow colour represents unreliable data because it has f0>10 Hz, and pink colour represents unreliable data because it has A0>2.

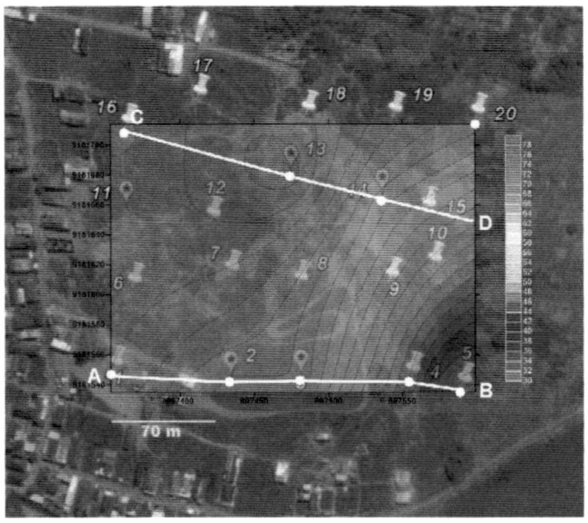

Figure 6. Engineering bedrock thickness map for Sedati Mud Volcano, Sidoarjo.

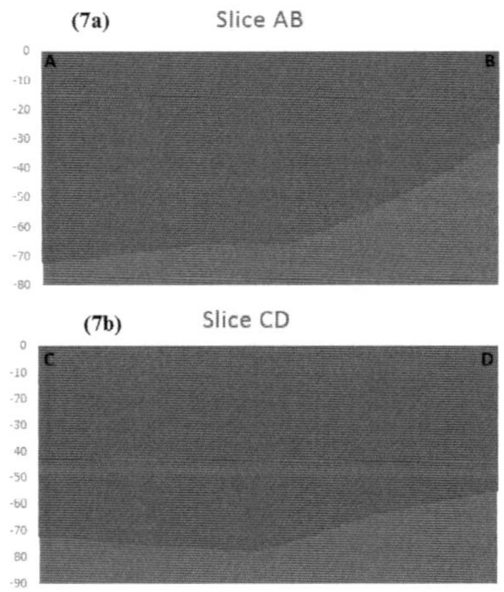

Figure 7. Slicing result for AB (7a) and CD (7b) line on figure 6, with lowest engineering bedrock thickness around ±30m and ±50m, respectively.

4. Conclusion

Through measurements of micro tremor conducted at Sedati mud volcano, we concluded:

- Duration of measurements on the mud volcano area should be more than 40 minutes as in point 2, 3, 11, 13, and 14 to obtain reliable and clear data.
- From the calculation of HVSR, the thickness of engineering bedrock layer of mud volcanoes Sedati is 30-70 m.
- Regional trend at depth of 68-70 m indicates the possibility of continuity of Watukosek fault with Southwest - Northeast trend direction in the research area of the mud volcano Sedati.

References

[1] Mazzini A, Svensen H, Planke S, Guliyev I, Akhmanov G G, Fallik T, Banks D 2009. When mud volcanoes sleep: Insight from seep geochemistry at the Dashgil mud volcano, Azerbaijan. *Marine and Petroleum Geology* **26** (9) 1704–1715.

[2] Istadi B, Pramono G H, Sumintadireja P, Alam S 2009. Simulation on growth and potential Geohazard of East Java Mud Volcano, Indonesia. *Marine and Petroleum Geology* **26** (9) 1724–1739.

[3] Irjan and Bukhori A 2001. *Pemetaan Wilayah Bencana Berdasarkan Data Mikroseismik Menggunakan TDS (Time Digital Seismograpph).* Jurnal Tugas Akhir. (Malang: Universitas Islam Negeri Maulana Malik Ibrahim).

[4] Nakamura Y 1989. A method for dynamic characteristics estimation of subsurface using microtremor on the ground surface. *Quarterly Report of Railway Technical Research Inst. (RTRI)* **30** 25–33.

[5] Sato T J, Saita, Nakamura Y 2008. Evaluation of the Amplification Characteristics of Subsurface using Microtremor and Strong Motion - the Studies at Mexico City. *Proc. 13th World Conference Earthquake Eng. Vancouver, B.C, Canada.* Paper No. 86213.

[6] Triwulan, Utama W, Warnana D D, Sungkono 2010. Vulnerability index estimation for building and ground using microtremor. *Aptecs 2nd International Seminar on applied Technology, Science and Arts.*

[7] Herak M 2008. ModelHVSR - a Matlab tool to model horizontal-to-vertical spectral ratio of ambient noise, *Comp. Geosci.* **34** 1514–1526.

[8] Arai H, and Tokimatsu K 2000. Effects of Rayleigh and Love Waves on Microtremor H/V

Spectra. *12th World Conference on Earthquake Engineering, Auckland, New Zealand.*

[9] Nakamura Y, 2000. Clear identification of fundamental idea of Nakamura's technique and its applications. *Proc. 12th World Conf. Earthquake Eng. New Zealand.* Paper No. 2656.

[10] Syaifuddin F, Bahri A S, Lestari W, Pandu J 2016. Microtremor study of Gunung Anyar mud volcano, Surabaya, East Java. *AIP Conference Proceedings* **1730**, 050004 (2016); DOI: 10.1063/1.4947400.

[11] SESAME European Research Project 2004. *Guidelines for the Implementation of the H/V Spectral Ratio Technique on Ambient Vibrations Measurements Processing and Interpretation.*

[12] Oubaiche E H, Chatelain J L, Bouguern A, Salem R B, Machane D, Hellel M, Khaldaoui F, and Guillier B 2012. Experimental Relationship Between Ambient Vibration H/V Peak Amplitude and Shear-wave Velocity Contrast. *Seismological Research Letters* **83** (6).

[13] Sawada Y, Taga M, Watanabe M, Nakamoto T, Nagumo H, Kudo K, Horike M, Sakajiri N, Masaki K, and Sasatani T 2004. Applicability of Microtremor H/V Method for KIK-NET Strong Motion Observtion Sites and Nobi Plain. *Proc. 13th World Conference Earthquake Eng. Vancouver, B.C, Canada.* Paper No. 855.

IOP Conf. Series: Earth and Environmental Science **62** (2017) 012011 doi:10.1088/1755-1315/62/1/012011

2-D Resistivity Assessment of Subsurface Characterization and its Engineering and Environmental Implications at SiLC

M M Nordiana[1], I N Azwin[1], Rosli Saad[1], Teoh Ying Jia[1], A B Anderson[1], Edy Tonnizam[2] and Muhamad Taqiuddin Zakaria[1]

[1]School of Physics, 11800 Universiti Sains Malaysia, Penang, Malaysia
[2]Department of Geotechnics and Transportation, Faculty of Civil Engineering, Universiti Teknologi Malaysia, Skudai, 81310 UTM Johor. Malaysia

E-mail: mmnordiana@usm.my

Abstract. The role of geophysics in Environmental Earth Sciences and Engineering is considered. In the developing era, geophysics has mainly contributed in investigation of new constructions such as tunnels, road, dams and high-rise buildings. This study was carried out to assess the foundation depths around a construction site in the Southern Industrial & Logistics Clusters (SiLC), Nusajaya, Johor using 2-D resistivity method. The 2-D resistivity method was carried out with a view to characterize different subsurface geological and to provide the engineering and environmental geophysical characterization of the study area. Measurements of eight 2-D resistivity profile using Pole-dipole array with 2 m minimum electrode spacing was taken with the use of ABEM Terrameter SAS4000 and ES10-64C selector. The results are presented as inversion model resistivity with the outline of the survey line. The inversion model resistivity from L1-L8 obtained is characterized by resistivity range of 1-8000 ohm-m. This range indicates the occurrence of silt, clay, sandy clay and sand whose ranges are; 10-100 ohm-m, 1-100 ohm-m, 100-800 ohm-m and 100-3000 ohm-m respectively. However, there was a boulder with range of >5000 ohm-m and saturated zone (1-20 ohm-m) which may indicate the weak zones of the study area. The 2-D resistivity method is not intended to replace borings, except in specific cases where information gathered would be sufficient to address the intended engineering and environmental purpose.

1. Introduction

The Southern Industrial & Logistics Clusters (SiLC), Nusajaya, Johor is a new site that is intended to be developed for industrial development with world-class managed, clean and green industrial park designed and planned to meet the current demand for environmentally sustainable development. Though there are existing structures on the site, however, there is need for the development of new structures to meet the population changes. For this reason, 2-D resistivity investigation was carried out for evaluation of site condition as an aid to geotechnical engineers.

Environmental problems will be definitely one of major social problems in this century. Many academic and engineering societies have paid their attention to the environmental issues in recent years. Environmental problems include many problems relating to human life such as natural disasters due to earthquakes, volcanic activities, landslides and various accidents due to existing underground constructions (collapse structures, karst void, cavity and sinkholes) as well as environmental pollutions.

Geophysical methods, particularly 2-D resistivity method have been extensively used for a wide variety of engineering and environmental problems [1,2,3]. This is due to the fact that, the 2-D resistivity survey is one of the simplest and less costly geophysical surveys employed. 2-D resistivity survey is relatively easy to perform and can be used to identify geological structures [4].

2. Site description

The area of investigation is located in the Southern Industrial & Logistics Clusters (SiLC), Nusajaya, Johor approximately at latitude 103° 35' 30" E and longitude 1° 28' 38" N (Figure 1 and 2). SiLC is a 1,372 acre (5.6 km^2) industrial park with close proximity to two international airports and cargo hubs as well as four seaports both in Johor and Singapore is a value-adding advantage. As an international gateway, this key logistical factor adds soaring value to the business activities of industry players.

Figure 1. Map of Nusajaya, Johor showing the study area.

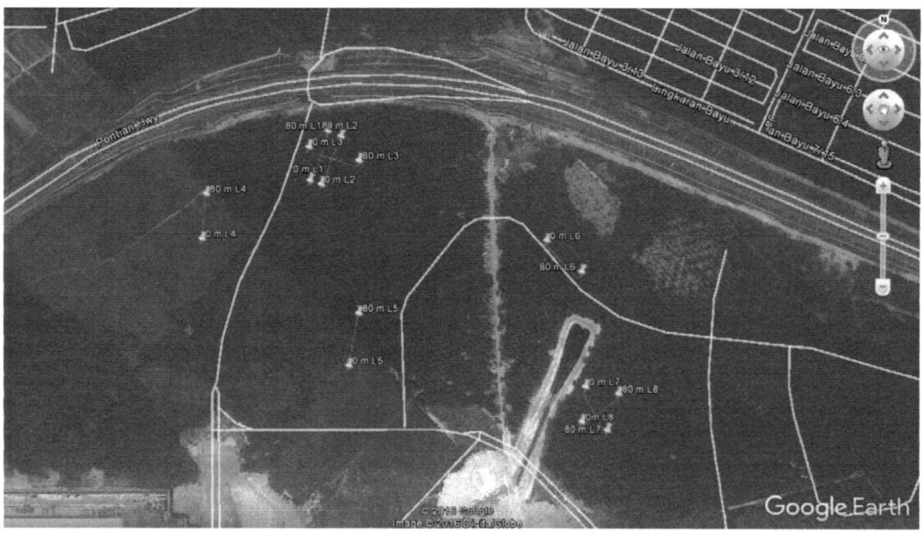

Figure 2. Map of survey line at SiLC, Nusajaya, Johor.

3. Geological area

Both [5] and [6] emphasized the variable lithology of the Older Alluvium. The most common variety is coarse feldspathic sand with occasional rounded phenoclasts, but gravelly clay, sandy gravel, sandy clay, silty clay, clayey sand and clay are also all well represented. The Older alluvium contains phenoclasts of vein quartz, quartzite, sandstone, siltstone, shale, hornfels, granite, granite porphyry, alaskite, aplite, rhyodacite, andesite and tuff. Bedding in it is generally only moderately well-developed, but close interbedding of the various lithological types can be seen in a few places. Where the formation is exposed in outctrops its feldspar content has in many places been weathered to a kaolinitic clay, and thus in the case of the common feldspathic sand type its surface appearance is that of a sandy clay or clayey sand [7].

4. Methodology

The 2-D resistivity survey was conducted using ABEM SAS4000 Terrameter, ES10-64C as a selector, electrode cables with 5 m takeouts and stainless steel electrodes. Eight survey lines were occupied in this study area. The survey used Pole-dipole array with space between electrodes are 2 m. The data were processed using Res2Dinv software for gridding, contouring and final presentation. The image from 2-D inversion model resistivity (L7) will be present with correlation of resistivity values and N-values from the borehole record. The results of the 2-D inversion model were combined to produce 3-D image using Surfer software.

5. Results and discussion

The inversion model resistivity from L1-L8 obtained is characterized by resistivity range of 1-8000 ohm-m (Figure 3-5). This range indicates the occurrence of silt, clay, sandy clay and sand whose ranges are; 10-100 ohm-m, 1-100 ohm-m, 100-800 ohm-m and 100-3000 ohm-m respectively. Borehole log records from each line were used as a control for the interpretation of the entire profiles obtained in this survey (Figure 6). The results of the inversion model resistivity assumed that the first layer is characterized by relatively high resistivity value (3000-8000 ohm-m) and thin thicknesses (1-3 m). This layer was considered to be reclaimed area; however their wide ranges of resistivity values are due to dry nature and percentage of sand. Low resistivity values (1-50 ohm-m) were interpreted as saturated zone. Boulder was found at L1, L2, L7 and L8 with resistivity values of 3000-8000 ohm-m. The borehole information starts at depth of 3 m from inversion model resistivity. The results of the borehole information and inversion model resistivity assumed that the second layer is topsoil which contain of sandy silt. It was observed that the maximum depth of sandy clay (100-800 ohm-m) is reached at 15-20 m with N-values of 14-22. The third layer can be said to be sand formation with gravel. It is characterized by relatively high resistivity value and at maximum depth of 21 m with N values of 22-50. Figure 7 shows the sedimentary topography of the study area base from the inversion model resistivity profiles.

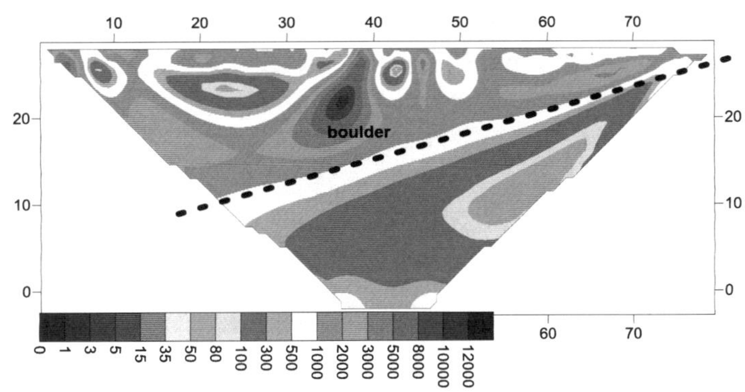

Figure 3. Inversion model resistivity of L1 at Nusajaya, Johor.

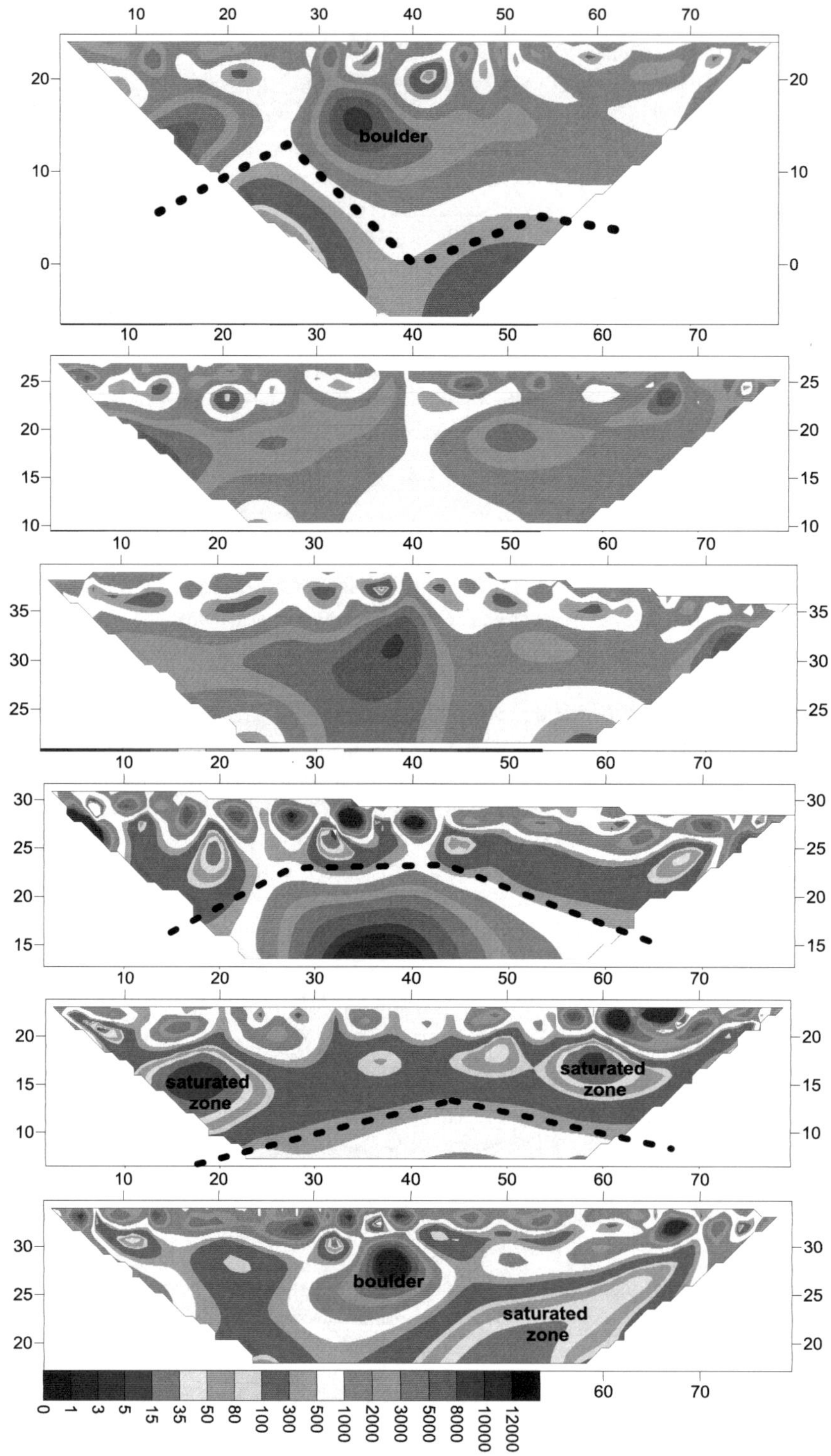

Figure 4. Inversion model resistivity of L2-L7 at Nusajaya, Johor.

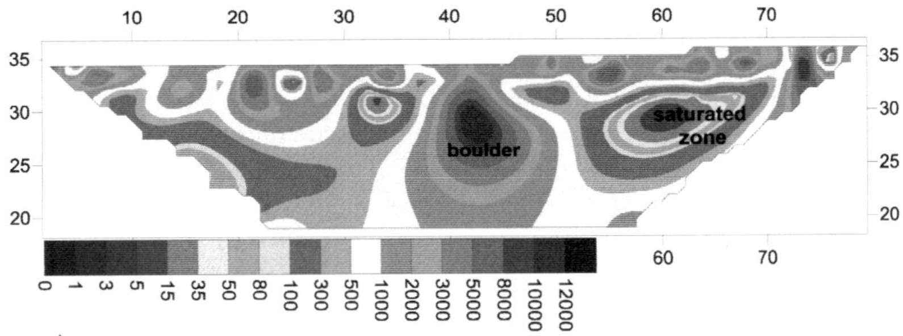

Figure 5. Inversion model resistivity of L8 at Nusajaya, Johor.

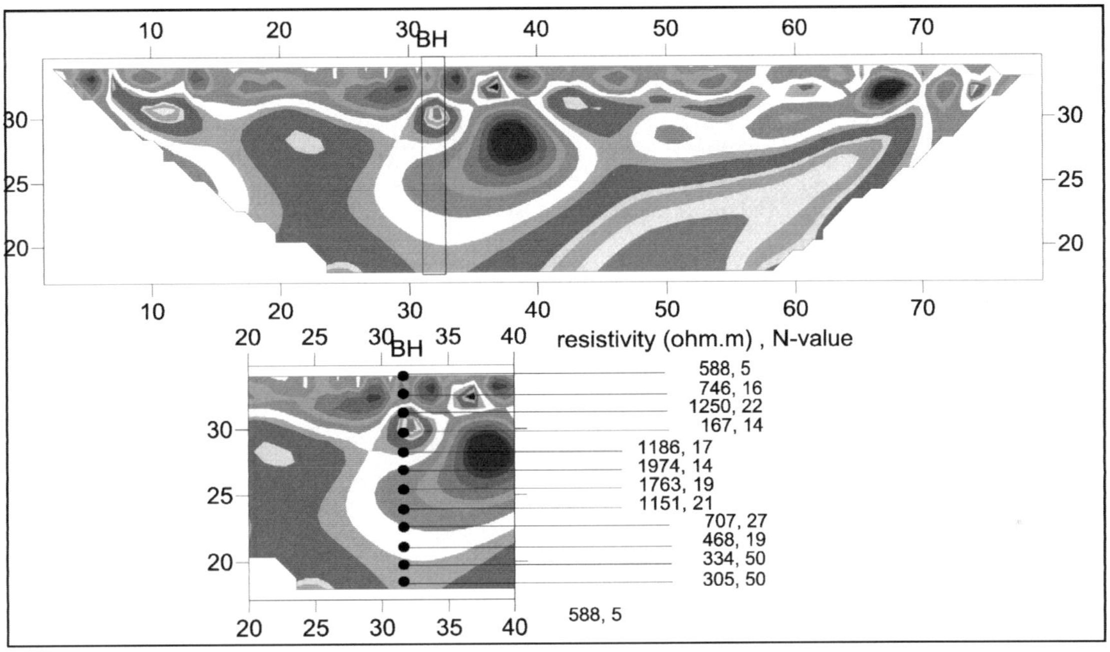

Figure 6. Inversion model resistivity of L7 with borehole at Nusajaya, Johor.

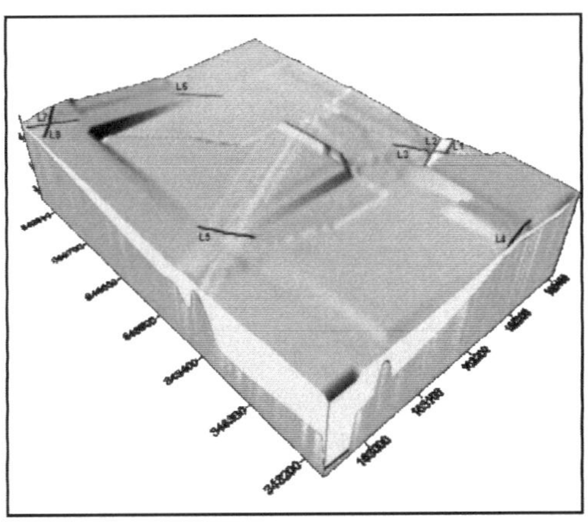

Figure 7. Sedimentary topography and survey lines of SiLC, Nusajaya, Johor.

6. Conclusion

A key to success in the development of a site is the understanding of the subsurface formation during the planning stages. Measurement of 2-D resistivity method is one of the most widely used geophysical survey methods in this respect. 2-D resistivity method can afford in determining inherent engineering and environmental effects of the subsurface variations. The inversion model resistivity from L1-L8 obtained is characterized by resistivity range of 1-8000 ohm-m. This range indicates the occurrence of silt, clay, sandy clay and sand whose ranges are; 10-100 ohm-m, 1-100 ohm-m, 100-800 ohm-m and 100-3000 ohm-m respectively. The investigation of the study area by using 2-D resistivity method was correlated with borehole log records and these show a sequence of topsoil, clay, sandy clay and sand. Therefore, the third layer to be the most competent for founding small to medium engineering structures due to the presence of sand thickness that varies from 10-25 m and resistivity values that range between 100-3000 ohm-m. The results of 2-D resistivity method could be used to detect subsurface problem and subsurface characterization supported by borehole records for engineering and environmental purpose.

References

[1] Pomposiello C, Dapeña C, Favetto A and Boujon P 2012 Application of geophysical methods to waste disposal studies in Municipal and Industrial Waste Disposal, X.-Y. Yu, Ed.

[2] Boyce J I and Kaseoglu B B 1996 Shallow seismic reflection profiling of waste disposal sites *Geoscience Canada* **23**(1) 9-21.

[3] Oyedele K F, Ayolabi E A, Adeoti L and Adegbola R B 2009 Geophysical and Hydrogeological Evaluation of Rising Groundwater level in the Coastal Areas of Lagos Nigeria *Bull. Eng. Geol. Environ.* **68** 137-143.

[4] Al-Sayed E A and El-Qady G 2007 Evaluation of Sea Water Intrusion using the Electrical Resistivity and Transient Electromagnetic Survey: Case Study at Fan of Wadi Feiran, Sinai, Egypt, EGM *International Workshop Innovation in EM, Grav and Mag Method: a new Perspective for Exploration Capri, Italy* April 15-18

[5] Willbourn E S 1928 The geology and mining industries of Johore *Malayan Br. Roy. Asiatic Soc. Jour.* **6** 5-35.

[6] Alexander F E S 1950 Report on the availability of granite on Singapore and the surrounding islands *Government Printing Office, Singapore*

[7] Burton C K 1972 Geology and Mineral resources, Johor Bahru- Kulai area, South Johore *Geological Survey of Malaysia Map Bulletin* **2** 56.

Acknowledgements

The authors thank to the technical staffs of the geophysics laboratory, School of Physics, Universiti Sains Malaysia for their cooperation in the geophysical field. The authors also would like to express their appreciation of Research University (RUI) Grantt, Universiti Sains Malaysia with the title Geophysical Application and Approaches In Engineering And Environmental Problems (1001/PFIZIK/811323) for their supports.

Enhancing Magnetic Interpretation Towards Meteorite Impact Crater at Bukit Bunuh, Perak, Malaysia

M. K. A. Nur Amalina[1*], M. M. Nordiana[2], Rosli Saad[2], Mokhtar Saidin[3]

[1]Geophysics Postgraduate Student, Geophysics Section, School of Physics, Universiti Sains Malaysia
[2]Geophysics Lecturer, Geophysics Section, School of Physics, Universiti Sains Malaysia
[3]Centre for Global Archaeological Research Malaysia, Universiti Sains Malaysia

E-mail: nuramalinamka93@gmail.com

Abstract. Bukit Bunuh is the most popular area of suspected meteorite impact crater. In the history of meteorite impact hitting the earth, Bukit Bunuh has complex crater of a rebound zone of positive magnetic anomaly value. This study area was located at Lenggong, Perak of peninsular Malaysia. The crater rim extended 5 km outwards with a clear subdued zone and immediately surround by a positive magnetic residual crater rim zone. A recent study was done to enhance the magnetic interpretation towards meteorite impact crater on this study area. The result obtained is being correlated with boreholes data to determine the range of local magnetic value. For the magnetic survey, the equipment used is Geometric G-856 Proton Precision magnetometers with the aids of other tools such as compass and GPS. In advance, the using of proton precision magnetometer causes it able in measures the magnetic fields separately within interval of second. Also, 18 boreholes are accumulated at study area to enhance the interpretation. The additional boreholes data had successfully described the structure of the impact crater at Bukit Bunuh in detailed where it is an eroded impact crater. Correlations with borehole records enlighten the results acquired from magnetic methods to be more reliable. A better insight of magnetic interpretation of Bukit Bunuh impact crater was done with the aid of geotechnical methods.

1. Introduction

Magnetic methods on meteorite impact have a long history in the meteorite impact study. Among the geophysical methods which had been previously applied in the meteorite impact crater investigation, magnetic field measurements around impact structures may have given the poorest response over the other methods such as seismic, electrical resistivity, and gravity. Magnetic low is a signature for some of the craters found, for example, Mien crater [1], Dellen crater [2] and Kärdla crater [3]. The dominant effect of magnetic low anomaly or subdued is closely related to the major impact structures [4][5], which generally distinguish as a truncation of the regional magnetic indication. However, the other impact structures on a strong local magnetic anomaly either positive or negative may be produced by the remanent magnetization. Hence, these types of magnetization are occurred due to the units of impact melt within the structure or also could be found by the uplift of more magnetic units from depth into the central uplift [6].

The magnetic method has seldom had been used in correlation with borehole data for interpretation compared with other geophysical method such as seismic and resistivity. It provides potential field measurement which does not relate directly to depth. A study of tectonic features and subsurface in the Thrace Basin using the integration of magnetic, borehole data, gravity, and seismic reflection had been done [7]. Sediment thickness in Thrace Basin is accurately determined, and magnetic response towards intrusion and metamorphic basement outcrops is identified. The exact depth is hard to obtain in the magnetic modeling where the anomalies are only caused by the lateral magnetization contrasts [8]. Geology survey and mapping in the area surrounded of Bukit Bunuh have found many block of rocks with angular and variety of sizes from 0.3 m till more than 9.0 m. Rock types found are classified into suevite, impact rock, and shock gneiss. Even though the original rocks have undergone an impact, the rocks found on the site were still comprised of clastic rock and igneous rock. Most of the impact rock is distributed at the west of Bukit Bunuh where a variety of sizes [9]. The distribution of rock sometimes concentrated with size in between 1 m to 2 m. Round shapes of rock indicate that the rock was produced from the meteorite impact. There is the distribution of some suevite rock, which covers less than 5 km^2 area and could be an indication of the diameter of meteorite impact not less than 2 km. The appearance of the rocks with a hole that resembles looks like vesicle present around the rocks due to the recrystallization occurs within the structures. The ejecta blanket that exists at the outer part of the crater is due to the fraction of rock that settles or falls after the impact, and it settled down as small sizes of the angular shape of rock [10]. Borehole records correlate well with magnetic residual value where the thicker sediment infilling has lower magnetic residual and vice versa. From the record, it is shown that the depth of bedrock at the crater rim and rebound area is shallow to only 3.00 m and the interior of crater is found with the depth of 18.0 m. The delineation of fault and fracture at this study area has been map by [11] with using magnetic response mapping and proven by the borehole records. Parameters like sediment thickness and range of magnetic residual values were extracted to correlate with the defined crater from previous studies.

2. General Geology

Bukit Bunuh was bounded between Titiwangsa Range and Bintang Range. The Lenggong Valley covered with a few of lithology which is alluvium, tephra dust, and granitic rock. Alluvium is the mostly found along the river area. Meanwhile, the granitic rock is from Jurassic to Carbonaceous were also dominated in the whole of Lenggong area which is originated from Bintang Range at the west of Lenggong [12]. In addition, the nearby area to the river also covered with Quaternary deposits such as clay and sand.

3. Study Area

This study was carried out in the area of Bukit Bunuh, located 1 km away from the town of Kota Tampan, Lenggong, Perak which is about $17\times 10\ km^2$. This study was dividing into two parts which are regional study and detailed study. The direction of the magnetic survey is towards the southwest. Figure 1 shows the map of Lenggong, Perak from front-view of Google map and pointed area represented the study area. There are about 18 boreholes were done around the rebound zone to reduce the ambiguity of magnetic data interpretation. Hence, crater interior and crater rim that defined from the magnetic results.

IOP Conf. Series: Earth and Environmental Science **62** (2017) 012012 doi:10.1088/1755-1315/62/1/012012

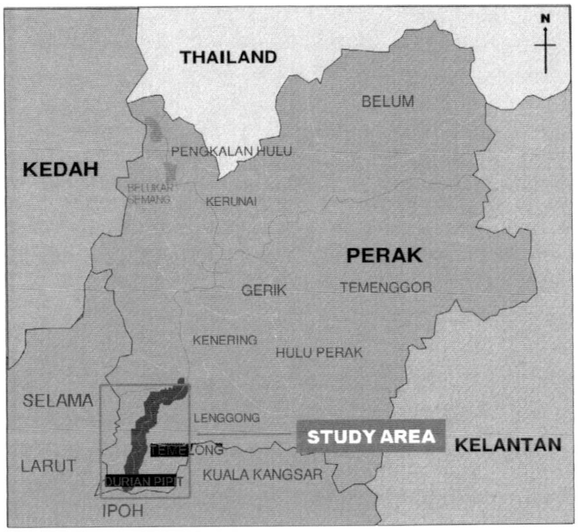

Figure 1. The study area at Lenggong, Perak.

4. Research Methodology

The magnetic surveys were conducted using Geometric G-856 Proton Precision magnetometers with the aids of some accessories such as GPS and compass. The moving station of this survey was set at 250 m to 500 m for regional study and 50 m for detailed study interval or in whichever distance permitted to cover the whole of the survey area. The location of the base station was set at the same spot throughout the survey and relatively far away from the magnetic disturbance. The reading of base station was set to auto function and taken for an interval of 1 minute. The magnetic data processing started with applying data reductions before deducting with the base reading to get magnetic residual values. Finally, a magnetic residual contour map is generated using Surfer 8 software. To supervise the interpretation, the records from four existing boreholes and addition of 14 new boreholes (Figure 2) are used to integrate with the magnetic response of Bukit Bunuh area.

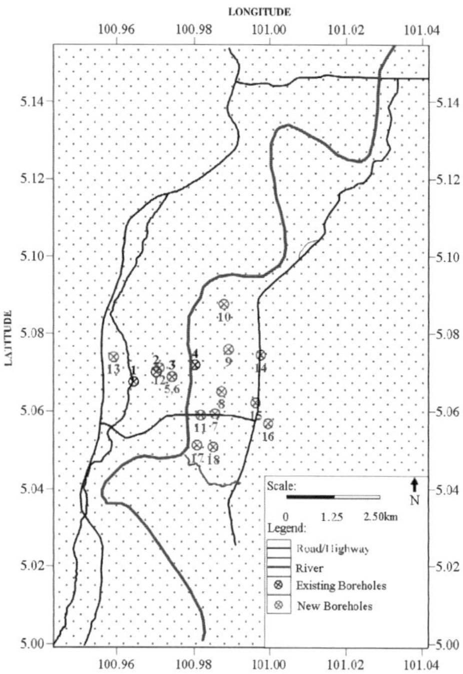

Figure 2. Existing and new boreholes located around Bukit Bunuh area.

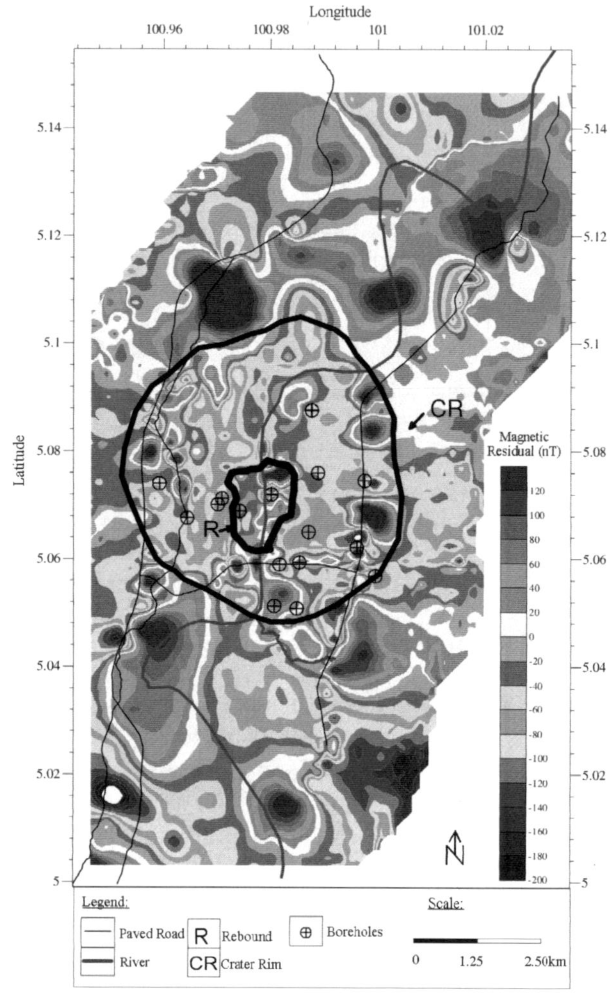

Figure 3. Map of Boreholes positions with crater rebound and crater rim identified.

5. Result and discussion

An additional of 14 boreholes which make the total number of 18 boreholes accumulates at the Bukit Bunuh impact crater area. The additional boreholes record will enhance the interpretation towards the subsurface utilizing the magnetic survey. The location of the boreholes is overlaid on top on the residual magnetic map of Bukit Bunuh impact crater (Figure 3) for better understanding and aid in interpretation at ease. Table 1 shows the sediment thickness and magnetic residual value for existing and new boreholes records.

Table 1 Sediment thickness and magnetic residual value for existing and new boreholes records.

Existing / New	Location	Borehole (BH)	Sediment thickness before reaches impact/weathered granite (m)	Magnetic residual value range (nT)
Existing	Crater rim	**1**	3.00	-20 to 0
Existing	Crater interior	**2**	18.00	-100 to -80
Existing	Rebound zone	**3**	3.00	0 to 20
Existing	Rebound zone	**4**	7.50	-60 to -40
New	Rebound zone	**5**	1.50	-20 to 0
New	Rebound zone	**6**	3.00	-20 to 0
New	Crater interior	**7**	12.50	0 to 20
New	Crater interior	**8**	13.00	-120 to -100
New	Crater interior	**9**	15.00	-60 to -40
New	Crater interior	**10**	12.00	-60 to -40
New	Crater interior	**11**	15.00	-60 to -40
New	Crater interior	**12**	12.00	-80 to -60
New	Crater rims	**13**	18.00	-60 to -40
New	Crater rims	**14**	13.50	-100 to -80
New	Crater rims	**15**	13.00	-20 to 0
New	Crater rims	**16**	13.60	-80 to -60
New	Crater rims	**17**	10.50	60 to 80
New	Crater rims	**18**	15.00	0 to 20

The suspected area of crater rim or the edge of the crater rim is located at the location of boreholes 1,13 – 18 with sediment thickness of 3 – 18 m and may residual values of -100 to 80 nT. Borehole 1 and 6 shows a characteristic of shallow sediment or also known as ejecta layer deposited of 3.00 m with the magnetic residual value of -20 to 0 nT. Boreholes 1, 17 and 18 showed a positive magnetic response while the other boreholes were in the negative magnetic response area. The inconsistent variation of sediments thickness with respect to magnetic residual value may be due to a few effects. Bukit Bunuh impact crater is located at the tropical region which exposes to extreme erosion condition due to weathering processes. At present, thick sediment was found as a result of the downwards gravity pull due to a collapse of an initially deep transient crater causing rock units near the rim slumping downwards. During the excavation event, the collision caused tensional stress exceeds the mechanical strength of the target rocks which cause them to be fractured and shattered. The original crater rim may not preserve well at these conditions. The low magnetic residuals value spotted around the crater rim is identified as the deeply eroded structure with thick sediments measured.

While borehole 2, 7 – 12 show sediment depth of 12.00 – 18.00 m to drill through before hitting impact/weathered granite and have the low magnetic residual value of -120 – 20 nT. These boreholes were in the suspected crater interior, which has the deepest sediment infilling and the authigenic breccia formation area. The magnetic response of this area is low due to the reduction of the magnetic intensity field by the sediments.

Boreholes 3 – 6 were identified with the depth of 1.5 – 7.5 m. Before reaching granite rock with the magnetic residual value range of -60 – 20 nT. Borehole 3, 4 and 5 located at the suspected rebound area that shows a characteristic of shallow bedrock depth. Negative anomaly zone was suspected as the zone of a fractured cause by the impact. Hence, borehole 4 shows negative anomaly and bedrock depth of 7.5 m.

Meteorite impact is a near-surface activity and the magnitude deformation and shock metamorphism associated with impact structures reducing with depth. Most of the impact structures found are relatively shallow, and subsurface rock metamorphism produced from the activity is comparatively thin units of the earth crust. The top soils of the target area are removed during the excavation stage and causes the impact structures are now uncovered and extremely vulnerable to erosion before the new layer of top soil. In the early stage of erosion, the near-surface ejecta deposits with the distinctively shocked and melted materials contain are preferred to be removed. Thus, the most convincing and comprehensive evidence of impact is rapidly destroying, and the crater is now camouflaged.

6. Conclusion

The additional boreholes data had successfully described in detail the structure of Bukit Bunuh the impact crater where it is an eroded impact crater. Correlations with borehole records enlighten the results acquired from magnetic methods to be more reliable.

7. Acknowledgements

The authors would like to express thanks to all of the lecturers, geophysics staff, geophysics postgraduate students and family for their support and willing to help in this project. In addition, many thanks are due to Centre for Global Archaeological Research (CGAR), University Sains Malaysia and Institutes of Postgraduate, Universiti Sains Malaysia Conferences Funds for sponsoring the project.

References

[1] Henkel H 1982 The lake Mien structure *Geol. Surv. Sweden Geophys. Dep* **8221**

[2] Henkel H 1992 Geophysical aspects of meteorite impact craters in eroded shield environment with special emphasis on electric resistivity *Tectonophysics* **216** 63–89

[3] Puura V A and Suuroja K Ordovician impact crater at Kardla, Hiiumaa Island, Estonia *Tectonophysics* **216** 143-156

[4] Dabizha A I and Fedynsky V V 1975 The Earth's "star wounds" and their diagnosis by geophysical methods *Zamlya Vselennaya* **3** 56-64

[5] Clark J F 1983 Magnetic survey data at meteoritic impact sites in North America *Geomagnetic Service of Canada, Earth Physics Branch* **83**(5) 1-32

[6] Hart R J, Hargraves R B, Andreoli M A G, Tredoux M and Doucouré C M 1995 Magnetic anomaly near the center of the Vredefort structure: Implications for impactrelated magnetic signatures *Geology* **23** 277–280

[7] Demir D, Bilim , Aydemir A and Ates A 2012 Modelling of Thrace Basin, NW Turkey using gravity and magnetic anomalies with control of seismic and borehole data *Journal of Petroleum Science and Engineering* 44-53

[8] Ormö J and Blomqvist G 1996 Magnetic modelling as a tool in the evaluation of impact structures, with special reference to the Tva¨ren Bay impact crater, SE Sweden *Tectonophysics* **262** 291–300

[9] Talib N K, Abdullah J and Saidin M 2009 Evidence of meteorite impact in Bukit Bunuh and its contribution to archaeology *Proceedings of the International Seminar on Sharing Our Archaeological Heritage, Johor Bharu, Malaysia*

[10] Talib N K 2013 Ekskavasi tapak Bukit Bunuh, Lenggong, Perak: sumbangan kepada pemahaman kebudayaan Paleolitik *M.A. Thesis Universiti Sains Malaysia*

[11] Saad R, Saidin M , Muztaza N M and Kiu YC 2015 Subsurface study Using 2-D Resistivity Imaging Method Impact at Bukit Bunuh, Perak *Electronic Journal of Geotechnical Engineering* **20** 4711-18

[12] Mokhtar S 1993 Kajian perbandingan tapak Paleolitik Kampung Temelong dengan Kota Tampan dan sumbangannya terhadap kebudayaan zaman Pleistosein akhir di Asia Tenggara *Msia. Museum J* **32**

Study of Sedimentary Outcrop of Semanggol Formation with the Correlation of Geology, Geotechnical and Geophysics Technique

A N Nordiana[1], M M Nordiana[1], Teoh Ying Jia[1], Hazrul Hisham[1], Nabila Sulaiman[1], Umi Maslinda[1], Z M Taqiuddin[1], M K A Nur Amalina[1], Muhamad Afiq Saharudin[1]

[1]School of Physics, 11800 Universiti Sains Malaysia, Penang, Malaysia.

E-mail: nordiana.nawawi@gmail.com

Abstract. The study location was at Bukit Kukus, Kuala Ketil, Kedah, Malaysia where the geological outcrop of this Semanggol Formation comprises of chert, mudstone, and volcanic tuff. The study was conducted using two geophysical methods, which are 2-D Resistivity and Ground Penetrating Radar (GPR). The objectives of the study are to correlate both of the geophysical methods through the value of conductivity and to identify the physical properties of rocks through the value of porosity and permeability. The data acquisition for both methods was conducted on the same line. For 2-D Resistivity method, the length of the line is 60 m with 1.5 m electrode spacing and the array used was Wenner-Schlumberger. For GPR method, the survey line was on top of the resistivity line, and the frequency of the antenna used is 250 MHz. A good correlation exists between both of the GPR signature and contour maps for resistivity from the surfer 10 software with the outcrop feature. Conductivity value from both GPR and Resistivity method was compared and the range value of conductivity obtained from GPR method almost equivalent with Resistivity method based on derivation and calculation for the sedimentary rocks, which are 0.037 to 0.574 miliSiemens per metre (mS/m) for chert and 0.186 to 10.142 miliSiemens per metre (mS/m) for mudstone. Two types of rock samples were taken, and several geotechnical tests were conducted, but only the value of permeability, K and porosity, ϕ of chert can be calculated, which are 1.95E-22 m^2 (original condition) and 2.27E-22 m^2 (dry condition) and 3 percent respectively as the sample of mudstone was damaged. The parameter of the 2-D resistivity method derived from Archie's law was used to calculate the porosity, ϕ_f value using the Formation Factor equation. The range values of porosity, ϕ_f for chert mostly in the range of 5 to 25 percent, which is 6.26 to 13.36 percent but slightly out of range for mudstone, which is 14.12 to 36.02 percent.

1. Introduction

Several studies on outcrop have been conducted for rocks corresponding in the subsurface that is rich in natural resources. Such continuous information on the outcrop and detailed information can be obtained, for example, petrology, facies, sedimentary structures, texture, grain types and diagenetic changes [1].

Mostly, the geotechnical event begins with a review of the task to define the required material properties. Then, site investigation of soil and rock properties was conducted at the area of interest to determine their engineering properties to proposed construction. Additional of site investigations are needed to gain an understanding of the area or which engineering will take place. It is a process collecting information and uses principles of soil mechanics and rock mechanics to investigate subsurface conditions and materials, which to determine the relevant physical or mechanical and chemical properties of these materials [2].

The geophysical survey can act as an important role in defining the subsurface geology and the associated parameters which are calculated by application of a variety of invasive and noninvasive techniques. Analysis of data using geophysical methods usually helps produce a more cost effective program in locating the target. Basically, the methods of geophysics are utilized to retrieve analog data in the absence of outcrops or to enhance those with petrophysical data in comparable to the subsurface of interest. For example, ground-penetrating radar (GPR) has long been used to for static imaging [1]. Of all the physical properties of rocks and minerals, electrical resistivity shows the greatest variation.

Recently, a fieldwork has been carried out at Kuala Ketil, at the coordinate N: 05.59100° E: 100.69568° until N: 05.59046° E: 100.69573° which to determine the value of resistivity and conductivity of the sedimentary rocks from the outcrops features. Thus, the application of 2-D Resistivity and Ground Penetrating Radar (GPR) methods plays an important role to determine the resistivity and conductivity values of those sedimentary rocks.

2. Study Area

The study area that was conducted at Bukit Kukus, Kuala Ketil, Kedah, Malaysia with the coordinate of N 05.59100° E 100.69568° until N 05.59046° E 100.69573°. Survey line length of 60 m was proposed on top of the outcrop, which at one specific area that has high possibilities to become the target of interest and produce expected a visionary result. Based on the observation of the geological features, the outcrop shows a contact of two different sedimentary rocks, which are chert and mudstone. According to Jasin and Harun [3], there are seven lithofacies have been identified that may represent the chert unit in ascending order, laminated black mudstone, interbedded sandstone and mudstone, interbedded tuffaceous sandstone and tuff with a paraconglomerate bed, interbedded tuffaceous sandstone, siliceous shale and chert, bedded chert, tuffaceous mudstone and interbedded chert and siliceous mudstone. However, only the presence of volcanic tuff can be identified at the study area (Figure 1(a)). Figure 1(b) is the presence of chert with iron stains, and lastly, Figure 1(c) shows laminated mudstone that has been identified in the study area and based on observation, it was formed through the compaction process.

(a) (b) (c)

Figure 1. (a) Volcanic tuff, (b) chert and (c) laminated mudstone.

3. Methodology

Regarding geology, the outcrop at Bukit Kukus, Kuala Ketil, Kedah was observed, and the characteristics of the facies distribution with its geological features were identified. The rock samples from the outcrop were taken using the geology hammer. There are two different types of materials of the rock samples taken which are chert and mudstone.

Two different geophysical methods were utilized in this study. For 2-D Resistivity survey, the data acquisition was conducted using Terrameter ABEM SAS4000 system and performed by Wenner-

Schlumberger array. A single line was laid out with minimum 1.5 m spacing between electrodes which make the total distance of 60 m. The data processing and interpretations were managed using the RES2DINV software, and the data were then outputted into Surfer 10 software for gridding, contouring, and final presentation.

For Ground Penetrating Radar (GPR), the antenna frequency used is 250 MHz antenna and only one survey line conducted at the study location which exactly on top of the resistivity line. For GPR processing, the raw data need to filter such for instrument noise or any irregularities in the data using the Ramac GroundVision software. The knowledge of radar attenuation crossed by EM waves introduces significance absorptive losses or decreases the intensity of electromagnetic allow calculating the electrical conductivity [4].

Based on the Radargram interpretation, the distance in meter (m) and time of reflection in the nanosecond (ns) are important in finding the velocity, v in meter per second (m/s) of the medium travel. Equation (1) used to calculate the value of attenuation, α based on the value of the velocity and the value of conductivity, σ obtain from the 2-D Resistivity method. The value of conductivity in Equation (2) is from the GPR method which was compared with the conductivity value from 2-D Resistivity method.

$$\alpha = 1.69 * \frac{\sigma}{\sqrt{k}} \tag{1}$$

$$\sigma = \omega\varepsilon \sqrt{\left[2\left(\frac{\alpha}{\omega\sqrt{\varepsilon\mu}}\right)^2 + 1\right]^2 - 1} \tag{2}$$

The resistivity of rock depends on number of factors such as the porosity and the water saturation [5]. Permeability and porosity test was conducted directly from the rock samples taken. Before the test begins, the rock samples undergo the coring process to form a cylindrical shape. Permeability test (Nitrogen Gas Permeability) was conducted in two conditions, which in original condition (after coring) and in dry condition. Flow rate, V in meter square per second (m^2/s) and area, A in meter square (m^2) was calculated through Equation (3) and (4) respectively, where D is the flowmeter diameter, H is the length head of flowmeter, T is average time, B is specimen diameter and L is the length of specimen. Intrinsic permeability, K in meter square (m^2) value was obtained through calculation from the Equation (5).

$$V = \frac{\left(D^2/4\right)\pi H}{T} \tag{3}$$

$$A = \frac{\pi B^2}{4} \tag{4}$$

$$K = 2P^2 \frac{VL * 1.7 * 10^{-16}}{A\left(P_1^2 - P_2^2\right)} \tag{5}$$

Porosity test (Water Absorption and Porosity Measured by Water Immersion under Vacuum) was conducted later where the weight of the rock samples was measured in original condition (*W4*). Then the rock samples undergo vacuum pressure process where the samples were left in the vacuum saturation apparatus for three hours. After three hours, the samples were immersed completely in water together with the vacuum pressure for another three hours. Then samples were left only in the water for another an hour. The samples were taken out, and the weight of the rock samples was recorded in water (*W3*) and in air (*W2*). Porosity, ϕ value was obtained in percentage, (%) through calculation from the Equation (6).

$$\phi = \frac{W2 - W4}{W2 - W3} \tag{6}$$

Archie's law link the electrical resistivity of a rock to its porosity, to the porosity of the water that saturates its pores and to the fractional saturation of the pore space with the water. Archie began by naming the ratio of the resistivity of the rock ρ_o to that of the pore water ρ_w the formation resistivity factor, *F*. Formation resistivity factor is the ratio of the resistivity of 100% water saturated rock to the resistivity of the water with which it is saturated [6].

Equation (7) is the calculus for the formation resistivity factor, where ρ_f is the resistivity value for the unconsolidated layer. The value of ρ_f was assumed on top of the layer of the outcrop, which at where the data was acquired, while for ρ_w is fluid resistivity which was assumed as the lowest resistivity value and using the Wenner-Schlumberger array the value of ρ_w is 55.57 ohm-m. Once the value of *F* is obtained, porosity, ϕ_f of each of the materials can be calculated using Equation (8), where the values of tortosity factor, *a* and cementation factor, *m* was referred from Archie's formula for different lithologies in Table 1. The value of *a* and *m* are fixed at 0.62 and 1.95 respectively as the outcrop is a sedimentary rock from the Paleozoic age.

$$F = {}^{\rho_F}\!/_{\rho_W} \quad (7)$$

$$F = a * \phi^{-m} \quad (8)$$

Table 1 Archie's formula for different lithologies (Keller, 1966)

Description of rock	*a*	*m*
weakly cemented detrital rocks, such as sand, sandstone and some limestones, with a porosity range from 25 to 45 % (Tertiary age)	0.88	1.37
moderately well cemented sedimentary rocks, including sandstones and limestones, with the porosity range from 18 to 35 % (Mesozoic age)	0.62	1.72
Well-cemented sedimentary rocks with a porosity range from 5 to 25 % (Palaeozoic age)	**0.62**	**1.95**
Highly porous volcanic rocks, such as tuff, aa and pahoehoe with porosity 20 to 80 %	3.5	1.44
Rocks with less than 4 % porosity, including dense igneous rocks and metamorphosed sedimentary rocks	1.4	1.58

4. Results and Discussion

Figure 2 shows the outcrop features of the study area consist of contact of chert and mudstone mark with a yellow dotted line. This zone could indicate fracture presence due to high weathering process at the contact. The properties of the rock could be confirmed and identify through the value of geophysical method that was conducted on top of red dotted line and geotechnical tests.

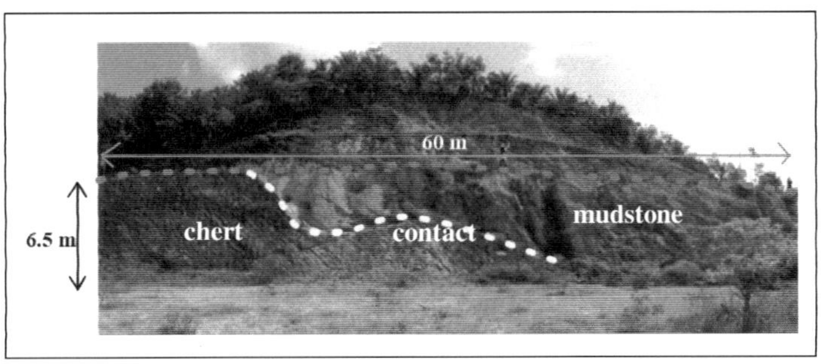

Figure 2. The outcrop with thickness of 6.5 m show contact of chert and mudstone.

In comparison with the outcrop in Figure 2, the result of the pseudosection for 2-D Resistivity method in Figure 3 for Wenner-Schlumberger array process from the Surfer 10 show a positive correlation. Based on the pseudosection map, the contact zone could dictate by the trend of the resistivity contour, which reflects the vertical resistivity distribution within the depth of 6.5 m that have been cut to correlate with the outcrop thickness. The value of resistivity for mudstone is the lowest compared to chert. This is because the majority of chemical sedimentary rocks of mudstone come from minerals left behind by evaporated water. Mudstone has a resistivity value of 12-900 ohm-m, while the chert exhibited a resistivity of 1700-130000 ohm-m respectively. For GPR interpretation, Figure 4 displays the result of the Radargram after filtered to obtain more perceive image. The white and black stripes indicate amplitude values alternately. The obvious reflected signal was identified, where the contact of both chert and mudstone are nearly at the same distance.

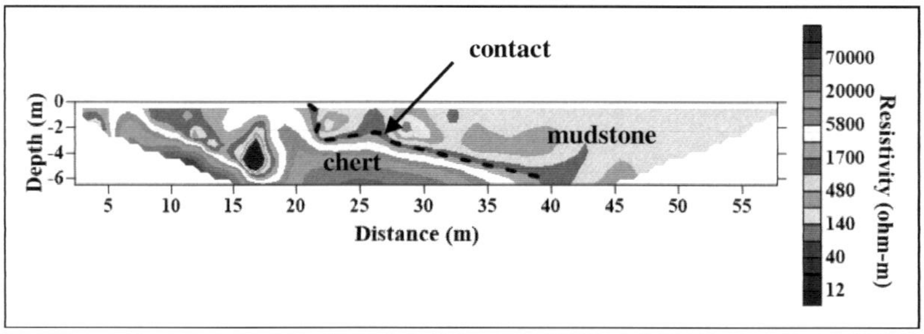

Figure 3. Pseudosection map result in using Wenner-Schlumberger array with depth of 6.5m.

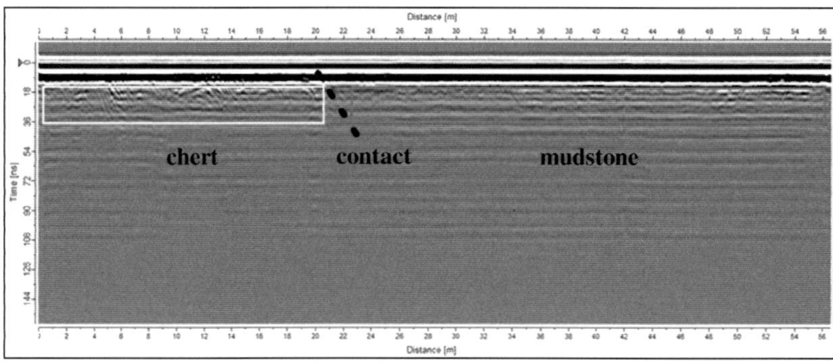

Figure 4. Amplitude contrast from the GPR Radargram results until depth of 7m.

At the different distance and depth give the different value of time in the Radargram, thus show the different velocity of materials. The knowledge of the radar energy attenuation is important to identify electrical properties of different materials in the ground, which in this case is to determine the value of conductivity for chert and mudstone at the study area. Generally, attenuation is the interruption of a radar beam by intense precipitation. Based on the calculation of average value, Table 2 shows attenuation value for chert and mudstone are in the range of 0-0.04 and 0.01-0.64 Neper per meter (Np/m) subsequently. Thus, conductivity value from GPR method could determine where for chert and mudstone are in the range of 0.037-0.574 and 0.186-10.142 miliSiemens per meter (mS/m) respectively. The conductivity value from both GPR and 2-D Resistivity method are almost equivalent. This implies that conductivity value of the outcrop can be obtained from GPR method and successfully correlated with the 2-D Resistivity method.

In general rock classification, permeability, K and porosity, ϕ is included in the basic rock properties. The permeability of rock is the measure of resistance to the flow of fluid through a rock. While for the porosity of rock could be simplified as a measure of its ability to hold a fluid. Based on the test of the geotechnical event, both permeability and porosity of the rock samples could be determined, and the values are organized as in Table 2. For permeability test (Nitrogen Gas Permeability), only permeability value of chert could be calculated, while the other sample which is mudstone is not available as the sample was damaged directly from the coring process. Using the Equation (5), the average time for the original condition is longer because it takes a lot of pressure to squeeze fluid through a rock compared to the dry condition which fluid passes through rock easily in shorter time. Thus, chert in original condition has the low permeability which is 1.95E-22 m², but high permeability after dry condition which is 2.27E-22 m². Water absorption, *A* for chert was measured by subtracted weight of the sample in original condition after dry in the oven, (*W4*) from the weight of the sample in the air after undergoes the porosity test procedure, (*W2*) and the product was divide with (*W4*). For porosity test (Water Absorption and Porosity Measured by Water Immersion under Vacuum), the water absorption, *A* for chert is 1.2. Thus, the porosity, ϕ was calculated when the product of (*W2* - *W4*) is divided by the product of the weight of the sample in the water after undergoes the porosity test procedure, (*W3*) subtracted from (*W2*) as in Equation (6). As a result, the porosity for chert is 3 percent. More interpretation for the porosity value for both samples of chert and mudstone could be done using the value of Formation Factor, *F*. Porosity value for both chert and mudstone can be calculated through resistivity method using both Equation (7) and (8). The porosity obtained from the Formation Factor, ϕ_f for chert is in the range of 5 to 25 percent referring to Table 1, however for mudstone, the porosity value, ϕ_f is slightly out of range.

Table 2 Permeability Test and Porosity Test

Sample	Attenuation, α (Np/m)	Conductivity, σ (mS/m) Resistivity	GPR	Conditions	Permeability, *K* (m²)	Porosity, (%) ϕ	ϕf
Chert	0 - 0.04	0.036 - 0.556	0.037 - 0.574	Original Dry	1.95E-22 2.27E-22	3	6.26 - 13.36
Mudstone	0.01 - 0.64	0.18 - 9.813	0.186 - 10.142	Original Dry	Not available Not available	Not available	14.12 - 36.02

5. Conclusions

This study area consists of chert and mudstone. The conductivity value obtains from GPR method was identified and compared with resistivity method, which is (0.037 – 0.574 mS/m) for chert and (0.186 – 10.142 mS/m) for mudstone. For permeability and porosity test, 1.95E-22 m^2 (original condition) and 2.27E-22 m^2 (dry condition) and 3 % respectively for chert sample only. Using Formation Factor, the porosity, ϕ_f obtain for chert is (6.26 – 13.36 %) and (14.12 – 36.02 %) for mudstone. The research on the porosity of clay rock should be improved to obtain a successful result and well correlation. The study on outcrop expands the interpretation and show good correlation in the view of Geology, Geophysics, and Geotechnical method.

6. Acknowledgements

Authors would like to extend sincere gratitude to all Geophysics Department Staff of School of Physics, Universiti Sains Malaysia for their assistance in acquiring the data and making the research a success. A special gratitude and appreciation are given to Institute of Postgraduate Studies, Universiti Sains Malaysia for their Conference Fund sponsorship and Universiti Sains Malaysia for providing Research University Grant (RUI) entitle Geophysical Application And Approaches In Engineering And Environmental Problems. (1001/PFIZIK/811323).

References

[1] Remke L V D, Joep E A S, Gerard T S, Alireza M, Jeroen A M K and Emanuele F 2015 Introduction to special section : *Geophysical imaging and interpretation of outcrops*
[2] Terzaghi K, Peck R B and Mesri G 1996 *Soil Mechanics in Engineering Practice* 3rd Ed
[3] Jasin B and Harun Z 2007 Stratigraphy and sedimentology of the chert unit of the Semanggol Formation. *Geological Society of Malaysia*, **53** 103–109

[4] Leucci G 2008 Ground Penetrating Radar: The Electromagnetic Signal Attenuation and Maximum Penetration Depth. *Scholarly Research Exchange* 1–7
[5] Loke M H 2014 Tutorial: 2-D and 3-D Electrical Imaging Surveys
[6] Glover P 2009 What is the cementation exponent A new interpretation. *The Leading Edge*, **28**(1) 82–85
[7] Keller G V and Frischknecht F C 1966 Electrical Methods in Geophysical Prospecting. Pergamon Press Inc Oxford

Aquifer Characterization using 2D Electrical Resistivity Imaging in Kidangpananjung, Cililin District, West Java

Dery Marsan[1*], Azizatun Azimmah[1], Dida Patera Adli[2], Muthi'a J M[1], Ethis Yuantoro[2], Fatkhan[3]

[1]Undergraduate Program of Geophysical Engineering, Institut Teknologi Bandung
[2]Undergraduate Program of Geological Engineering, Institut Teknologi Bandung
[3]Exploration and Engineering Seismology Research Group, Institut Teknologi Bandung

Email: derymarsan@gmail.com

Abstract. Water shortage is a big problem for those who live in the region with unsustainable water resource like in Kidangpananjung - a village in Cililin district West Java. With elevation of 1070 m above mean sea level, Kidangpananjung stands on crest of a hill with the geographical coordinate of 113.5^0 BT, 6.97^0 LS. Based on geological survey, the outcrop which found in Kidangpananjung indicates that this region consists of pyroclastic rock such as fresh tuff and andesitic breccia. Four springs are found in the foothills with elevation of ± 1040 m above mean sea level which indicates the location of water table. To map the groundwater distribution more precisely and understand the aquifer rock more accurately, geo-electrical approach was conducted. This method is chosen based on the principle that the survey target, the water saturated rock, would give a relatively low resistivity contrast than its surrounding rocks. The target aquifer is considered as confined aquifer at 30m - 40m beneath Kidangpananjung. The data acquisition was designed with two lines of wenner-alpha arrays with 235 meters length each. Two lines of profiling were chosen in order to map the underground layer and its resistivity and thicknesses. The resistivity measurements were carefully interpreted by using least-square inversion technique by using RES2DINV program. The purpose of this research is to understand the characteristics and depth of Kidangpananjung aquifer. Therefore, it can be used to be a reference in groundwater drilling in order to improve the living of the inhabitants of Kidangpananjung
Keywords: Aquifer Characterization, 2D Resistivity Imaging, confined Water

1. Introduction

Clean water crisis has become a major problem occurring in Indonesia. This problem mostly happens not because of water scarcity issues but because of the difficulties in reaching one. Finding a more reachable new water source is needed in order to solve such problem. This water source could be found beneath the ground and often called groundwater. Groundwater is water located beneath the ground surface in soil pore spaces and in the fractures of lithological formations. A unit of rock or an unconsolidated deposit is called an aquifer when it can yield a usable quantity of water [1]. One of the advantages of groundwater is that its sustainable existence when all the surface water source has become diminished due to any season change. Resistivity method as one of geophysical methods is proven to be suitable for being applied in investigating groundwater potential existence [2] and determination of depth, thickness and boundary of an aquifer [3, 4]. Furthermore, this method possesses some useful advantages such as relatively easy field logistic and instrumentation using, and its data processing which could be done immediately even if when we are still in the field.

The investigation site for this research took place in Kidangpananjung Village, Cililin District, West Java, Indonesia. This research area is located at 113.5^0 BT, 6.97^0 LS with the involved investigation area of ± 90.000 m^2. The procurement of research permission was done in 2 weeks and the data acquisition was conducted on May 7th, 2016 and June 11th, 2016 using ARES resistivity meters as the instrument and Wenner-alpha as the measurement configuration. We used two measurement lines which had been arranged to cross to each other (Figure 1) with the length of 235 meters each and electrode spacing of 5 meters. Each line has a total amount of 360 datum points. These lines were chosen based on geological observation at the location of springs in the investigation area.

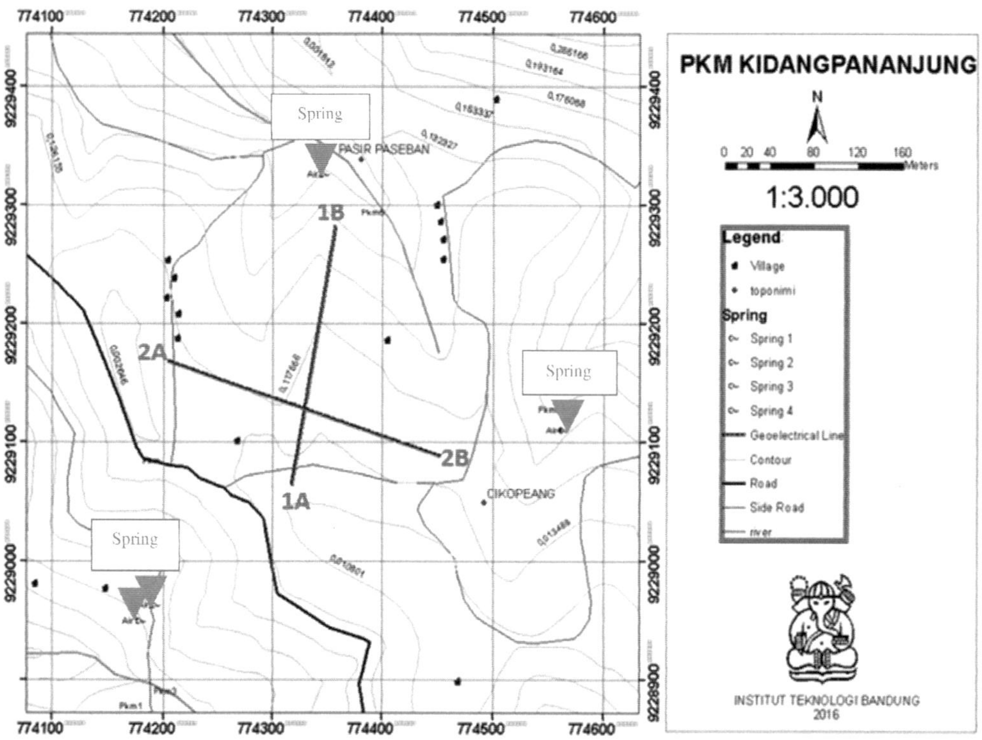

Figure 1. Springs and measurement line locations shown in topography map of Kidangpananjung area.

2. Method

Geological survey was done by observing the rock's outcrops and the springs around obsrvation area. These locations were then plotted in coordinate points on topography map in order to estimate the possible resistivity acquisition lines and the target depth of aquifer rock. The four springs found in the area of investigation are located in valley-like regions so that the depth of aquifer could be estimated. The investigation site is dominated by weathered rocks originated from volcanic eruption such as pyroclastic rock and andesitic breccia. The rocks found in the site of investigation are pyroclastic rock known as reddish brown weathered tuff with relatively good porosity, making it suitable to be an aquifer (Figure 2), and also andesitic breccia which is in fresh condition, dark colored, aphanitic, containing hornblende and plagioclase minerals, and might be considered to be bedrock and cap rock of aquifer (Figure 3).

Figure 2. Pyroclastic tuff rock in Kidangpananjung area

Figure 3. Andecitic breccia rock in Kidangpananjung area

In order to determine subsurface resistivity distribution, apparent resistivity needs to be known to estimate the true resistivity. Geo-electrical data are usually expressed as apparent resistivity:

$$\rho_a = \frac{\Delta V}{I} K \tag{1}$$

Where ΔV is the measured potential, I denotes the transmitted current, and K is the geometrical factor.

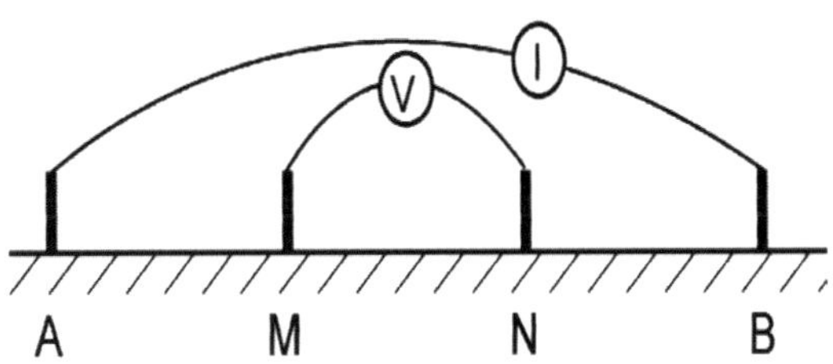

Figure 4. Wenner array electrode configuration. The letters A and B denote the current electrodes and the letters M and N denote the potential electrodes [5]

To determine the true subsurface resistivity, the inversion process must be performed with the input of measured apparent resistivity data. The data processing and inversion were done by using RES2DINV software with the apparent resistivity as the input to obtain the 2 dimensional (2D) subsurface resistivity distributions. Inversion process could also calculate the RMS error for each line. The inversion technique used in this research is referred to the algorithm of Loke and Barker, the least-square inversion [6]. It is done with the maximum iteration of 6 times for line 1 and 7 times for line 2.

3. Result

The inversion results in RMS error of about 2.8% for line 1 and 3.4% for line 2 which indicates that the misfit between the true models and inversion results is relatively small. Therefore, the inverted model results can be assumed to be close to the subsurface geology and proper to be interpreted as long as the model is plausible as the geologic features model. Based on the selected lines in (Figure 1), we presented the results of resistivity distribution profiles along 1A-1B profile (figure 5) and 2A-2B profile (figure 6) where the horizontal axes represents the distance from 0 point and the vertical axes represents the elevation from mean sea level:

1A 1B

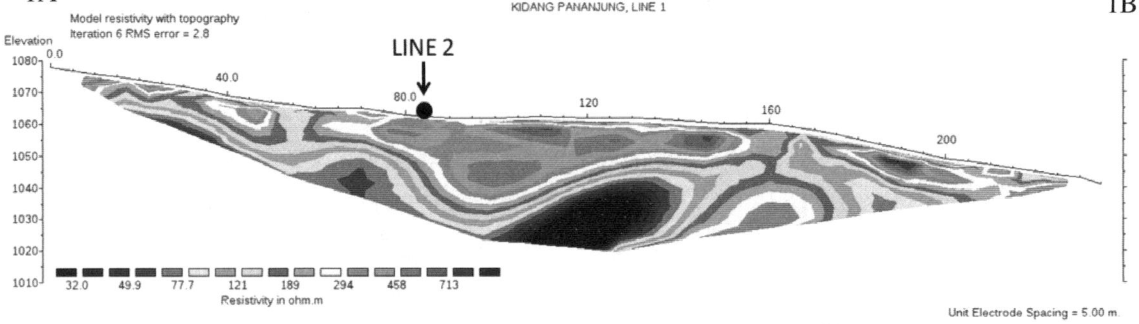

Figure 5. Resistivity profile of line 1 (1A-1B profile)

Line 1 was arranged in the orientation towards the north – south, crossing the topography contours and the line 2 at the electrode position of 85 meters. The length of line 1 is 235 meters where the northern region is lower than the southern region. At the position of 80 meters to 160 meters of line 1, the region is considered to be a plain region with the altitude of ± 1065 masl. At the position of 0 meter to 80 meters of line 1, there is a slope of about 10^0, while at 160 meters to 235 meters, the slope is about 15^0. The range of resistivity values resulted by the measurement in line 1 ranged from 32 Ωm to 890 Ωm with low resistivity values were represented in dark blue colour, and the high resistivity values were represented in dark red colour. The ground resistivity is related to various geological parameters such as the mineral and fluid content, porosity and degree of water saturation in the rock [4]. Based on subsurface resistivity distribution, the layer is dominated by high resistivity rock which indicates massive rock with the lack of conductivity material properties. At an elevation of ± 1040 m or 30 meters below the surface, low resistivity anomalies below 30 Ωm are spotted and suspected to be water saturated layer or aquifer zone.

2A 2B

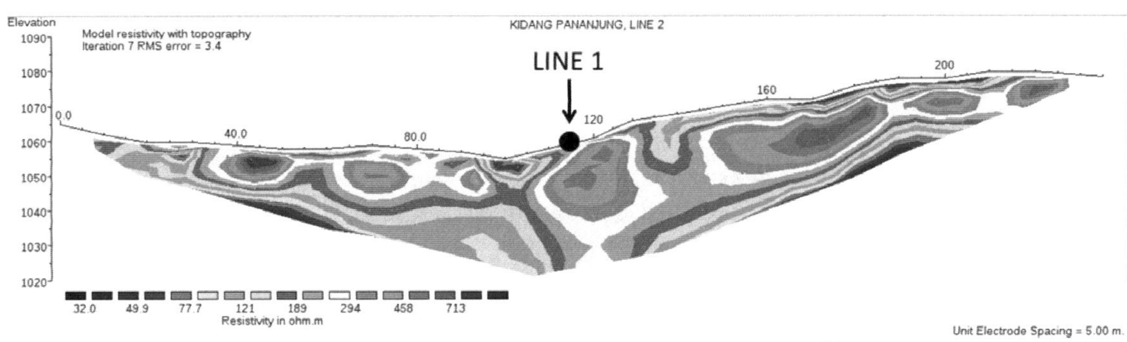

Figure 6. Resistivity profile of line 2 (2A-2B profile)

Line 2 was arranged in the orientation towards west - east, crossing the line 1 at the electrode position of 115 meters. This line is 235 meters long where the western region is relatively flatter than the east region with a slope of about 15^0. The range of resistivity values resulted by the measurement in line 1 ranged from 32 Ωm to 890 Ωm which high resistivity characteristic correlated with line 1 at elevation ± 1040 meters, where there is also found a low resistivity anomaly which is similar to the line 1 and suspected to be water saturated layer zone.

Based on geological survey and interpretation results of both profiles, the predicted aquifer layer is formed by tuff pyroclastic deposits. This layer was saturated by water at a depth of 30 meters beneath the surface as the blue or low resistivity zone was observed at 30 meters beneath the surface and the elevation of springs are also at 30 meters below the crossing of Line 1 and Line 2. The result of this study was confirmed by a nearby well which drilled after this study. The depth of water which emerged from that well is ± 30 m beneath the surface. The high resistivity values in both profiles indicate the andesitic breccia layer acting as the cap rock because of its impermeable characteristic. The table below shows the rocks types that have been found in Kidangpananjung along with their resistivity characteristics and descriptions:

Table 1. Rock properties and characteristics in Kidangpananjung area

Rock Types	Resistivity Range (Ω m)	Thickness (m)	Characteristics
Water Saturated Tuff	32 – 95	10 – 30	Low resistivity, high porosity, permeable
Tuff	95 - 300	10 -45	Medium resistivity, medium porosity, permeable
Andesitic Breccia	300 - 890	5 – 25	High resistivity, low porosity, impermeable

The profiles derived from 2D geo-electrical surveys have clearly delineated the variations and distribution of resistivity laterally. The high and low resistivity zones are identified from the 2D profiles. The inversion results are able to show the water saturated zone in aquifer layer. The proposed point for the next well drilling to fulfill the need of water in Kidangpananjung is at at a point of 115 meters on line 1 where the ground water is estimated to be found at a depth of ± 30 meters beneath the subsurface (figure 7). This recommendation point lies at 774335 and 9229167 (the 48M zone UTM) where the predicted aquifer found at an elevation of 1036,25 masl. The proposed descriptions and characteristics of rocks which would be found beneath this recommendation point are shown in figure 8.

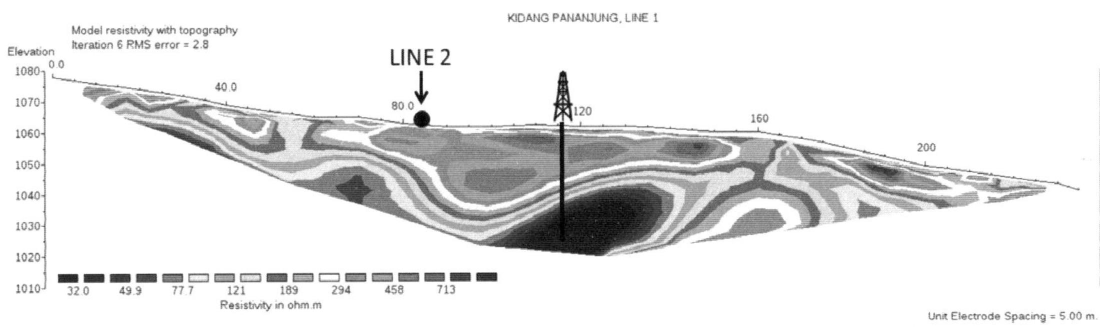

Figure 7. Proposed point of borehole drilling shown in resistivity profile of line 1.

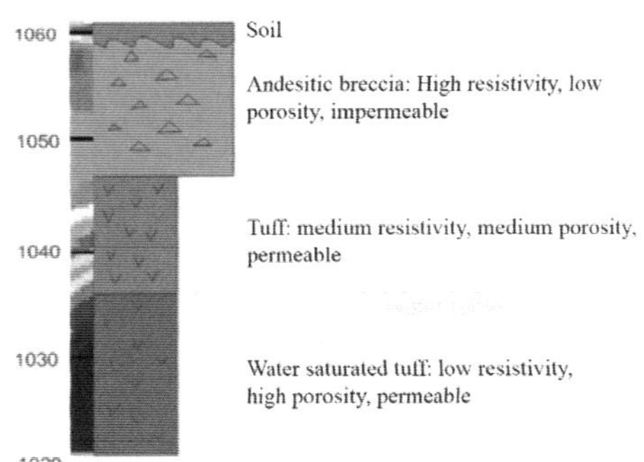

Figure 8. Proposed descriptions and characteristic of rocks found underneath the recommendation point.

4. Conclusion

To conclude, in this study, 2D resistivity method has been proved to be able to locate groundwater potential as the depth of groundwater have been confirmed by a nearby well and able to determine the rocks characteristics in volcanic area while provided minimum error during investigation. The recommendation point for well drilling in Kidangpananjung is lies at 774335 and 9229167 (the 48M zone UTM) where the predicted aquifer found at ± 30 m beneath the surface. Furthermore, it is considered that the aquifer in this area belongs to the kind of confined aquifer with andesite breccia as the cap rock and pyroclastic rock or tuff as the volcanic rock aquifer.

References

[1] Alabi A A, Bello R, Ogungbe A S, and Oyerinde H O 2010 Determination of groundwater potential in Lagos State University, Ojo; Using geoelectric methods (Vertical electrical sounding and horizontal profiling) *Report Opinion* **24**: 68-75

[2] Oseji J O, Atakpo E A, and Okolie E C 2005 Geoelectric investigation of the aquifer characteristics and groundwater potential in Kwale, Delta State, Nigeria. *J. Applied Sci. Environ. Mgt.* **9**: 157-160

[3] Ismailmohamaden M I 2005 Electric resistivity investigation at Nuweiba Harbour Gulf of Aqaba, South Sinal, Egypt. *Egypt. J. Aquatic Res.* **31**: 57-68

[4] Omosuyi G O, Adeyemo A, and Adegoke A O 2007 Investigation of groundwater prospect using electromagnetic and geoelectric sounding at afunbiowo, near Akure, Southwestern Nigeria. *Pacific J. Sci. Technol.* **8**: 172-182

[5] BurVal Working Group 2006 Groundwater resources in buried valleys - a challenge for geosciences (R. Kirsch, Ed.). *Leibniz Inst. for Appl. Geosc* 77-78

[6] Loke M H and Barker R D 1996 Rapid least-squares inversion of apparent resistivity pseudosections by a quasi-Newton method. *Geophysical Prospecting* **44**: 131-152

IOP Conf. Series: Earth and Environmental Science **62** (2017) 012015 doi:10.1088/1755-1315/62/1/012015

The Identification of Seulimeum Fault System in Iejue Aceh Besar (Indonesia) Using 2-D Resistivity Imaging Method

Z M Taqiuddin[1*], M M Nordiana[1], S Rosli[1]

[1]Geophysic Section, School of Physics, 11800 Universiti Sains Malaysia, Penang, Malaysia

[1*]Email's: taqiuddinzakaria92@gmail.com

Abstract. 2-D resistivity imaging method is widely used to delineate fault zones. The fault referred as a rock fractures or discontinuity in rock volume due to relative displacement caused by the earth movement. In this study, 2-D resistivity imaging survey conducted in Iejue, Aceh Besar locality with the aim to delineate the Seulimeum fault system for this locality. Two survey lines L1 and L2, with a total length of 1200 m each, conducted across the suspected fault using ABEM SAS4000 system with Pole-dipole array and 10 m minimum electrodes spacing. 2-D resistivity imaging profile shows the exploration depth of >320 m with resistivity range of 1-1800 Ωm. The contrast in resistivity values for both lines L1 and L2 indicates the suspected fault at distance of 550-600 m and 500-550 m respectively. Correlation of 2-D resistivity imaging analyses with a geological map of the area shows that the fault is trending in the NW-SE direction with fault lineament approaching the resistivity result. Two distinct resistivity zones identified for each survey line; a saturated zone starting at distance of 600 m toward the end of the line and high resistivity zone at distance of 0 m to 550 m. This saturated zone interpreted as a ground fluid, while the high resistivity zone represents as existing alluvium, sandstones, and volcanic sediments.

1. Introduction
The Sumatra fault recognized as an active fault zone which consists a part of oblique convergence subduction zone between the Indo-Australian and Eurasian plates [1;2]. Strike-slip faulting occurs a hundred kilometers to the east and involves slip directed parallel to the trench which is pass through the Sumatra Island. Sumatra was accommodating from the northern edge of Australia during Triassic to early Jurassic and become stable continental margin after the subduction in Cretaceous ages [3]. Active volcanoes of Sumatra located along the Barisan range as the backbone of Sumatra which composed of pre-Tertiary rocks covered by Cenozoic volcanic rocks. In 2000, the Sumatra fault was map according to the geomorphic expression on a topographic map and aerial photographs [4]. The geological characterization of Sumatra fault is clearly observed based on contrasted in geological rocks along the fault. The fault classified into three-part which is southern, central, and northern [5]. The Sumatra fault recorded as right lateral strike-slip with the length of 1900 km and consist about 20 major segments with length about 60-200 km [6].

2. General Geology
Aceh province located in the northern region of Sumatra island and overlain by Tertiary sedimentary and volcanic rocks from northeast to the southwest which transacted by Sumatra fault with the

orientation of NW-SE direction [7;8]. The study area, Iejue located near to Banda Aceh with dominated by Lam Tuba volcanic which composed of andesitic to dacitic volcanic, pumiceous breccia, tuffs, agglomerate and volcanic ash flow with the composition of tuffaceous and calcareous sandstones, conglomerates, and mudstones [9]. The geomorphology for this area a due to a topographic depression with occupied of alluvial flat and formed of flat-topped hills [10;11]. Following closely the crest of Barisan Range is a continuous system of axial valley which mark as the outcrop of Sumatra fault system [12]. Figure 1 shows the geological map of study area.

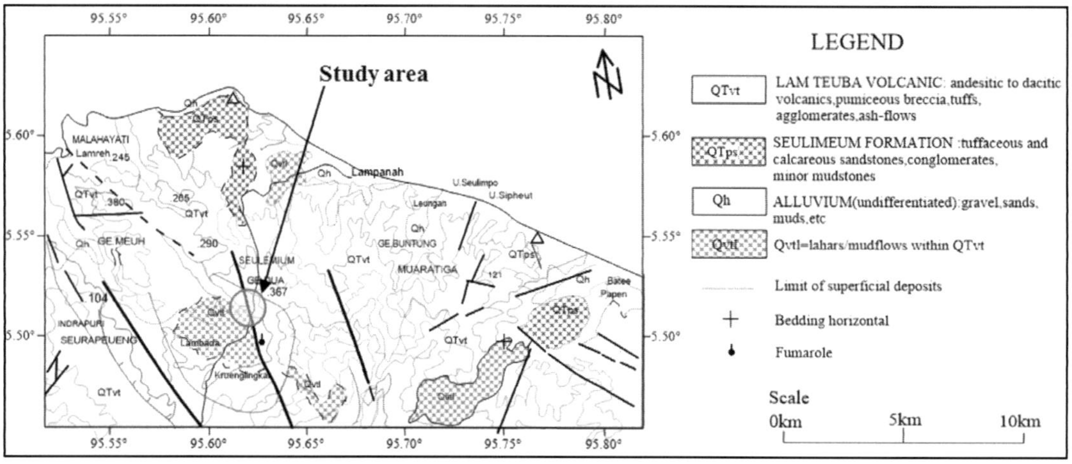

Figure 1. Geology map of Banda Aceh Quadrangle, Sumatra (modified: [9]).

3. Resistivity Theory

The electrical method utilizes direct or low-frequency alternating current to investigate the electrical properties of the subsurface. In resistivity method, the electric current injected into the ground which resulting in a potential difference [13]. Resistivity method used to investigate the horizontal and vertical discontinuity of resistivity in overburden material. The apparent resistivity, ρ_a is calculate based on Equation 1.

$$\rho_a = k\frac{V}{I} \tag{1}$$

where;

V: voltage (V)
I : current (A) or (mA)
k : geometric factor depends on electrode arrangement

In 2-D resistivity imaging survey, k values play an important role in determining the deepest depth of penetration with respect to the electrodes spacing. Electrode arrays such as Pole-dipole capable of providing a detailed information in vertical resolution for subsurface exploration. Changes in composition, layer or contaminant levels will give variation in electrical resistivity which also dependent on the degree of fracturing [14].

4. Methodology

2-D resistivity imaging survey with Pole-dipole array conducted at Iejue, Aceh, Indonesia and its vicinity. Two survey lines, L1 and L2 length of 1200 m (Figure 2) conducted crossing the suspected fault using ABEM SAS4000 system with 10 m electrode spacing. The data processed using Res2Dinv and display in Surfer10 software for interpretation and discussion.

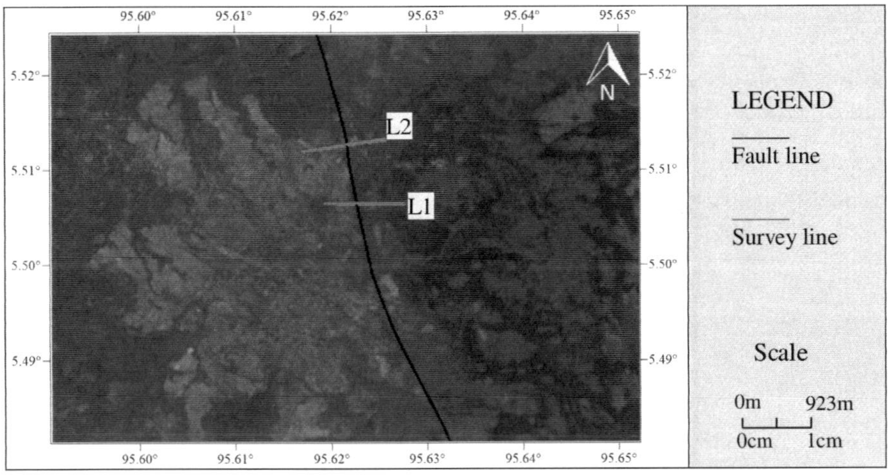

Figure 2. The study area of Iejue locality [15].

5. Result and Discussion

Figure 3 shows the inversion model 2-D resistivity for line L1 with exploration depth >320 m and resistivity values cover a range of 1-1800 Ωm. From the result, the resistivity values of 1-12 Ωm and 15-50 Ωm represent as the contrast zones of high and low values. High resistivity values from 100-500 Ωm dominated for topsoil underlain with a lower resistivity at a depth up to 40 m refer as a presence of alluvium, sandstone, and volcanic tuffs. Low in resistivity values at 1-7 Ωm indicated the present of ground water which expected consist of geothermal fluid. The contrast in resistivity with low and high values represent the fault lineament at distance of 550-600 m shown in Figure 3.

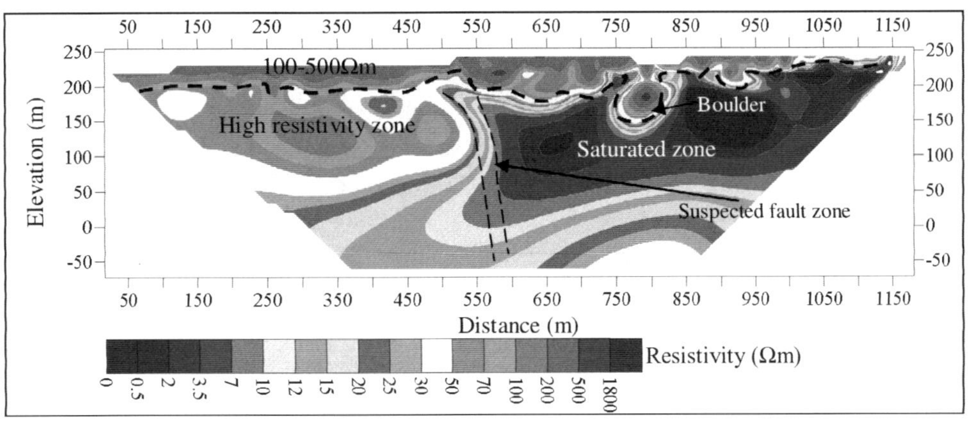

Figure 3. Inversion model 2-D resistivity of line L1.

Figure 4 shows inversion model 2-D resistivity of line L2 with the depth of exploration up to 320 m and resistivity values cover from ranges of 1-1800 Ωm. High resistivity values from 100-500 Ωm dominated the topsoil underlain with low resistivity zones. High resistivity zone indicated the existence of alluvium soil and present of volcanic rocks at a topsoil with depth about 40 m. The result shows the presence of two distinct resistivity zones with values of 1-10 Ωm for the low (saturated) and 15-50 Ωm at high resistivity region. Low resistivity region from 30-50 Ωm indicated the presence of clay soil while at 1-7 Ωm refer the presence of ground water. The boulders identified with resistivity values of (100-200) Ωm at a depth of 50 m. The contrast of resistivity zone at distance of 500-550 m indicated the present of suspected fault line in this study area as shown in Figure 4.

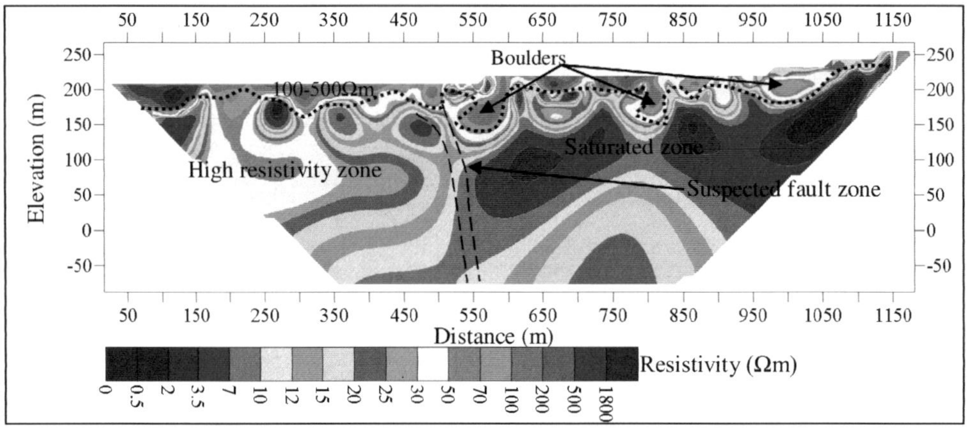

Figure 4. Inversion model 2-D resistivity of line L2.

6. Conclusion

The 2-D resistivity survey delineates the fault zone of Iejue area due to the contrast of high and low region with resistivity values from 1-1800 Ωm. Low resistivity region identified at values of 1-7 Ωm while high resistivity region identified at values of 100-500 Ωm. The contrast of resistivity zone at distance of 550-600 m for line L1 and a distance of 500-550 m for L2 indicates the presence of fault. The correlation with the geological map shows that the Seulimeum fault trending from NW-SE direction (Figure 5).

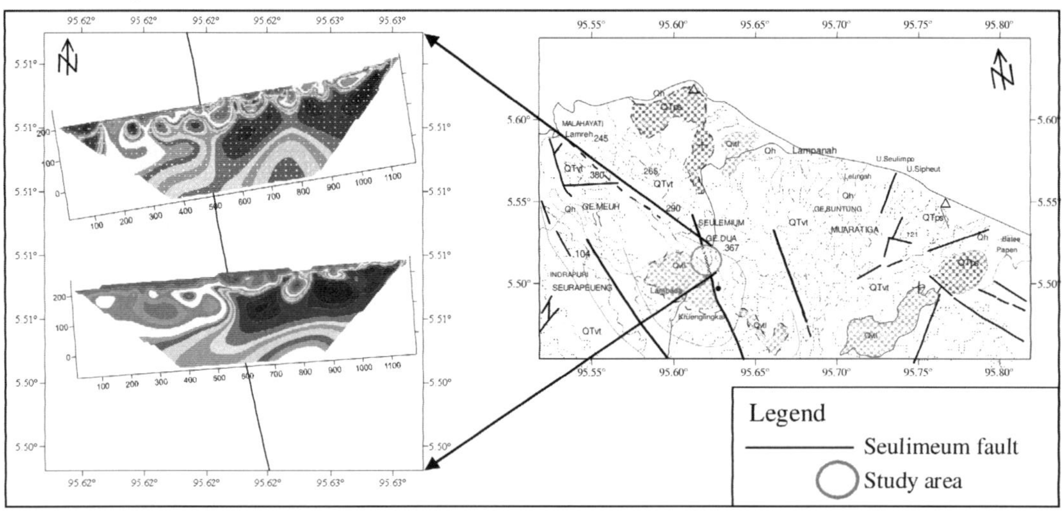

Figure 5. Correlation of inversion model 2-D resistivity with the geological map of Seulimeum fault.

References

[1] Bellier O, Sébrier M, Pramumijoyo S, Beaudouin T, Harjono H, Bahar I, and Forni O 1997 Paleoseismicity and Seismic Hazard Along the Great Sumatran Fault (Indonesia) *Journal of Geodynamics* **24(1-4)** pp 169–83

[2] Syukri M, Saad R, Nordiana MM, and Azwin IN 2014 Preliminary Study of Sumatera Fault Using 2-D Resistivity Imaging Method *Electronic Journal Geotechnical Engineering (EJGE)* **19** pp 971-979

[3] Hamilton W B 1979 *Tectonics of the Indonesian Region* (US Govt Print Off)

[4] Sieh K and Natawidjaja D H 2000 Neotectonics of Sumatran Fault Indonesia *Journal of Geophysical Research* **28** pp 295-326

[5] Aydan Ö 2010 *A Reconnaisance Report on the Pariaman-Padang Earthquake of September 30 2009* Japan Society of Civil Engineers Japan pp 1–36

[6] Natawidjaja D H, and Wahyu T 2007 The Sumatran Fault Zone-From Source to Hazard *Journal of Earthquake and Tsunami* **1(01)** pp 21-47

[7] Barber A J, and Crow M J 2003 An Evaluation of Plate Tectonic Models for the Development of Sumatra *Gondwana Research* **6(1)** pp 1–28

[8] Nordiana MM, Taqiuddin ZM, Saad R, Azwin IN, and Syukri M 2016 Exploring The Seulimeum Fault in Aceh, Indonesia Using Magnetic Method *Jurnal Teknologi* **78(8-5)** pp 89-95

[9] Bennett J D, Bridge McD D, Cameron N R, Djunuddin A, Ghazali S A, Jeffrey D H, Kartawa W, Keats W, Rock N M S, Thomson S J and Whandoyo R 1981 Geologic map of the Banda Aceh Quadrangle Sumatra

[10] Katili J and Hehuwet F 1967 On The Occurrence of Large Transcurrent Faults on Sumatra Indonesia Osaka Japan *Osaka City University Journal of Geoscience* **10** pp 1-17

[11] Page B G N, Bennett J D, Cameron NR, Bridge D, McC Jeffery D II, Keats W and Thaib J 1979 A Review of The Main Structural and Magmatic Features of Northern Sumatra *Journal of Geological Society of London* **1** 36 pp 569-79

[12] Nordiana MM, Syukri M, Saad R, Marwan, and Kamaruddin NA 2014 The Identification of Fault Zones in Krueng Raya, Aceh Besar (Indonesia) Using Magnetic Method *Electronic Journal Geotechnical Engineering (EJGE)* **19** pp 1-9

[13] Kearey P, Brooks M and Hill I 2002 *An Introduction to Geophysical Exploration* (Blackwell Science Ltd Oxford)

[14] Loke M H 1999 *Electrical Imaging Surveys for Environmental and Engineering Studies. A practical guide to 2-D and 3-D surveys* pp 1-57

[15] Google Earth Iejue 5.510332 N and 95.921692 E 18 January 2015 2 September 2015

Acknowledgements

The authors wish to thank lecturers, technical staffs, and all member of Geophysics group, Universiti Sains Malaysia (USM), Penang, Malaysia, all students and staffs of Faculty of Science, Universitas Syiah Kuala(UNISYIAH), Banda Aceh (Indonesia) for their assistance during the geophysical field data acquisition. The author also wishes to thank Institute of Postgraduate Studies Universiti Sains Malaysia (IPS USM) for providing the IPS Conference Funds for this conference and USM for providing the Research University Grant (RUI) entitled of Geophysical Application and Approaches in Engineering and Environmental Problems (1001/PFIZIK/811323).

How to Build New Interpretation Concept using Dynamic Data: A case Study in Carbonate of Upper Cibulakan, North West Java Basin, Indonesia

Dimas Pramudito[1], Tania Meidiana[1], A D Alfianto[2], Dhea Rizki Nurhadi[2]

[1] Geoscience in Geology and Geophysics Department, Pertamina EP Asset 3, Indonesia
[2] Engineer Support in Geology and Geophysics Department, Pranalika Energi Nusa, Indonesia

E-mail: dimas.pramudito@pertamina.com

Abstract. Bravo Structure is one of the Pertamina EP gas producing field in North West Java Basin region. This structure located at Subang about 10 km southeast from City Subang. The objective of this structure is a vertical compartment of carbonate due to diagenetic effect and transgression. This evident cause shale deposited and divide multi-layer carbonate reservoir. The carbonate lies at Cibulakan Formation and has three reservoirs: Z-14, Z-15, and Z-16 limestone. Since each reservoir has different pressure data and CO_2 contain, so it becomes challenges to make new interpretation and to know those compartments between reservoirs. To look up geometry distribution detailed, this structure is reinterpreted using 3D seismic (2014) and new concept depositional environment based on wells correlation and core analysis. Generating seismic interpretation using seismic inversion and attribute. Amplitude envelope and instantaneous frequency are calculated to obtain sweetness. Both of them are built by 3D volume seismic. High magnitude from amplitude envelope is used to characterize geometry distribution of carbonate while instantaneous frequency determines low frequency because of gas distribution. The result of this study suggests vertical and lateral carbonate distribution characteristics Meanwhile sweetness attribute can determine gas contain in each layer Z-14 and Z-15. Vertical distribution of Z-14 layer about 140 m (porosity 7-11%) and Z-15 about 70 m (porosity 8-13%). Finally, interpreting carbonate both Z-14 and Z-15 as shelf margin that new concept gives a chances to develop this structure and optimize reservoir management.

1. Introduction

Bravo structure is one of PT Pertamina EP gas producer field in North West Java Basin region. It is part of Subang Field which located southeast from Subang city. Based on Regional Geology, it lies at Pamanukan High formed as result of *bed rock* lifting with a north-south orientation and it is part of Pasir Putih sub-basin system. This high is surrounded by fault system which separates Kepuh low at East and Pasirbungur low at West [1].

This structure has three main wells which are: Bravo-01, Bravo-02, and Bravo-03. Among those wells data, the objective of this structure is vertical compartment of carbonate due to diagenetic effect and transgression. Those processes cause shale deposited and divides into multilayer carbonate reservoirs. In this case, these carbonate reservoirs that lies in Upper Cibulakan Formation divided into three reservoirs; those are Z-14, Z-15, and Z-16 layer. Interpretation and analysis for that carbonate vertical compartment using the different pressure and high CO2 contain of each reservoir (Figure 1).

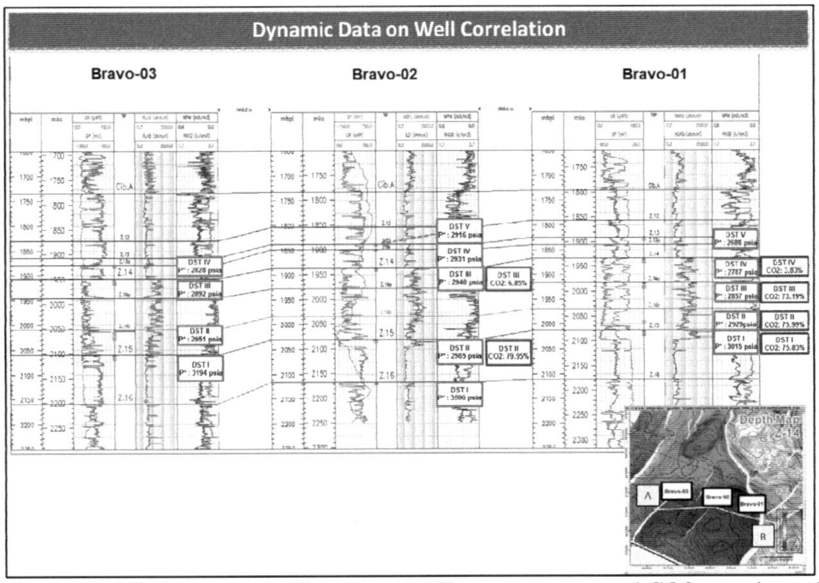

Figure 1. Well data correlation shows different pressure and CO2 contain each layer.

Table 1. Core analysis results Well Bravo-01

Depth (Metres)	Permeability (mDarcies)	Porosity (%)	Res. Liq. Sat Oil	Res. Liq. Sat Total Water	GD	Lithological Description
1897	1.01	15.44	5.96	47.15	2.637	SD: gy, med hd, vgy, sbrnd, w strd, tr.calc, sli foss, sli arg, patchy brn flu, slow lt yell-lt brn uniform cut
1935.5	13	20.70	1.02	63.46	2.707	LM: (Packstone) gy-lt brn, hd, l.foram, algae, coral, pp-mott, mott lt yell flu, slow lt yell-milky uniform cut
1992	3.98	11.34	0	69.97	2.707	LM: (Wack-Packstone) gy, hd, l.foram, algae, styl, no flu, no cut
2061	3.63	12.70	0	65.14	2.698	LM: (Granstone) lt gy-wht, hd, l.foram, algae, pp lt yell flu, slow lt yell-milky strw cut
2081.5	21	11.03	0	67.48	2.722	LM: (Packstone) gy, hd-vhd, foram, l.foram, moldic, no flu, no cut

First, to look forward geometry distribution and depositional environment of the reservoir, it is reinterpreted using seismic 3D Post Stack Time Migration data (the year 2014), well correlation, petrophysical analysis and core analysis. Core analysis seen in Table 1. Then next seismic interpretation is doing acoustic impedance inversion and sweetness attribute.

The results of the processing conducted are the conceptual geological model built from acoustic impedance volume and sweetness. Acoustic impedance inversion can show carbonate characteristic associated with the porosity of reservoir. Sweetness succeeds to confirm stratigraphy feature. The high magnitude of amplitude envelope can describe the character of geometry distribution of carbonate reservoir. Instantaneous frequency expected to describe fluid distribution that trapped in the carbonate reservoir. The relation between acoustic impedance and sweetness used for creating a conceptual geological model of carbonate build up that refers to Bubb & Hatlelid concept [2].

2. Methodology

2.1. Seismic Inversion Acoustic Impedance

Acoustic impedance inversion is a process to get reflection coefficient value from the seismic record that will be used to determine quantitative rock property such as acoustic impedance value. Inversion method for this study is the model based. It needs an early impedance model (initial model) from well log data, which are velocity log and density log [3, 4]. Parameters used for inversion are background value 14,806 gr/cc.m/s, 40% model weight and 40x iteration in the inversion process.

2.2. Sweetness Attribute

Sweetness attribute is the empirical algorithm to identify reservoir existence in an area based on seismic data that can reveal high amplitude value and low frequency. The ratio between amplitude envelope and the root of instantaneous frequency can build it. This attribute will identify feature which has general energy change of seismic data [5].

3. Results and Discussion

Figure 2 shows carbonate reservoir characteristic, where deposited carbonate Z-14 above carbonate Z-15. Porous carbonate distribution is presented by low acoustic impedance value for carbonate Z-14 at 1935-1960 m and shows up again in carbonate Z-15 at 2080-2120 m. The evenly spread of high acoustic impedance value associated with tight carbonate occur at less than 1960 m to 2080 m. The area on those depths is the platform of carbonate Z-14 growth. Finding out acoustic impedance cut-off value that separates porous zone and tight zone shown in cross plot between acoustic impedance log and density log (Figure 3)

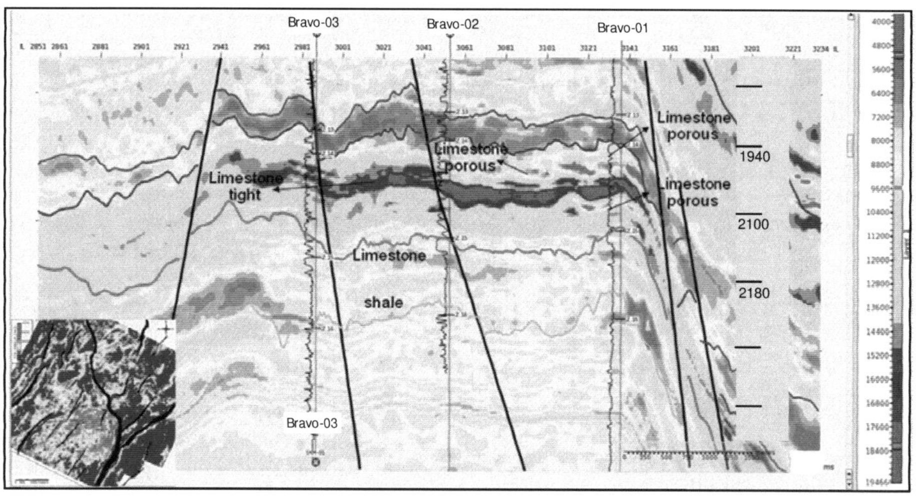

Figure 2. Acoustic impedance inversion shows that low acoustic impedance value (red color) associated with porous carbonate, whereas high acoustic impedance value (blue-violet color) associated with tight carbonate

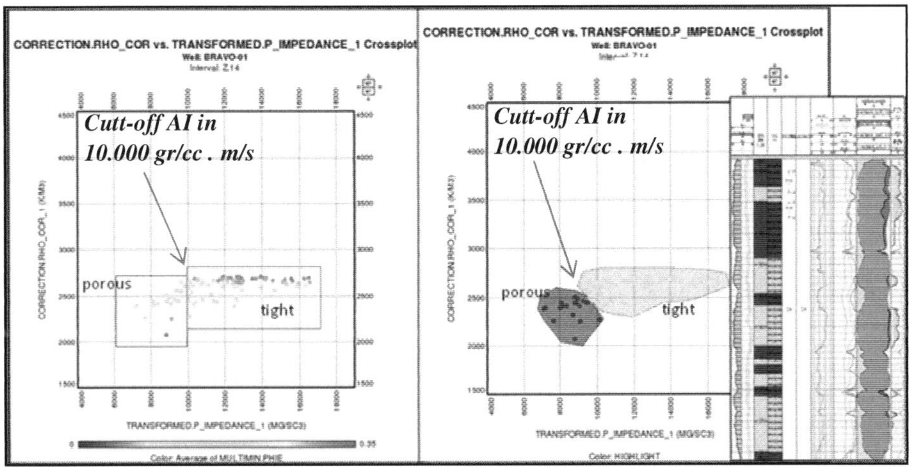

Figure 3. Sensitivity analysis with cross plot between acoustic impedance log and density log using effective porosity log as the color scale.

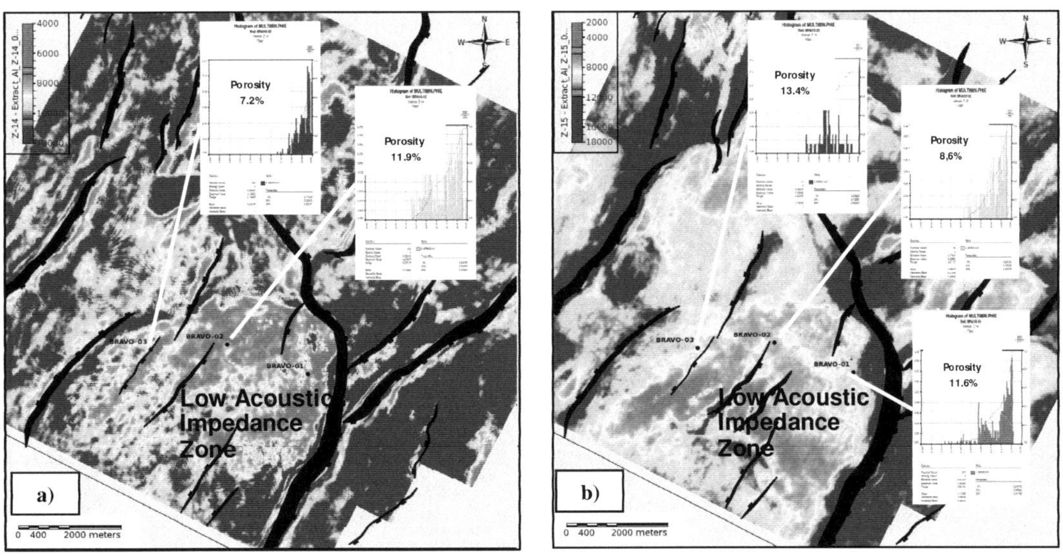

Figure 4. Carbonate reservoir properties distribution derived from acoustic impedance using window 10ms; a) Carbonate Z-14; b) Carbonate Z-15

To acquire acoustic impedance value distribution pattern laterally for carbonate Z-14 and Z-15, slicing with window width 10ms along the horizon Z-14 and Z-15 is done (Figure 4). The result can be used for facies interpretation of build up carbonate depositional environment based on rock property value, which is porosity. Low acoustic impedance value distribution that relates with porous carbonate has porosity value around 7-11% for Z-14 and 8-13% for Z-15. Refers to core analysis data (Table 1), depth 1935.5 m (Z-14) and depth 2080 m (Z-15), lithology description shows that carbonate facies is packstone with coral, algae, and foram contains. So that interpreted low acoustic impedance zone associates with shelf-margin facies (Figure 7).

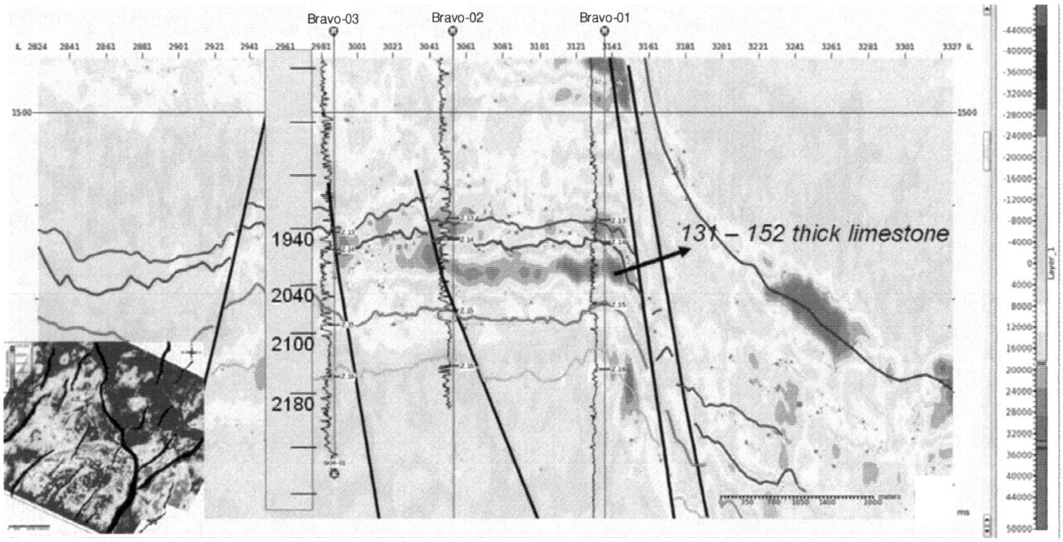

Figure 5. Comparison between sweetness attribute and gamma ray on well, sweetness applied to predict reservoir carbonate. High sweetness can show 131-152 m thick of carbonate at 1935 m depth, while carbonate under depth 2080 m can't present good correlation with sweetness

Figure 5 shows high sweetness can describe 131-152 m thick of carbonate at depth 1935 m (Z-14), while carbonate under 2080 m can't shows good correlation with sweetness. Sweetness distribution pattern laterally is sliced along the Z-14 top horizon with window width 10 ms, and supported by frequency and envelope amplitude slicing with the same window (Figure 6). The high magnitude of envelope magnitude can show the body of carbonate Z-14 associates with lithological change among rock beds around or boundary depositional environment change of carbonate Z-15. Then low-frequency distribution can show carbonate Z-14 trap that contains fluid (gas). The relation between them generates high sweetness value that can describe stratigraphy feature as a patch reef and shelf-margin. Facies open marine clearly presented with low sweetness value (Figure 7).

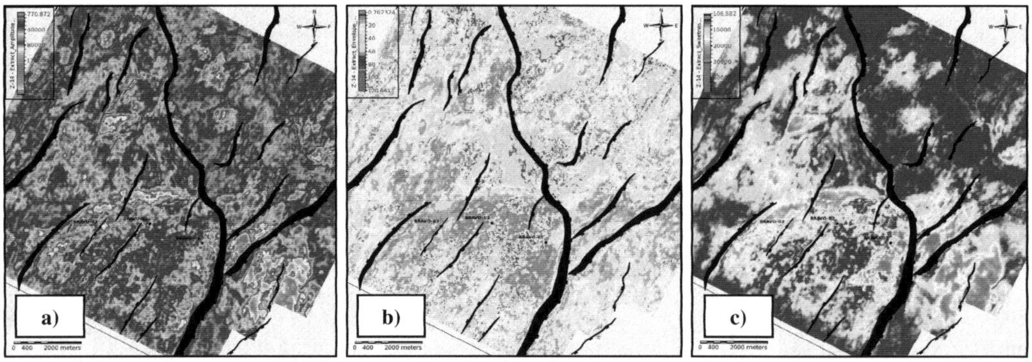

Figure 6. Same feature on RMS extract to calculate Sweetness attribute; a) feature on RMS extract amplitude envelope attribute with window length 10 ms; b) feature on RMS extract instantaneous frequency attribute with window length 10 ms; c) feature on RMS extract sweetness with window length 10 ms.

Finally, interpretation result using acoustic impedance value and sweetness distribution, the conceptual geological model built for describing the geological condition (Figure 6). Acoustic impedance value distribution overlayed with sweetness distribution supported by porosity data from petrophysical analysis and core analysis in the model while building that model. Shelf-margin facies generated based on acoustic impedance value distribution, petrophysical analysis, and core analysis. Then patch reef, shelf and open marine facies produced by sweetness distribution. Conceptual geological model refers to carbonate build up conceptual geological model belongs to Bubb & Hatlelid [2]

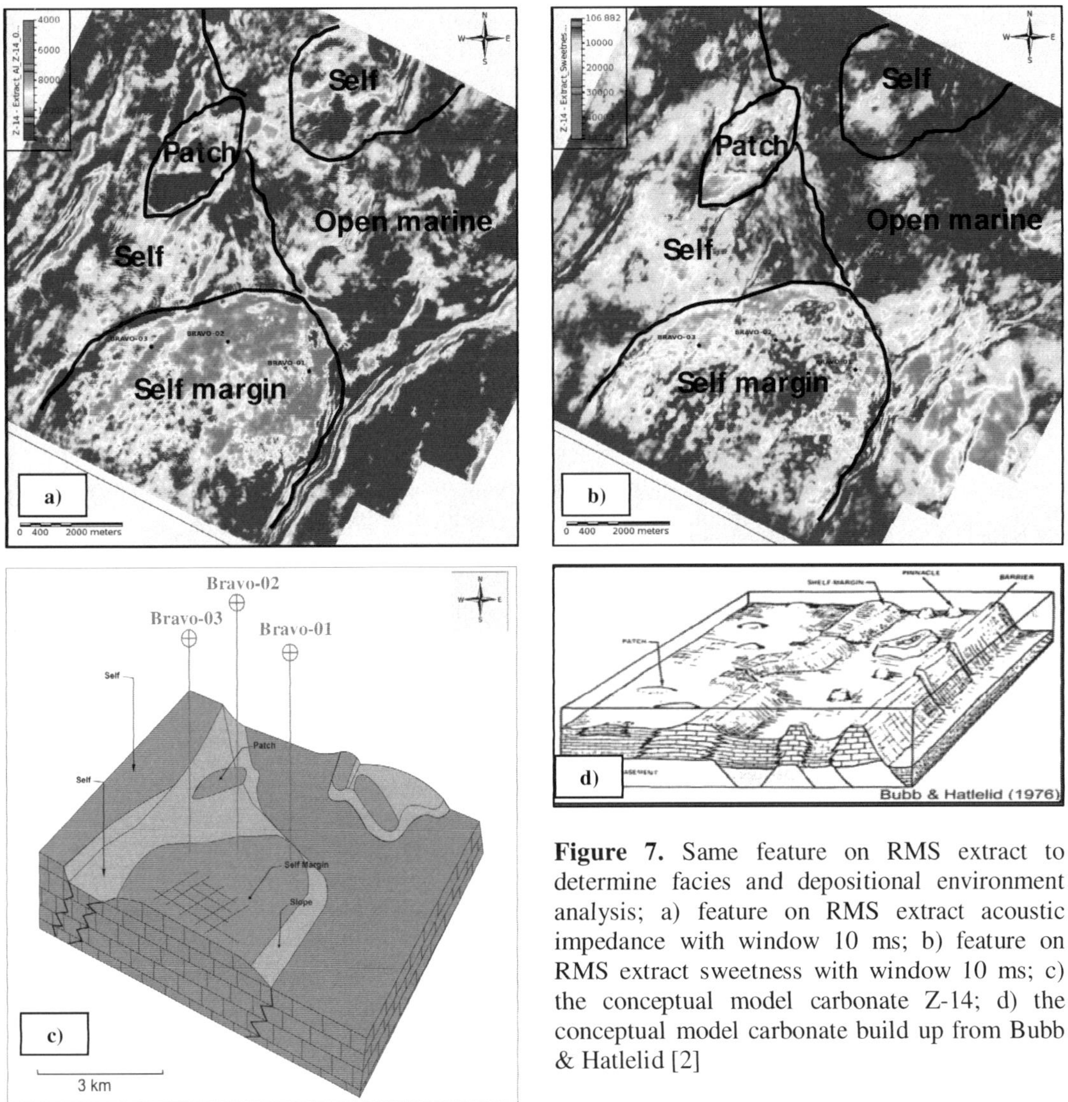

Figure 7. Same feature on RMS extract to determine facies and depositional environment analysis; a) feature on RMS extract acoustic impedance with window 10 ms; b) feature on RMS extract sweetness with window 10 ms; c) the conceptual model carbonate Z-14; d) the conceptual model carbonate build up from Bubb & Hatlelid [2]

4. Conclusions

Based on dynamic data correlation and seismic characterization, we can build the new concept facies, diagenesis, and also environment depositional carbonate. Acoustic impedance inversion can explain reservoirs character of carbonate Z-14 and Z-15. Porous carbonate shown by low acoustic impedance at 1935-1960 m depth and 2080-2120 m depth. While at 1960-2080 m depth, interpreted as the platform of carbonate Z-14 growth because it is tight that shown by high acoustic impedance.

Sweetness attribute can describe a carbonate trap that contains fluid (gas) at 1935 depth with reservoirs thickness around 131-152 m.

Acoustic impedance distribution, sweetness, petrophysical analysis and core can build a conceptual geological model of carbonate Z-14. Shelf-margin facies generated from acoustic impedance value distribution, petrophysical analysis, and core, whereas patch reef, shelf, and open marine facies determined by sweetness value distribution.

Acknowledgements
We would like to thanks to PT. Pertamina EP Asset 3 as the company that owns Bravo structure

References

[1] Martodjojo 1984 *Evolusi Cekungan Bogor* Bandung: Institut Teknologi Bandung
[2] Bubb J N and Hatlelid W G 1976 *Recognition of Carbonate Buildups on Seismic Section* Proceedings of the Carbonate Seminar
[3] Russel B 1988 *Introduction to Seismic Inversion Methods* S.N: Domenico Series Editor Course Notes Series Volume 2 (An Seg Continuing Education Short Course USA)
[4] Richardson M and Zandt G 2003 *Inverse Problems In Geophysics* Department of Geosciences University of Arizona, pp 2-6
[5] Hart B 2008 *Channel detection in 3-D seismic data using sweetness* American Association of Petroleum Geologists Bulletin, AAPG, 92(6):733-742

IOP Conf. Series: Earth and Environmental Science **62** (2017) 012017 doi:10.1088/1755-1315/62/1/012017

Drilling exploration design controlled by pore pressure prediction from 2D seismic and well data: case study of South Sumatra Basin

A Haris, T Mulyawan and A Riyanto

Master Program for Reservoir Geophysics, Department of Physics, Faculty of Mathematics and Natural Sciences, Universitas Indonesia, Depok 16424, Indonesia

E-mail: aharis@sci.ui.ac.id

Abstract. To have safe and economical in drilling design, an information of formation pore pressure is required. Pore pressure can be estimated from seismic data using a velocity to pore pressure transform. The objective of this paper is proposing the drilling exploration design for the case study of South Sumatra field, which is controlled by predicted pore pressure. The pore pressure is predicted by using Eaton method that used velocity from 2D seismic and was validated with well log data. The predicted pore pressure is used to design exploration drilling including casing depth and mud weight. Eaton parameter (N =1.1), shear stress (Ko= 0.6), Gardner (A = 0.198 and B = 0.268), which is used in this works, is gained from existing well data. The velocity model is derived from RMS velocity that should be converted into interval velocity. In addition, this velocity should be validated with the sonic log from existing well. The Normal Compaction Trend (NCT) from interval velocity that was combined with generated previous parameter is used for predicting pore pressure and fracturing pressure. Our experiment shows that based on pore pressure prediction, the drilling exploration design is divided into three sections. i.e. section 17-1/2", 12-1/4" and 8-1/2" and four casing sections, i.e. Casing 20", K-55, 90 ppf at 160 ft, casing 13-3/8", K-55, 54.5 ppf at 1400 ft with mud weight 8.8 - 13.7 ppg, casing 9-5/8 ", K-55, 40 ppf at 4000 ft with mud weight 9.5 - 14.0 ppg and casing 7", L-80, 26 ppf at 5500 ft with mud weight 10.4 - 14.6 ppg.

1. Introduction

The successful in drilling design is dependent on the knowledge of subsurface pore pressure. Inaccurate prediction of the subsurface pore pressures can cause slow penetration, excessive consumption of drill bit, increase costs and risk in drilling activity [1]. Pore pressure can be predicted from seismic data using a velocity to pore pressure transform. Given seismic velocities with sufficient spatial resolution, we can predict subsurface pore pressure. Existing approaches include the empirical methods of Eaton [2] which are widely used in the industry. In this paper, we proposed pore pressure prediction to design exploration well drilling by using Eaton method that used velocity from 2D seismic and was validated with well log data of G-field of South Sumatra basin.

G Field is an exploration field in South Sumatra Basin, precisely in the Middle Palembang Sub-Basin. Exploration target in this region is Talang Akar Formation. South Sumatra Basin is a tertiary back-arc basin that has the NW-SE direction, bounded by the Bukit Barisan in the southwest and Sunda Shelf in the northeast. South Sumatra Basin separated from Sunda Basin on the east by

Tinggian Palembang and Lampung, which was filled by Tertiary sediments with a thickness up to 4000 meters.

One of exploration wells (G-1 well) has been drilled in this field, which undergone a kick in the depth of 5132 ft. This kick occurred due to lack of pore pressure information, so the drilling design is not optimal. To overcome this problem, predrill pore pressure prediction has been carried out to propose the next drilling activities in term of location and drilling design.

In this study, the pore pressure is predicted from seismic interval velocity, which is validated with Sonic from well log data. The Eaton coefficient parameter (N), Shear Stress (Ko) and Gardner coefficient were obtained from well log data. The predicted pore pressure will be used for drilling design including; setting the number of well trajectories, casing seats and mud weight.

2. Pore pressure prediction and casing design

Pore pressure estimation, which is developed by Eaton, used seismic interval velocity and pressure information from previous drilling result. Eaton method explained that the majority of the pressure in the subsurface is influenced by overburden activity the so called primary overpressure [2]. Relation between overburden pressure and pore pressure described by Eaton written on the following equation:

$$P_p = P_{ovb} - (P_{ovb} - P_h)\left[\frac{V_i}{V_n}\right]^N, \text{ for seismic data} \quad (1)$$

$$P_p = P_{ovb} - (P_{ovb} - P_h)\left[\frac{\Delta t_n}{\Delta t_i}\right]^N, \text{ for sonic data} \quad (2)$$

Where P_p is pore pressure (psi), P_{ovb} is overburden pressure (psi), P_h is hydrostatic pressure (psi), N is Eaton exponential coefficient, V_n is interval velocity from NCT (ft/s), V_i is interval velocity from seismic (ft/s), Δt_n is transit time from NCT (μsec/ft), and Δt_i is measured transit time from well log (μsec/ft). NCT is determined on a layer, which has a low permeability value, such as shale. NCT for acoustic travel time in the depth domain is an exponential trend which is plotted in semi-log, whereas NCT for seismic velocity in the depth domain is a linear relation [3].

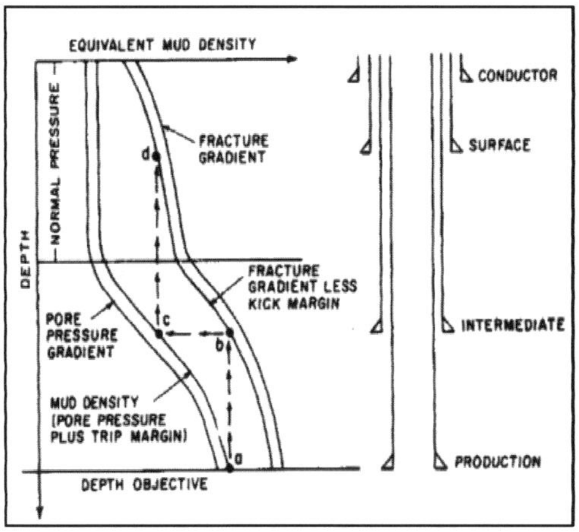

Figure 1. Illustration of casing design that was guided by pore pressure and fracturing pressure [4].

To achieve the optimal solution, the design engineer must consider casing as a part of a whole drilling system. A knowledge of formation properties such as pore pressure, fracture pressure, strength, and chemical stability/sensitive shale (mud type and exposure time) involved in the design process is required. Mud weight is applied to prevent the flow of formation water penetrate into well, which is also known as "kick." Moreover, the casing must be fitted to the depth where the amount of mud weight is equal to the fracturing pressure to prevent formation damaged, the illustration is shown in Figure 1.

In general, pore pressure prediction process is divided into three stages; well evaluation, analysis of seismic interval velocities and application of Eaton method to predict pore pressure. The availability data in this study consist of 2D seismic data with the length of 15.68 km (Figure 2), VSP, well log data (GR, Density, and Sonic), Repeat Formation Test (RFT), Drilling Report of G-1 well and Formation Integrity Test (FIT).

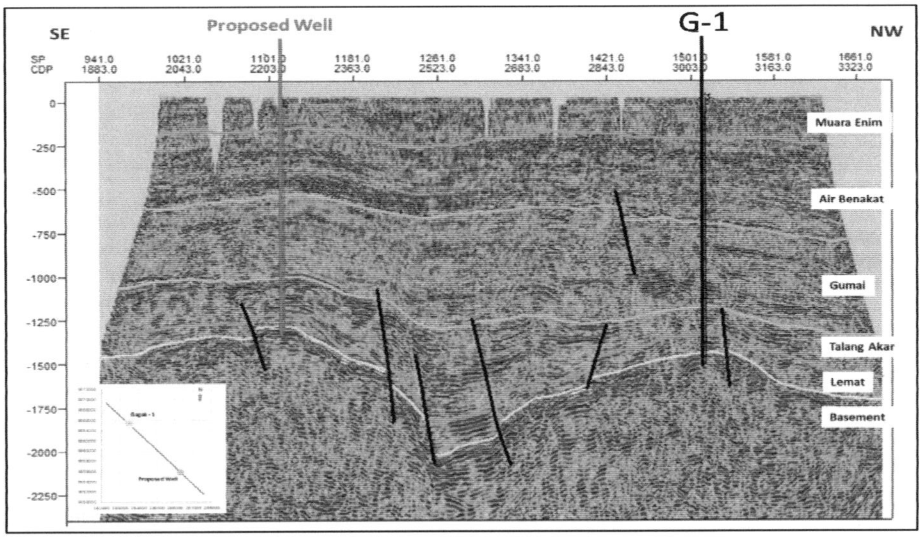

Figure 2. The seismic section in the study area, which is passing through the existing Well G-1 (black line) and proposed well (red line).

3. Result and discussion

Our evaluation is performed on G-1well to obtain several parameters such as NCT, Eaton pressure Parameter (N), shear factor (Ko) and Gardner coefficient. Calculation of NCT is carried out to G-1 well by using sonic of the shale data as shown in Figure 3(a). Gardner coefficient A and B determined using a cross-plot of density and sonic log as illustrated in Figure 3(b). From the cross-plot, Gardner coefficient was obtained as follows A=0.198 and B =0.268.

The best Eaton coefficient produces pore pressure values which have the best fit with RFT. Our exercise to various values, the best Eaton coefficient d is 1.1 as shown in Figure 3(c). Similar to Eaton coefficient, shear factor (Ko) is determined by comparing the fracturing pressure with FIT. From this process, it was obtained that shear factor is 0.6. The plot of pore pressure and fracturing pressure that has the best fit with RFT and FIT is displayed in Figure 3(c).

Root Mean Square (RMS) velocity model was obtained from velocity analysis in the seismic data processing. This kind of velocity must be converted to interval velocity by using Dix's equation. The interval velocity model is then converted to depth domain for adjusting the well. The result of the interval velocity in the depth domain can be seen in Figure 4(a). The predicted pore pressure, which is based on the interval velocity, is shown in Figure 4(b).

Pore pressure prediction in this study used the interval velocity and the parameters that have been obtained from the well analysis. Before the interval velocity is being used it must be corrected by the sonic log. The process was done by cross plot between the interval velocity and the sonic log as shown in Figure 5(a).

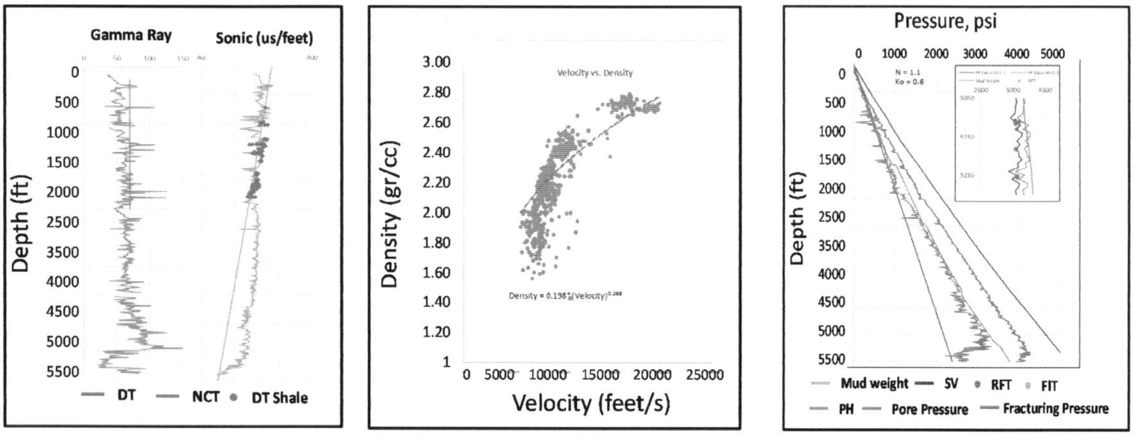

(a) (b) (c)

Figure 3. (a) The plot of Gamma Ray log (track one) and Sonic log (track two), which is overlaid with predicted normal compaction trend from the sonic of shale (red line). (b) The plot of Density versus Sonic to determine Gardner coefficient. (c) The plot of pore pressure (brown line) and fracturing pressure (red line) for determining Eaton constant (N) and shear factor (Ko) in Well G-1.

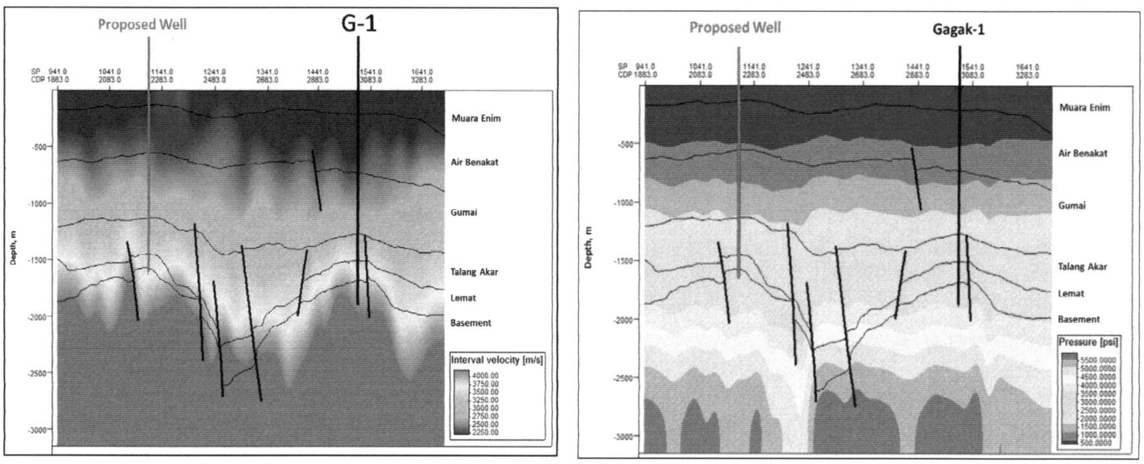

(a) (b)

Figure 4. The section of the interval velocity model (a), and the predicted pore pressure (b) in the depth domain (hot colors represent high magnitude).

NCT prediction of the proposed well is calculated using linear trend of shale velocity. Whereas the density log is determined using the Gardner equation using the coefficient, which was already obtained in the well log evaluation. These two parameters of the proposed well are shown in Figure 5(b). The last step is calculating pore pressure and fracturing pressure on the location of the proposed well using Eaton coefficient and shear factor. Detailed analysis of predicted pore pressure is shown in Figure 5(c). It can be observed that the pore pressure in the prospect area has relatively normal pore pressure. However, at depths below 4200 ft, it has increased pressure. This phenomenon is possible due to the presence of oil or gas content.

The predicted pore pressure and fracturing pressure were then used to design the casing seat and mud weight in the well drilling planning. Pore pressure and fracturing pressure are converted to equivalent mud weight. After converting the predicted pore pressure to equivalent mud weight, the trajectory well design is proposed, which is divided into three following sections:

- Section 17-1/2" has a normal pore pressure or equal to the hydrostatic pressure.
- Section 12-1/4" has pore pressure range 9.0 - 10.0 ppg or 0.67 - 1.67 ppg above hydrostatic pressure.
- Section 8-1/2" has pore pressure range 10.0 - 11.0 ppg or 1.67 - 2.67 ppg above hydrostatic pressure.

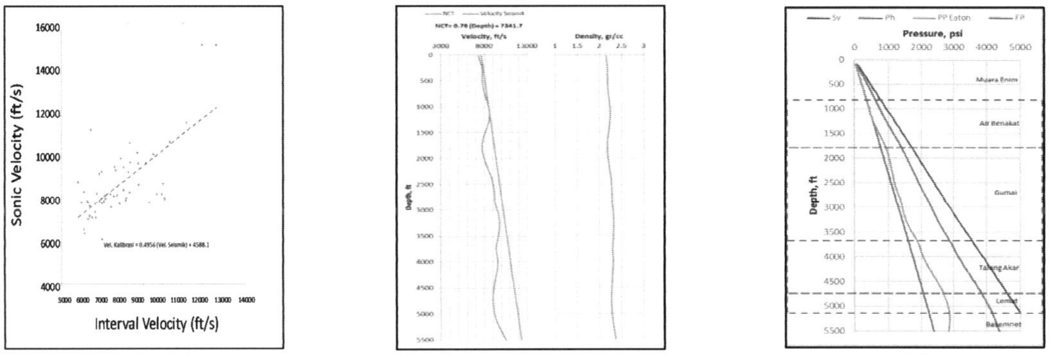

(a) (b) (c)

Figure 5. (a) Linear regression in interval velocity calibration, (b) Track one is the result of NCT (red indicates NCT and blue indicated seismic velocity), and track two is density log, (c) Predicted pore pressure (brown line) and fracturing pressure (red line).

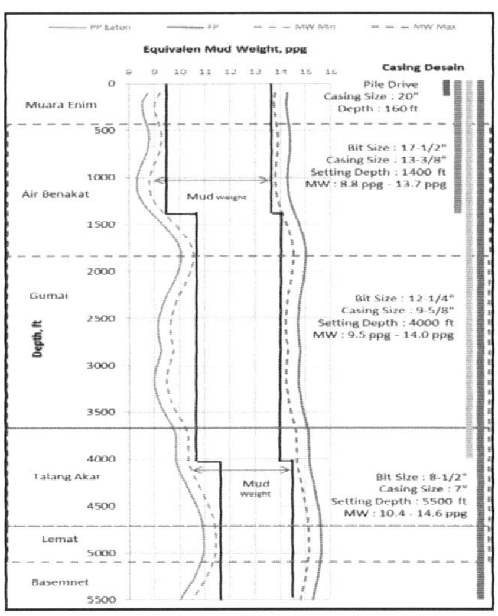

Figure 6. Drilling design on proposed well which contains drilling trajectory, casing design and mud weight.

Figure 6 shows the trajectory well design, which is constrained by pore pressure and mud weight. In general, the designed well trajectory, which is illustrated in Figure 6, can be described as follows:

- Section 1: Casing 20" with grade K-55, 84 ppf serves as surface casing, which is piled up to a depth of 160 ft or bedrock. This casing is intended to protect the aquifer of fresh water in a shallow layer, and avoiding the washout in the surface layer, which can lead to the foundation of the drilling rig and the installation of a diverter system during the drilling. In this section, the estimated pore pressure is equal to the hydrostatic pressure.

- Section 2: Casing 13-3/8" with a grade K-55, 54.5 ppf is installed up to a depth of 1400 ft which reach Muara Enim and Air Benakat formation. This casing is a conductor casing which is drilled with bits 17-1/2" and the mud weight range of 8.8 - 13.7 ppg. The function of this casing is as the holder of the Blow-out Preventer (BOP), the wellhead and the protector of the freshwater aquifer.
- Section 3: Casing 9-5/8" with a grade K-55, 40 ppf is installed up to a depth of 4000 ft which close to lower Air Benakat, Gumai, and Top Talang Akar Formation. Section 3 is drilled with bits 12-1/4" and mud weight range of 9.5 - 14.0 ppg. The purpose of section 3 is to avoid problems such as clay/shale swelling due to exposure of drilling mud which can lead to the collapse of the borehole wall that causes the pipe pinched.
- Section 4: Casing 7" with grade L-80, 26 ppf is installed up to 5500 ft (Total Depth) that close to the Talang Akar, Lemat Formation, and Basement. Section 4 drill with bits 8-1/2" and mud weight range of 10.4 - 14.6 ppg. This section is production casing that penetrates the objective zones which are estimated at a depth of 4200 ft.

4. Conclusions

The pore pressure and fracturing pressure is the most important parameter in determining the reliability and success of a casing design. Therefore, the pore pressure and fracture prediction by considering all available parameters are required. For G Field case, we have proposed the trajectory well design, which is constrained by pore pressure and mud weight, which is divided into three sections that are section 17-1/2", 12-1/4" and 8-1/2". In general, the designed well trajectory in G Field case is proposed into four sections casing step with respect to drilling depth. Casing 20" is piled up to a depth of 160 ft, casing 13-3/8" installed up to a depth of 1400 ft, casing 9-5/8" installed up to a depth of 4000 ft and casing 7" installed up to 5500 ft (Total Depth) that close to the Talang Akar, Lemat formation, and Basement.

References

[1] Mukerji T, Dutta N, Prasad M and Dvorkin, J 2002 Seismic detection and estimation of overpressures Part I The rock physics basis *CSEG Recorder* **27** 34-57

[2] Eaton B A 1975 The equation for geo-pressure prediction from well logs *Proc. In Fall Meeting of the Society of Petroleum Engineers of AIME Society of Petroleum Engineers*

[3] Zhang J 2011 Pore pressure prediction from well logs: Methods, modifications, and new approaches *Earth-Science Reviews* 108 **1** 50-63

[4] Bourgoyne A T, Millheim K K, Chenevert M E and Young F S 1986 *Applied drilling engineering*

IOP Conf. Series: Earth and Environmental Science **62** (2017) 012018 doi:10.1088/1755-1315/62/1/012018

Shale Hydrocarbon Potential of Brown Shale, Central Sumatera Basin Based on Seismic and Well Data Analysis

Abdul Haris, Husein Agil Almunawwar, Agus Riyanto and Andang Bachtiar

Reservoir Geophysics Program, Department of Physics, Faculty of Mathematics and Natural Sciences, Universitas Indonesia, Depok 16424, Indonesia.

E-mail: aharis@sci.ui.ac.id

Abstract. The development of unconventional shale hydrocarbon is really depending on integrating approach of wide range disciplines. The integrated approach for analysing organic-rich shale reservoirs involves calibration of core and well-log data, building petrophysical and rock-physics models, and finally characterizing the key reservoir parameters (TOC, porosity, and natural fractures) and mechanical properties evaluation from seismic data. In this research, integrated approach of geochemical, geomechanical, mineralogy, petrophysical, and geophysical analysis are carried out in Brown Shale, Central Sumatera Basin. Total Organic Carbon (TOC), maturity, and brittleness index are the main parameters used in this study to analyse the shale hydrocarbon potential. The result of geochemical analysis shows that the maturity level of shale in the interest zone in oil window, which means it can generate shale oil in early mature phase at depth of 6400 ft. Quantity of shale hydrocarbon potential is indicated by the TOC value of 0.5-1.2 wt. % (fair to good), with average of shale thickness for over 50 ft. The result of geomechanical analysis shows that brittleness index of interest zone for over 0.48 and rock strength below 10000 Psi.

1. Introduction

Indonesia has shale hydrocarbon potential within selected marine-deposited formations, as well as more extensive shale resources within non-marine and often coaly shale deposits. The best overall potential appears to be mostly oil-prone, lacustrine deposited shales within the Central and South Sumatra basins, which sourced the prolific nearby conventional oil and gas fields [1]. Central Sumatra basin is situated in a tectonically active region as a result of subduction of the Indian - Australia plate. Central Sumatra basin is located between North Sumatra Basin and South Sumatra basin, surrounded by Asahan uplift in the north and northwest, Tigapuluh uplift in the southeast, Sunda Shelf in the east, as well as the Bukit Barisan in the west and southwest [2]. Brown Shale formations are known as a main source rock in the presence of oil and gas in Central Sumatra Basin [1]. Depositional environment of this formation was formed from Lacustrine with lithology rock consists of laminated shale, brown colour, rich in organic material, which indicates the depositional environment with calm water conditions.

Previous studies of shale hydrocarbon in Indonesia mostly analysed in terms of geochemistry, which focus on the characteristics of shale for the content value of Total Organic Content (TOC), the type of kerogen, and level of maturity. Geomechanics analysis, which is determined from brittleness index is also the important parameter to define the economically production.

Hydrocarbon basin in Sumatra has a wide distribution of shale, particularly in Central Sumatra Basin. Brown Shale formation is the main source rock for oil and gas in Central Sumatra Basin. According to published paper [3], TOC content of Brown Shale Formations reach 3-5% with a thickness of 300-1000 feet. The maturity level of this formation is on the mature phase (oil-gas window). This research will be conducted integrated study of the results of geochemical analysis, geomechanics, mineralogy, petrophysical and geophysics in Brown Shale Formation, Central Sumatra Basin. By looking at the relation between TOC, maturity and the brittleness index on seismic data and wells, we could be able to identify the characteristics of potential shale hydrocarbon in Brown Shale Formation, Central Sumatra Basin.

2. Methods and Analytical Procedures

Characteristics of oil shale is almost similar to shale gas, the only difference is from the hydrocarbon content in the shale, kerogen type and level of maturity. Jarvie (2012) stated that the value of vitrinite Reflectance (Ro) for oil shale ranged from 0.6% to 1.1% (oil window), but it could be up to 1.4%. Then for the kerogen type it could be Type I, II or Type III with Ideally type II [4].

The integrated approach for analysing organic-rich shale reservoirs involves calibration of core and well-log data, building petrophysical and rock-physics models, and finally characterizing the key reservoir parameters (TOC, porosity, and natural fractures) and geomechanical were carried out in Brown Shale, Central Sumatera Basin to achieve the aims of this study.

3. Geochemical and Petrophysical Analysis

A number of selected sample of the Pematang Formation have been analysed to obtain organic geochemical data and pyrolysis values. Results of the present analysis were compared with those of the published work [4] in order to better understand geochemistry of the source system throughout the region.

In this work, geochemical analysis on the core data were carried out to determine TOC, mineralogy, Tmax, and kerogen type. Geochemical analysis of the source rock was used to analyse the quality, quantity and level of maturity. The parameters used are the parameters of Peter [5]. Then to determine the brittleness index value it could be divided into two ways, by mineralogy or geomechanics. The difference is, in mineralogy is based on the minerals composition in the shale sample and using formula that often used in determining the brittleness index values as mineralogy initiated by Jarvie et al [6] and Wang & Gale [7]. But in terms of geomechanics used equation was initiated by Grieser and Bray for calculating brittleness index value using value of Young's modulus and Poisson's Ratio [8,9].

Our assessment to potential Brown Shale of wells Afaf-1 and Fahed-2, in term of maturity, it can be explained based on cross-plot between Tmax versus depth and Ro, which is shown in Figure 1a. Those, it can be classified into early mature phase with Tmax of 435°C and Ro of 0.55%, which is located at a depth of 6400 ft. In term of keogen type, the modified van Krevelen diagram (Figure 1b) showing the HI vs. OI plot based on the Pematang Brown Shale pyrolytic data, which is taken from wells Afaf-1 and Fahed-2, indicating that the source section is gas prone and type III kerogen.

However, our observation on diagram Ro vs. depth shows that the sample data is mostly in oil window, therefore it could be concluded that this Pematang Brown Shale Formation potentially generated oil. It has also been confirmed with the results of cross-plot between Hydrogen Index (HI) with Tmax in Figure 1c. From the results cross-plot between HI to Tmax, the shale formation Pematang Brown Shales entered phase early to peak maturity, which means could generate oil and has type III kerogen. It was also confirmed from mud log data at Fahed-2 well, which has oil show at a depth 8500ft.

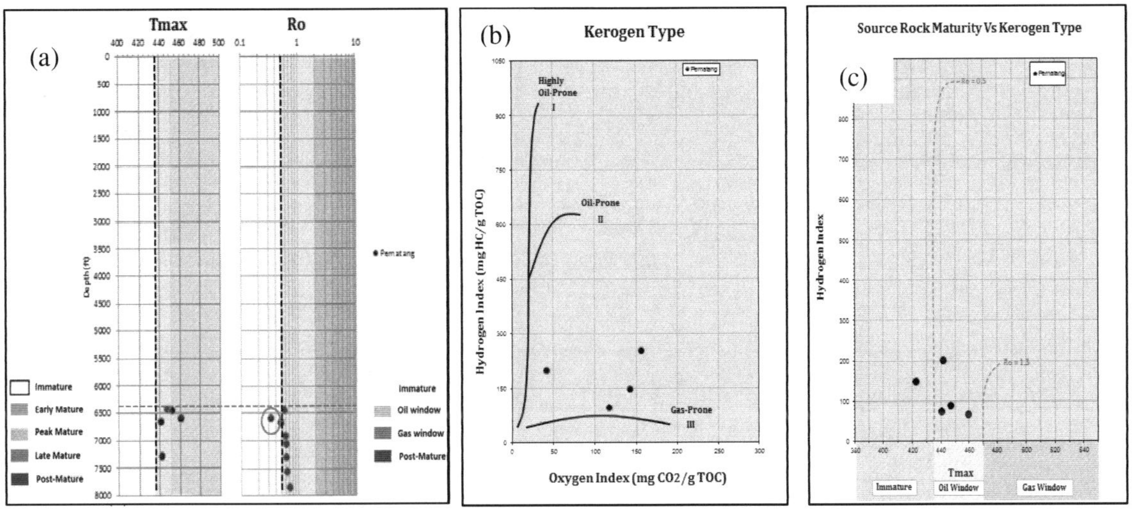

Figure 1. Pore (a) Cross-plot of depth versus Tmax (left) and Cross-plot of depth versus Ro (right). (b) Plot of Van Krevelen Diagram and (c) Plot of Hydrogen Index versus Tmax of Afaf-1 and Fahed-2 well.

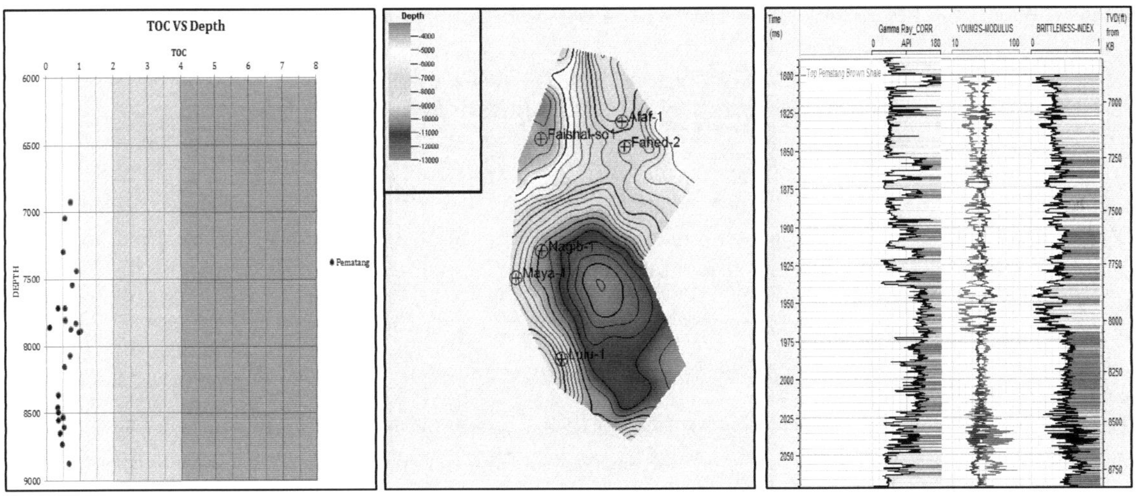

(a) (b) (c)

Figure 2. (a) TOC vs. depth for Brown Shale Formation, (b) Depth structure map and (c) Brittleness Index calculation at Pematang Brown Shale Formation using Geomechanical method [9] equation of Fahed-2 well.

The TOC content of the selected samples of Brown Shale ranges from 0.5 to 1 wt. %, which means the quantity of shale into the medium category (fair). This condition is because of the well location Fahed-2 is on the edge of the sub-basin (Figure 2a) so gives the low value of TOC. If the well located at the centre of the basin (Figure 2b), the TOC value will be higher due to the central basin sedimentation rate is slower and gives rich organic material. In contrast, the edge of the basin areas tends to have a high sedimentation rate so rarely found in shale with rich in organic material.

The next assessment to the potential Pematang Brown Shale is intended to Brittleness Index (BI), which is based on mineralogy content. Table 1 shows the test results of X-Ray Diffraction (XRD) of the sample data Fahed wells core-2. The result of X-Ray Diffraction (XRD) on wells Fahed-2 is used to determine the value of Brittleness Index (BI). In common, the BI is estimated by using Jarvie (2007) and Wang (2009) approximation. Figure 2c shows Brittleness Index calculation of Pematang Brown Shale Formation using Geomechanical method Grieser equation [9] of Fahed-2 well.

Table 1. Mineralogy of X-Ray Diffraction (XRD) test for Fahed-2 well.

TABLE OF X-RAY DIFFRACTION ANALYSIS RESULTS FOR FAHED-2 WELL (CORE SAMPLES)																
			CLAY MINERALS (%)				CARBONATE MINERALS (%)			OTHER MINERALS (%)				TOTAL (%)		
NO.	Well	Lithology	SMECTITE	ILLITE	KAOLINITE	CHLORITE	CALCITE	DOLOMITE	SIDERITE	QUARTZ	K-FELDSFAR	PLAGIOCLASE	PYRITE	CLAY	CARBONATE	OTHER
1	Fahed-2	Sand	-	3	10	-	-	-	tr	87	-	-	-	13	tr	87
2	Fahed-2	Shale	-	30	45	-	-	-	tr	25	-	-	-	75	tr	25
3	Fahed-2	Sand	-	3	24	-	-	-	1	72	-	-	-	27	1	72
4	Fahed-2	Sand	-	7	14	5	-	-	-	74	-	-	-	26	-	74
5	Fahed-2	Sand	-	4	7	-	-	-	-	89	-	-	-	11	-	89
6	Fahed-2	Sand	-	2	3	-	-	-	-	95	-	-	-	5	-	95

Since the limitation of sample number, therefore we only have a few TOC data, which is not cover for the whole depth. In order to patch the lack of this data, we need to estimate the TOC information as function depth by applying Multi-linear Regression. The following is derived equation from Multi-linear Regression as function of GR, Density, NPHI, Sonic and Resistivity.

$$\begin{aligned} TOC = -14.8534 + (-0.0168xGR) + (3.9498xDensity) + (4.7925xNeutron\ porosity) \\ + (0.0751xSonic) + (0.0651xResistivity) \end{aligned} \quad (1)$$

Our estimation based on equation (1), the TOC value for for potential of Pematang Brown Shale is identified as fair to good (0.5-1.2 wt.%) with thickness of 50 ft and brittle brittleness index (BI > 0.48). The depth of target zone is in interval of 8450-8660 ft. Our assessment on the sensitivity analysis, shale is not significantly separated with sand because the shale thickness is low and intercalated with sand.

4. Geophysical analysis

The geophysical analysis is performed to transform the seismic data, which has lateral resolution, into the organic geochemical data. This strategy is very helpful in mapping the potential of hydrocarbon shale laterally. In order to transform seismic data to organic geochemical data, we first applied the seismic inversion to produce the AI section. Further, we did sensitivity analysis as a reference to define shale potential zones. Based on sensitivity analysis, the shale potential hydrocarbon zones was indicated by having AI higher than 43750 ft/s*g/cc, TOC of 0.5-1.2 w.t % (*fair - good*) and Rock Strength value less than 10000 Psi.

Figure 3a shows AI inversion result at seismic line 4777. The resulted AI is then distributed to all seismic line guided with top and bottom Pematang Brown Shale Formation to produce TOC section, which is illustrated in Figure 3b. The same step is done to define rock strength map and illustrated in Figure 3c. After having all 2D seismic line was transformed into TOC section, then we interpolate TOC and rock strength section to make TOC and rock strength map. Figure 3d shows a map of Rock Strength, which is overlaid with TOC structure, dark area indicates sweet spot potential brown shale that has Rock Strength less than 10000 psi and TOC value is more than 1%. Furthermore geological structure of this area is a basin structure which possibly makes sediment turns mature phase by sedimentation in high temperature.

IOP Conf. Series: Earth and Environmental Science **62** (2017) 012018 doi:10.1088/1755-1315/62/1/012018

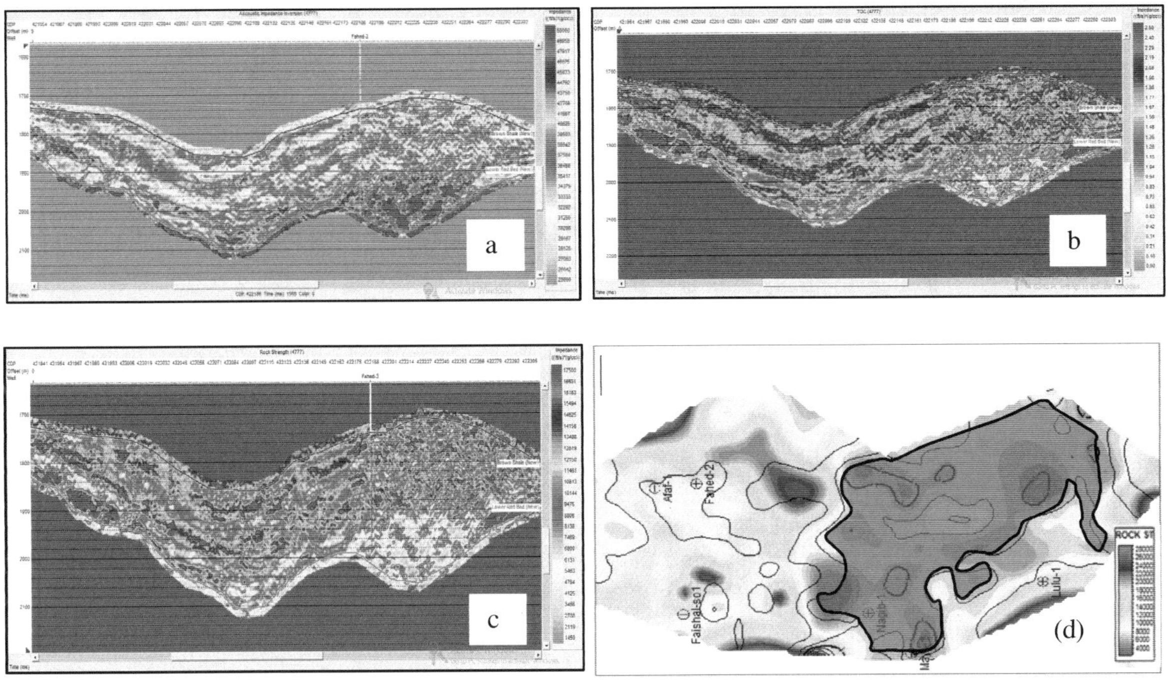

Figure 3. (a) AI section of inverted seismic line 4777, (b) TOC section for seismic line 4777, (c) Rock strength section for seismic line 4777, (d) Rock Strength map, which is overlaid with contour of TOC for delineating sweet spot of potential shale hydrocarbon.

5. Conclusions

The integrated approach for analyzing organic-rich of Pematang Brown Shale Formation has been completely carried out for characterizing the key hydrocarbon shale parameters geochemical (TOC, Ro, and kerogen type) and mechanical properties evaluation from seismic and well data. The potential of Pematang Brown Shale Formation show fair to an excellent hydrocarbon potential level, and classified into phase early to peak maturity, which means could generate oil and has type III kerogen. It was also confirmed from mud log data at Fahed-2 well, which has oil show at a depth 8500ft.

Our delineation by integrating geochemical and geomechanical analysis show that ppotential shale hydrocarbon distribution was indicated by having following parameters such as TOC - 1.2 0.5 wt.% (Fair - good), a thickness up to 50 ft., Brittleness Index above 0.48 and Rock Strength value below 10000 Psi. From the results of acoustic impedance inversion, seen that the shale with hydrocarbon potential has acoustic impedance value above 43,750 ft/s*g/cc, Directions deployment zone of potential shale hydrocarbon is in the southern part sub-basin around wells Nagib-1, Maya-1 and Lulu-1.

References

[1] U.S. Energy Information Administration 2013 Technically Recoverable Shale Oil and Shale Gas Resources: An Assessment of 137 Shale Formations in 41 Countries Outside the United States, Washington, DC

[2] Heidrick T L and Aulia K 1993 A Structural and Tectonic Model of The Coastal Plains Block, Central Sumatera Basin, Indonesia *Proceedings 21st Indonesia Petroleum Association* 285-317.

[3] Camell A, Butterworth P, Hamid B, Livsey A, Badon J, and Bates C 1998 The Brown Shale of Central Sumatra: A Detailed Geological Appraisal of a Shallow Lacustrine Source Rock *Proceedings Indonesian Petroleum Association, Twenty-Sixth Annual Convention* 51-69

[4] Sarjono S and Sardjito 1989 Hydrocarbon source rock identification in the South Palembang subbasin *Proceeedings 18th Annual Convention IPA* 427-467

[5] Peters K E and Cassa M R 1994 Applied Source Rock Geochemistry The Petroleum System – From Source to Trap, ed. Magoon L B and W G Dow *AAPG Memoir* **80** 93 – 120

[6] Jarvie D M, R J Hill, T E Ruble, and R M Pollastro 2007 Unconventional shale-gas systems: The Mississip-pian Barnett Shale of North-Central Texas as one model for thermogenic shale-gas assessment *AAPG Bulletin* **91** 475–499

[7] Wang F P and Gale J 2009 Screening Criteria for Shale Gas *System Gulf Coast Association of Geological Societies 59th Annual Convention, 27-29 September, Shreveport, LA*

[8] Jarvie D M 2012 Shale resource systems for oil and gas: Part 2—Shale-oil resource systems, in J. A. Breyer, ed., Shale reservoirs—Giant resources for the 21st century *APG Memoir* **97** 89–119.

[9] Grieser B and J Bray 2007 Identification of Production Potential in Unconventional Reservoirs *SPE Production and Operations Symposium #106623*

Estimating anisotropy parameter by shear wave splitting of crosswell seismic data: a case study on inter-bedded sand-shale layers

T Suroso[1,2], H Laksono[3], W Triyoso[1] and A Priyono[1]

[1]Graduated Program of Geophysical Engineering, Faculty of Mining and Petroleum Engineering, Institut Teknologi Bandung, Ganesha 10, Bandung 40132, Indonesia
[2]Pertamina Upstream Technology Center, Medan Merdeka Timur 6, Jakarta 10110, Indonesia
[3]Elnusa Geosains Indonesia, TB Simatupang Kav 1B, Jakarta 12560, Indonesia

E-mail: tsuroso.itb@gmail.com

Abstract. This paper describes a method to determine anisotropy parameter by shear wave splitting of crosswell seismic data. The purpose of this work is to present a section of anisotropy to be used in detecting hydrocarbon. Splitted S-waves were detected by picking the traveltimes of S-fast and S-slow on SH and SV components, respectively. Polarization method was done on the 3-C crosswell data using hodogram analysis, so that from these three components we can observe P-, SH- and SV-waves. Tomography of SH- and SV-wave of crosswell seismic were run to produce SH and SV tomograms. Due to the azimuth randomness of the 3-C receivers, an azimuth and inclination correction were implemented on the data, and some preconditioning were applied to enhance the quality of the firstbreaks including deconvolution, bandpass filtering and fx-decon. The traveltimes were picked on SH and SV components, furthermore to be inversed using traveltime tomography algorithm. In this work, a case study was carried out to the seismic data collected on inter-bedded sand-shale layers. The results of this work are SH- and SV-wave tomograms and the anisotropy section. We conclude that this method is effectively presenting an anisotropy section to be used for reservoir recognition.

1. Introduction

Anisotropy has been widely used in seismic seismic exploration work for imaging, fracture detection, lithology discrimination and reservoir characterization [1,2,3]. Anisotropy has been successfully improved seismic image, such as subsalt exploration targets in Gulf of Mexico, even it was considered as mild anisotropy [4]. Anisotropy parameter η estimated from surface seismic whereas in a relatively low resolution [5]. There is relationship between apparent anisotropy with lithology [6] and there is a correlation between hydrocarbon and anisotropy [7]. It was obviously observed the effect of "hockey-stick" at the CMP gather of surface seismic dataset in sand reservoirs containing gas (gas sand), while CMP gather in the same layer that does not contain gas (wet sand) did not show any symptoms of anisotropy [7]. One method in estimating anisotropy parameter is Shear Wave Splitting (SWS) [8,9,10,11]. When passing through fractures, S-wave is splitted into perpendicular components. One component is parallely polarized to the fracture plane propagating as faster velocity, and the other component is perpendiculary polarized to the fracture plane as slower velocity. This phenomenon is shear wave splitting (SWS).

Although anisotropy can be estimated from surface seismic data, however it can be estimated in higher resolution utilizing crosswell seismic data. Crosswell seismic survey is technically conducted to obtain detailed interwell profiles. Crosswell seismic provides higher resolution inter-well images compared to surface seismic. In our study area, some geological challenges are presence in high undulating topography, unconsolidated weathering layer, shallow coal streaks, complex structure and inter-bedded sand-shale layers. Field evaluation using conventional surface seismic survey is very challenging due to the limitation of data quality. Therefore, crosswell seismic survey was conducted in this area to overcome those geological challenges and data limitation.

The main data used in this study was a set of crosswell seismic data recorded using three orthogonal sensors (3-C) and Z-track tool as borehole source producing P- and S-wavefields with the source sweep frequency was defined at 30 – 200 Hz. One profile of crosswell data involved two boreholes; one borehole was assigned as source-well and another borhole was assigned as receiver-well. Each receiver component recorded P- and S-waves, including the converted waves. The P- and S-waves are mixed up on each component. A wavefield separation technique was applied to separate the P- and S-waves from the gather. The separation method used in this work was two passes rotations to produce the P-, SH- and SV-waves. In this work, we utilized the traveltime differences of SH- and SV-waves to estimate the anisotropy parameter. A Tomographic inversion of SH- and SV-waves were implemented to produce SH and SV velocities.

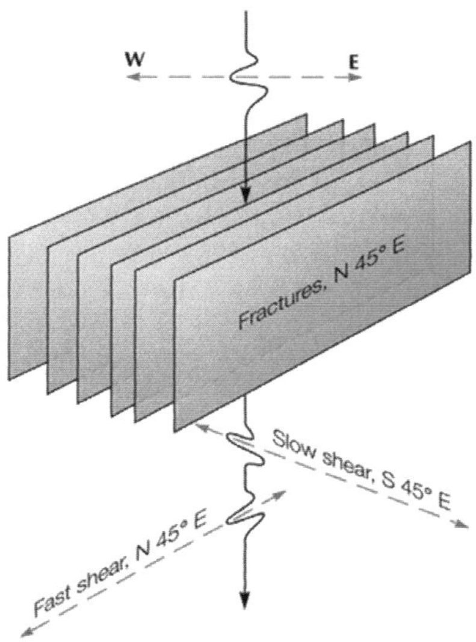

Figure 1. The ilustration of S-wave passing through fractures. S-wave is splitted into perpendicular components. One component is parallely polarized to the fracture plane propagating as faster velocity, and the other component is perpendiculary polarized to the fracture plane as slower velocity [8].

The goal of this work is to present a section of anisotropy estimated by SWS method to be used in detecting hydrocarbon. A petrophysical data (composite logs) was used to validate and evaluate the presence of hydrocarbon.

2. Methodology

The methodology used to achieve the objectives was as follows; 1) Tomography parameterization using synthetic data; 2) Wave separation on the 3-C gather using a rotational technique with hodogram analysis to obtain SH- and SV-waves; 3) Traveltimes measurement by picking the direct wave at each gather dominated by SH- and SV-waves; 4) Traveltime inversion by non-linear traveltime tomography to obtain SH and SV velocities; 5) Calculating anisotropy parameter by SWS; 6) Quick evaluation on layers indicating hydrocarbon presence.

A set of synthetic data was produced to evaluate tomography parameters to be implemented on real data. A wedge shape was used as true model in forward modeling. The geometry of the synthetic data was set exactly the same as the geometry of the real data (table 1).

Table 1. Crosswell seismic acquisition geometry.

Parameter	Profile
Receiver well	left
Source well	right
Inter well distance (m)	686.2
Receiver interval (m)	7.5
Source interval (m)	7.5
Sample rate (ms)	0.5
Sweep frequency (Hz)	30-200
Listening time (s)	1.6
Receiver depth start (m)	1917.5
Receiver depth end (m)	125.0
Source depth start (m)	2028.0
Source depth end (m)	220.5
Total number of shot	242
Total number of receiver	240

Wave separation was performed by rotation based on hodogram analysis. Two passes rotational were implemented to correct the azimuth and the inclination of each receiver component, followed by traveltimes picking on SH- and SV-waves. Some preconditioning were applied to enhance the quality of the firstbreaks including deconvolution, bandpass filtering and fx-decon. The traveltimes measured then inverted with tomography parameters tested on the synthetic data. The output were velocity of SH- and SV-waves (V_{SH} and V_{SV}). Assuming the medium is VTI, the SH-wave is roled as faster S-wave and the SV-wave is roled as slower S-wave.These two types of velocities were used to calculate the anisotropy parameter (K) by SWS method with equation (1), where V_{fast} is V_{SH}, V_{slow} is V_{SV} and V is the average velocity of V_{SH} and V_{SV}.

$$K \approx \frac{(V_{fast} - V_{slow})}{V} \tag{1}$$

Well data was needed to evaluate the correlation between anisotropy and the presence of hydrocarbon. The well data were mud log and composite log.

3. Results and discussion

Real field data was inverted using tomography parameters tested on synthetic data. A wedge model was used to evaluate the resolution. The true model and the tomography results are shown in figure 2. The inversion result using the best tomography parameters was closely similar to the true model, either the structures shape or the velocities. The best parameters was used in tomography is shown in tabel 2. This parameter was then implemented on the real field data with changing in number of iteration. Number of iteration used in tomography of the real field data was 30. The tomography results of the real field data, i.e. V_{SH} and V_{SV} were used to estimate anisotropy parameter (K).

Table 2. Tomography parameters.

Inversion Parameters	
Grid size	7.5 m
Smoothing control	10
X/Z smoothing ratio	40
Minimum velocity	1900 m/s
Maxmimum velocity	4000 m/s
Number of iteration	20

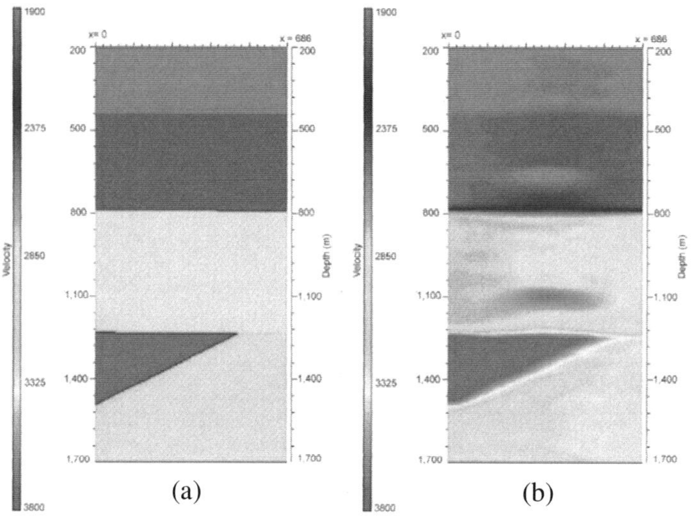

Figure 2. Depth velocity model. (a) true model and (b) tomography result of the synthetic data. The tomography result is similar to the true model, either in structures shape or the velocities.

Seismic anisotropy is the variation of velocity in the direction of wave propagation, either P- or S-waves, or the velocity difference in accordance with the direction of polarization of the S-wave [12]. In other words, anisotropy is the ratio between the maximum and minimum velocity in different directions. Quick detection of anisotropy was observed by analizing the traveltime differences between maximum and minimum traveltime. Figure 3 shows the delay time of SH- and SV-waves. Zone with the depth below 1000 m showed the larger delays than the shallower zone. It indicates that below 1000 m potentialy has stronger anisotropy.

Tomography was implemented to invert the traveltime of SH- and SV-waves producing tomogram SH (V_{SH}) and tomogram SV (V_{SV}). The tomography parameters suitably worked on the real field data. The tomograms were verified using the seismic reflection images. The seismic images were depth migrated crosswell seismic (figure 4). Verification was done by overlying the velocities on top the reflection seismic image of the S-wave. The velocity anomalies were conformable with the S-wave reflections image. High contrast on the tomograms were correlable with the high magnitude reflection on the seismic image. The tomograms provided resolution inter-well velocites as high as the crosswell seismic reflection image.

IOP Conf. Series: Earth and Environmental Science **62** (2017) 012019 doi:10.1088/1755-1315/62/1/012019

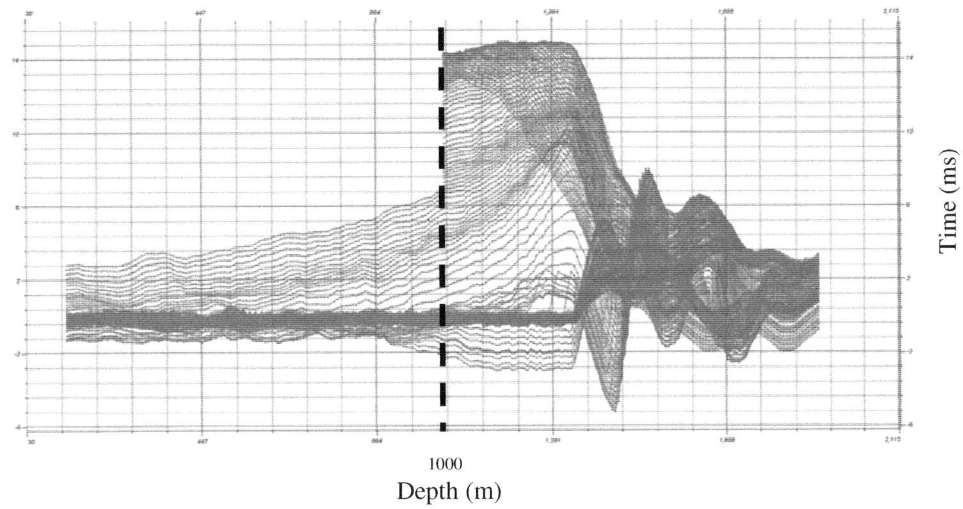

Figure 3. Delay time of SH- and SV-waves. Zone with the depth below 1000 m shows the larger delays than the shallower zone.

The velocities were used to calculate anisotropy parameter (K) with equation (1), where V_{fast} is V_{SH}, V_{slow} is V_{SV} and V is the average velocity of V_{SH} and V_{SV}. Figure 5 shows the tomogram SH and SV and the calculated anisotropy parameter (K). The quick indication by time delays of SH and SV (figure 2) was conformable with anisotropy calculated by SWS (figure 5.c). Anomalies were observed on the zone below 1000 m.

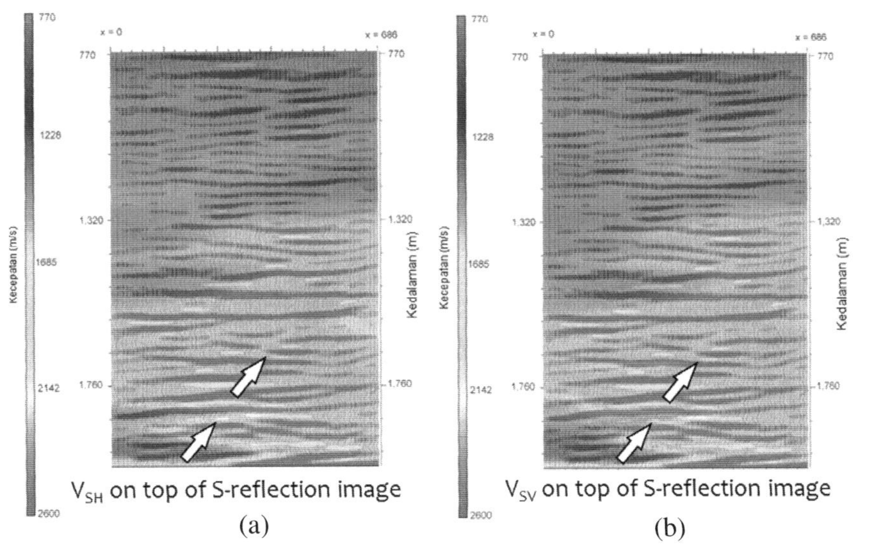

Figure 4. Tomography verification by overlying the tomograms on top of the seismic reflection image. (a) V_{SH} on top of S-reflection image; (b) V_{SV} on top of S-reflection image.

Well data were used to evaluate the estimated anisotropy parameter K. Well data were used to investigate any correlation between the parameter K with the presence of hydrocarbon. From mud log data, it can be observed that the gas is started read by the depth of 1000 m. And from composite log, the water saturation (S_W) is start showing low percentage by 1000 m. In this work, the presence of hydrocarbon indicated from well data correlable with the anisotropy anomaly K.

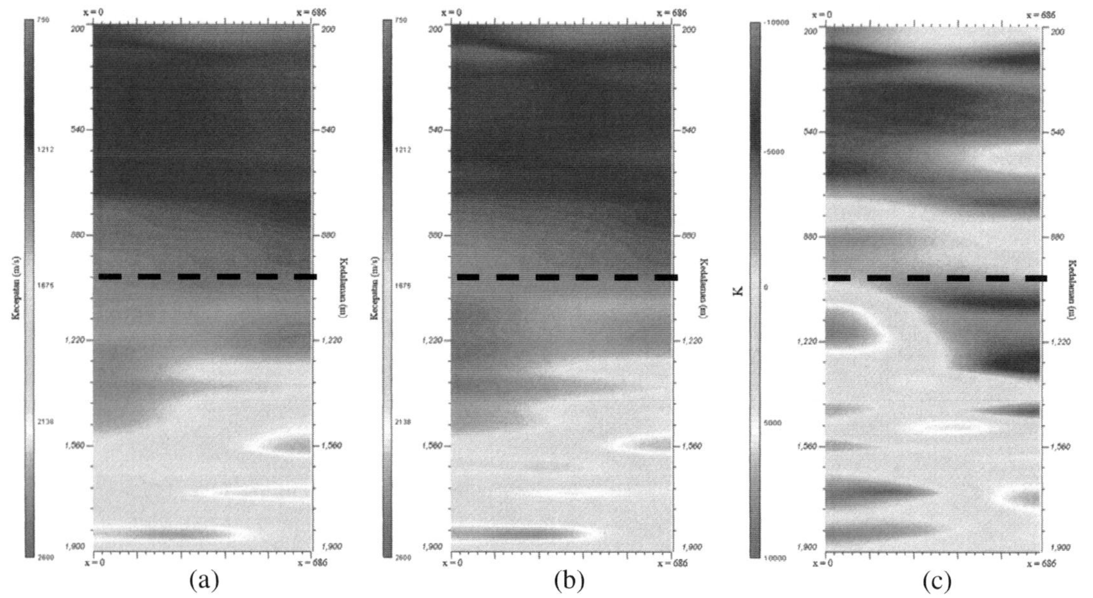

Figure 5. Tomographic inversion result and parameter anisotropy K calculated by SWS. (a) V_{SH}; (b) V_{SV} and (c) K.

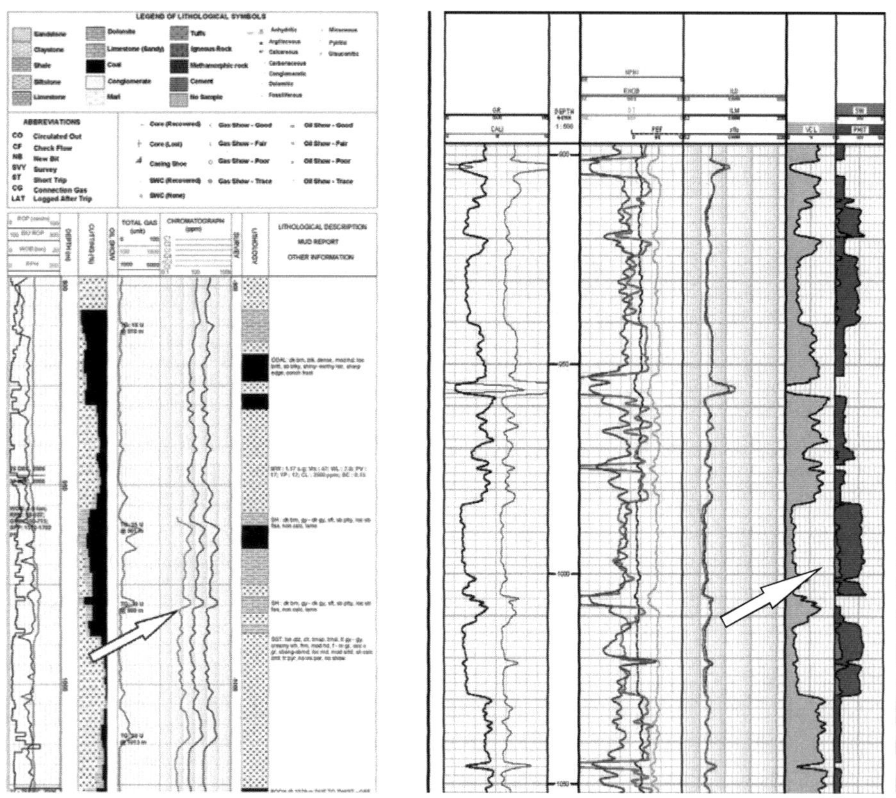

Figure 6. Validation with well data. (a) mud log; (b) composite log. The two well data indicate the presence of hydrocarbon start at around 1000 m and it is correlable with the anisotropy anomaly K.

4. Conclusions

Anisotropy parameter *K* calculated by SWS shows anomalies below 1000 m with the resolution is as high as the reflection iamge. In this work, the anisotropy anomaly *K* is correlable to the presence of hydrocarbon indicated from well data.

5. Acknowledgement

We acknowledge Pertamina Upstream Technology Center sponsor for their support in this geophysical exploration research.

References

[1] Cheng A C H dan Zhang J 2001 Imaging complex structures with crosswell seismology *SEG Int'l exposition and annual meeting* San Antonio Texas

[2] Dong Q Marion B Meyer J Xu Y dan Xu D 2005 Imaging complex structure with crosswell seismic in Jianghan oil field *The Leading Edge*

[3] Jessop J A Friedel M J Jackson M J dan Tweeton D R 1992 Fracture detection with seismic crosshole tomography for solution control in a stope *Proceeding of the Symposium on the Application of Geophysics to Engineering and Environmental Problems* EEGS

[4] Tsvankin I Gaiser J Grechka V Baan M V D dan Thomsen L 2010 Seismic anisotropy in exploration and reservoir characterization An overview *Geophysics* **75** 75A15-75A29

[5] Alkhalifah T and Rampton D 2001 Seismic anisotropy in Trinidad A new tool for lithology prediction *The Leading Edge*

[6] Byun B S Corrigan D dan Gaiser J E 1989 Anisotropic velocity analysis for lithology discrimination *Geophysics* **54** 1564-1574

[7] Supriyono 2011 Velocity variation with angle or offset VVO A new tool for direct hydrocarbon indicator case studies in low and high impedance clastic reservoirs *Doctoral dissertation* 120p Institut Teknologi Bandung

[8] Anderson B Bryant I Luling M Spies B dan Helbig K 1994 Oilfield anisotropy Its origins and electrical characteristics *Oilfield Review* October 1994

[9] Liu J Zeng X Xia J dan Mao M 2012 The separation of P- and S-wave components from three-component crosswell seismic data *Journal of Applied Geophysics* **82** 163-170

[10] Winterstein D F dan Paulsson B N P 1990 Velocity anisotropy in shale determined from crosshole seismic and vertical seismic profiling *Geophysics* **55** 470-479

[11] Zhang H Liu Y Thurber C dan Roecker S 2007 Three-dimensional shear-wave splitting tomography in the Parkfield California region *Geophysical Research Letters* **34**

[12] Sheriff R E 2002 Encyclopedic dictionary of applied geophysics *4th edition Geophysical References Series* **13** 376p Society of Exploration Geophysicists

[13] Aoki Y Kawanaka T Mitsunaga J Ishii Y Rokugawa S dan Miyazaki 1991 Cross-well tomography experiment in a highly fractured area *SEG*

Hydrocarbon Prospect Derived from Attributes Analysis on Low-Frequency Passive Seismic Survey: a Case Study from Kalimantan, Indonesia

Billy Sugiartono Prabowo[1*], Rexha Verdhora Ry[2], Andri Dian Nugraha[3], Katrine Siska[1]

[1]Volcanology and Geothermal Laboratory, Geophysical Engineering, Faculty of Mining and Petroleum Engineering, Institut Teknologi Bandung, Indonesia
[1]Graduate Program of Geophysical Engineering, Faculty of Mining and Petroleum Engineering, Institut Teknologi Bandung, Indonesia
[3]Global Geophysics Research Group, Faculty of Mining and Petroleum Engineering, Institut Teknologi Bandung, Indonesia

*Corresponding author's email: billysprabowo@gmail.com

Abstract. Hydrocarbon Microtremor Analysis is a low-frequency passive seismic method which derives a quick look estimates new hydrocarbon reservoir prospect area. This method based on the empirical study which investigated an increasing of spectra anomaly between 2 – 4 Hz above the reservoir. We determined five attributes on low-frequency band of microtremors including Power Spectral Density integral of vertical component (PSD-IZ), Power Spectral Density (PSD) on 3 Hz frequency, frequency shifting, the spectral ratio of vertical and horizontal components (V/H) maximum and integral of spectral ratio of vertical and horizontal components (V/H). We deployed 105 points of measurement spreading in our suspect area. We used time series data that recorded from particle velocity of three components with 80 minutes duration and 100 Hz of the sampling frequency. The noise identification analysis in each station data set has been made from the measurement location, considering the suspect area had different local cultural noise. We proceed attributes for each data acquired from all station then used the interpolated map using a standard kriging algorithm spatially. As a result, each attribute analysis and spatial attribute map are combined to identify and estimate a good prospect of the hydrocarbon reservoir.

1. Introduction

Recent studies of low-frequency passive seismic survey or called Direct Hydrocarbon Indicator (DHI) in the range of frequency of 2-6 Hz indicates that the frequency content of the seismic background noise changes over the reservoir of hydrocarbons [2, 3, 5, 10]. Analysis of spectrum amplitude of the passive seismic data accumulated from reservoir hydrocarbon shows anomaly spectrum at a frequency of approximately 1-5 Hz with signal peaks around 3 Hz as observed in the oil fields of Abu Dhabi [5, 2], Brazil Mossoro field and field Voitsdorf Austria [6]. Furthermore, [10] give some good results about hydrocarbon reservoir related passive seismic attributes maps with the well drilled after the passive seismic survey, confirmed hydrocarbon prospect in the exploration area maps.

Based on that experience, the passive seismic survey was conducted over the green area located onshore in the Central Kalimantan. We deployed 105 points of measurement spreading with grid about 1 km node spacing in our suspect area approximately 400 km^2 using four station seismometers. We used time series data that recorded from particle velocity of three components with 80 minutes duration and 100 Hz of the sampling frequency. The noise identification analysis in each station data set has been made from the measurement location, considering the suspect area had different local cultural noise.

2. Methodology

Signal processing was used to calculate the passive seismic microtremors data raw taken from three components seismometer. We choose 1 minutes of time windowing which sampled the data with good signal to noise ratio by using STA/LTA algorithm. These data selection resulted into a component in every station. For every window, we calculated Fast Fourier Transform (FFT) spectra and a smoothing moving average of 30 data were applied to derived spectrum anomaly from each station.

Dangel et al. [3] suggested the primary goal identify low-frequency energy anomalies that expected with hydrocarbon reservoir-related microtremor (i.e., approximately 1-6 Hz) could be determined by attribute related to the strength of spectral peaks [10]. In this study, we determined five spectral attributes on the low-frequency band of microtremors including:

2.1. Power Spectral Density integral of vertical component (PSD-IZ)

The integral value under the power spectral density curve of the vertical component [6]. Figure 1a illustrated the integration to quantify low-frequency anomalies visible in the power spectral density plot.

2.2. Power Spectral Density (PSD) on 3 Hz frequency

The peak amplitude of 3 Hz frequency vertical spectrum component (Figure 1b) [5].

2.3. Frequency Shift of Maximum Spectral Peak

Determine frequency value on maximum spectral peak (Figure 1b). In this study, we use a range of 2.3 – 3.7 Hz. [3, 2, 10] as the DHI indicator range.

2.4. The Spectral Ratio of Vertical and Horizontal components (V/H) maximum

This attribute represents to reduce the instability between vertical and horizontal component related to the time of measurement. The procedure is to measure the maximum of V/H ratio spectra between 1-6 Hz (Figure 1d) [6].

2.5. Integral of Spectral Ratio of Vertical and Horizontal components (V/H)

This attribute represent to reduce the instability between vertical and horizontal component related to the time of measurement. The procedure is to measure integral of V/H ratio spectra between 1-6 Hz above V/H=1 [10]. Figure 1d illustrated the integration to quantify in V/H ratio spectra plot.

Figure 1a show the 60 minutes continuous recorded waveform at station PS082, ground particle velocity (m/s) versus sample number of 100 Hz sampling frequency for three components, Up-Down (Vertical U-D) (above), north-south (horizontal N-S) (middle) and east-west (horizontal E-W). Green color are indicates set of a time window which was used for further processing for each component.to get the amplitude spectrum analysis. Furthermore, smooth amplitude spectrum of three component, black line represent up-down (vertical U-D), red line represent east-west (E-W horizontal) and blue line represents north-south (N-S horizontal) component (Figure 1b). Blue area is quantified an anomaly of the power spectral density of the vertical component in the low-frequency (approximately between 1.7 and 3.7 Hz), value of PSD-IZ (blue area) attribute is 421.3545 arbitrary units, the amplitude spectrum value on frequency 3 Hz (PSD 3Hz) is 266.3544 (red dash line), and frequency shifting in 2.5 Hz (Figure 1c). Blue area is a quantifies an anomaly of the spectral ratio V/H attribute (approximately between 1 – 6 Hz), the value of integral V/H ratio spectra is 1.7779 arbitrary unit, with maximal peak 1.5 (Figure 1d).

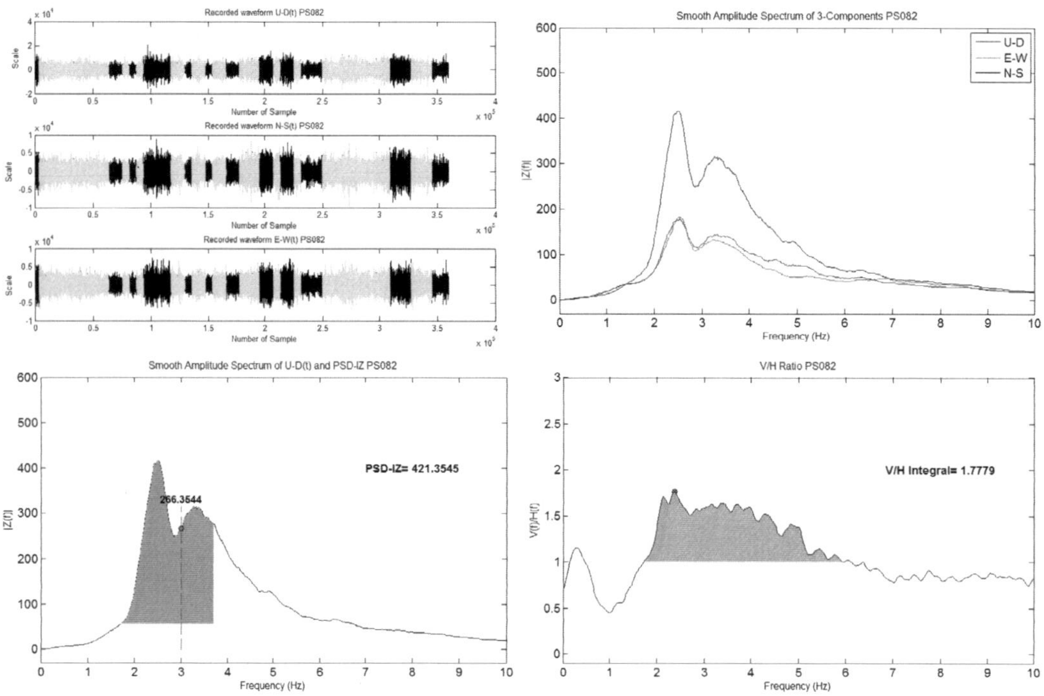

Figure 1. Spectral attributes of the microtremor waveform used in this study (from measurement PS082, as an example). (a) Raw data 3 component (b) Smooth spectrum of 3 components (c) Smooth amplitude spectrum of the vertical component and the Power Spectral Density of the vertical component (PSD-IZ) attribute (d) The spectral ratio V/H (Vertical/Horizontal) attribute.

3. Results and Discussion

This processing procedure is applied to the all passive seismic data of each measurement stations points. As an output, all spectral attribute maps were generated base on a specific attribute. Attributes values were mapped spatially about the station location (Figure 2). Combined standard kriging interpolated algorithm attribute map shows slicing area which derived new hydrocarbon reservoir prospect (orange area). Red line represent PSD-IZ attribute values, thick red line represents cut off values more than 100; Green line represent PSD-3Hz attribute values, thick green line represent cut off values more than 100; Blue line represent contour map frequency shifting attribute values, with thick line represents value 2.3 – 3.7; Magenta line represent contour map of spectral ratio V/H attribute which thick line represent value more than 1; Purple line represent contour map integral of spectral ratio V/H attribute which thick line represent value more than 0.6. Green point represent interesting new hydrocarbon reservoir prospect; yellow point represent middle interesting and red point represent not interesting, based on statistically score and value combined.

This comes into our consideration that the DHI range of spectrum attribute values needs to be adapted from noise effect near survey site or any other seismic source. The adaptation can be made by considering noise identification and controlling the DHI range using calculated attribute values from known hydrocarbon production stations [4, 8, 1, 7], that aspect can increase or decrease the amplitude and frequency of hydrocarbon microtremors.

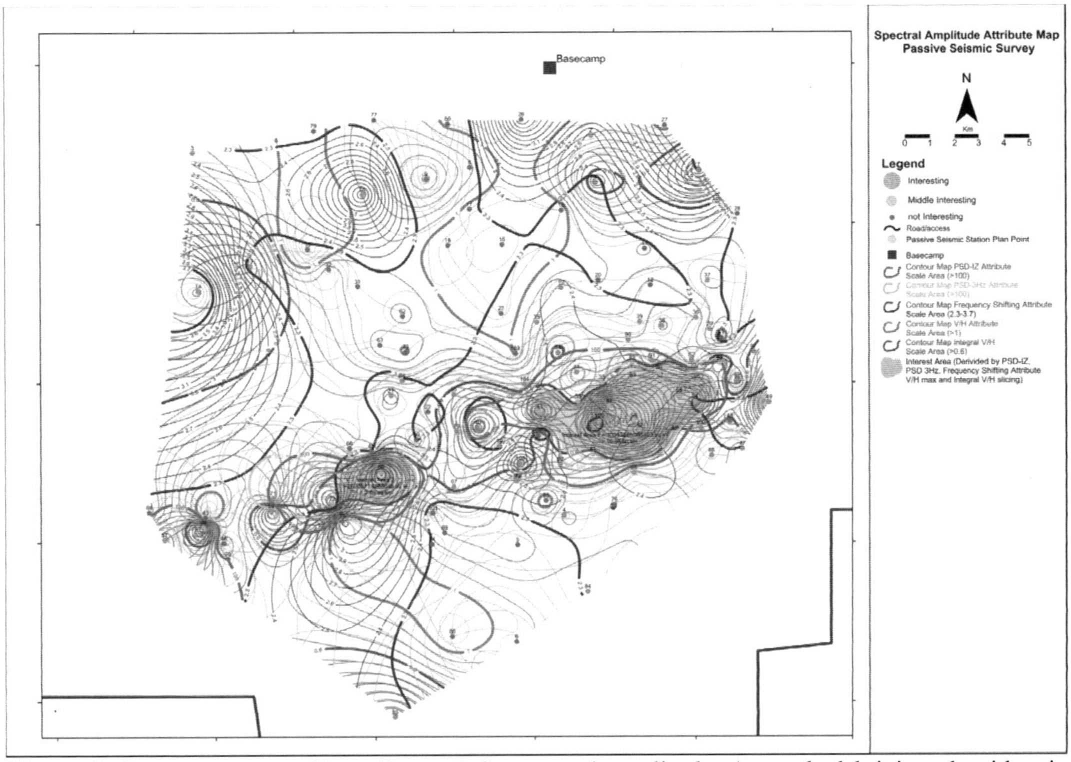

Figure 2. Survey map a multi attribute of the spectral amplitude. A standard kriging algorithm is applied. New hydrocarbon reservoir prospect area is represented by slicing five spectral amplitude attribute map area which has statistically cut off value on each attribute (orange color boundary in eastern and central-western side).

4. Conclusions

Based on our study, we can summarize DHI range for 5 attribute value mention before as follows: PSD IZ: >100; PSD 3Hz: >100; Frequency Shift: 2.3-3.7 Hz; Integral V/H: > 0.6; and V/H Maximum: >1. Each attribute analysis and spatial attribute map are combined to identify and estimated a good prospect of a hydrocarbon reservoir. We derived that orange area (Figure 2) is the most prospective hydrocarbon area, based on five spectral attributes and to be suggested for the exploration well placement.

References

[1] Akbar M F, Afnimar, Nugraha A D, Ramadhan D, and Mulyanegara G 2013 Noise Identification in Passive Seismic Hydrocarbon Reservoir Microtremor Data *Proc. Indonesia Petroleum Association 37th Convention & Exhibition*
[2] Ali M Y, Berteussen K A, and Barkat B 2009 *Geophysical Prospecting* **58**(5) 875-899
[3] Dangel S, Schaepman M E, Stoll E P, Carniel R, Barzandji R, Barzandji O, Rode E D, Singer, J M 2003 *Journal of Volcanology and Geothermal Research* **128** 135-158
[4] Gerivani G, Haghshenas E, Moghaddas N H, Ghafoori M, Lashkaripour G R 2012 *Journal of Geophysics and Engineering* **9** 632–641
[5] Holzner R, Eschle P, Dangel S, Frehner M, Narayanan C, Lakehal D 2006 Hydrocarbon

Microtremors Interpreted as Oscillations Driven by Oceanic Background Waves *Communications in Nonlinear Science and Numerical Simulation* **14** (1) 160-173

[6] Lambert M A, Schmalholz S M, Saenger E H, Steiner B 2008 *Geophysical Prospecting* **57** 393–411

[7] Nugraha A D, Ramadhan D, Afnimar, Akbar M F, and Mulyanagara G 2013 Identification of New Hydrocarbon Prospect Area Using Multi-Attribute Analysis of Hydrocarbon Microtremor Data *Proceedings of the 11th SEGJ International Symposium, Yokohama, Japan* 10.1190/segj112013-047.

[8] Ramadhan D, Nugraha A D, Afnimar, Akbar M F, and Mulyanagara G 2013 Detection of new hydrocarbon reservoir using hydrocarbon microtremor combined attribute analysis *AIP Conf. Proc.* **1554** 277 doi: 10.1063/1.4820339.

[9] Ramadhan D, Nugraha A D, Afnimar, Akbar M F, and Mulyanagara G 2013 Multi-Attribute Analysis of Low-Frequency Passive Seismic Method for Hydrocarbon Indicator Prospecting Case Study in 'HFD' Region *Proc. Indonesia Petroleum Association 37th Convention & Exhibition*

[10] Saenger E H, Schmalholz S M, Lambert M A, Nguyen T T, Torres A, Metzger S, Habiger R M, Müller T, Rentsch S, and Méndez-Hernández E 2009 *Geophysics* **74** (2)

Pore pressure prediction using probabilistic neural network: case study of South Sumatra Basin

A Haris, R J Sitorus and A Riyanto

Master Program for Reservoir Geophysics, Department of Physics, Faculty of Mathematics and Natural Sciences, Universitas Indonesia, Depok 16424, Indonesia

E-mail: aharis@sci.ui.ac.id

Abstract. Pore pressure prediction in the planning of the drilling well commonly carried out using seismic stacking velocity and Normal Compaction Trend (NCT) analysis with Eaton's equation. There are other parameters that correlate to pore pressure, i.e. density, P-impedance, S-impedance, and Vp/Vs ratio. The aims of this study are to predict pore pressure distribution from 2D pre and post-stack seismic data of South Sumatera field by applying the Probabilistic Neural Network (PNN). The pre-stack seismic inversion, which resulted in the elastic parameters such as Density (ρ), Vp/Vs ratio, P-impedance (Zp), S-impedance (Zs), is used as input for PNN training. In another hand, the post-stack seismic data, which resulted in the following parameters such as the average frequency, absolute integrated amplitude, apparent polarity, and dominant frequency, is also used to predict the lateral distribution of pore pressure. Our data training using PNN with pre-stack seismic data provided the best correlation up to 98% compared with the post-stack seismic data. Our prediction, in general, provides the pore pressure model and in detail provides over-pressure. The advantage of PNN shows vertical resolution as good as seismic resolution and provides more helpful information for a further drilling operation.

1. Introduction

An understanding of pore pressure distribution and its mechanisms are important in the planning of drilling, both in term of security aspect and cost. The pore pressure knowledge can also be a consideration in the exploration strategy, as an example of consideration for drilling on the flank rather than on the peak of the structure. The use of velocity to pore pressure transformation by utilizing conventional stacking velocity has been commonly used to pore pressure prediction in the planning of drilling [1]. There are problems in pore pressure prediction using stacking velocity. First, conventional seismic stacking velocities are commonly not appropriate for pore pressure prediction since stacking velocity is not the real wave velocity in rocks [2]. Second, stacking velocity has a lower resolution in the depth that cannot detect changes in the pressure in short intervals.

There are other parameters that are affected by the overpressure zone. The less granules contacts in overpressure zone, which is low effective stress area, gives more influence on the Vs rather than Vp [3]. They expressed that Vp/Vs ratio is a suitable parameter for pore pressure prediction. The elastic parameter such as S-impedance, P-impedance, and density has a strong correlation to the pore pressure. These parameters are used to identify the possible presence of overpressure. In this paper, we predict the pore pressure by applying the PNN from 2D different seismic data, i.e., pre and post-stack seismic data.

IOP Conf. Series: Earth and Environmental Science **62** (2017) 012021 doi:10.1088/1755-1315/62/1/012021

The area of study is located in the western part of South Sumatra Basin as shown in Figure 1, which is indicated by yellow shadow. Geologically, the South Sumatra Basin is a basin formed by the interaction of Asian plate with Indo-Australian plate in the Middle Mesozoic. Extension phase during the Palaeocene to Early Miocene formed half graben filled by clastic sediments and carbonate where old tertiary and precipitated were not aligned above the basement (igneous, metamorphic, volcanic and limestone) which was formed pre-Eocene. South Sumatera Basin was separated by Central Sumatera Basin with Tigapuluh mountains, separated by Sunda Basin with Palembang/Lampung high and in the western part of Barisan mountains. This basin was divided into five sub-basins: Jambi Sub-basin, North Palembang Sub-basin, Central-Palembang Sub-basin, South-Palembang Sub-basin and Bandar Jaya Sub-basin [4]. The study area is located in South-Palembang Sub-basin that has the potential for over-pressure.

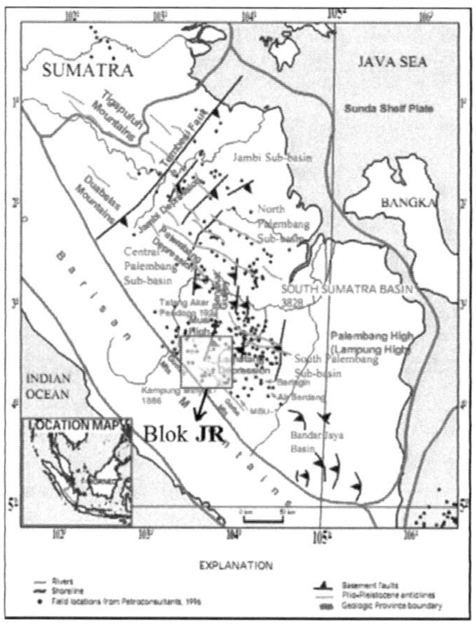

Figure 1. Map of South Sumatera Basin, where the yellow shadow is study area [4]

Our observation to the study area, which is based on the existing well data, shown overpressure zone particularly in Air Benakat and Gumai formation. The identified overpressure is indicated on the pore pressure changes with exceeding the value of the normal pressure. Therefore, further study is needed to map the distribution of pore pressure and delineate the over-pressure in the study area.

2. PNN pore pressure prediction

Neural networks have been widely used for several years in geophysical studies [5]. One of neural network algorithms is a probabilistic neural network (PNN). PNN is an interpolation scheme in mathematics that uses neural network implementation. By implementing PNN, the approach of pressure log can be estimated with multi attributes as training inputs without having to build models of Normal Compaction Trend (NCT) in advance. PNN is more superior to the Multilayer Feedforward Neural Network (MLFN), where the mathematical formulas of PNN are more understandable than MLFN [6].

In this work, PNN is applied to two different types of 2D seismic data, i.e., pre-stack and post-stack seismic data from South Sumatera field. Both processes including training and validation are carried out in predicting pore pressure distribution. The availability of data in the present study consists of 2 wells data and 14 line of land 2D seismic data, where two seismic line (JR-01 and JR-02) are pre-stack seismic data. Two pre-stack seismic line (Figure 2) crossing to the wells, in which seismic line of JR-01 passes through XX-1 wells and seismic line of JR-02 passes through ZZ-1 wells.

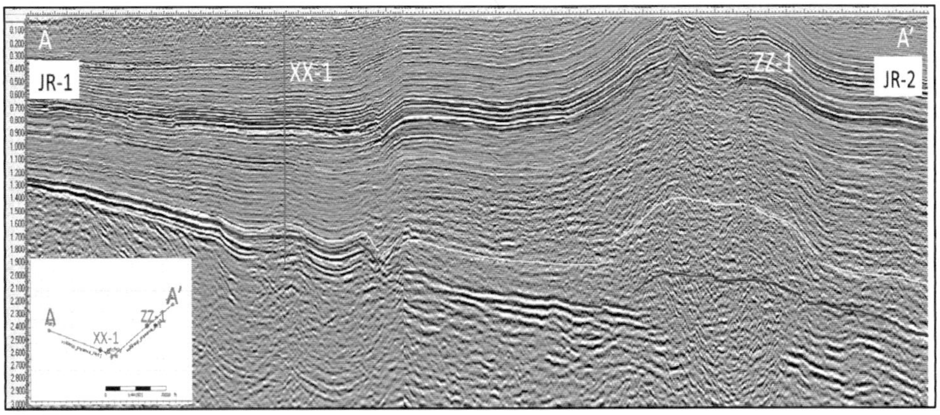

Figure 2. Seismic section of JR-01 and JR-02, which is passing through 2 wells (XX-1 and ZZ-1)

Before further use of seismic data, a pre-conditioning of seismic data is performed in advances such as the amplitude balancing and time shifting. The XX-1 well contains density, Vs, Vp, pore pressure and fracture gradient, while the ZZ-1 well contains only density, pore pressure, and fracture gradient. The same treatment to seismic data, we applied the pre-conditioning to the well data including the environmental correction to caliper log. Figure 3 shows the well log data of XX-1 and ZZ-1 well. These two well data show over-pressure condition with a mechanism of disequilibrium compaction.

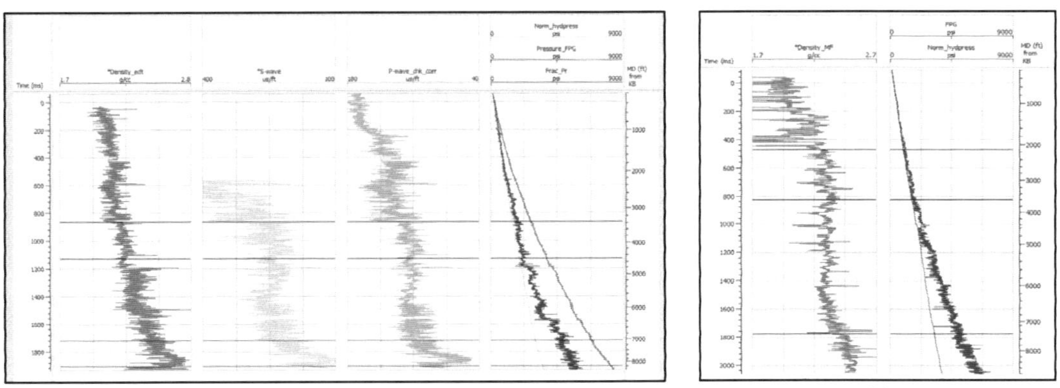

(a) **(b)**

Figure 3. From left to right: a) density, S-wave, P-wave and pressure log of XX-1 well. (b) Density and pressure log of the ZZ-1 well. The pressure log consists of hydrostatic pressure (green), pore pressure (blue) and fractures pressure (red).

In the first step, PNN pore pressure prediction is applied to the pre-stack seismic data, where pre-stack seismic inversion is performed to extract elastic parameter such as Density (ρ), Vp / Vs, P-wave impedance (Zp), S-wave impedance (Zs). The extracted elastic impedance is then used as input of PNN for data training. In the second step, PNN pore pressure prediction is applied to the post-stack seismic data, where seismic attributes such as average frequency, dominant frequency, absolute integrated amplitude, and apparent polarity. In this study, the training process is done by using XX-1 well and validation by using ZZ-1 well.

3. Result and Discussion

PNN pore pressure prediction by using the extracted elastic parameter resulted in different sensitivity to the predicted pore pressure. Our analysis to the predicted pore pressure from the elastic parameter is shown in Table 1. The input of density produces more accurate pore pressure prediction compared to the other elastic parameters, which is indicated by the highest correlation.

Table 1. The correlation between predicted pore pressure with pressure log from different elastic parameter.

No	Target	Elastic Parameter	Correlation
1	Pressure	Density	0.94
2	Pressure	P-impedance	0.89
3	Pressure	S-impedance	0.88
4	Pressure	Vp/Vs ratio	0.75

The important step in this work is validation, which is performed by using one of the existing wells (XX-1 well). The validation aims to see the level of confidence of PNN predicted pore pressure. Figure 4 shows PNN predicted pore pressure and its validation. PNN predicted pore pressure is indicated by the red line and the pore pressure log of XX-1 well is indicated by the black line. Our observation concludes that PNN predicted pore pressure shows good agreement with the pore pressure log, which is indicated by 98% correlation coefficient. In term of vertical resolution, it can be seen that the predicted pore pressure shows similar resolution by mean that we can observe the sudden change pore pressure distribution.

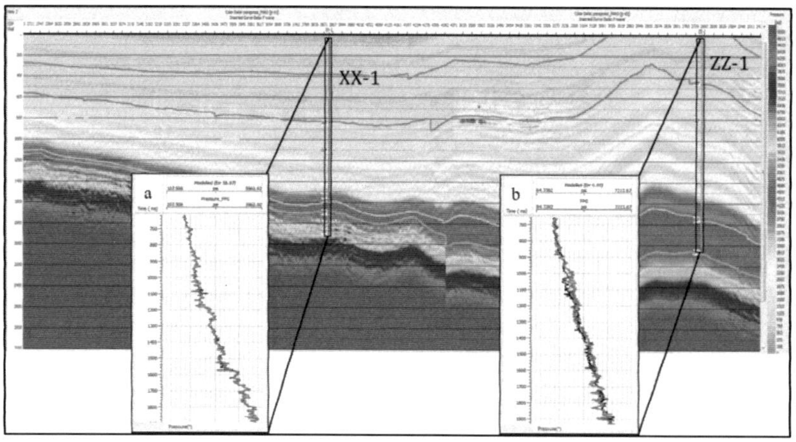

Figure 4. PNN predicted pore pressure section from pre-stack seismic data, which is overlaid with (a) ZZ-1 well for data training and (b) XX-1 well for validation.

After having the validated prediction, the PNN pore pressure is then applied to all seismic section. PNN predicted pore pressure by using the extracted seismic attribute (average frequency, dominant frequency, integrated absolute amplitude, and apparent polarity) is shown in Figure 5. This work is carried out to have a dense distribution of predicted pore pressure since the availability of post-stack seismic data is denser compared to the pre-stack seismic data that is only represented by two seismic line. It can be seen that PNN predicted pore pressure generated by post-stack data have a strong lateral variation than PNN predicted pore pressure generated by using pre-stack data. This result can be understood since the PNN predicted pore pressure produced by using pre-stack data is strongly controlled by the well, which is guided by the horizon during the inversion process. Although there are no controls on the horizon during the PNN pore pressure prediction by using post-stack data, we can observe that PNN predicted pore pressure shows a trend to follow the pattern of the structure. This means that predicted lateral pore pressure distribution is also controlled by the pattern of structure.

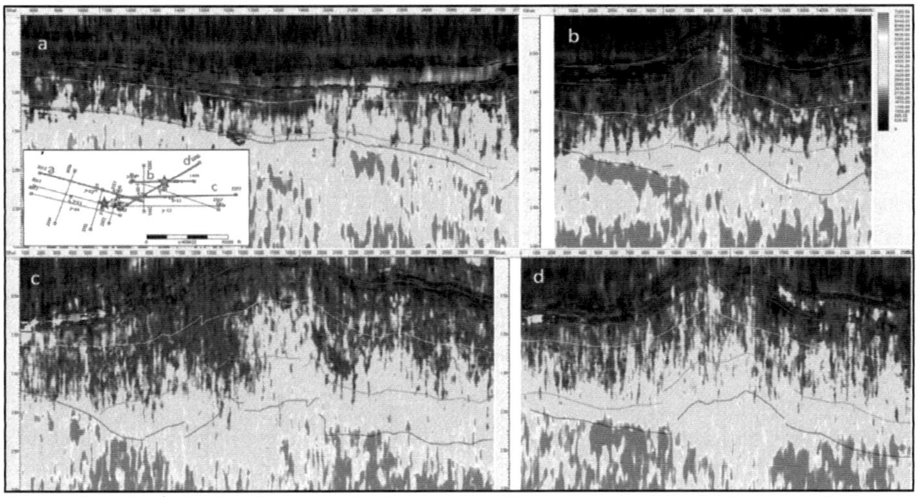

Figure 5. PNN predicted pore pressure section from post-stack seismic data from the different 2D seismic line, which is indicated in index map.

Roughly speaking to the mechanisms of over-pressure, it can be understood that overpressure can be caused by loading (associated with sediment loading) and unloading mechanisms that influences decreasing the effective stress. Our analysis to over-pressure potential in this study area can be caused by disequilibrium compaction occurs in the thick shale formation, which is Gumai formation. This formation was deposited during the Lower to Middle Miocene, which was composed of fine clastic sediments in the form of shale, marl, and calcareous claystone. This formation represented a phase of maximum transgression in South Sumatra Basin and deposited in deep-sea environments. The over-pressure mechanism in this study area can be understood by the relationship between density and sonic log from XX-1 well, where the shale compaction in the smectite zone strongly control to the over-pressure. So that it can be summarized that over-pressure in the study area occurs as burial process, which occurs so fast and pores fluid did not drain out from the formation.

Figure 6 shows section and map of overpressure, where green is the zone with normal pore pressure. The zone which has a deviation of less than 1500psi is marked in yellow and areas with greater than 1500psi deviation is shown in purple. Top of overpressure zone is commonly located in the upper Gumai formation, but there are anomalies that occur in Air Benakat formation, as in Figure 6 marked with an ellipse with the red dotted line.

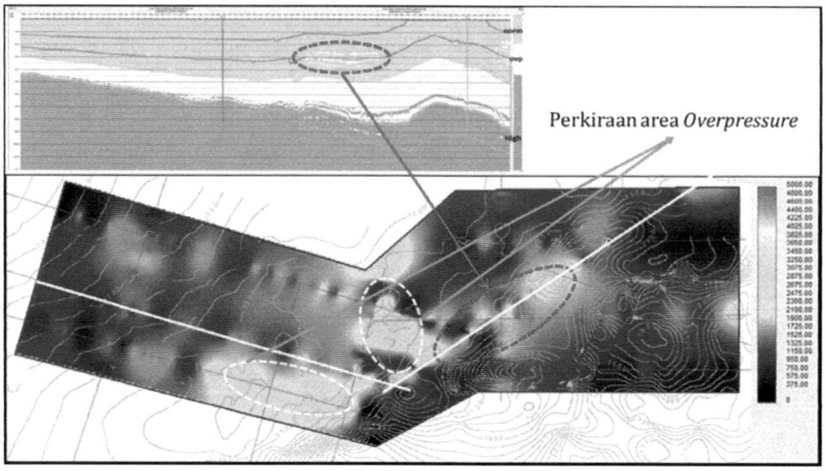

Figure 6. Section and map of predicted over-pressure at the top of Air Benakat Formation

4. Conclusions

PNN Pore pressure prediction has been successfully applied to two different 2D seismic data i.e., pre-stack and post-stack seismic data. PNN predicted pore pressure generated by post-stack data have a strong lateral variation than PNN predicted pore pressure generated by using pre-stack data. This result can be understood since the PNN predicted pore pressure produced by using pre-stack data is strongly controlled by the well, which is guided by the horizon, during the inversion process. The predicted pore pressure produces the same vertical resolution as seismic resolution, which is very important to evaluate the pore pressure change at intervals of 100 ft.

The over-pressure potential in this study area is identified at the upper Gumai formation, but there are small anomalies that occur in Air Benakat formation. The over-pressure at the upper Gumai formation can be caused by disequilibrium compaction occurs in the thick shale formation. This formation was deposited during the Lower to Middle Miocene, which was composed of fine clastic sediments in the form of shale, marl, and calcareous claystone. This formation represented a phase of maximum transgression in South Sumatra Basin and deposited in deep-sea environments.

References

[1] Eaton B A 1972 Graphical method predicting pressure worldwide *World Oil* **185** p 51-56
[2] Al-Chalabi M 1994 *First Break* **12** p 589-596
[3] Xu S and Keys B 1999 Study of the coupled effect of pressure, frequency and fluid content on P-and S-wave velocities *Proc. in 1999 SEG Annual Meeting. Society of Exploration Geophysicists*
[4] Sumuyut S and Sardjito 1989 Hydrocarbon source rock identification in the South Palembang Sub-basin *Proc. Indonesian Petroleum Association, 18th Annual Convention* p 427-467
[5] McCormack M D 1991 *The Leading Edge* **10** p 11-15
[6] Hampson D P, Schuelke J S and Quirein J A 2001 *Geophysics* **66** p 220-236

IOP Conf. Series: Earth and Environmental Science **62** (2017) 012022 doi:10.1088/1755-1315/62/1/012022

Introduction to Digital Rock Physics and Predictive Rock Properties of Reservoir Sandstone

Handoyo[1]*, Fatkhan[2], Suharno[3], and Fourier DEL[4]

[1]Undergraduated Program of Geophysical Engineering, Institut Teknologi Sumatera
[2]Exploration and Engineering Seismology Research Group, Institut Teknologi Bandung
[3]Geophysics Engineering, University of Lampung
[4]Earth Science, Institut Teknologi Bandung

*Corresponding author's email: handoyo.geoph@itera.ac.id

Abstract. Rock properties analysis (porosity, permeability, elastic modulus, and wave velocity) of the rock is important to note as one of the methods to determine the characteristics of the reservoir rock. Rock properties can calculated in conventional (laboratory), indirect (inversion of seismic waves), and digital computation (Digital Rock Physics). This paper will introduce and discuss the digital calculation/simulation and empirical equation to predict the value of the rock properties from reservoir sandstone. The data used is the samples of the data sandstone core (reservoir) subsurface in an oil field. The research method is to combine the data from a thin layer, a digital image of rocks in three-dimensional (μ-CT-Scan), and empirical approaches of the equations of permeability on rocks and Lattice Boltzmann equation. Digital image of a scanned using μ-CT-Scan used to determine value rock properties and pore structure at the microscale and visualize the shape of the pores of rock samples in 3D. The method combined with rock physics can be powerful tools for determining rock properties from small rock fragments.

1. Introduction

Reservoir rocks possess a certain characteristic which can be identified by several physical parameters. Among many of the physical parameters are the porosity and permeability. The characterization is a process of elaborating its characteristic qualitatively and quantitatively by using the available data. Characteristics provided as information regarding physical parameters from reservoir rock is essential to understand the reservoir better.

The characterization of reservoir rock can be conducted directly in-situ, or also in the laboratory. In this paper, we utilized analysis of digital images (Digital Rock Physics; DRP) to obtain information regarding the characteristic of the reservoir rock [1, 2, 3, 4]. This method utilizes the absorption of X-ray to obtain the digital image from the rock. The goal of this research is to estimate porosity, permeability and elastic properties of the reservoir sandstone using computer-based simulation and empirical method. Digital rock physics (DRP) aims at providing qualitative and quantitative understanding of flow transport units as well as geometrical properties of rocks. Some of the rock properties are extremely difficult, if not impossible, to measure in the laboratory. Thus, DRP in conjunction with laboratory measurements, will compliment and complete well log analysis with not only detailed information but also with new kinds of insights. Such well logs enhance analysis of reservoirs and will open a way for economic exploration and better recovery of hydrocarbon [5].

2. Data and Basic Theory

The sample data used in this research is a core of reservoir sandstone. Name sample is Ngrayong-54 which diameter 5 cm, length 15 cm, porosity 0.24 and density 2.3 gram/cm^3. Subsequent after acquiring the digital images, porosity is calculate as follows:

$$\phi = \frac{V_{pore}}{V_{total}} = 1 - \frac{V_{matriks}}{V_{total}} \quad (1)$$

V_{pore} is volume of pore and V_{total} is the volume rock. The simple bounds for an isotropic linear elastic composite, defined as giving the narrowest possible range without specifying anything about the geometries of the constituents, are the simple effective medium theory is the Voigt bound (illustrated by Figure 1). The Voigt bound calculated effective elastic modulus M_V, from the volume fraction N phase f_i, and elastic modulus N phase fraction M_i as follows:

$$M_V = \sum_{N=1}^{N} f_i M_i \quad (2)$$

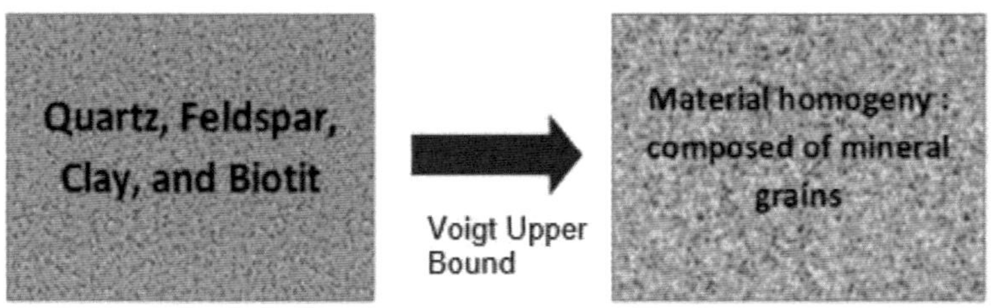

Figure 1. Effective medium theory simple illustration using Voigt upper bound.

The empirical calculation to determine elastic properties is the P-wave velocity, S-wave velocity, Bulk Modulus and Shear Modulus. P-wave velocity (V_p) and S-wave velocity (V_s) is calculated as follows [6]:

$$V_p = (1-\phi)^2 V_m + \phi V_f$$
$$V_s = (1-\phi)^2 V_m \sqrt{\frac{(1-\phi)\rho_m}{(1-\phi)\rho_m + \phi\rho_f}} \quad (3)$$

V_m is the Velocity in the mineral, V_f is the velocity in the fluid, ρ_m is the density of mineral and ρ_f is the density fluid. Bulk modulus (K) and Shear Modulus (G) is calculate as follows:

$$K = \rho V_p^2 - \tfrac{4}{3} G$$
$$G = \rho V_s^2 \quad (4)$$

One of the characteristics of the reservoir rock is the acoustic impedance (AI). Acoustic impedance is calculated using equation [6]:

$$AI = \rho V_P \tag{5}$$

Then, the rock ability to pass through the pore fluid is expressed by the permeability. Absolute permeability appears in Darcy's law as a constant coefficient relating fluid, flow, and the porous medium parameters as follows:

$$\frac{Q}{A} = -\frac{k}{\mu}\frac{\Delta P}{L} \tag{6}$$

To calculate permeability, we simulate fluid flow using Lattice Boltzmann method that is as follows [7]:

$$k = \frac{Q\mu}{A}\frac{1}{dP/dL} = \frac{\mu\langle v\rangle}{dP/dL} \tag{7}$$

Where Q is the fluids debit, A is the surface area, μ is the viscosity, and dP/dL is the differential of pressure by the length.

3. Rock Imaging

The process of DRP begins with an image of a large rock sample acquired at a relatively coarse resolution to cover a large field of main cube. At this stage, rock fabrics larger than the image resolution are resolved while smaller ones are unresolved (sub-cube). Information concerning the unresolved rock fabrics is analysis from additional images acquired at a finer resolution and smaller field of view. The information from resolved and unresolved rock fabrics are average from the small-scale image and back into the large-scale image. In general, the stage of the DRP analysis shown in Figure 2. Multi-scale DRP provides a promising method to characterize rocks. They can be roughly summarized as follow:

Digital rock imaging using μCT-Scan: the sample is later converted into digital images μCT-Scan device SkyScan 1173 which is installed in the BSC-A building, Institut Teknologi Bandung

Image segmentation: digital image result is separated between the pore and solid matrix rock. This process is called thresholding.

Rock imaging at the finer resolution and smaller field of view: Thresholding result made in the form of 2D and 3D (cube). Furthermore, the main cube made into the small cube (sub-cube) to determine variation the physical properties of the rock.

DRP analysis, 2D and 3D analysis: the next step is to calculate the percentage of pore section and solid rock matrix to determine the value of porosity every sub-cube. Permeability values were calculated using Lattice Boltzmann equation. The volume fraction of each constituent mineral rock calculated to determine effective medium theory. Then, calculation elastic properties of sandstone using empirical equation.

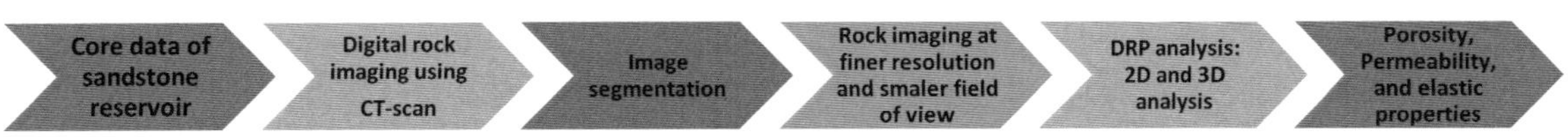

Figure 2. DRP data processing stages on sandstone sample. The stage begins with digital image processing to the calculation porosity, permeability, and elastic properties using empirical equations.

A variety of imaging and detection techniques have been used to gain insights into rocks. Ideally, the image resolution being used should resolve all significant rock features and provide a large field of view. Due to limitations in imaging technology both image resolution and image scale are overly compromised. Figure 3 imaging of sandstone sample. Figure 3a is the original image slice (2D), has a dimension 300×300 pixels. They have a gray-scale color image. Black is the pore and gray is the solid matrix of rock. Black image shows pore (air), the gray color shows solid matrix low density, and bright white color shows solid matrix higher density. Figure 3b is the image made in the thresholding process from Figure 3a. Black is the pore fraction and white are the solid matrix of the rock. Figure 3c and 3d is the sub-cube from the original image. The Figure 3e and Figure 3f are the 3D image from the sub-cube.

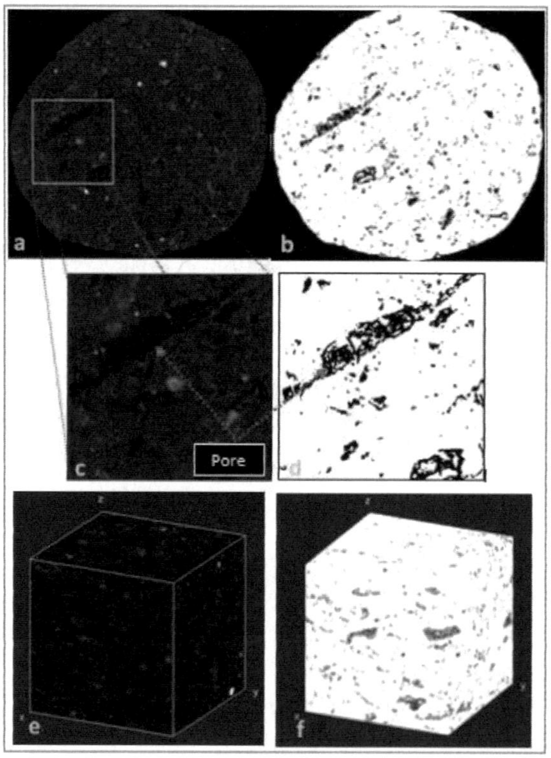

Figure 3. Digital rock physics was processed of sandstone. (a) 2D original image (gray scale), (b) image made in thresholding, (c) sub-cube from the original image, (d) threshold image from c, (e) 3D image from c by dimension 300×300×300 pixels, and (f) 3D threshold image from e.

4. Result and Discussion

Calculation of porosity values of obtained from Figure 3d. Porosity values obtained from 16 sub-cube to determine variations. Then, permeability values obtained from Lattice Boltzmann equation from 16 sub-cube. Elastic properties values obtained using equation (3), (4), and (5). The volume fraction of the constituent mineral rock consisting of biotite (4%) and feldspar-quartz-clay (96%) from the solid matrix of rock. The calculation results are shown in Table 1. The Chart on the calculation of permeability, elastic modulus, seismic wave velocities, and acoustic impedance as a function of porosity is shown in Figure 4. In Figure 4a, the permeability values are increasing with increasing porosity values. Permeability value varies between 160-6000 mD with the average permeability is

2700 mD. This value indicates that the sandstone sample can accommodate and passes fluid well. Thus, this sandstone sample can be a good reservoir rock.

Table 1. Summary of the results calculation values of the porosity, permeability, and elastic properties of the sample sandstone.

Sub Cube	Porosity	Permeability (mD)	V_p (km/s)	V_s (km/s)	Shear Modully (GPa)	Bulk Modully (GPa)	Acoustic Impedance (Kg/m³ Km/s)
1	0.241	1989	3.21	2.00	10.89	13.56	8353.17
2	0.211	824	3.47	2.16	12.72	15.82	9026.42
3	0.332	6420	2.49	1.55	6.52	8.15	6470.62
4	0.159	160	3.94	2.46	16.42	20.41	10255.21
5	0.287	3909	2.84	1.76	8.47	10.56	7371.54
6	0.263	2880	3.03	1.89	9.67	12.06	7876.04
7	0.302	5022	2.72	1.69	7.78	9.71	7064.71
8	0.284	3760	2.86	1.78	8.62	10.74	7433.69
9	0.215	835	3.44	2.14	12.45	15.50	8932.61
10	0.326	5820	2.53	1.58	6.76	8.44	6587.35
11	0.274	3195	2.94	1.83	9.10	11.34	7638.98
12	0.208	486	3.50	2.18	12.90	16.05	9090.23
13	0.295	4350	2.77	1.73	8.10	10.11	7209.20
14	0.188	380	3.68	2.29	14.27	17.74	9560.25
15	0.253	2459	3.11	1.94	10.21	12.72	8091.17
16	0.226	867	3.34	2.08	11.75	14.63	8677.94

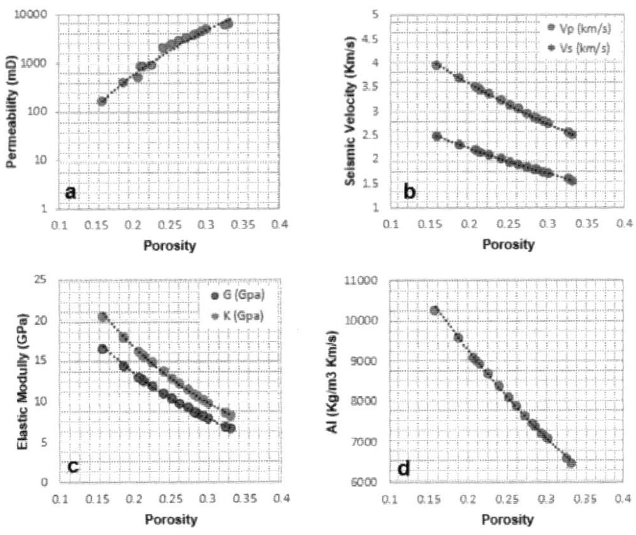

Figure 4. The calculation of permeability and elastic properties of sandstone sample. (a) estimation of permeability, (b) estimation of V_p and V_s, (c) estimation of Bulk modulus and shear modulus, and (d) estimation of acoustic impedance.

Figure 4b shows the variation of the seismic wave velocities decreases with increasing porosity value exponentially. The red marker is the *P*-wave velocity and blue is the *S*-wave velocity. *P*-wave velocity value varies between 2.49 km/s – 3.94 km/s with the average value is 3.12 km/s. *S*-wave velocity value varies between 1.55 km/s – 2.46 km/s with the average value is 1.94 km/s. Figure 4c shows the variation of the elastic modulus decreases with increasing porosity value exponentially. The red marker is the bulk modulus and blue is the shear modulus. Bulk modulus value varies between 8.15 Gpa – 20.41 Gpa with the average value is 12.97 Gpa. Shear modulus value varies between 6.52

Gpa – 16.42 Gpa with the average value is 10.41 Gpa. Figure 4d shows the variation of the acoustic impedance decreases with increasing porosity value exponentially. Acoustic impedance value varies between 6450 (Kg/m^3 Km/s) – 10250 (Kg/m^3 Km/s) with the average value is 8100 (Kg/m^3 Km/s).

5. Conclusion

The conclusions of this paper are: (1) μ-CT image is able to visualize the micro structure of rocks such as the pore and density matrix, (2) μ-CT image combined with digital computation can deliver the result of pore visualization, porosity value prediction, permeability value prediction, fraction of high-low density rock matrix, and elasticity of sandstone samples, (3) The rock properties such as porosity, permeability, and elastic properties have done calculated using digital imagining analysis and empirical equation, and (4) The method combined with rock physics can be powerful tools for determining rock properties from small rock fragments.

6. Acknowledgments

Thanks to supporting for this work was partly provided by LPPM of Institute Technology Sumatera, Geophysical Engineering Department of Institute Technology Bandung, Department of Institute Technology Bandung, and Geophysical Engineering Department of University Lampung.

References

[1] Arns C H and Knackstedt M A 2002 *Journal of Geophysics* **67** (5) 1396-1405
[2] Andra H, Combaret N, Dvorkin J, Glatt E, Han J, Krzikalla F, Lee M, Madonna C. Marsh M, Mukerji T, Ricker S, Saenger E H, Sain R, Saxena N, Wiegmann A and Zhan X 2012 *Journal Computers & Geosciences ELSEVIER* **50** 25-32.
[3] Wiegman 2012 Predicting Effective Elastic Properties with Elastodict *Fraunhofer ITW* Germany
[4] Handoyo, Fatkhan, and Fourier D E L 2014 Digital Rock Physics Application: Structure Parameters Characterization, Materials Identification, Fluid Modeling, and Elastic Properties Estimation of Saturated Sandstone *HAGI Proceeding 2014*
[5] Sungkorn 2015 Multi-Scale And Upscaling Of Digital Rock Physics With A Machine That Can Learn About Rocks *Ingrain Inc.* 3733 Westheimer Rd., Houston, Texas, 77027, U.S.A.
[6] Mavko G and Nur A 2009 The Rock Physics Handbook, Second Edition Tools for Seismic Analysis of Porous Media *Cambridge University Press*
[7] Fourier D E L 2014 Analysis of Permeability and Tortuosity of Fontainebleau Sandstone and Its Models Using Digital Rock Physics Approach *Bandung Institute of Technology*
[8] Raymer L L, Hunt E R, and Gardner J S 1980 An improved sonic transit time-to-porosity transform *Trans. Soc. Prof. Well Log Analysts, 21st Annual Logging Symposium*

IOP Conf. Series: Earth and Environmental Science **62** (2017) 012023 doi:10.1088/1755-1315/62/1/012023

Integrated Approach for Characterizing Unconventional Reservoir Shale Hydrocarbon: Case Study of North Sumatra Basin

A Haris, B Seno, A Riyanto and A Bachtiar

Master Program for Reservoir Geophysics, Department of Physics, Faculty of Mathematics and Natural Sciences, Universitas Indonesia, Depok 16424, Indonesia

E-mail: aharis@sci.ui.ac.id

Abstract. Integrated analysis of geochemical, rock mechanic and geophysics were carried out to characterize and map the unconventional reservoir shale hydrocarbon in Baong field, North Sumatera Basin, which is proven with the large potential hydrocarbon, particularly in the sand reservoir. The new challenge of this field is the present of thick shale layer, which is offering the new concept of an unconventional reservoir. The shale layer has the double role as source rock and reservoir. In this work, we performed geochemical analysis on the core data to determine the Total Organic Carbon (TOC), mineralogy, Tmax, and Kerogen type. In term of rock mechanic, the Rock Strength was calculated to determine the hardness and brittleness index. While for petrophysical analysis, we performed multi-linear regression of log data to estimate TOC relationship with the seismic attribute. In term of geophysics, we carried out seismic inversion to produce acoustic impedance, which is useful to map shale distribution. Our analysis shows that the target of shale layer has TOC range from 2 up to 3.5 wt.% with brittleness index of 0.48. Based on the predicted Tmax, this shale layer is categorized into early mature phase and classified into II Kerogen type, which means it has a potential to produce oil. The shale layer was indicated by the result of acoustic impedance inversion which has a value for over 25000 ft/s *g/cc and Rock Strength less than 3000 Psi.

1. Introduction

In term of energy resource exploration, the generation potential of shale hydrocarbon from organic-rich shale has been considered intensively as new and unconventional energy resources. The hydrocarbon shale has long been of an interest for indirect energy resource and has unique petroleum system where shale was playing a role as a source rock and reservoir hydrocarbon. Finding other alternative energy resources of fossil fuel origin requires a serious effort for geoscientists to overcome the continuing decline of petroleum supplies, accompanied by increasing costs of petroleum-based products may present opportunities for hydrocarbon shale to produce some of energy needs. Thereby, the hydrocarbon shale potential in Indonesia, which have been abundantly identified, can be explored to meet the energy needs in the future [1].

The previously published paper discussed the hydrocarbon shale potential, which was identified in some parts of Sumatra, including Baong filed in North Sumatra. Those shales have preliminary been studied, however, their potential and quality has not been known well yet. These hydrocarbon shale beds seem to superimpose the oil, gas, and coal seam basin, and thereby the basin setting is similar to the tectonic configuration of the clastic depositional system [1][2].

Tectonically, North Sumatra Basin is a back-arc basin which covers an area of 60,000 square km in offshore and onshore northern Sumatra Island. The Basin is a prolific hydrocarbon basin in western Indonesia and produces hydrocarbons [3]. Exploration history began in the late 19th century in which the oil field was first discovered in the Basin of North Sumatra that field Telaga Said (1885), followed by the field of Land (1889), Perlak (1990), Serang Jaya (1926), Rantau (1929), Gebang (1936), and the west Palu (1937). Arun field is a large field discovered in 1971 with reserves at that time is estimated around 17TCF.

Stratigraphically, Baong shale is younger than Bampo shale in North Sumatera but the availability of well data only up to Baong formation. Baong formation was formed in the deep lacustrine environment and consists of darks shale. In term of stratigraphy, Baong Shale is divided into three sections, Lower, Mid and Upper Baong [4]. In mid-section, shale was intercalated with sand then act as a reservoir. The regional structure is high in the southwest and north due to the Bukit Barisan high and the Asahan High [5].

The integrated analysis including geochemical, rock mechanic, and geophysics reported in the present paper aimed to characterize and map the unconventional reservoir shale hydrocarbon in term of potential source rocks, and thermal maturity of the rock succession in Baong field, North Sumatera Basin. The availability of data consists of 2D seismic data and 3 well will be used to characterize and map shale hydrocarbon, which is guided by AI, TOC, and Rock Strength.

2. Methods and analytical procedures

To achieve the goal of this study, we used core laboratory and geophysical analysis. Primary laboratory work activity includes accurate determinations, observations, and measurements on organic matter characteristics, and detailed seismic interpretation containing shale of Baong Formations, North Sumatra Basin. Afterward, we generated some approximation model from the limited data to gain the relation between the field (samples) and calculated data for geochemical and organic petrographic analyses. The samples are gained from a freshest relatively selected representative core of shale and mudstones.

3. Geochemical and petrophysical analysis

The quantities of organic matter in shale samples were determined based on the total content of organic carbon (TOC). In this work, geochemical analysis on the core data was carried out to determine TOC, mineralogy, Tmax, and Kerogen type. Geochemical data are collected from cutting sample for each well. Geochemical data of shale sample in the form of Rock-Eval Pyrolysis from each well were taken from Baong Formation.

Geochemical analysis of 3 existing wells is shown in different color point where blue is well A; red is well B, and green is well C. The location of well A and B are in close each other, well B located about 10 km from well A and C. Figure 1a shows the distribution of Tmax from each well. Figure 1b shows that well A and B are categorized as Kerogen type II which potentially generate oil but has different amount of Hydrogen Index (HI). Figure 1c shows that Well A and B shows temperature less than 435°C which is categorized immature and on the other side Well C indicates temperature in early mature phase. The distribution of TOC data from each well has variation 0.5 – 4 wt.%.

TOC collected from geochemical analysis at certain depth while TOC data is needed for entire depth. Therefore TOC data is derived from mathematical relationship to gain TOC as a function of depth (in the range of shale formation) by using Multilinear Regression method. The mathematical relationship of TOC data will be the function of GR, Density, Neutron Porosity, Sonic, and Resistivity is given by equation (1) as follow:

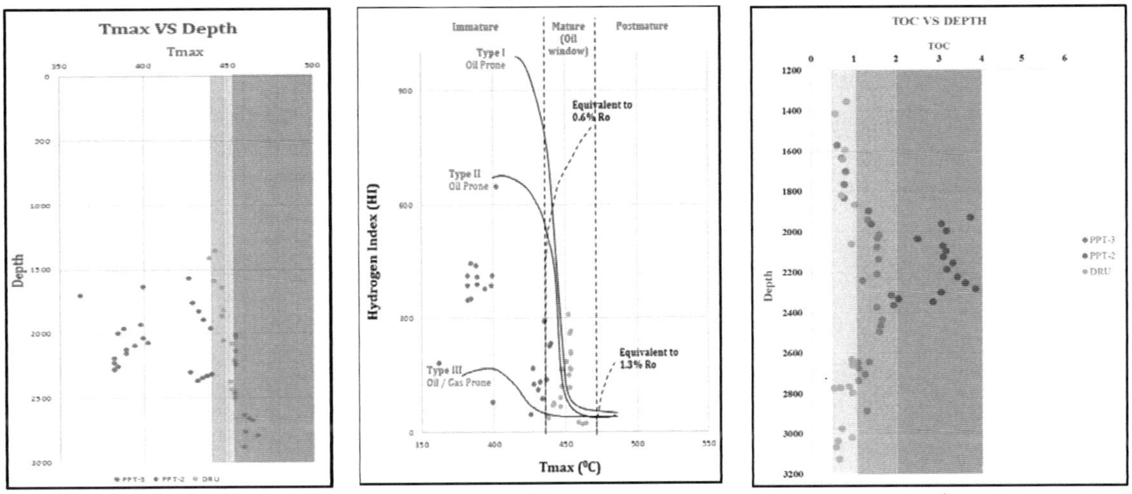

a) b) c)

Figure 1. Geochemical analysis of three available well a) Diagram plot Tmax vs. depth, b). The relationship of Tmax and HI, and c). Diagram plot TOC vs. depth.

$$TOC = 1.8994 - 0.GR + 1.0176xDensity + 20.893xNeutron\ Porosity - 0.0488xSonic + 0.321xResistivity \quad (1)$$

The correlation between the TOC lab and the calculated TOC using Multi-linear regression, which is illustrated in equation (1), shows good correlation and the derived equation will be used to distribute the calculated TOC along the seismic section, which is guided by Acoustic Impedance (AI).

In term of petrophysical analysis, we determine the relationship between AI and TOC in the way of cross plotting analysis to see the separation of sand on shale. Figure 2 shows that shale was well separated from sand in a certain range. Shale hydrocarbon potential is indicated by AI of 25000 ft/s*g/cc and GR around 70 GAPI. The Rock Strength parameter is determined from well data using Horsrud [6], which is derived from sonic log [6]. The relation between Rock Strength and brittleness index is inversely proportional, where brittleness was determined from the percentage of reversible strain.

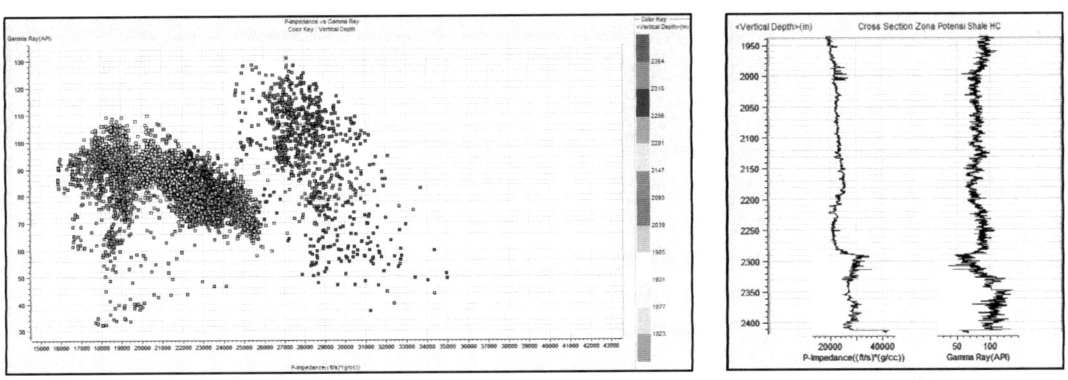

a) b)

Figure 2. Cross-plotting analysis to see the separation of sand with respect to shale a), and cross section as a function of depth b), which is represented by AI and GR.

In term of mineralogy analysis, we applied the equation of Brittleness Index (BI) that has been defined by Jarvie et al. and modified by Wang and Gale [7][8]. They modified the equation by adding dolomite fraction into their equation which could increase brittleness index value. In this study area, the presence of mineral plagioclase is dominant. It was shown that presence of plagioclase around 2-6 %. Our observation to the both calculated BI from Jarvie and Wang and Gale, show similar trends of BI. Figure 3 shows the relationship of BI and Rock Strength, where more brittle shale the Rock Strength decreases. Our assessment of the total potential of shale rock refers to more brittle rock, less Rock Strength, and higher TOC value. In general, the potential of shale rock has TOC range from 2 up to 3.5 wt.% with brittleness index of 0.48 and Rock Strength less than 3000 Psi.

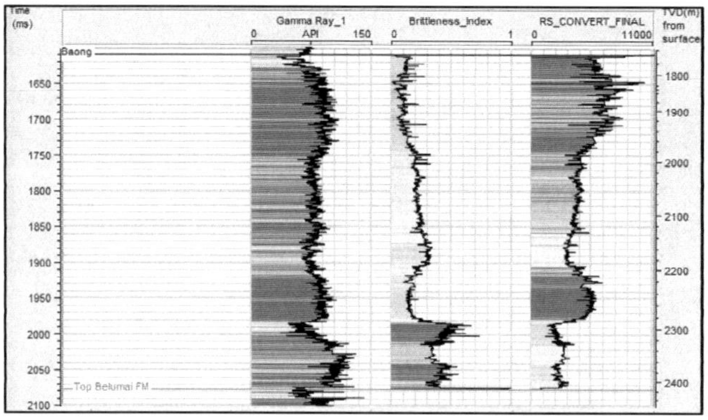

Figure 3. The panel from left to right is illustrating Gamma Ray, Brittleness Index, and Rock Strength. More brittle shale will have less strength.

4. Geophysical analysis

The geophysical analysis is performed to transform the seismic data, which has the lateral resolution, into the organic geochemical data. This strategy is very helpful in mapping the potential of hydrocarbon shale laterally. The available seismic data has different vintage so that seismic data should be applied conditioning such as amplitude matching and balancing.

To transform seismic data to organic geochemical data, we first performed the seismic inversion to produce the AI section. Further, we did sensitivity analysis as a reference to define potential shale zones. Based on sensitivity analysis, the shale hydrocarbon zones was indicated by having AI of above 25000 ft/s*g/cc, TOC of 2-3.5 wt.% (good to very good quality) and Rock Strength less than 3000 Psi. The prospect zone is marked by the black ellipse. Figure 4a show AI inversion result at seismic line 77 through well A. The resulted AI is then distributed to all seismic line guided with top and bottom Baong Formation to produce TOC section, which is illustrated in figure 4b. The same step is conducted to define Rock Strength section as displayed in figure 4c. After having all 2D seismic line were transformed into TOC, then we interpolate TOC and Rock Strength section to make TOC and Rock Strength map.

Figure 4d shows TOC distribution, where high TOC was shown at the south-eastern part of the study area. Thus overlay between Rock Strength map and TOC shows the potential sweet spot where the color legend indicates Rock Strength and contour indicates TOC distribution. The area which has high TOC and low Rock Strength is shown in the south-eastern area.

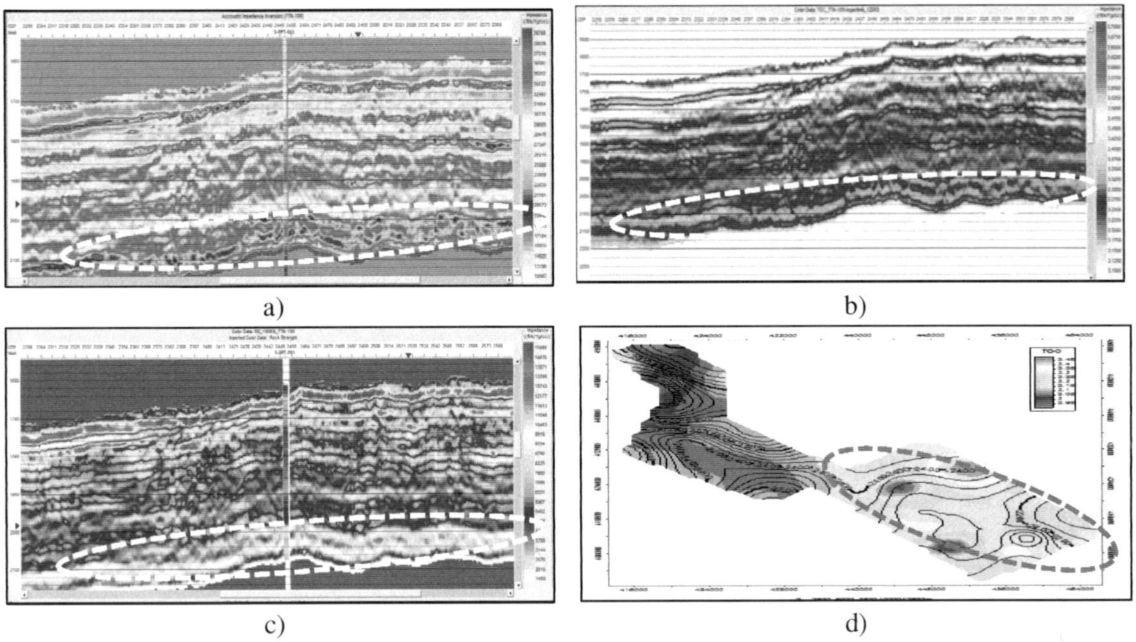

Figure 4. a) AI section b). TOC section and c) Rock Strength section, and d) TOC surface of Baong Formation. The dashed-line is the sweet spot of potential shale.

5. Conclusions

Integrated analysis of geochemical, rock mechanic and geophysical was successfully applied to characterize and map the unconventional reservoir shale hydrocarbon in Baong field, North Sumatera Basin. The shale hydrocarbon of Baong Formation shows fair to very good hydrocarbon potential level, with Kerogen contained is classified into Type II at depth of 1500m. They can be categorized as gas-prone source-rock, with thermal maturity significantly varies from immature to early mature level, dominated by immature stage. Our delineation by integrating geochemical and geomechanical analysis show that sweet spot can be identified in the south-eastern part of the study area. Based on the integrated analysis, we identified that potential shale layer has TOC range from 2 up to 3.5 wt.% with brittleness index of 0.48. The shale layer was indicated by the result of acoustic impedance inversion which has a value for over 25000 ft/s *g/cc and Rock Strength less than 3000 Psi.

References

[1] Iqbal M, Suwarna N, Syafri I and Winantris W 2014 *Indonesian Journal on Geoscience* **3** 135-49
[2] Kusumahbrata Y and Suwarna N 2003 Characteristic of the Keruh Formation Oil Shale: Its Implication to Oil Shale Resource Assessment *Prosiding Kolokium Energi dan Sumber Daya Mineral* 362-70
[3] Cockcroft P, Anli J, and Duignan J 1998 EOR Potential of Indonesian Reservoirs, *Proc. of 17th Annual Convention of Indonesian Petroleum Association* **2** 73-188
[4] McCaffrey R 2009 The Tectonic Framework of the Sumatran Subduction Zone. *Annual Review of Earth and Planetary Sciences* **37** 345-66
[5] Doust H, Noble R A 2008 Petroleum Systems of Indonesia *J. Marine and Petroleum Geology* **25** 103-129

[6] Horsrud P 2001 Estimating Mechanical Properties of Shale from Empirical Correlations *SPE Drilling & Completion* **16** 68-73
[7] Jarvie D M, Hill R J, Pollastro R M, Wavrek D A, Bowker K A, Claxton B L, Tobey M H 2003 Evaluation of Unconventional Natural Gas Prospects: The Barnett Shale Fractured Shale Gas model. *In 21st International Meeting on Organic Geochemistry*
[8] Wang FP and Gale JF 2009 Screening Criteria for Shale-gas Systems Gulf Coast *Association of Geological Societies Transactions* **59** 779–793

The Preliminary Results of GMSTech: A Software Development for Microseismic Characterization

Maman Rohaman[1*], Cahli Suhendi[1], Rexha Verdhora Ry[1], Billy Sugiartono Prabowo[2], Sri Widiyantoro[3], Andri Dian Nugraha[3], Tedi Yudistira[3], Bambang Mujihardi[4]

[1]Master Program of Geophysical Engineering, Faculty of Mining and Petroleum Engineering, Institut Teknologi Bandung, Indonesia
[2]Volcanology and Geothermal Laboratory Faculty of Mining and Petroleum Engineering, Institut Teknologi Bandung, Indonesia
[3]Global Geophysics Research Group, Faculty of Mining and Petroleum Engineering, Institut Teknologi Bandung, Indonesia
[4]Pertamina UTC, Indonesia

E-mail: rohamanmaman@gmail.com

Abstract. The processing of microseismic data requires reliable software for imaging the condition of subsurface related to occurring microseismicity. In general, the currently available software is only specific for certain processing module and developed by the different developer. However, the software with integrated processing modules will give a better value because the users can use it easier and faster. We developed GMSTech (Ganesha Microseismic Technology), a C# language-based standing-alone software consisting several modules for processing of microseismic data. Its function is to solve a non-linear inverse problem and imaging the subsurface. C# library is supported by ILNumerics to reduce time consumption and give good visualization. In this preliminary result, we will present four developed modules: (1) hypocenter determination, (2) moment magnitude calculation, and (3) 3-D seismic tomography. In the first module, we provide four methods for locating the microseismic events that can be chosen by a user independently: simulated annealing method, guided grid-search method, Geiger's method, and joint hypocenter determination (JHD). The second module can be used for calculating moment magnitude using Brune method and to estimate the released energy of the event. At last, we also provided the module of 3-D seismic tomography for imaging the velocity structures based on delay time tomography. We demonstrated the software using both a synthetic data and a real data from a certain geothermal field in Indonesia. The results for all modules are reliable and remarkable, reviewed statistically by RMS error. We will keep examining the software using another set of data and developing further modules of processing.

1. Introduction

Observation of micro-seismic events is very important to characterize geothermal field or volcano-tectonic activity. Our previous works in geothermal characterization [1, 2, 3], the reservoir can be characterized for example for detecting fracture and permeability zone, Vp/Vs anomaly in water liquid replacement, attenuation structure, shear wave splitting to image anisotropy distribution, etc. In volcano-tectonic [4, 5, 6], its activity is imaged by a distribution of the event's locations. By observing that distribution, we can map active fault on a shear zone, as well as fluid compression.

The processing of micro-seismic data requires reliable software for imaging the condition of subsurface related to occurring micro-seismicity. In general, the currently available software is only specific for certain processing module and developed by the different developer. Two examples of these softwares are VELEST, GAD (Geiger's method with Adaptive Damping) and SIMULPS. In 1984 E. Kissling and W. L. Ellsworth modified VELEST, the program based on Fortran. They used it to calculate a Minimum 1-D model (i.e., a well-suited 1-D velocity model of earthquake location and 3-D seismic tomography) of Long Valley area, California [7]. In 2005, K. Nishi developed hypocenter calculation software GAD based on Fortran. This software uses least square optimization to determine hypocenter location [8]. Originally, in 1983 Thurber developed SIMULPS to calculate tomography inversion. It provides alternative forward solution methods (ray tracer), approximated ray tracing plus pseudo-bending (ART-PB) and 3D shooting method using paraxial rays and perturbation theory (RKP) [9].

The software with integrated processing modules will give a better value because the users can use it easier and faster. We developed GMSTech (Ganesha Micro-seismic Technology), a C# language-based standing-alone software consisting of several modules for micro-seismic data processing. Its main function is to solve the non-linear inverse problem and to image the subsurface. C# library is supported by ILNumerics to reduce time consumption and give better visualization. The compiled program is an executable file which is compatible with Windows Operating System.

2. Methods

2.1. Ray Tracing

2.1.1. Straight Method

This method uses straight line connecting the source to the receiver. It is so simple using the assumption that the medium is homogenous.

2.1.2. Shooting Method

In wave propagation, this method uses Snell's Law to determine incident and transmission angles at the boundary layer. These angles affect total path length from the source to the receiver. This method will calculate root mean square of error then choose minimum error to determine the best location. The Snell's Law used follows bellow equation:

$$\frac{V_1}{V_2} = \frac{\mathrm{Sin}\theta_1}{\mathrm{Sin}\theta_2} \qquad (1)$$

2.1.3. Pseudo-Bending Method

Pseudo-bending is an approach in minimization of travel time based on Fermat's Principle by giving small perturbations gradually on ray paths. Equation bellow is used to calculate travel time (T).

$$T = \int_{source}^{station} \frac{1}{v} dl \qquad (2)$$

Where l is ray path and v is velocity. Travel time is calculated by sum all of Ray segment. The pseudo-bending method calculates it using perturbation scheme. Initial definition is defined by 3 points which are interpolated linearly. The middle point is an estimation of initial trajectory. Furthermore, it will be perturbed iteratively using wave equation to calculate optimum travel time (the wave propagates using the fastest travel time). The segments of initial trajectory are separated to calculate travel time which is more optimum. Zhao et al. modified the equation using Snell's law to determine the middle point to calculate more optimum travel time [10].

2.2. Hypocenter Determination

2.2.1. Simulated Annealing Inversion

One method of guided random search is simulated annealing method (SA) [11, 12]. In the inversion process, SA is based on the thermodynamic formulation of substance's crystal annealing. The important step is to define a prior model. The model is defined by choosing the interval of the minimum value and the maximum value of the model parameter (earthquake location). The model parameter will be determined randomly in that its value following equation:

$$Model_i = Model_i^{min} + R\,(\,Model_i^{max} - Model_i^{max}\,) \tag{3}$$

Where R is a uniform probability.

The annealing process in thermodynamic is adopted in the inversion problem solving. It uses the model parameter to define system configuration and uses the objective function as energy. Temperature factor is controller factor that has the same unit as the objective function. The model perturbation probability is following equation:

$$P(\Delta E) = \exp(-\frac{\Delta E}{T}) \tag{4}$$

Where ΔE is the objective function change or the misfit change caused by the model perturbation.

2.2.2. Grid Search Inversion

The global approximation is an alternative to obtain a solution of non-linear inversion. This method evaluates value of the objective function for each model in space model. This systematic evaluation is the easiest way to obtain the solution of the non-linear solution. Basic calculation of the objective function is a forward modeling.

2.2.3. Geiger's Method Inversion

Geiger's method is one of single event determination (SED). It is an iterative processing using least square optimization to determine hypocenter. Its principle is solving the objective function that gives smaller residual square cumulative of all stations [8].

2.2.4. Joint Hypocenter Determination

This method adapts Geiger's Method to relocate hypocenter. The difference is hypocenter determination calculated using all events in one matrix.

2.3. Moment Magnitude Calculation

In this calculation, we adapt Brune's method. The classic Brune model of a shear source is used to fit the far-field displacement spectra, computed by integrating the series in time or frequency domain the velocity particle motion. That method is used to optimized the model, to obtain low-frequency level, and corner frequency that will be used to calculate seismic moment [13]. Seismic moment can be estimated from low-frequency plateau. Seismic moment is converted to moment magnitude following the equation below [14]:

$$M_w = \frac{2}{3}Log(M_0) - 6.07 \quad (5)$$

2.4. Delay Time Tomography

Delay time tomography is the method that was developed in this research. The delay time is the difference of travel time observation and travel time calculation. Travel time observation is obtained in field observation. Another one is calculated using velocity model which is known [15]. For resolving tomography inversion, least square method was implemented iteratively to minimize the differences between observed and calculated travel times.

3. Results and Discussions

3.1. Graphical User Interface

Ganesha Micro-seismic Technology (GMSTech) is a software for micro-seismic processing to determine hypocenter location, magnitude, 1D and 3D seismic velocity structure in geothermal, oil , and gas exploration field. There are four modules in this software such as *hypocenter determination, moment magnitude calculation,* the *1D velocity model,* and *tomography*. Each module has specific parameter and file required to be loaded. Graphical user interface of GMSTech is shown in Figure 1.

Figure 1. Graphical user interface of GMSTech for micro-seismic processing

3.2. Hypocenter Determination

In general, this module requires *arrival time data, location of station data, and 1D velocity.* Furthermore, this module requires ray tracing method to obtain ray path and calculated travel time. In the calculation, ray tracing of straight method or shooting method can be used. Straight method is used for contras medium which is not extreme or homogenous medium. For high contras medium, shooting method is available by calculating incident angle and transmission angle at each layer. Like the straight method, shooting method can be used in homogenous medium but requires more time due to angle calculation. Figure 2 shows the histogram of Simulated Annealing Inversion result applied for synthetic and real data. The number of event of synthetic data and real data respectively are ten events and 482 events, respectively. In this case, inversion results using ray tracing Shooting Method is better than inversion results using ray tracing Straight Method (Figure 2). These ray tracing will be applied for all method of hypocenter determination.

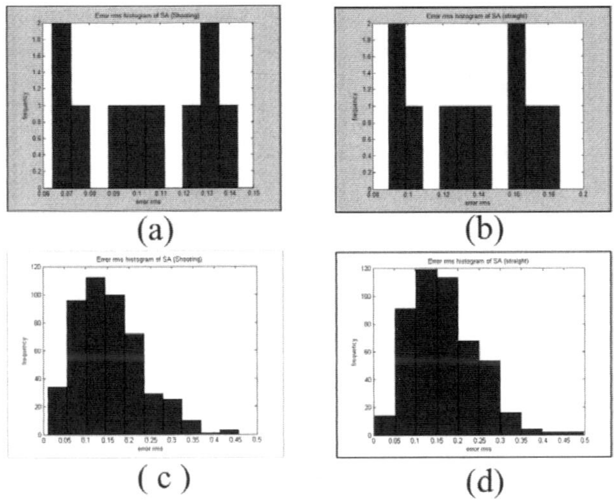

Figure 2. Histogram of Simulated Annealing Inversion results using (a) Shooting Method and (b) Straight Method applied for synthetic data; using (c) Shooting Method; and (d) Straight Method applied for real data

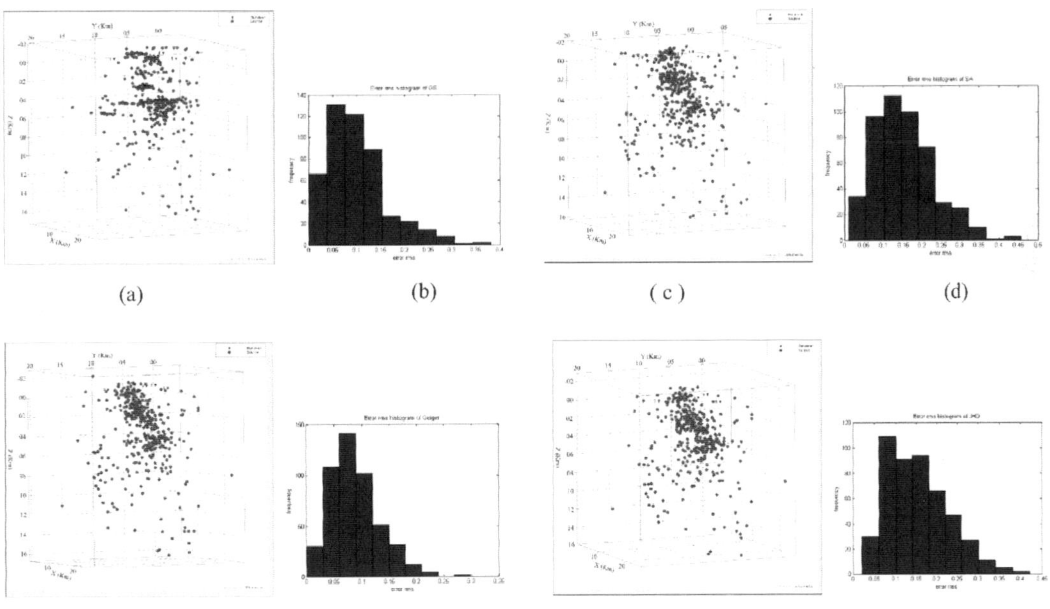

Figure 3. Hypocenter determination results using (a) Grid Search Inversion, (b) Simulated Annealing, (c) Inversion Geiger's Method Inversion, and (d) Joint Hypocenter Determination; and histogram of RMS error (e) Grid Search Inversion, (f) Simulated Annealing Inversion, (g) Geiger's Method inversion, and (h) JHD

Grid Search Inversion calculates the hypocenter by trying at every grid [16]. The calculation is processed for each event and each station. The optimation of this method uses error value and maximum iteration as iteration boundary. The optimum result will be defined as the final result. This method has a limitation that the result may be trapped in minimum local. The calculation that process in every grid makes this method requires much time. Figure 3 (a) and (e) respectively show the result of hypocenter determination using Grid Search Inversion.

Simulated Annealing Inversion can solve Grid Search Inversion limitation by finding minimum global. The optimation using temperature function reduces time consumption. Figure 3 (b) shows the result of hypocenter determination using Simulated Annealing Inversion and (f) its histogram of error rms.

Geiger's Method Inversion is single event determination like Grid Search Inversion and Simulated Annealing Inversion, but it doesn't process for each station. An event recorded in several stations is calculated in a matrix. The inversion uses Least Square Inversion guided by error value and maximum iteration. Actually, this method requires initial hypocenter. In this software, there are two choices to use initial hypocenter or not. The choice without initial hypocenter uses the station that has the fastest velocity as initial hypocenter. The result using initial hypocenter is better than the result without initial hypocenter.

Joint Hypocenter Determination adapts Geiger's Method Inversion. The difference is the calculation for all events and all stations. It requires *initial hypocenter* like Geiger's Method Inversion.

3.3. Moment Magnitude Calculation

The required data for moment magnitude calculation are *the location of the station, arrival time data, hypocenter data, velocity model* and *waveform list*. We use the same format for *the location of the station, arrival time data,* and *hypocenter data* as we used in hypocenter determination process. *The waveform list* is list of events, the station recording related events, and the waveform file for each direction.

Waveform loaded is processed by converting from velocity domain to displacement domain, calculating signal to noise ratio, spectral fitting, calculating moment seismic and then converting it to moment magnitude. In spectral fitting, we optimize to fit the spectral and amplitude and to determine corner frequency and low-frequency level. Figure 4 (b) shows spectral fitting for P wave and S wave, event one recording by station bor3. Figure 4 (a) shows the magnitude of all events.

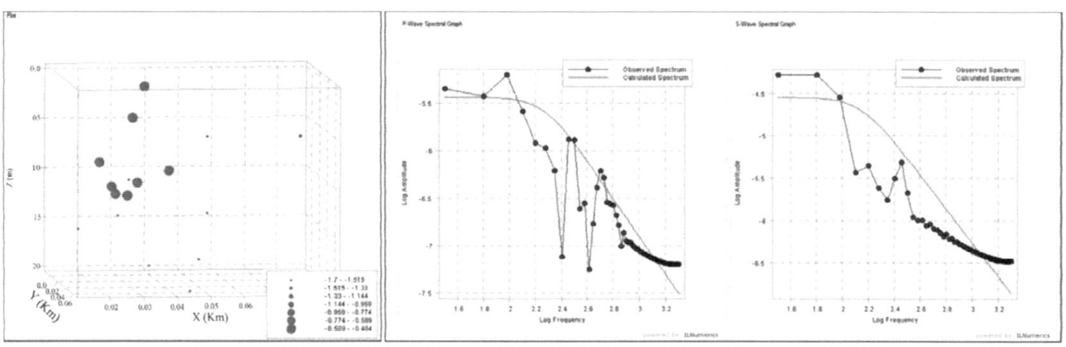

(a) (b)

Figure 4. (a) The result of magnitude calculation and (b) spectral fitting between observed and modeled spectrum for P wave and S wave

3.4. Tomography

The required data in this module are the event catalog, the location of the station and 1-D velocity model. Furthermore, the definition of space model and the grid size are important. The space model accommodates event location and the location of the station. The other important parameters are the number of bending points, maximum perturbation number, normal and gradient damping value, the maximum number of iteration and RMS error. The results of this processing are checkerboard and velocity perturbation. The checkerboard images the resolution (shown in Figure 5). We can draw a line to establish a section. Figure 5.(a) and 5.(b) images the different line and its section.

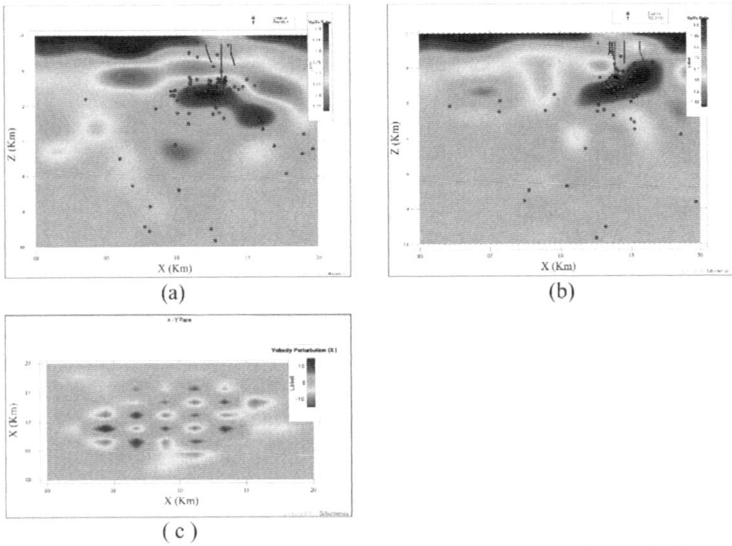

Figure 5. Tomographic inversion display. (a) & (b) Vertical section of Vp structure and (c) checkerboard resolution test displays.

4. Conclusions

We present the data storage format in ASCI for all modules. Hypocenter Determination module is available to use four methods, Simulated Annealing, Grid Search, Geiger's method, and JHD Inversion. The results are plotted in the 3D display that can be rotated in any direction. Magnitude calculation using Brune's method shows spectral fitting and plotting data. Delay time tomography produces 3D seismic velocity structure and the model resolution is evaluated by checkerboard resolution test.

5. Reference

[1] Ry R V and Nugraha A D 2016 Reservoir characterization around Geothermal Field, West Java, Indonesia derived from 4-D seismic tomography *IOP Conf. Series: Earth and Environmental Science* **29** (1) doi: 10.1088/1755-1315/29/1/012001.

[2] Nugraha A D, Syahputra A, Fatkhan and Sule R 2013 Seismic velocity and attenuation structures in the geothermal field *AIP. Conf. Proc.* **1554** 238-241 doi: 10.1063/1.4820329.

[3] Hasanah M U, Nugraha A D and Sule R 2013 Attenuation tomography using microearthquake (MEQ) data in the a geothermal field AIP. Conf. Proc. **1554** 273-276 doi: 10.1063/1.4820338.

[4] Nugraha A D, Widiyantoro S, Gunawan A, Suantika G 2013 *Journal of Mathematical and Fundamental Sciences* **45** (1) 17-28 doi: 10.5614/j.math.fund.sci.2013.45.1.2.

[5] Ry R V, Priyono A, Nugraha A D, and Basuki A 2015 Seismicity study of volcano-tectonic in and around Tangkuban Parahu active volcano in West Java region, Indonesia *AIP Conf. Proc.* **1730** 020004

[6] Firmansyah R, Nugraha A D and Kristianto 2015 The preliminary results: Internal seismic velocity structure imaging beneath Mount Lokon *AIP. Conf. Proc.* **1658** 050012 DOI: 10.1063/1.4915051.

[7] Kissling E, Ellsworth W L and Cockerham R S 1984 Three-dimensional structure of the long valley caldera, California region by geotomography. In Active Tectonic and Magmatic Processes beneath Long Valley Caldera, Eastern California, *Open-le Report 84-939. U. S. Geol. Survey*, Menlo Park.

[8] Nishi K 2005 Hypocenter calculation software GAD (Geiger's method with Adaptive Damping) *Manual. JICA*

[9] Thurber C H 1983 *J. Geophys. Res.* **88** 8226-8236.

[10] Um J and Thurber C H 1987 *Bull. Seism. Soc. Am.* **77** 972–986.

[11] White S R 1984 Concepts of scale in simulated annealing *Proc. IEEE Int. Conference on Computer Design, Port Chester* pp.646–651.

[12] Weber Z 2000 *Elsevier Physics of Earth and Planetary Interiors* **199** pp.149-159.

[13] Brune J N 1970 *J. Geophys. Res.* **75** 4997–5009.

[14] Kanamori H and Hanks T C 1979 *J. Geophys. Res.* **84** (B5)

[15] Widiyantoro S, Gorbatoc A, Kennett B L N and Fukao Y 2000 *Geophys. J. Int.* 141 747-758

[16] Ry R V and Nugraha A D 2015 Improve earthquake hypocenter using adaptive simulated annealing inversion in regional tectonic, volcano-tectonic, and geothermal observation *AIP. Proc. Conf.* **1658** 030004 DOI: 10.1063/1.4915012.

IOP Conf. Series: Earth and Environmental Science **62** (2017) 012025 doi:10.1088/1755-1315/62/1/012025

Improved Location of Microseismic Events in Borehole Monitoring by Inclusion of Particle Motion Analysis: a Case Study at a CBM Field in Indonesia

Rexha Verdhora Ry[1], **T. Septyana**[1,3], **S. Widiyantoro**[2], **A. D. Nugraha**[2], **A. Ardjuna**[4]

[1]Graduate Program of Geophysical Engineering, Faculty of Mining and Petroleum Engineering, Institut Teknologi Bandung, Ganesha 10, Bandung 40132, Indonesia
[2]Global Geophysics Research Group, Faculty of Mining and Petroleum Engineering, Institut Teknologi Bandung, Ganesha 10, Bandung 40132, Indonesia
[3]BP Indonesia
[4]SKK Migas Indonesia

E-mail: rexha.vry@gmail.com

Abstract. Microseismic monitoring and constraining its hypocenters in and around hydrocarbon reservoirs provides insight into induced deformation related to hydraulic fracturing. In this study, we used data from a single vertical array of sensors in a borehole, providing measures of arrival times and polarizations. Microseismic events are located using 1-D velocity models and arrival times of P- and S-waves. However, in the case of all the sensors being deployed in a near-vertical borehole, there is a high ambiguity in the source location. Herein, we applied a procedure using azimuth of P-wave particle motion to constrain and improve the source location. We used a dataset acquired during 1-day of fracture stimulation at a CBM field in Indonesia. We applied five steps of location procedure to investigate microseismic events induced by these hydraulic fracturing activities. First, arrival times for 1584 candidate events were manually picked. Then we refined the arrival times using energy ratio method to obtain high consistency picking. Using these arrival times, we estimated back-azimuth using P-wave polarization analysis. We also added the combination of polarities analysis to remove 180° ambiguity. In the end, we determined hypocenter locations using grid-search method that guided in the back-azimuth trace area to minimize the misfit function of arrival times. We have successfully removed the ambiguity and produced a good solution for hypocenter locations as indicated statistically by small RMS. Most of the events clusters highlight coherent structures around the treatment well site and revealed faults. The same procedure can be applied to various other cases such as microseismic monitoring in the field of geothermal and shale gas/oil exploration, also CCS (Carbon Capture and Storage) development.

1. Introduction

Microseismic monitoring involves the passive seismic recording of acoustic emissions from certain event source known as microearthquake. Microearthquake is defined as a small magnitude earthquake, and it often is referred to as microseismic events. Microseismic events are associated with naturally occurring of artificially induced fracture movements, resulting in inelastic geomechanical fracture deformation. Due to it, in the hydraulic fracturing treatment, microseismic monitoring is the powerful technology to image stimulated hydraulic fractures [1].

Typically, the passive recordings are used to detect microseismic events and to constrain its hypocenter location and attribute. Accurate event locations are important to study the induced fractures. Most methods used to determine the location assume a velocity model and minimize the difference between the observed and predicted arrival times of P- and S-wave phases at some receivers, e.g. [2,3,4,5]. Several factors control this location accuracy; one of them is source-receiver geometry [6]. Geophone array deployed in a single near-vertical borehole has often been used for microseismic monitoring of hydrocarbon reservoir and hydraulic fracturing. This geometry may lead to ambiguous solution due to the ambiguity of misfit function in the horizontal direction. The use of a 1D velocity model and a near-vertical array of sensors also leads to an 180° ambiguity where the event can be either side of the monitoring well. In contrast, these ambiguities do not occur in monitoring as the stations are located on the surface.

In this paper, we apply a location procedure using the inclusion of P-wave particle motion to constrain the initial source location. The information of P-wave polarization is required to constrain the direction of the source location from receiver [7,8]. We also added the combination of polarities analysis to remove 180° ambiguity. Then, we determined the hypocenters location using guided-grid-search method. This procedure was applied to a dataset acquired during fracture stimulation of particular coalbed methane (CBM) field in Indonesia.

2. Event Location Workflow

The data processing for locating microseismic events consists of 4 steps: 1) manually picking of the P- and S-wave arrival times, 2) hodogram analysis of P-wave polarization to estimate back-azimuth, 3) removal of 180° ambiguity using the combination of polarity analysis, 4) inversion process using guided grid search method.

2.1. Arrival time picking

One of the major contributors to errors in event location is an error in arrival time phase picking [6]. An automated procedure can be applied in this picking by comparing a short-term average and a long-term average of the signal [9]. This automated procedure can help to overcome the problems of consistency and time-consuming. However, a disadvantage of the automated procedures is that they lack consistency in where the pick is placed with respect to the waveform [7].

In this paper, we picked the phase of P- and S-wave manually, although it is very time-consuming. We insist that manually picking is the better way to reduce error in event location, as long we perform the refinement of arrival time picking. We refined the arrival times using modified energy ratio (MER) method to reduce a human error while picking and obtain high consistency picking. The MER algorithm is an extension of STA/LTA in which the pre- and post-sample windows are of equal size. The energy ratio (ER) at the i-th time sample is given by [10],

$$MER(i) = \left(\frac{\sum_{j=i}^{i+w} x_j^2}{\sum_{j=i-w}^{i} x_j^2}\,|x_i|\right)^3 \quad (1)$$

where w is the window length and x is the input series. Because the energy ratio function is computed using post- and pre-sample windows, the time index associated with the maximum of the MER represents the arrival time pick.

2.2. P-wave polarization

We need to estimate P-wave polarization to constrain the direction of source location from a receiver. Since the P-waves are vertically and radially polarized, the vector of P-wave amplitude can be used to calculate the back-azimuth to the source. We performed a hodogram analysis of P-wave polarization by cross plotting the signal amplitude on the horizontal components, e.g. [8,11]. This hodogram depicts a graphical movement of the ground motion (eg., fig. 1b).

First, we cut the very first signal of P-wave after it arrived to obtain a sharp P-wave linear polarization (shown in fig. 1a). We applied principal component analysis on the relative signal amplitude to calculate a principal direction of horizontal components: A_N for the North-South component and A_E for the East-West component. We then calculate the signal polarization azimuth $\emptyset$ [12] by

$$\emptyset = tan^{-1}\frac{A_E}{A_N}. \quad (2)$$

2.3. *Combination of Polarity Analysis*

There is an ambiguity of 180° where the event can be either side of the monitoring well. It occurs due to the limitation of the solution from arctangent. Therefore, the solution will be limited by

$$-90 \leq \emptyset \leq 90. \quad (3)$$

Therefore, we should add 180° to the azimuth we obtained based on the combination of polarities. The first polarity can be up (positive) or down (negative) so it can be used to get the correct back-azimuth. If the first motion on the vertical component of the P is upward, then the radial component of P is directed away from the source. The opposite is true if the P-polarity is negative [12]. This combination is shown in Table 1.

As an example in fig. 1, we could determine the first polarity for every component. The first polarity for vertical, N-S, and E-W respectively is positive, positive, and negative. The back-azimuth calculated from P-wave polarization is -50° or 310°. However, based on the combination of polarities, we must add 180° to our solution. So, the back-azimuth of source location from the receiver is 130°.

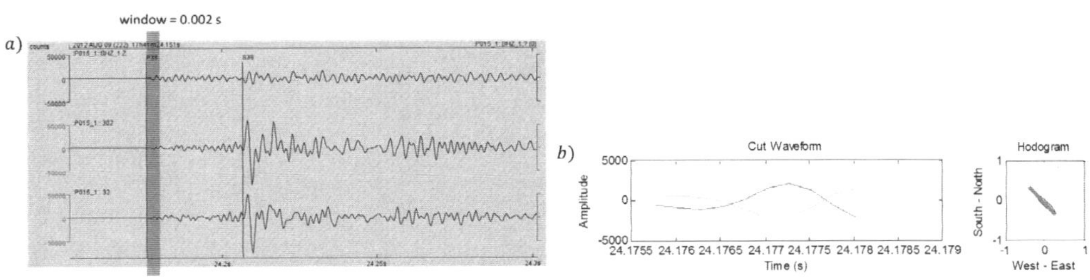

Figure 1. An example of P-wave polarization analysis: a) the recorded signal waveform and very first part of P-wave windowing; b) the cut waveform and the hodogram for horizontal component (the black line shows the cross plotting and the magenta line shows principal direction).

Table 1. Combination of polarities and cases where 180° has to be added to the calculation

Z	+	−	+	−	+	−	+	−
N	+	+	−	−	−	−	+	+
E	+	+	+	+	−	−	−	−
Add	180	0	0	180	0	180	180	0

2.4. *Event location*

The last step of our procedure is the inversion process to obtain the location of the microseismic event. We used guided grid search to produce accurate hypocenter location. Grid search method work to minimize an objective function by searching trough all points systematically in the model space we set [13]. The objective function used is the misfit function, the difference between the observed and calculated arrival times of P- and S-wave phases at a number of receivers.

Since the origin time of the microseismic event is an unknown parameter, we used another way of incorporating differential arrival times and their associated calculated travel times into the hypocenter determination problem. We used master station method [13,14] which is minimizing all combinations of arrival times and sensors. The objective function for the minimization of all combinations of arrival times and sensors is given by

$$E(i) = \sum_{j=1}^{N-1}\sum_{k=1+1}^{N}\left[\left(t_j^{obs} - t_k^{obs}\right) - \left(T_{ij}^{cal} - T_{ik}^{cal}\right)\right]^2 \quad (4)$$

which is the generalization of minimizing the misfit. In equation (4), t_j^{obs} is the observed arrival times at receiver *j*; t_k^{obs} is the observed arrival times at receiver *k*; T_{ij}^{cal} is the calculated travel time from source *i* to receiver *j*; and T_{ik}^{cal} is the calculated travel time from source *i* to receiver *k*.

From step 2 and 3, we already had the back-azimuth of source location from a receiver. Using this back-azimuth, we could constrain the model space in our grid search inversion process. We can also set a tolerate angle towards the back-azimuth. We set the model space to search systematically in that direction (shown in fig 3). So, we can reduce the ambiguity of misfit distribution using this technique.

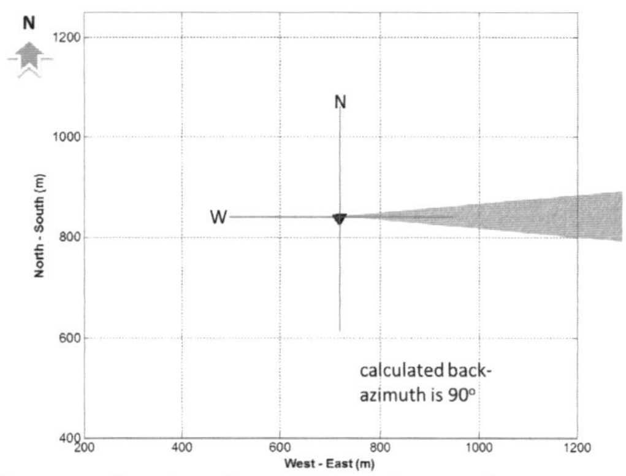

Figure 2. The schematic diagram showing the grid search area from a receiver (gray zone), which is constrained by back-azimuth and 5° tolerate angle.

3. Application to The CBM Field

Hydraulic fracture treatment was conducted in particular coalbed methane (CBM) field in Indonesia. The production stage in this field is still on development stage. Therefore, the purpose of this treatment is to develop and to increase the permeability of the reservoir. The fracture stimulations were divided into 11-days of injection stages, located in the treatment well (Fig. 3a). In this paper, the procedure was applied to a dataset acquired during 1-day monitoring which had the most seismicity. The 1D velocity model was derived from sonic logs recorded in the treatment well. The logs were smoothed to interface blocked model (Fig. 3). Every single block represents a lithological group in this area study. The thin layered block with low velocity represents coal-bed where the methane gas lied.

The microseismicity was recorded using eight downhole seismometers with sampling rates of 4000Hz or 0.25 milliseconds (Fig. 3a-b). The earlier step is the event identification. We identified the event based on STA/LTA analysis [9]. During the 1-day monitoring, we had identified up to 1700 microseismic events. This number of events is considered as a high seismicity if compared with the other days. Then, the first step of the procedure is the manual picking of P- and S-wave arrival times. However, not all identified events had a good signal to noise ratio. We only selected the events with the high confidence of arrival times picking. So, the number of microseismic events is reduced to 1584 events.

The next step is the estimation of back-azimuth using P-wave polarization. Using the hodogram analysis, we calculated the polarization direction in every station for certain event. Fig 4 shows an example of the polarization that recorded in every station. The refinement of arrival times picking is helpful to obtain the good consistency of polarization direction in every station. The back-azimuth for this certain event is defined as the average value of these polarization directions that have been corrected using the combination of polarities.

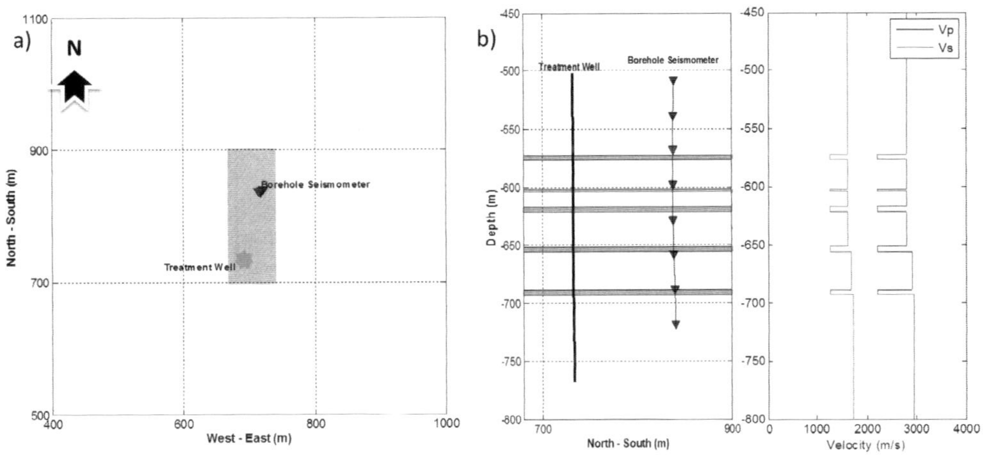

Figure 3. (a) Map view of borehole geometry at the study area. (b) The geometry of downhole seismometer and the one-dimensional velocity model for P- and S-wave which used in this study.

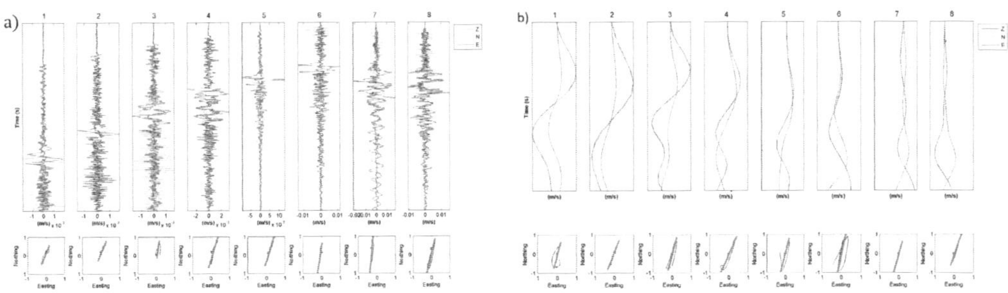

Figure 4. An example of recording waveform and hodogram analysis for every station, in a display of (a) the long waveform and (b) the cut waveform according to the window.

4. Result and Discussion

The totals of 1584 events were successfully located using guided grid search (Fig. 5). The microseismic events occurred at a depth between 570 to 580 meters. It is well correlated with the hydraulic fracturing location. The injection during that day located in the first layer of coal-bed (depth of 572 meters), so the distribution in depth comparison is reasonable.

Since the sensor geometry used is single vertical borehole array, the possibility of miss interpretation exists in a horizontal plane. From the microseismic event distribution in a horizontal plane, we can divide them into 2 clusters: (1) South cluster, where the events distributed in SE-NW direction started from treatment well; (3) East cluster, where the events distributed in SE-NW direction like the first cluster, but they have a gap from treatment well.

The South cluster is the most reasonable cluster in this result. The previous study inferred that there is a fault with a Southeast-Northwest direction around the treatment well. In this cluster, fractures were generated around treatment well. As more fluid was injected, the fractures continue to widen and grow outward from the well. Most of the fractures generated during hydraulic fracturing could have followed the same preexisting fractures, opening them, causing a slip on them, and getting filled with some proppant, e.g. [16].

The East cluster, nevertheless, is the most interesting distribution. The microseismic events occurred in the Southeast-Northwest direction like the South cluster but located not along that. There is a gap in the middle that separated this cluster. The possibility is that triggered seismic events occurred in this area. The triggered seismicity can occur if a fault is close to failure and requires comparatively less pressure change to cause an earthquake [1]. Other study inferred that there is an ancient fault in this area. However, due to the lack of the data, it is not confirmed yet. Furthermore, we need to confirm them by their source characterization like magnitude and focal mechanism.

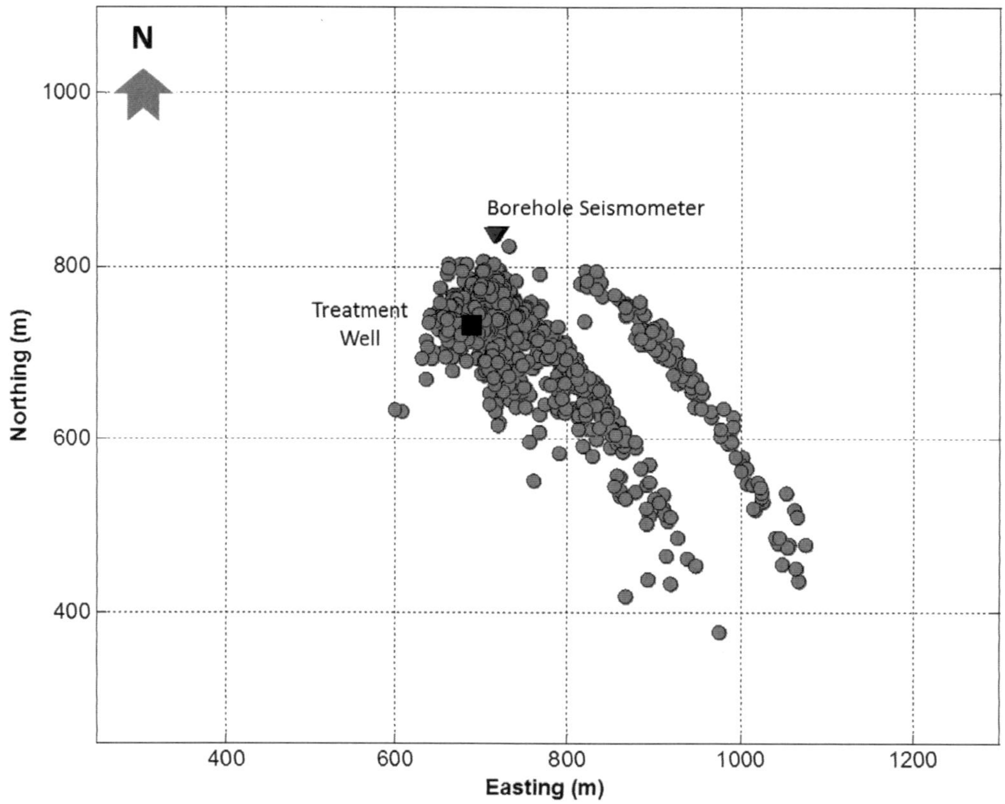

Figure 5. Map view of microseismic event locations (red dot) at certain coalbed methane (CBM) Field in Indonesia, stimulated by hydraulic fracturing treatment during 1-day monitoring. The black square is the treatment well, and the blue reverse triangle is monitoring well.

5. Conclusions

We have applied a procedure to improve the microseismic event location and remove 180° ambiguity inflicted by using a single vertical borehole array. We found that the event distribution in East cluster is unusual but yet still questionable. The next step of our research is to confirm this cluster by analyzing its source characterization like magnitude and focal mechanism.

The distribution of microseismic events is very reliable in both depth and horizontal plane for the South cluster. This events cluster highlights coherent structures around treatment wells and inferred faults. However, the reliability of this procedure will depend on the quality of arrival times picking. The same procedure can be applied to various other cases such as microseismic monitoring in the field of CCS (Carbon Capture and Storage), geothermal, and shale gas/oil exploration development.

Acknowledgement

The authors greatly acknowledge SKK Migas Indonesia for providing data in this study; Laboratory of Near Surface Geophysics, Faculty of Mining and Petroleum Engineering, Institut Teknologi Bandung (FTTM ITB) and Indonesia Endowment Fund for Education (LPDP) for supporting this study.

References

[1] Maxwell S 2014 Microseismic imaging of hydraulic fracturing: Improved engineering of unconventional shale reservoirs *Publication of SEG.*
[2] Foulger G R 1982 *Geothermics* **11** 259-268
[3] Nishi K 2005 Hypocenter Calculation Software GAD (Geiger's Method with Adaptive Damping) ver 1. *JICA report*
[4] Ry R V and Nugraha A D 2015 Improve earthquake hypocenter using adaptive simulated annealing inversion in regional tectonic, volcano tectonic, and geothermal observation *AIP Conference Proceedings* **1658** 030004
[5] Ry R V, Priyono A, Nugraha A D, and Basuki A 2015 Seismicity study of volcano-tectonic in and around Tangkuban Parahu active volcano in West Java region, Indonesia *AIP Conference Proceedings* **1730** 020004
[6] Pavlis G L 1986 *Bull. of the Seismological Society of America* **76**(6) 1699–1717
[7] Jones G A, Raymer D G, Chambers K, and Kendall J M 2010 *Geophysical Prospecting* **58** 727–737
[8] Moriya H 2008 *Geophysical Prospecting* **56** 667–676
[9] Earle P S and Shearer P M 1994 *Bull. of the Seismological Society of America* **84**, 366–376.
[10] Akram J and Eaton D W 2016 *Geophysics* **81** 67-87
[11] Alessandrini B, Cattaneo M, Demartin M, Gasperini M, and Lanza V 1994 *Annali di Geofisica* **37**
[12] Havskov J and Ottemöller L 2010 Routine data processing in earthquake seismology *Springer*
[13] Jones G A, Kendall J M, Bastow I D, and Raymer D, 2013 *Geophysical Prospecting* **62** 34–49
[14] Zhou H W 1994 *Journal of Geophysical Research* **99**(B8) 15439–15455
[15] Moriya H 2008 *Geophysical Prospecting* **56** 667–676
[16] Rodriguez-Pradilla G 2015 *Leading Edge* **34**(8) 896-902

An Integrated Method 3D Velocity Model and Fuzzy Clustering for Fracture Characterization

Erlangga Ibrahim Fattah[1], Andri Dian Nugraha[2], Rachmat Sule[3]

[1]Graduate Program of Geophysical Engineering, Faculty of Mining and Petroleum Engineering, Institut Teknologi Bandung, Ganesha 10, Bandung 40132, Indonesia
[2]Global Geophysics Research Group, Faculty of Mining and Petroleum Engineering, Institut Teknologi Bandung, Ganesha 10, Bandung 40132, Indonesia
[3]Applied Geophysical Research Group, Faculty of Mining and Petroleum Engineering, Institut Teknologi Bandung, Ganesha 10, Bandung 40132, Indonesia

E-mail: erlanggaif@gmail.com

Abstract. Enhanced Geothermal system assessment conducted to have a better understanding of characteristics of fracture caused by a fluid injection in the geothermal reservoir. Fluid injection may cause microseismic occur. These events allow us to map and characterize fractures in the geothermal system. Fractures play important role at the geothermal system due to its ability to increase permeability for fluid movement within the reservoir. In this study, we combined tomographic inversion method and fuzzy clustering to identify fracture characteristics at the EIF Geothermal Field. Tomography helped in delineating fluid-filled fractures with high permeability area which shown by lower velocity Vs anomaly than Vp and higher Vp/Vs ratio. Fuzzy clustering allowed us to map microseismic movement and estimate suitable locations for the future well injections or productions.

1. Introduction

Microseismic occurs due to several parameters, such as pore-pressure increases, temperature and/or volumetric changes due to fluid withdrawal or injection, and chemical alteration of fracture surfaces. Fracture characterization in the geothermal system is conducted by monitoring microseismic events. These events correlate with the migratory of fracture network and indicate connectivity within fracture [1,2]. In this study, we implement the combination of tomographic inversion and fuzzy clustering methods to have a better understanding of fracture characteristics at the EIF Geothermal Field. SimulPS12 code [3] utilizes travel time tomography and classifies microseismic events regarding Vp, Vs and Vp/Vs characteristics. The previous study of tomographic inversion around geothermal field in Indonesia region showed a prominent anomaly in seismic and attenuation structure around geothermal field [4, 5, 6].

We use microseismic data from a period of October 2012 until Mei 2013 which are recorded at least by four station observations. Catalog data contains of 1,116 micro-seismic events and 11,433 P- and 11,433 S- wave phases and is monitored by 35 stations observation. First, we relocate hypocenter catalog data and update velocity model simultaneously by using coupled hypocenter – velocity method from VELEST program code [7]. The relocated hypocenter and updated 1D velocity model are applied as input for tomographic inversion by applying SimulPS12 program code [3] resulting three-dimensional seismic velocity model and a precise hypocenter from the codes. These results can determine fracture characterization and also predict potential high permeability zone.

Second, we clustered hypocenter and three-dimensional velocity structures of Vp, Vs and Vp/Vs using fuzzy clustering method. The movement of cluster center over time can represent the fracture propagation. We used a time window of three months since October 2012 – May 2013.

2. Methodology

2.1. Coupled hypocenter – velocity method
Minimum 1D velocity model and hypocenter location utilized as the input parameters in tomographic inversion SimulPS12. These two factors were obtained by coupled hypocenter – velocity method from VELEST program code. The relocated hypocenter, updated 1D velocity model, and stations correction were calculated simultaneously until minimum RMS misfit of the solution reached. The minimum RMS misfit, which lied between 0.02 – 0.2 s, showed the most reliable solution.

2.2. Inversion of Tomographic
Three-dimensional velocity structure obtained from SimulPS12 program code. This program calculated hypocenter, updated three-dimensional velocity model (Vp and Vp/Vs ratio) and station correction simultaneously. Initial velocity model obtained from VELEST inversion program. Damping parameters were determined from trade – off curve between variance of both data and model. The most applicable damping were chosen by the minimum value of those variances. We used damping parameter of 10 for both Vp and Vp/Vs. Grid velocity model with a dimension of 2 x 2 x 1 Km^3 applied to get three-dimensional velocity structures of Vp and Vp/Vs ratio with good resolutions.

2.3. Fuzzy clustering
Fuzzy clustering is one of the statistical analysis technique. Data was classified into several groups consist of similar characteristic using fuzzy logic. Fuzzy logic ‘membership’ degree pointed out the similarity of data for each clusters. The degrees of similarity lied between 0 to 1 which indicated the lowest and the highest similarity of data within each clusters. In this process, microseismic events were clustered as a function of time window from October 2012 – May 2013. To obtain optimum partition data, we validated the number of clusters using validity index of Xie and Beni [8]. The optimum partition data determined by using two clusters.

3. Results
Figure 1 depicts tomogram section of Vp, Vs and Vp/Vs ratio at elevation of 0 to 1 Km BSL. The area within black dashed line had good resolution. The observation at the elevation of 1 Km showed lower Vp and Vs anomaly with higher Vp/Vs ratio.These features were interpreted as fluid-filled fracture zones. Tomogram images at elevation of 0 to 1 Km could not show significant anomaly at red dashed line area because hypocenter at this area concentrated at the depth between -6 to -10 Km. We presumed that microseismic events at this area might not correlate with geothermal activities. Hence, we clustered microseismic event only at black dashed line area.

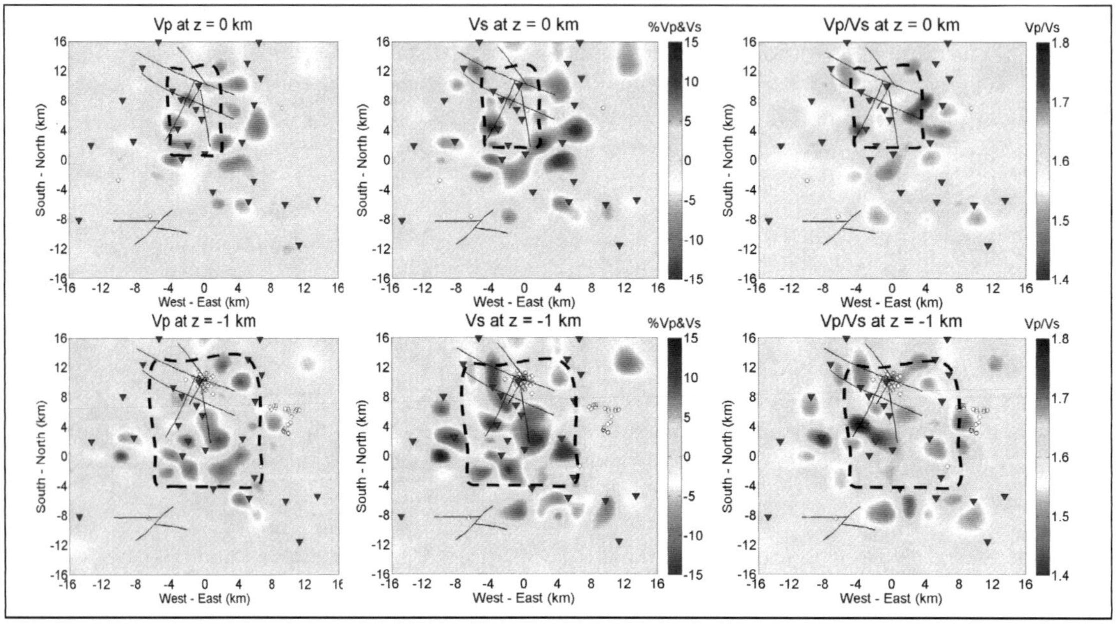

Figure 1. Depth section of tomogram Vp, Vs, and Vp/Vs ratio at elevation of 0 to 1 Km BSL. Black dash line indicated area with good resolution, fault(black line), station(blue triangle) and mountain(black triangle). Red color indicated low anomaly meanwhile blue indicated high anomaly.

To estimate microseismic movement, we clustered microseismic event over time with the time window of three months from Oct 2012 until May 2013. The cluster center represented a connected fracture network. The result of fuzzy clustering which depicted in Figure 2 were located at the center of microseismic distribution and presumed as the center of connected fracture network. Then, the microseismic movement represented by the cluster center over time. The cluster center showed movement with orientation of NW from Oct 2012 to May 2013.

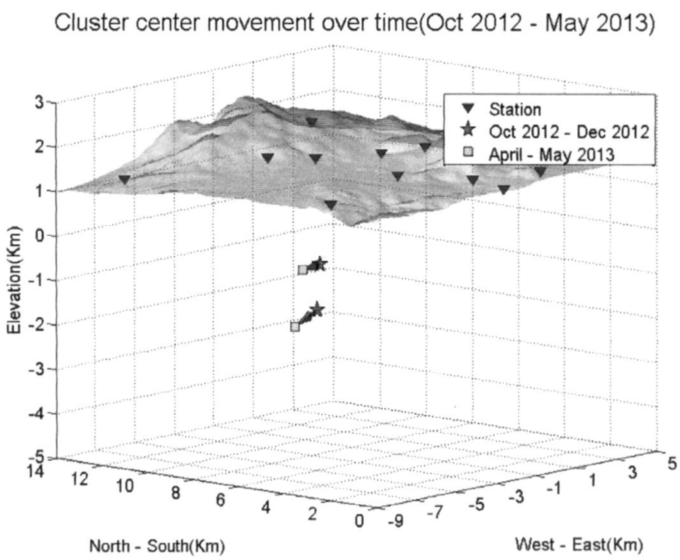

Figure 2. Cluster center movement with time, star symbol is cluster center at microseismic event in October – December 2012, square symbol is cluster center at microseismic event in April – May 2013.

4. Conclusions

The combination method of fuzzy clustering and tomographic inversion could map microseismic movement and characterize fracture at EIF geothermal field. The three-dimensional of velocity structure delineates area with high permeability shown by low Vp and Vs anomalies with high Vp/Vs ratio. This anomaly were interpreted as fluid-filled fractures. The movement of microseismic event fuzzy cluster centers movement as a function of time. Cluster center position located at the center of microseismic distribution may correlate as center of fracture network. The cluster center showed movement with the orientation of NW from October 2012 to May 2013. After overlaying fuzzy cluster center and three dimensional velocity, the cluster center located around high permeability zone and this location considered as the best location for the next injection or production wells for enhancing the geothermal system.

5. Acknowledgements

The author would like to acknowledge Star Energy Geothermal Limited for micro-seismic waveform data used in this research, GDP Lab, Faculty of Mining and Petroleum Engineering ITB for supporting this study and LPDP Scholarship.

References

[1] Aminzadeh, Fred, Tayeb A Tafti and Debotyam Maity 2013 *Computer and Geosciences* **54** 39 – 49.
[2] Aminzadeh, Fred, Tayeb A Tafti and Debotyam Maity 2010 *GRC Transactions* **34** 1193 – 1198.
[3] Evans, J R Eberhart – Phillips D and Thurber C H 1994 *California; U.S. Department of the Interior, U.S. Geological Survey.*
[4] Ry R V, Nugraha A D 2016 *IOP Conf. Series: Earth and Environmental Science* **29** **(1)** 012001 DOI: 10.1088/1755-1315/29/1/012001.
[5] Nugraha A D, Syahputra A, Fatkhan, Sule R 2013 *AIP. Conf. Proc.* **1554** 238-241 DOI: 10.1063/1.4820329.
[6] Hasanah M U, Nugraha A D, Sule R 2013 *AIP. Conf. Proc.* **1554** 273-276 DOI: 10.1063/1.4820338.
[7] Kissling E, Ellsworth W L, Eberhart – Phillips D and Kradolfer U 1994 *J. Geophys. Res* **99** 19635 – 19646
[8] Xie, Xuanli Lisa and Gerardo Beni 1991 *IEEE Transactions on Pattern Analysis and Machine Intelligence.* **13**

IOP Conf. Series: Earth and Environmental Science **62** (2017) 012027 doi:10.1088/1755-1315/62/1/012027

Steam and Brine Zone Prediction around Geothermal Reservoir Derived from Delay Time Seismic Tomography and Anisotropy Case Study: "PR" Geothermal Field

Kadek Hendrawan Palgunadi[1*], A D Nugraha[2], R Sule[3] and T Meidiana[1]

[1]Undergraduate Program of Geophysical Engineering, Faculty of Mining and Petroleum Engineering, Institut Teknologi Bandung
[2]Global Geophysics Research Group, Faculty of Mining and Petroleum Engineering, Institut Teknologi Bandung
[3]Exploration and Engineering Seismology Research Group, Faculty of Mining and Petroleum Engineering, Institut Teknologi Bandung

[*]Email: hendrawan.palgunadi@gmail.com

Abstract. Development of geothermal production can be conducted in several ways, one of them analyses the fracture or crack and structure within the reservoir. Due to low permeability and porosity value within the reservoir in geothermal field. This crack or fracture provide porosity for fluid storage and permeability for fluid movement and play a major role in production from this kind of reservoir. Structure and polarization direction can be derived from anisotropy parameter and seismic velocity parameter in geothermal field. In this study, we used micro-earthquake data of 1,067 events that were recorded by the average of 15 stations during almost 1-year measurement. We used anisotropy parameter using 3-D shear-wave splitting (SWS) tomography method to represent the distribution of anisotropy medium around the geothermal field. Two parameters produced from the S-wave analysis, which is polarization direction and delay time between fast S-wave and slow S-wave. To determine SWS parameters, we used a rotation of horizontal seismogram including N-S component and E-W component. Furthermore, we used short-time fourier transform (STFT) to calculate lag time and time window based on wave periods. Two horizontal components have been rotated from azimuth 0° to 180° with an increment of 1°. Cross-correlation coefficient used every azimuth of two horizontal components based on delay time with predetermined time window obtained by STFT. When cross-correlation coefficient is high, the corresponding value of delay time and azimuth are chosen as the polarization direction and delay time of SWS. Normalized time different divided by total ray length was used to determine the distribution of crack density. Through correlation of seismic velocity model, crack density, and 3-D anisotropy tomography, we can delineate a geothermal reservoir model. Our results show, high degree of anisotropy and crack density occur in the northern and eastern part of "PR" geothermal field for further development. The result had good correlation with high production of geothermal activities.
Keyword: micro-earthquake, tomography, SWS, geothermal

1. Introduction

The propagation of shear-wave in anisotropic medium leads to the splitting of waveform and arrival time. Anisotropy of the medium, shown by shear-wave splitting (SWS) in local earthquake seismogram and had been proven toward upper crust [1]. The mechanism divided into two main categories, which are stress induced anisotropy and structural control anisotropy [2]. In the earth crust, the split of shear-wave mainly caused by pressure that parallel with vertical micro-crack [3], the density of crack and the direction of crack can be estimated from the delay time among two separated shear-wave and the polarization given by the fast wave.

Anisotropy is the variation of seismic velocity on its direction [2]. The medium is anisotropy if the value of the physical properties vector is also variated along the direction. In general, there are three major factors responsible for seismic anisotropy: 1) interlayer of lithology; 2) preferred orientation of minerals; and 3) stress-induced fractures and micro-crack that show preferred alignment [3].

The previous study regarding shear-wave splitting using anisotropy parameter (K) has been conducted by Zefeng Li et al [4] similar method had done by Haijiang Zhang et al [5]. This study is the development of a shear-wave splitting tomography method to distribute the spatial distribution of anisotropy in the upper crust using the backpropagation of delay time shear-wave along the ray-path in 1-D and 3-D velocity models at North Anatolian Fault (NAF) northern Turkey.

Through the previous study, shear-wave splitting method using the delay time of SWS may be used in the development of 3-D tomography as a function of anisotropy parameter percentage (K) and determine the crack density in the geothermal field with the assumption of delay time just on the ray-path and horizontal anisotropy is neglected.

In this study, to obtain the value of anisotropy parameter in the positive value, we used non-negative least square (NNSL) method [6]. Furthermore, this method does not including the distance of anisotropy in inversion scheme, so that the ray-path was calculated by pseudo bending principle.

2. Regional Geology

The "PR" Geothermal field is located in the southern area of West Java, approximately 40 km southern of Bandung, Indonesia (figure 1).

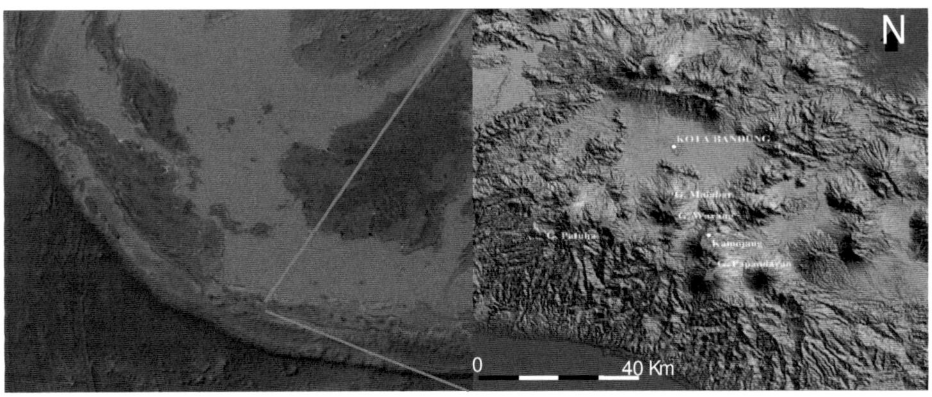

Figure 1. The location of "PR" Geothermal Field. Located 40km to the southern part of Bandung City. This field is associate with south mount cluster of west Java Mountain [13]

Also, it is located in the active tectonic area, which is the main structural pattern are on W-NW; NW; and NE direction. These structures are the result of both regional tectonic activities and local volcanic area.

3. Theory

Using the assumption of delay time that accumulated along the ray-path [2] and applying a rotation into two seismograms, then the delay time of SWS can be obtained. Characterization of anisotropy can be obtained using isotropic slowness (U) trough equation of anisotropy coefficient (K) expressed in [5]. It shows a linear relation between the anisotropy percentage (K) and the delay time SWS. Therefore, the nonlinear approximation of iteration is unnecessary. By using this scheme, the calculation does not rely on the initial model.

4. Delay Time Tomography

The data were recorded in January 2015 to 7 December 2015. The microearthquakes were recorded using 15 three-component seismometers that were distributed in the research area. The tomography was calculated using an SIMULPS12 software [3]. The advantage of 3-D velocity model is that the data of P phases and S phases were relatively close. Thus the result of velocity model and hypocentre is improved.

Determining velocity model requires the parameterization of the model. Parameterization is constructed by grid block. The amount and dimension of grid model determined by three parameters: 1) distribution of microearthquakes; 2) depth of microearthquakes: and 3) the zone with high density of ray-path. The grid dimension 1km x 1km x0.5km was chosen from elevation 2km up to 2km at depth (figure 2).

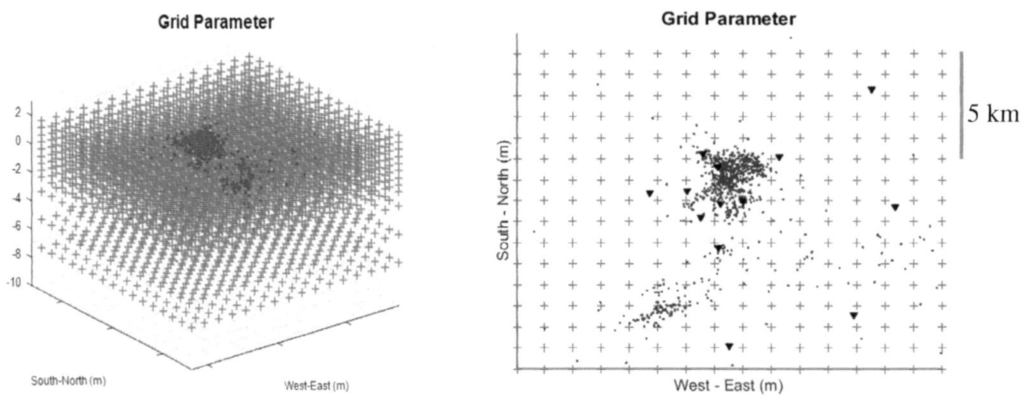

a) b)

Figure 2. Parameterization of grid data. a) Grid model shows in 3-D section; b) Grid model parameter of horizontal slice, the invert triangles are the station distribution, blue dots are the micro-earthquakes.

4.1. *Checkerboard resolution test*

Resolution checkerboard test was experienced on the data to show the area that had a good inversion result and minimum grid space that could be accommodated by the available data. The resolution test used synthetic data to verify the best inversion parameters. Figure 3 shows seismic velocity model perturbation with ±10% of initial 1-D model.

4.2. *The Result of Tomography 3-D*

Color bar tomography used in this research is red for describing low anomaly of VP, Vs, and ratio of Vp/Vs, whereas blue color indicates high anomaly. Color scale on Vp and Vs is the deviation value through the 1-D initial model. Meanwhile, color scale on Vp/Vs ratio shows the absolute values. Also, the result of seismic velocity inversion, throughout the period of January 2015 to December 7th, has been conducted with 6,776 for both P phases and S phases.

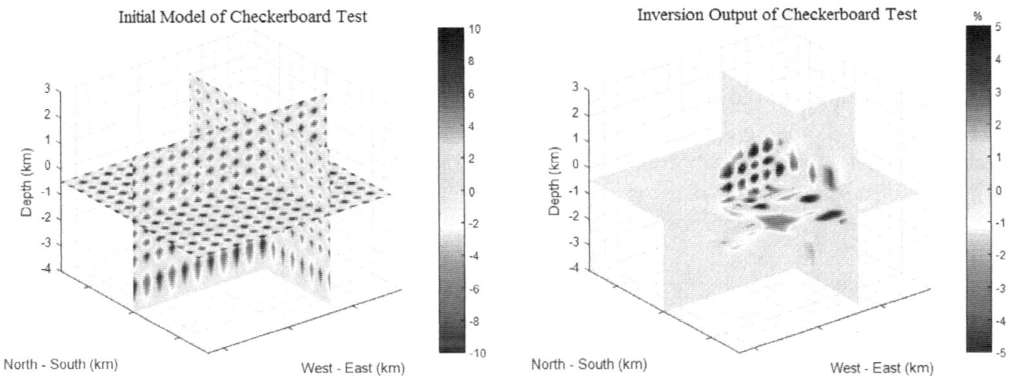

a) b)

Figure 3. Checkerboard resolution test. a) The initial model with 10% perturbation that used to create synthetic travel time. b) Inversion result of the synthetic data, the velocity model recovered about 50%. Only the area inside the checkerboard result that would interpret.

5. Anisotropy

5.1. Determining Window of Shear-wave Splitting

We used short-time fourier transform (STFT) to determine the dominant frequency in the selected time window around picking of S-wave arrival time. The dominant frequency used to calculate the length of waveform's period. Furthermore, this period is used as pale in cross-correlation between two waveforms.

5.2. Shear-wave Splitting Parameter

Determining SWS parameter, there are the polarizations of S-wave and delay time SWS that were analyzed using cross-correlation between two waveforms with rotated seismogram to radial and tangential coordinate. Figure 4 show an example of seismogram that had been rotated and correlated until it can produce a contour value of cross-correlation between radial and tangential direction. The rotation angle is used from 0^0 to 180^0 with the increment of 1^0. The coordinate of the highest value of lag time and polarization direction are assumed as SWS parameters.

5.3. Crack Density

The previous study had been done by Vlahovic et al [8], showed crack density had been obtained by normalizing the delay time along the ray-path. Afterward, the normalization on delay time plotted as illustrated in figure 5.

The crack density value is in the range of 4ms/km to 11 ms/km. Dense crack density is located at the eastern part of the study area and vice versa at the opposite side. It is also followed by moderate to high value of crack density. However, the weakness of this method is not able to delineate the anisotropy in the certain depth. Therefore, analyzing the anisotropic area using tomography anisotropy is need to be done.

5.4. Tomography Anisotropy

Since SWS parameters obtained, thus the delay time and polarization direction proceed to acquire the percentage of anisotropy. Percentage of the anisotropy derived in [9].

$$[\mathbf{dt}_{SWS}] = [\mathbf{A}][\cos\alpha][\Delta\mathbf{U}] \quad (1)$$

To obtain ΔU in the equation (1), is applied a non-negative least square inversion, which is the solution of the minimum deviation error that resulted in the positive value. The parameter of K is the value of percentage constant of anisotropy. The difference of slowness between fast S-wave and slow S-wave velocity is the output that produced by non-negative least square whereas the S-wave velocity model derived from 3-D tomography using the similar data. Figure 6 shows the result of 3-D SWS tomography that illustrated the distribution of anisotropy in "PR" Geothermal field.

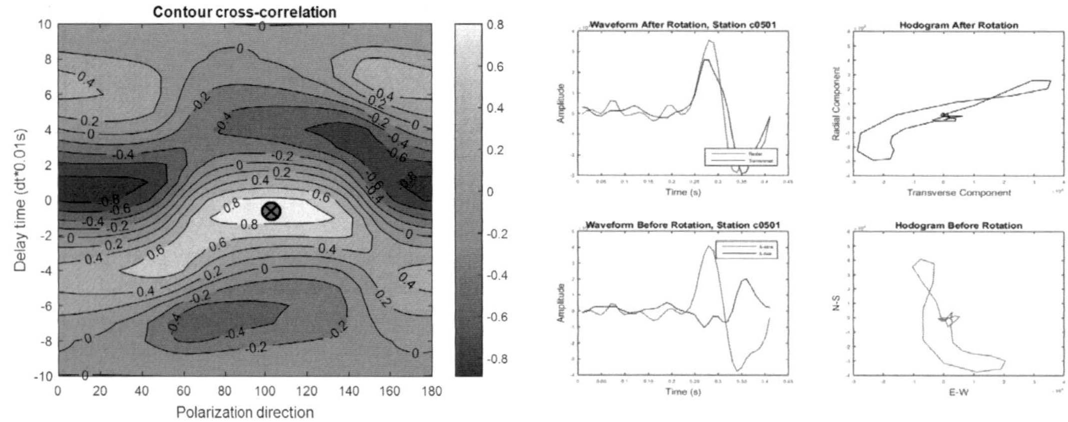

a) b)

Figure 4. a) Example of cross-correlation result among radial and tangential component. Contour with maximum value indicates by red dot. b) Comparison of hodogram before and after rotation.

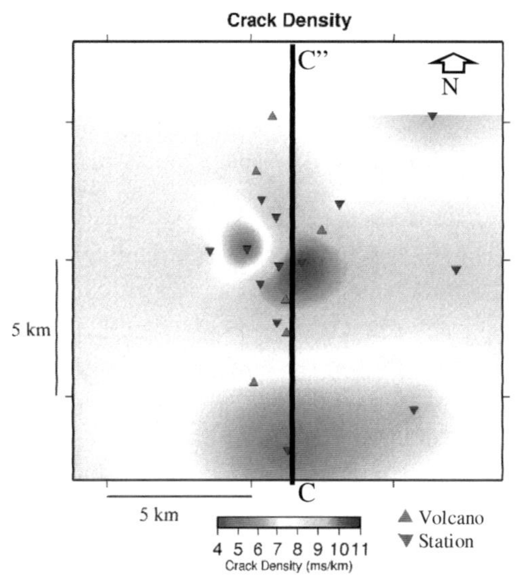

Figure 5. Crack density map in the study area. Blue colour indicates denser crack density than red

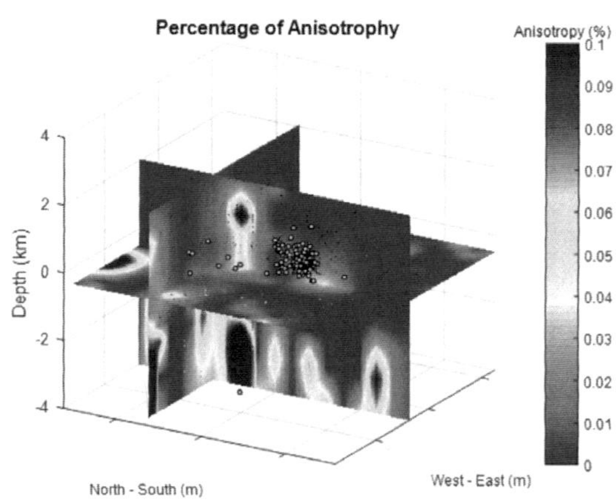

Figure 6. 3-D slice of tomography anisotropy. The blue colour shows high anisotropy while red colour is low anisotropy.

6. Discussion

Output distribution of tomography anisotropy is compared to the perturbation of Vp, Vs, Vp/Vs ratio, and crack density to delineate the reservoir in the geothermal field (see figure 7). At depth -1.0 km to

0.5 km anisotropy occupied in the production area, however, the higher degree of anisotropy is on the eastern part of production area. High degree of anisotropy lied in the range of 0.3% to 1.5%, supported by the low perturbation of Vp, Vs, and Vp/Vs ratio. The features described as the saturated steam zone in "PR" Geothermal Field. And it is supported by dense crack density in the eastern part and could be interpreted as high permeability area.

Furthermore, at depth -0.5 km to 1 km anisotropy not only lied in production area but also in the injection area. It is caused by intensive fluid injection activities and occupies to the high degree anisotropy in the area. At mean sea level illustrated, on the southern part of the study area, high level of anisotropy associated with Vs lower than Vp and high Vp/Vs ratio is expected as a water zone due to fluid injection activities. Latitude cross section shows toward on the southern part the reservoir becomes deeper, the top of reservoir lied at depth -1 km. The expected water reservoir lied at depth -0.5 km on the northern part and become 0.5 km on the southern part; it is also supported by high degree of anisotropy in the production area that is associated with rapid microearthquakes. On the middle of two mountains, there is occupied a high anisotropy in vertical section that is associated to high Vp/Vs ratio. This anomaly is expected as discharge area with alteration rock, probably the manifestation comes from this features. Generally, percentage of anisotropy was following the regional structure in the study area which is NNW-SSE and NE-SW. The schematic model based on interpretation of Vp, Vs, Vp/Vs ratio, crack density, and tomography anisotropy shown in figure 8.

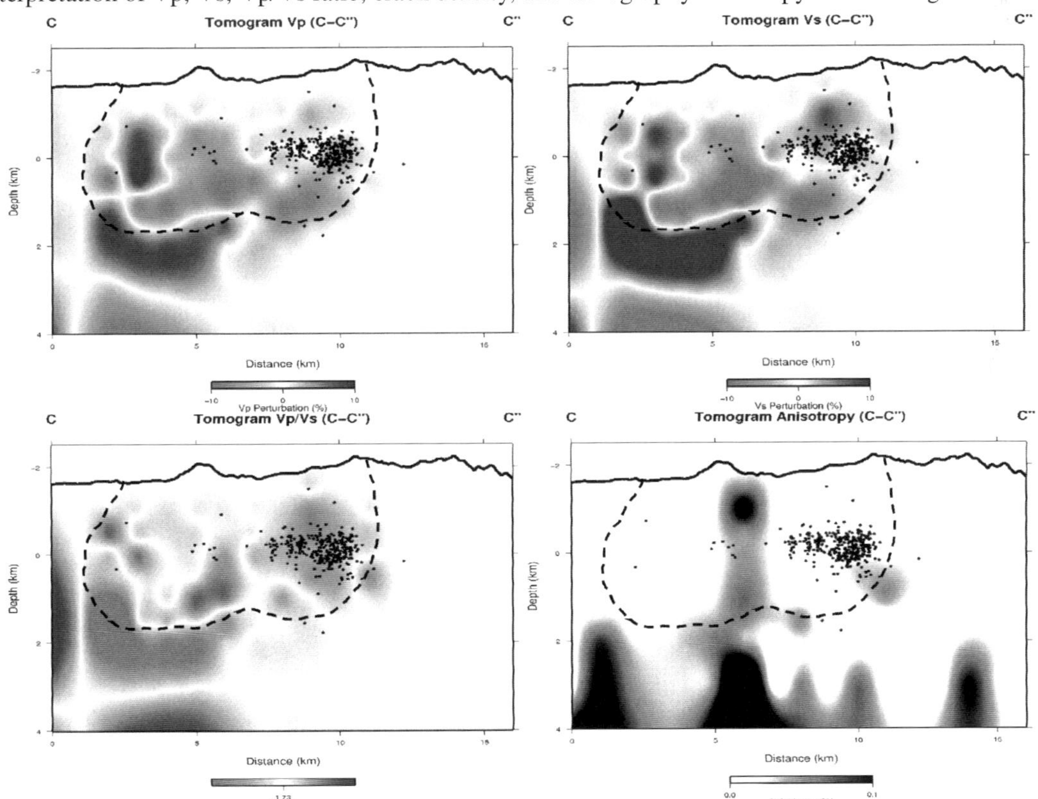

Figure 7. The comparison of Vp, Vs, Vp/Vs ratio, and anisotropy in the cross-section C-C" (figure 5). The dashed line bordered the best resolution derived by checkerboard resolution test.

7. Conclusion

Based on Vp, Vs, and ratio Vp/Vs map, the steam zone lies on the northern side of the study area and visible at depth -1 km to 0km. Steam zone spread out to the eastern part with 3km wide associated with low Vp/Vs ratio by 1.5-1.68. Water zone lies exactly below the steam zone, however it spreads with high Vp/Vs ratio within 1.78-1.8.

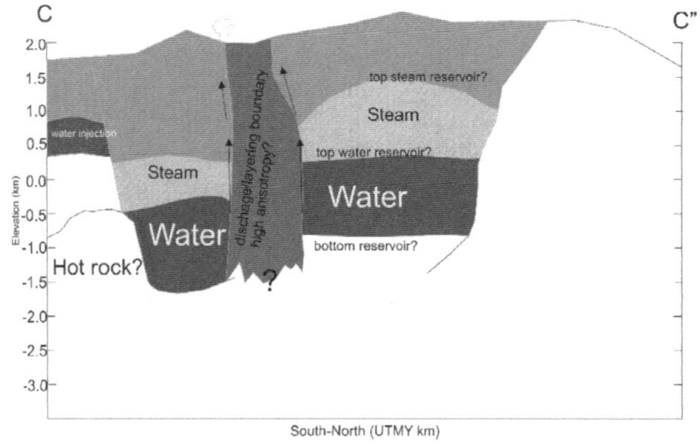

Figure 8. The schematic model of reservoir derived from Vp, Vs, Vp/Vs ratio, anisotropy, and crack density.

Commonly, dense crack density lies in the eastern part of the study area and vice versa on the opposite side within the range of 4-11 ms/km. Therefore, the area with high permeability can be interpreted as a reference for geothermal development.

The output of SWS parameter depends on the maximum value of cross-correlation. The delay time of SWS has a range of 0.01 to 0.2 second.

High degree of anisotropy and Vp/Vs ratio are possibly caused by interaction among fluid and rock; this interaction leads the occurrence of alteration which is recognized as high anisotropy. Thus, the vertical views indicate the discharge area that supports with manifestation around the location.

References

[1] Crampin S and Y Gao 2006a *Physics of the Earth and Planetary Interior* **156** 1-14
[2] Crampin S 1991 *Geophys. J. Int* **107** 611-623
[3] Crampin S 1994 *Geophysics* **57(5)** 727-735
[4] Li Zefeng, Haijiang Zhang, and Zhigang Peng 2014 *Earth and Planetary Science Letter* **391** 319-326
[5] Zhang H, Liu Y, Thurber C, and Roecker s 2007 *Geophysical Research Letter* **34**
[6] Lawson C L and Hanson R J 1974 Solving least square problems *Prentice-Hall Eaglewood Cliffs N.J* 340
[7] Evans J R, Eberhart-Phillips D and Thurber H H 1994 User's manual for SIMULPS12 for imaging Vp and Vp/Vs: A derivative of the "Thurber" tomographic inversion SIMUL3 for local earthquakes and explosions, *U.S. Geological Survey Open-File Report* 94-431
[8] Vlahovic, Gordana, Maya Elkibi, and Jose A. Rial 2005 Temporal Variation of Fracture Direction and Fracture Densities in the Coso Geothermal Field From Analysis of Shear-Wave Splitting *Proceedings Twenty-Seventh Workshop on Geothermal Reservoir Engineering. Stanford University, Stanford California, January 28-30 2002* **SGP-TR-171**
[9] Ry R V and Nugraha A D 2016 Reservoir Characterization around Geothermal Field, West Java, Indonesia Derived from 4-D Seismic Tomography *IOP Conference Series Earth and Environmental Science* **29(1):012001**
[10] Hasanah M U, Nugraha A D, and Rachmat Sule 2013 Attenuation tomography using

microearthquake (MEQ) data in the "A" geothermal field *AIP Conf. Proc.* **1554** 273
[11] Nugraha A D and Jim Mori 2006 *Geophysical Research Letter* **33**
[12] Nugraha A D, Ahmad Syahputra, Fatkhan, and Rachmat Sule 2013 3-D seismic velocity and attenuation structures in the geothermal field *AIP Conf. Proc.* **1554** 238
[13] https://www.earth.google.com (accessed 8 February 2016)

Stable Iterative Methods for Inversion of Magnetotelluric and Transient Electromagnetic Data

Ridho Nanda Pratama[1], Widodo[2]

[1]Undergraduate Program of Geophysical Engineering, Institut Teknologi Bandung, Indonesia
[2]Applied Geophysics and Exploration Research Group, Faculty of Mining and Petroleum Engineering, Institut Teknologi Bandung, Indonesia

email: ridhonpratama@gmail.com

Abstract. The joint inversion of geophysical data can reduce the ambiguity of model parameters. Joint Inversion of Magnetotelluric (MT) and Transient Electromagnetic (TEM) data was performed to get detailed information of subsurface structure. Whereas TEM data is sensitive to describe in the shallow structure, while the deeper structure is related to MT data. We derived the joint inversion scheme from the second order of Marquardt algorithm using singular value decomposition (SVD). The detailed analyses of the model parameters are performed by damping factors, V-matrix, damped error bounds and importances value. The joint models parameters with damping factor and importances value above 0.9 indicate that the associated eigen parameters are well-resolved.

1. Introduction

Modeling and inversion of geophysical data are crucial for the meaningful interpretation of the processed field response. For this purpose, a sequence of steps based on well-defined physio-mathematical formulations is implemented. These steps include the computations of responses for the assumed geoelectrical model referred as modeling in geophysical literature. Solutions to geophysical inverse problems are generally non-unique. It is usual to reduce the non-uniqueness by restricting the complexity of the Earth models. The restricted class of the earth models consists of horizontally layered, isotropic media, with constant resistivity in each layer.

The analysis of the problem applies to the geophysical inverse problem in which the partial derivatives of the data respect to the model parameters can be calculated which are classified as Important, Unimportant, and Irrelevant. The classification is based on the Singular Value Decomposition [1] of the Jacobian matrix, which also allows the visualization of which combination of earth parameters was most effectively resolved and allocated the confidence intervals for the parameters. Many inverse problems in geophysics are ill-posed (unstable) that is small changes in the data can lead to large changes in the solution model. The unstable nature and consequent numerical instability of the inverse problem are largely contained in Unimportant and Irrelevant parameters.

The detailed analyses of the model parameters are performed by damping factors, V-matrix, damped error bounds, and importances. The damped error bounds measure the expected variation in the well-resolved (Important) parameters, in response to small variation in the data. The bounds attributed to the Unimportant parameters are damped, and only measured the variation due to their correlation with the well-resolved parameters. The greatest expected variations occur for parameters which are near the threshold or dividing line between the Important and Unimportant parameters.

In this paper, the methods are applied to synthetic data sets of Radiomagnetotelluric (RMT) and Central Loop Transient Electromagnetic (TEM). Joint inversion can increase the numbers of Important parameters of a model to include some which the methods cannot resolve separately. Whereas RMT data is very sensitive to changes in the upper few meters, the TEM data is more sensitive to deeper variations. It is essential to note that this effect cannot be achieved merely by increasing the number or accuracy of data values of a single method. Joint inversion is a means of eliminating some defects inherent in the individual methods.

2. The analysis and solution of the inverse problem

2.1. General notation and the Jacobian matrix

The forward problem generates a set of model data for each setting of restricted earth model can generally denoted by

$$\mathbf{d} = g(\mathbf{x}) \tag{1}$$

With d is data that can be apparent resistivity values (ρ_a) or phase (Φ) and $g(\mathbf{x})$ the forward operator of model parameters (x) while in our examples d_i is the induced voltage at the i'th period for the TEM data; and the apparent resistivity or phase at the i'th period for MT data. In 1D case, models can be defined by the subsurface layer resistivity and thickness. Jupp and Vozoff [1] suggest the value of a model in logarithmic value to avoid a negative value in the inversion result.

The inverse problem determines values of $\mathbf{x}$ such that $g(\mathbf{x})$ matches $\mathbf{d}$ in some sense, which in this paper is the minimum of the familiar Root Mean Squared (RMS) relative error between model and data,

$$F(x) = \left\{\frac{1}{N}\sum_{i=1}^{N}\frac{(d_i-g_i)^2}{d_i^2}\right\}^{\frac{1}{2}} \tag{2}$$

The iterative method can improve a current model until the error measure is small, and the parameters are stable on reasonable changes in the model. When the partial derivates of the model data on their parameters can be obtained accurately and relatively easily, the Gauss method [1] is an attractive iterative inverse method. The Gauss method iteratively changes the current $\mathbf{x}$ by an amount δx calculated by solving the linear least-squares problem. We expand $g(\mathbf{x})$ about $\mathbf{x}$ in a Taylor expansion to derive it

$$g(x+\delta x) = g(x) + J\delta x \tag{3}$$

where J is the Jacobian matrix of the vector function $g(\mathbf{x})$ which is a by-product of linearizing a non-linear problem. Its elements are the partial derivatives of the calculated data on the model parameters.

$$J_{ij} = \left.\frac{\partial g_i}{\partial x_j}\right|_{\substack{i=1,M \\ j=1,N}} \tag{4}$$

The Jacobian matrix is also called parameter sensitivity matrix that explains its importance in the course inversion. Statistically, sensitivity values show whether the layer parameters are seen individually in the measured data or not. They qualitatively display which parameter or combinations of parameters are resolved by the data. The model parameter is only well-resolved by the measurements if its sensitivity matrix has large entries in the corresponding column. On the other hand, the normalized Jacobian values or relative sensitivities give an idea of each data point's sensitivity to a change of an individual parameter [1].

We drop the weighting factor W which are diagonal matrices corresponding to the mean relative error measure, $W = (1/d_i)|_{i=1,N}$. When $J^T WJ$ is non-singular, we find that

$$\delta x = (J^T WJ)^{-1} JW\big(d - g(x)\big) \quad (5)$$

The MT and TEM data recorded from field contain noise and hence, are inconsistent. Therefore, the inverse problem is ill-conditioned, and solution is normally ill-posed (unstable). Hence, regularized inversion techniques are used to obtain the solution of the inverse problem [1]. In this paper, we have used Marquardt-Levenberg scheme for the inversion and also Jupp and Vozoff approach for the joint inversion.

2.2. *The singular values of J*

Singular Value Decomposition (SVD) method has been used for the detailed analysis of the inverse problem. Stabilized solution of the ill-posed inverse problem is obtained by truncating the small eigen values of the matrix. The system matrix is decomposed into a product of three matrices as

$$\mathbf{J} = \mathbf{USV}^{\mathbf{T}} \quad (6)$$

Where **U** is a NxM matrix containing data space eigenvectors of J in its columns, **V** is a MxM matrix which contains the parameter space eigenvectors, and **S** is a MxM diagonal matrix with eigenvalues (s_i) as its diagonal elements. The s_i's are the non-negative square root of eigenvalues of $J^T J$, which are arranged in decreasing order as

$$s_1 \geq s_2 \geq s_3 \cdots s_m \geq 0 \quad (7)$$

By retaining only significant eigenvalues, a generalized stable inverse solution can be formulated. The damping factors act more like thresholds at each iteration. They are generally measures the '*importances*' of the corresponding layer parameters and vary between 0 and 1 to represent the minimal and maximal parameter resolution respectively. The SVD approach [1] implemented in EMUPLUS code has been applied for detailed analysis of RMT and TEM data in the following sections.

2.3. *Marquardt-Levenberg inversion scheme*

Levenberg-Marquardt Algorithm is an algorithm that applying the minimization of model perturbation to the Gauss-Newton solution. It can be done by minimizing the objective function F

$$\mathrm{F} = \big(\mathbf{d} - \mathrm{g}(\mathbf{x_0} + \delta\mathbf{x})\big) + \lambda||\delta\mathbf{x}||^2 \quad (8)$$

The solution now becomes

$$\boldsymbol{\delta}\mathbf{x} = [\mathbf{J}^{\mathbf{T}}\, \mathbf{W}\, \mathbf{J} + \lambda\mathbf{I}]^{-1}\, \mathbf{J}^{\mathbf{T}}\, \mathbf{W}\, (\mathbf{d} - \mathrm{g}(\mathbf{x})) \quad (9)$$

With the parameter λ is showing of the effect of the model perturbation. Small λ value will make the last equation becomes equal to Gauss – Newton solution in function F. The parameter λ is initialized with a large initial value. If the solution after iteration is worse than the previous iteration λ value is increased, otherwise it is reduced. To ensure the calculation stability the last equation is modified to

$$\boldsymbol{\delta}\mathbf{x} = [\mathbf{J}^{\mathbf{T}}\, \mathbf{W}\, \mathbf{J} + \lambda\, \mathrm{diag}(\mathbf{J}^{\mathbf{T}}\, \mathbf{W}\, \mathbf{J})]^{-1}\, \mathbf{J}^{\mathbf{T}}\, \mathbf{W}\, (\mathbf{d} - \mathrm{g}(\mathbf{x})) \quad (10)$$

The method effectively controls the instability caused by the existence of zero or very small eigen value of the system matrix.

2.4. *Parameter classification and 'error bounds'*

Eigenstructure (singular values) of J can be used to classify the parameters as Irrelevant, Unimportant, and Important [1].

i) Irrelevant parameters have no influence on the model data. They correspond to layers which are the range of the measurements, or as in the case of thin resistive layers, to parameter combinations that cannot be resolved from the data.

ii) Unimportant parameters have the only small influence on model data. Inversely, large changes in these parameters can occur for only marginal improvement in fit to the actual data. For this reason, they must be either neglected, or altered only marginally during the inversion process.
iii) Important parameters correspond to the well-resolved, and often gross features that are well represented in the data

2.5. Joint Inversion

To have a good fitting and avoid ambiguity that is inherent to the individual methods, we can use joint inversion. Joint inversion is a method for inverting two or more data sets from different geophysical methods and using the same model. Application of joint inversion was first introduced by Vozoff and Jupp [1] and first applied by Meju [2] for EM sounding.

The Jacobian matrix **J** determines how the model should be altered to improve the fitting:

$$J_{ij} = w_i \frac{\partial g_i}{\partial x_j} \tag{11}$$

where the weight w_i and N-layer earth model of resistivity and thickness are given as

$$\begin{cases} \rho_i \geq 0 & i = \overline{1, N+1} \\ h_i \geq 0 & i = \overline{1, N} \end{cases} \tag{12}$$

The elements of the matrix **J** can be written as

$$J_{ij} = w_i \frac{\partial g_i}{\partial x_j} = \frac{1}{g_i} \frac{\partial g_i}{\partial \log \rho_j} = \frac{\rho_j}{g_i} \frac{\partial g_i}{\partial \rho_j} \; or \; \frac{1}{g_i} \frac{\partial g_i}{\partial \log h_j} = \frac{h_j}{g_i} \frac{\partial g_j}{\partial h_i} \tag{13}$$

In this scheme, the Jacobian matrix is made scale free. Since the relative error will be proportional to the model, the two kinds of the data will equally influence the correction that improves the current model. Theoretically, each of them can compensate for the weaknesses of the other, and therefore the overall benefit will be a reduction of ambiguity [3].

3. Applications to TEM and RMT synthetic data

3.1. Synthetic model generated

Synthetic data sets were generated for RMT at the frequency range $10^4 - 10^6$ Hz and TEM data were generated for time range $10^{-5} - 1$ seconds. To make synthetic data close to the field conditions, 20 percent random Gaussian noise were added to each data. For each data set, Marquardt inversion was carried out. The three layers model with conductive layer sandwiched between two resistive layers ($\rho_1 > \rho_2 < \rho_3$) was used to generate the synthetic data. The results of single inversion are more dependent on the starting model, so then we choose the best influence of the starting model that have been observed before. Joint inversion effectively combines the capabilities of each method.

3.2. Detail analysis for the inversion

The damping factors, the V-matrix, calculated error bounds and parameters *importance* can be used to judge the performance of the inversion. By limits of damping factors to judge the resolution of the model parameter suggested, EP1, EP2, EP3 and EP4 are well-resolved; EP5 is unresolved for single RMT inversion and EP5 is shaky for single TEM inversion. However, in joint inversion case EP1, EP2, EP3, EP4, and EP5 are well-resolved. The structure of V-matrix is convenient as significant parameter combinations in resistivity are often products and ratios. The factors $\rho_i, h_i, \rho_i h_i, h_i/\rho_i$ and ρ_{i+1}/ρ_i may all appear as transformed parameters and interpret the V-matrix regarding these combinations.

The V-matrix analysis of single RMT inversion shows that EP1 has relatively large entries in the first rows, it consists of the resistivity product of the top layer. EP2 consists of the thickness product of the first layer. Since EP1 and EP2 have damping factors of 1.0, they are well-resolved. EP3 consists of resistivity of the second layer which is well-resolved since EP3 has damping factor more than 0.9. EP4 and EP5 have very low damping factor, consequently, the thickness of second layer and the basement is not resolved. Error bounds and parameters importances for single RMT inversion also indicates each parameter as the V-matrix and damping factors analysis as they are. The first layer parameters are well-resolved in comparison to the second and third layer parameters.

Similar analysis has been performed for the TEM inversion. EP1 is the third layer resistivity and has damping factor 1.0 while it means the basement is well-resolved. EP2 and EP3 show that the thickness and resistivity of the first layer, and are well-resolved since they have damping factors more than 0.9. EP4 consists of the third layer resistivity and the second layer thickness ratio and EP5 consists of the resistivity thickness ratio of the second layer. Since EP4 has damping factor more than 0.9 so the resistivity of the third layer is completely well-resolved that associated with EP1 but EP5 has damping factor between 0.8 and 0.9 so the thickness of the second layer is not resolved as well or shaky, then exactly for the resistivity of the second layer. Hence, the thickness of the second layer cannot be estimated independently. Error bounds and parameters importances for single TEM inversion indicates that the basement is well-resolved in comparison to the first and the second layer parameters.

The V-matrix analysis for joint inversion shows the better resolution. EP1, EP2, EP3, EP4, and EP5 have damping factors 1.0 respectively while it means they are well-resolved at all. EP1 shows the resolution of the second layer parameters, EP2, EP3, and EP4 show the resolution of the first layer parameters. EP5 is the basement resistivity which is resolved as well. Similar results are also indicated in error bounds and importances.

Table 3.1 Synthetic data and single and joint inversion of RMT and TEM results.

Parameter	True Model	Starting Model	Single RMT inverted model	Single TEM inverted model	Joint Inversion model
$\rho_1(\Omega m)$	100	400	142.6958	186.129	117.8884
$\rho_2(\Omega m)$	10	30	43.7311	29.0225	11.4097
$\rho_3(\Omega m)$	1000	800	737.1805	1152.911	993.5697
$h_1(m)$	30	20	30.0475	31.3023	30.9257
$h_2(m)$	10	20	31.1785	26.4486	11.6399

Table 3.2 V matrix and damping factor for single RMT inversion.

	EP1	EP2	EP3	EP4	EP5
PAR1	0.689	-0.369	0.204	-0.104	-0.280
PAR2	-0.014	0.103	-0.833	0.324	-0.163
PAR3	-0.002	-0.018	-0.004	-0.124	0.883
PAR4	0.064	0.864	0.175	-0.300	-0.182
PAR5	0.021	0.178	0.421	0.876	0.056
Normalized singular value	1.000	0.257	6.21E-02	2.42E-03	2.17E-03
Damping Factor	1.000	1.000	0.998	0.033	0.001
Parameter Combination	ρ_1	h_1	$1/\rho_2$	$1/h_2$	ρ_3

Table 3.3 Error bounds and importance for single RMT inversion.

Layer resistivities – 68 percent confident interval (damped)				
	RO(I)	*BOUND(1)*	*BOUND(2)*	*IMPORTANCE*
1	142.6958	121.102	168.1399	0.8135
2	43.7311	25.1377	76.0775	0.8911
3	737.1805	637.338	852.6639	0.1176
Layer thickness – 68 percent confident interval (damped)				
	H(I)	*BOUND(1)*	*BOUND(2)*	*IMPORTANCE*
1	30.0475	20.6738	43.6712	0.9275
2	31.1785	11.0028	88.3502	0.1366

Table 3.4 V matrix and damping factor for single TEM inversion.

	EP1	EP2	EP3	EP4	EP5
PAR1	0.422	-0.042	-0.813	-0.439	0.188
PAR2	0.162	-0.450	0.142	0.273	0.768
PAR3	0.660	-0.248	0.411	0.530	-0.176
PAR4	0.373	0.657	-0.281	0.120	0.008
PAR5	0.435	0.167	-0.252	-0.483	-0.509
Normalized singular value	1.000	0.718	0.430	0.281	0.232
Damping Factor	1.000	1.000	0.951	0.908	0.875
Parameter Combination	ρ_3	$1/h_1$	$1/\rho_1$	ρ_3/h_2	ρ_2/h_2

Table 3.5 Error bounds and importance for single TEM inversion.

Layer resistivities – 68 percent confident interval (damped)				
	RO(I)	*BOUND(1)*	*BOUND(2)*	*IMPORTANCE*
1	186.129	177.867	194.7747	0.8746
2	29.0225	28.2094	29.8591	0.7885
3	1152.911	1118.295	1188.599	0.9511
Layer thickness – 68 percent confident interval (damped)				
	H(I)	*BOUND(1)*	*BOUND(2)*	*IMPORTANCE*
1	31.3023	28.7802	31.9048	0.9198
2	26.4486	25.3215	27.6259	0.8712

Table 3.6 V matrix and damping factor for joint inversion.

	EP1	EP2	EP3	EP4	EP5
PAR1	0.280	-0.703	0.253	-0.603	0.003
PAR2	0.664	0.160	-0.152	0.069	-0.179
PAR3	0.108	0.027	0.030	0.039	0.982
PAR4	-0.03	-0.224	-0.955	-0.155	0.060
PAR5	-0.681	-0.181	0.004	-0.090	-0.022
Normalized singular value	1.000	0.783	0.503	0.271	9.84E-02
Damping Factor	1.000	1.000	1.000	1.000	1.000
Parameter Combination	ρ_2/h_1	$1/\rho_1$	$1/h_1$	$1/\rho_1$	ρ_3

Table 3.7 Error bounds and importance for joint inversion.

Layer resistivities – 68 percent confident interval (damped)				
	RO(I)	*BOUND(1)*	*BOUND(2)*	*IMPORTANCE*
1	117.8884	103.3135	134.5194	1.0000
2	11.4097	3.1497	41.3314	0.9923
3	993.5697	548.668	1799.231	0.9995
Layer thickness – 68 percent confident interval (damped)				
	H(I)	*BOUND(1)*	*BOUND(2)*	*IMPORTANCE*
1	30.9257	25.7304	39.6127	0.9998
2	11.6399	3.1284	43.3084	0.9919

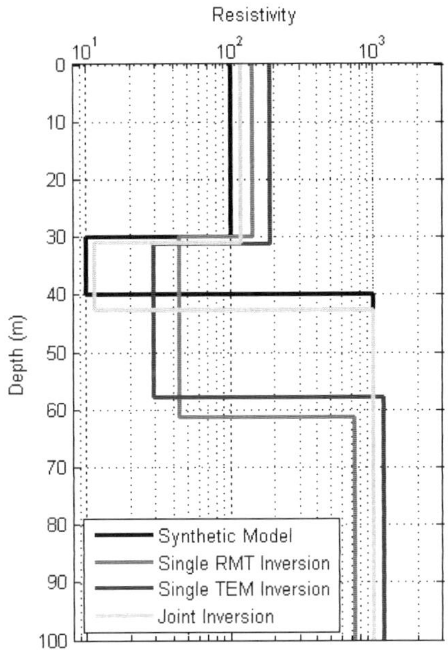

Figure 1. The resistivity and thickness of the synthetic model (black), single RMT inversion model (red), single TEM inversion model (blue), and joint inversion model (green).

3.3. Conclusion

From the above example, which has been demonstrated that joint inversion produces the better results in the equivalence problem in comparison with single inversion of RMT and TEM. However, if the data contain no information about the model parameters, the no inversion scheme is capable for delineation of the parameters.

Starting from the synthetic model with $\mu = 0.01$ and with data perturbed to the 20 per cent relative error level, we obtained the final model of Fig. 3.1. The residual error was 5 per cent, which was mainly the original error, but shows the model fits some of the original error at the expense of its truth. Table 3.2 to 3.7 records the singular values, the damping factors, the V matrix, error bounds, and parameters importances for single and joint inversion of RMT and TEM respectively, all of which have been interpreted as described in Section 3.2. This method provides a basis for its extension to the more realistic situation.

References

[1] Jupp D L B and Vozoff K 1975 Stable Iterative Methods for the Inversion of Geophysical Data *Geophysical Journal of the Royal Astronomical Society* **42** Issue 3: 957 –976.

[2] Meju M A 1996 Joint inversion of TEM and distorted MT sounding: Some effective practical considerations *Geophysics* **61** 56–65

[3] Widodo 2016 Multidimensional Interpretation of Near Surface Electromagnetic Data Measured in Volvi Basin, Northern Greece *Journal of Enviromental and Engineering Geophysics* **21** 121-133

Central Loop Time Domain Electromagnetic Inversion Based on Born Approximation and Levenberg-Marquardt Algorithm

I B S Yogi[1], Widodo[2]

[1] Master Program of Geophysical Engineering, Institut Teknologi Bandung, Indonesia
[2]Applied Geophysics and Exploration Research Group, Faculty of Mining and Petroleum Engineering, Institut Teknologi Bandung, Indonesia

E-mail: suanandayogi@gmail.com

Abstract. Time Domain Electromagnetic (TDEM) method uses electromagnetic wave to detect resistivity or conductivity difference of lithology in the subsurface, measured in the time domain. TDEM method has been developed in decades. There are forward modeling and inversion programs have been made. The purpose of making the forward modeling programs is to calculate TDEM response so that data acquisition parameters can be chosen correctly. The inversion program is for synthesizing geological model from measured TDEM data. Several TDEM programs have been made, but the procedures require heavy and complex computation, which need high computer specification and lot of calculation time. Nowadays program that used less complex computation and faster calculation is needed to match field data acquisition productivity. To achieve faster and more accurate process, we use Born Approximation of the apparent conductivity for the forward modeling program and Levenberg-Marquardt algorithm for the inversion program. We use many circumstances in testing our programs, from one layer to multi layers by varying the resistivity or thickness of lithology and compared with a validated program, EMUPLUS. At the end of the test, inversion of real data is taken as confirmation that the program can be used to process real TDEM data. Inversion results of both synthetic and real TDEM data show pleasant results. These test results indicate that our program can be used as daily forward modeling program for determining data field acquisition parameters and do the inversion procedure for synthetic and real TDEM data.

1. Introduction

Time Domain Electromagnetic (TDEM) method uses electromagnetic wave to detect resistivity or conductivity difference of lithology in the subsurface, measured in the time domain. Inversion program gives geological parameters from measured data of the Time Domain Electromagnetic. Previous forward modeling procedures require heavy and complex computation, which need high computer specification and lot of calculation time. To achieve a faster and more accurate process, we use approximation procedures in our program. The approximation procedure has been done by using Born Approximation [1,2]. This study will present modification of Levenberg-Marquardt algorithm that is suitable with the forward modeling algorithm. Forward modeling and inversion algorithms in this study will be used in our pyTEM1D program that is based on Python programming language.

2. Time Domain Electromagnetic

Central Loop Time Domain Electromagnetic measured secondary vertical magnetic field (H_z) and the time derivative of vertical magnetic field $\left(\frac{\partial H_z}{\partial t}\right)$ [3]. The vertical magnetic derivative for homogeneous half-space can be expressed in the following equation.

$$\frac{\partial H_z}{\partial t} = -\frac{I}{\mu_0 \sigma a^3}\left[3erf(\theta a) - \frac{2}{\sqrt{\pi}}\theta a(3 + 2\theta^2 a^2)e^{-\theta^2 a^2}\right] \quad (1)$$

The transmitter radius is symbolized by a, transmitter current is I, and σ is conductivity of the homogenous earth. Generally, in TDEM measurement, apparent resistivity can be distinguished as early time and late time resistivity, although in geophysical prospecting, late time resistivity is used:

$$\rho_a \approx \frac{I^{2/3}\mu_0 a^{4/3}}{20^{2/3}\pi^{1/3}t^{5/3}}\left(\frac{-\partial H_z}{\partial t}\right)^{-2/3} \quad (2)$$

3. Forward Modeling

Forward modeling method in this research is Adaptive Born Forward Mapping [2]. The apparent resistivity can be calculated as.

$$\sigma_a(t_i) = \sum_{j=1}^{L} \sigma_j . \mathcal{F}_{ij} \quad (3)$$

$\mathcal{F}_{ij}$ is integral of Frechet kernel which is a function of depth (z). $\sigma_a(t_i)$ is apparent conductivity from subsurface layer in time t_i, i is i^{th} measurement time, and σ_j is synthetic conductivity models for j^{th} layer $(j = 1,\ldots,L)$. The linear approximation of the integral of Frechet kernel can be expressed as:

$$\mathcal{F}\left(z_j, t_i, \sigma_a(t_i)\right) = \begin{cases} \frac{z_j}{D_i}\left(2 - \frac{z_j}{D_i}\right), & for\ z_j \leq D_i \\ 1, & for\ z_j > D_i \end{cases} \quad (4)$$

With

$$D_i = \sqrt{\frac{c\, .t_i}{\mu_0 \sigma_a(t_i)}} \quad (5)$$

c is a constant that affect the sensitivity of Frechet kernel, which in this study used a value of 1.2. Later using the known apparent conductivity in a certain time, the time derivative of second magnetic field can be calculated using the analytic equation as in equation 1 and will give the time derivative of secondary magnetic field response in time domain.

4. Inversion

Inversion method is used to find target models or parameters from data. In the inversion process, data and desired model can be simplified as a matrix [4]. Measured data with N number data can be defined as (***d***). Further M number desired models can be stated as (***m***) matrix.

One of the methods to solve non-linear problem is Levenberg-Marquardt algorithm. It is known that inverse matrix cannot be calculated if the matrix singular value is zero, while sometimes become unstable when the model change very small [5]. To overcome this problem, Levenberg-Marquardt equation is solved by using singular value decomposition as explained in [4].

$$\Delta \boldsymbol{m}_n = \boldsymbol{V}\left[diag\left(\frac{Q_i}{\lambda + Q_i^2}\right)\right]\boldsymbol{U}^T(\boldsymbol{d} - \boldsymbol{g}(m_n)) \quad (6)$$

Where Jacobian $(\mathbf{J})$ matrix which is partial derivative of forward model function and models, is defined as singular value decomposition matrix

$$\mathbf{J} = \boldsymbol{UQV}^T \quad (7)$$

$\boldsymbol{U}$ is NxM matrix, $\boldsymbol{V^T}$ is MxM matrix and $\boldsymbol{Q}$ is diagonal matrix of singular value with MxM dimension. λ is damping factor, which at the beginning of the iteration the is big enough. Because of this, the equation can control the result so that the updated model is not overshoot. After getting a better model, the damping factor is reduced so that the equation can give the best result faster.

5. Forward Modeling Test

The forward modeling process is done by using approximation and analytic equation from previous sections. Here, three basic models and two three layers models that will be processed (Figure 1). Time domain electromagnetic (TDEM) central loop transmitter has 1 Ampere current, 25 m radius and sampling time 1 μs – 1 s, with 20 sampling numbers.

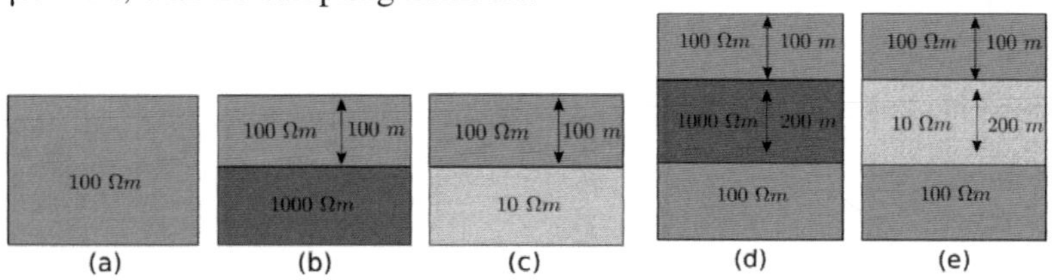

Figure 1. Synthetic models for forward modeling. Three basic models: **(a)** A, **(b)** B, and **(c)** C. Three layers models: **(d)** D and **(e)** E

The time derivative of magnetic field responses from pyTEM1D and EMUPLUS are similar as shown in Figure 2. The homogenous model has a constant trend of decay. Model with a more resistive and conductive second layer has the exact same value for earlier part of the curve, but in 0.1 ms the curve is bent to the lower value of time derivative of magnetic field for the model B and bigger value for the model C.

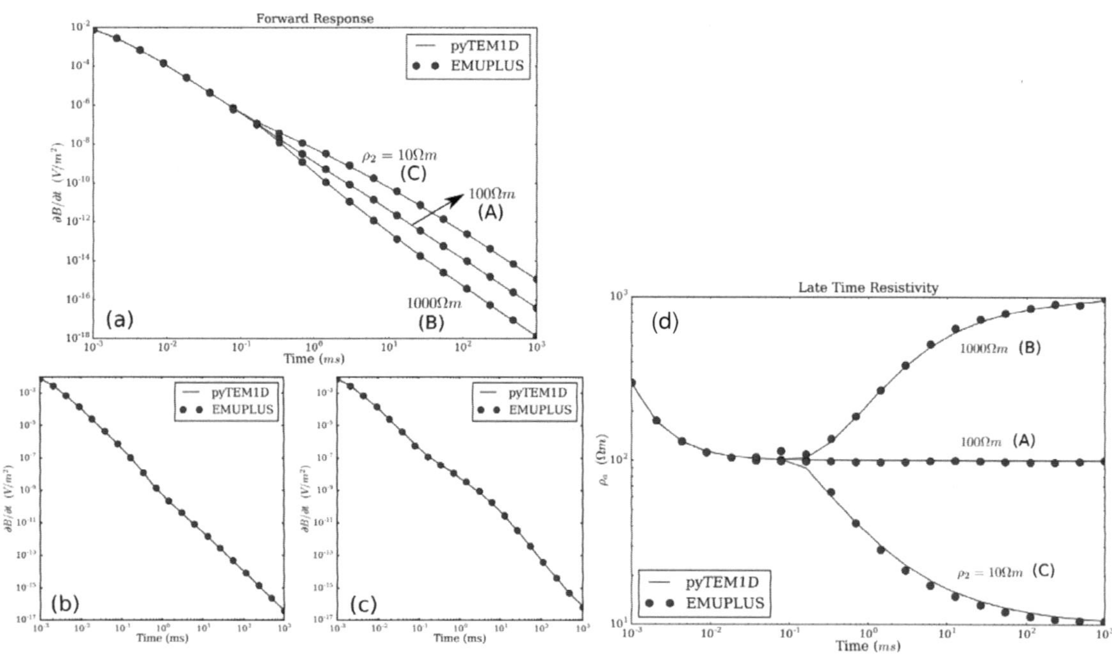

Figure 2. The time derivative of magnetic field of **(a)** three basic models: A, B, and C. Three layers models: **(b)** D and **(c)** E. **(d)** Late time apparent resistivity of three basic models

The time derivative of magnetic field from Figure 2 then transformed to become apparent resistivity at late time condition by using equation 2. The transformation process produces apparent resistivity for the late time in Figure 2. The late time apparent resistivity responses in time earlier than 0.1 ms give greater value than it has supposed to be (100 Ωm). This time range is usually called as early time condition [6].

6. Inversion Test

In this section, the inversion algorithm will be applied to the synthetic models from the forward modeling process, model B, C, D, and E (Figure 1). The inversion algorithm is Levenberg-Marquardt that has been explained. Starting models for this process are four starting models, 3-6 layers, 10 Ωm and thickness every layer is 50 m. Figure 3a is time derivative of the magnetic field response of the model E and its inversion results from four different starting models. In this figure, all inversion responses give similar to one another and match with the sytnetic respones of the model E. In determining the best result, a magnification is done to the Figure 3a (Figure 3b). In magnification figure, it appears that the curves of the inversion results are separated from each other.

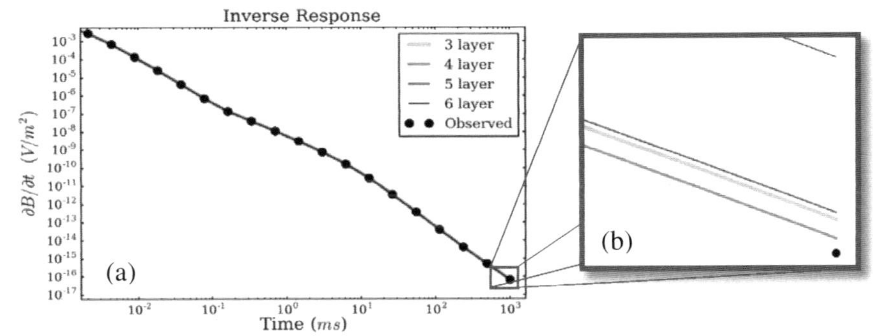

Figure 3. **(a)** The time derivative of magnetic field of the model E and its inversion results using four different starting models; **(b)** The magnification of the inversion results, shows separated curves for each starting model.

Figure 4 shows the inversion results of the model E using four starting models. Generally, the four inversion results show a model similar to one another. Resistivity value of each layer can be approached with pleasant results for all starting models. Starting model with more layers than the synthetic model can produce inversion models close to the synthetic model by combining thickness and resistivity of the layers.

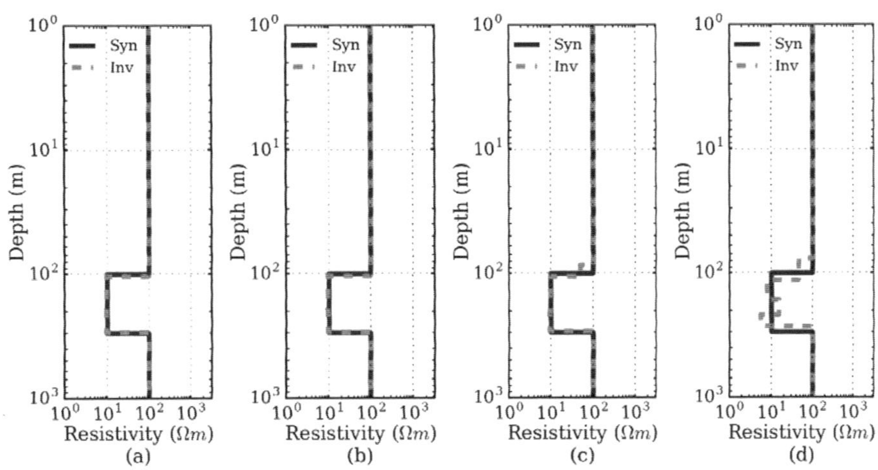

Figure 4. Inversion results of model E by using starting model **(a)** 3, **(b)** 4, **(c)** 5, and **(d)** 6 layers. Dash line is inversion results and full line is synthetic models.

7. Real data inversion

This section will use real TEM data for testing the inversion and forward modeling algorithm. The TEM data were measured in Volvi basin. The Volvi basin, a neotectonic graben structure, is located northeast of the city of Thessaloniki in Northern Greece. The research was done by Widodo, with a combination of radio magnetotelluric and transient electromagnetic to analyze the structure. In this research, three data are chosen for further inversion processes, which are TEM 1, 2, and 3 [7].

The results of this inversion show that the models of three layers staring model have not been able to produce the appropriate model, this refers to the high RMS error (Figure 5). Inversion of four layers model give best RMS error.

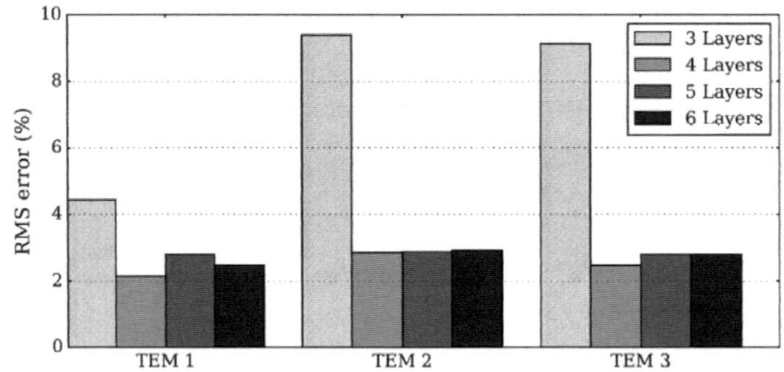

Figure 5. RMS error graph of inversion results for data TEM 1, 2 and 3 by using four synthetic models by using four different layers starting models.

Figure 6a is the time derivative of magnetic field response of TEM 1 and its inversion results from four different starting models. The synthetic and inversion responses are combined in order to compare it qualitatively. With the objective of determining the best results, magnification is done to the Figure 6a (Figure 6b). In the magnification figure (Figure 6b), it appears that the curves of the inversion results are separated from each other. Further inversion result from starting model with four layers is the closest to the synthetic response.

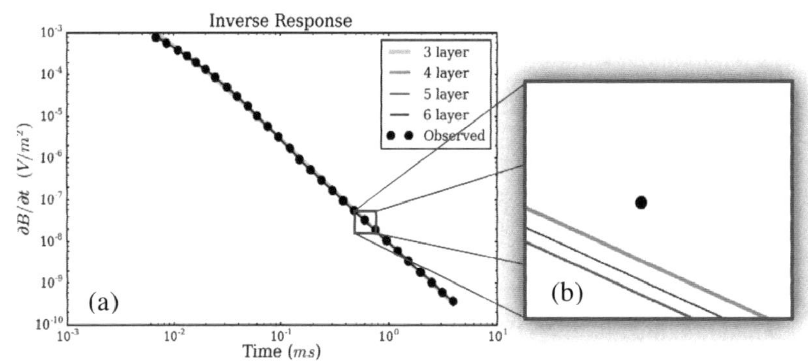

Figure 6. (a) The time derivative of magnetic field of data TEM 2 and its inversion results using four different starting models; **(b)** The magnification of the time derivative of magnetic field inversion results, shows separated curve for each starting model.

Through qualitative analysis using the RMS error (Figure 5) and qualitative analysis, it can be specify that the four layers of inversion result is the best reisitivity model for data TEM 1, TEM 2, and TEM 3 (Figure 7).

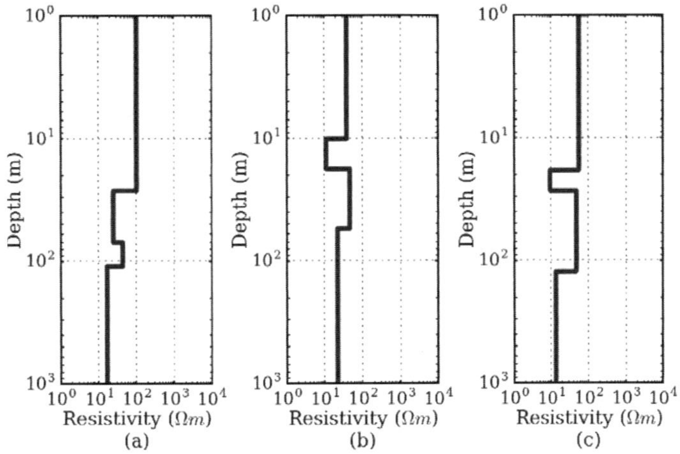

Figure 7. (a) The best inversion model for data TEM 1; **(b)** data TEM 2; and **(c)** data TEM 3.

8. Conclusion

Central loop time domain electromagnetic (TDEM) or transient electromagnetic (TEM) is a method to detect lithology layer by its conductivity properties. TDEM forward modeling program based on Born approximation is a simple method and very fast in the computation. This method is acceptable, as gives similar results with EMUPLUS's. For the inversion algorithm, Levenberg-Marquardt approach also gives pleasant results, though different layers, thickness and resistivity of starting models are used. So far, the inversion algorithm has stable inversion for solving synthetic data or real TEM data, but good starting model must be used in order to get good inversion results.

References

[1] Christensen N B 1997 *Geophysics* **18** 477–510
[2] Christensen N B 2002 *Geophysics* **67** 438-447
[3] Ward S H and Hohmann G W 1988 Electromagnetic Theory for Geophysical Applications, in Nabighian, M. N., Electromagnetic Methods in Applied Geophysics Vol I Tulsa *Society of Exploration Geophysic*
[4] Zhdanov M S 2002 Geophysical Inverse Theory and Regularization Problems, Methods in Geochemistry and Geophysics *Elsevier*
[5] Widodo and Saputera D 2016 Improving Levenberg-Marquardt Algorithm Inversion Result Using Singular Value Decomposition *Earth Science Research* **5** No. 2
[6] Spies B R and Eggers D E 1986 *Geophysics* **51** 1462-1471
[7] Widodo 2012 Multidimensional Interpretation of Near Surface Electromagnetic Data Measured in Volvi Basin, Northern Greece *Dissertation, Institut fur Geophysik und Meteorologie, Universitatzu Koln*

Analysis of Ground Penetrating Radar's Capability for Detecting Underground Cavities: A Case Study in Japan Cave of Taman Hutan Raya, Bandung

Azizatun Azimmah[1], Widodo[2]

[1]Undergraduate Program of Geophysical Engineering, Institut Teknologi Bandung, Indonesia
[2]Applied Geophysics and Exploration Research Group, Faculty of Mining and Petroleum Engineering, Institut Teknologi Bandung, Indonesia

Email: azizatunazimmah@gmail.com

Abstract. Underground cavities or voids detection is essential especially when it comes to building construction. By knowing the presence of void lying underground, one could consider whether the subsidence is likely to be prevented or not. Ground penetrating radar is a high-frequency electromagnetic sounding technique that has been developed to investigate the shallow subsurface using the contrast of dielectric properties. This geophysical method is suitable to be used to detect and locate voids beneath the surface especially those that lie in shallow depth. This research focused on how GPR could be implemented as void detector using model simulation or forward modelling. The models applied in the forward modelling process are to be made as similar as the real condition in the case study location which took place in Tahura Japan Cave, Bandung, Indonesia. Forward modelling needs to be done so in the future, we might use the modelling results as the references in measuring real GPR data in the location. We used three models that we considered fairly representative to prove that GPR is capable of detecting and locating voids underneath the ground. This research resulted in the different amplitude region around the considerably homogeneous region. The different amplitude region is characterized having an arc shape and is considered to be air which is known as the key component of voids.

1. Introduction

Underground voids could happen naturally as well as artificially like underground mining, tunnel and also cavities. Voids detection holds an important role in urban planning and construction. In such projects, especially those in the civil engineering field, it is necessary to identify every single void which lies underneath the construction site because it would bring unwanted effects such as subsidence and even total collapse to the ground surface [1][2]. Some geophysical methods commonly applied for detecting underground voids are GPR, seismic, gravity and also electrical measurement [3][4]. Ground Penetrating Radar (GPR) is one of non-destructive geophysical method most recommended to be used in studying shallow subsurface imaging [5]. GPR method has been used for many purposes due to its effectiveness in mapping the shallow depth subsurface including its application for detecting human buried body [6]. Its data sensitivity to dielectric permittivity variation make GPR method could be applied to detect any void (natural or anthropogenic), for the reflection caused by the interface between soil-air and air-soil could be easily identified. This research aims to prove that GPR could be applied to detect underground voids by imaging subsurface structure where its voids positions have already been known. Hence, we chose Goa Jepang Taman Hutan Raya (Japan Cave of Tahura) as the area of the investigation.

2. Method

2.1. Ground Penetrating Radar in Void detection

GPR works by transmitting EM waves as a source that penetrates medias beneath the site of investigation. Any reflected wave caused by material interfaces in the subsurface is then detected by receiver antenna.

When a propagating electromagnetic wave encounters a discontinuity in electric, magnetic or conductive properties, part of the electromagnetic energy is reflected, and the reflection strength being proportional to the magnitude of change. For a perpendicular incident wave, the strength of a GPR reflection is a function of the contrast in relative dielectric constants across the reflecting boundary, and the reflection coefficient RC can be expressed as [7]:

$$\boldsymbol{RC} = \frac{\sqrt{\boldsymbol{\varepsilon_{r2}}} - \sqrt{\boldsymbol{\varepsilon_{r1}}}}{\sqrt{\boldsymbol{\varepsilon_{r2}}} + \sqrt{\boldsymbol{\varepsilon_{r1}}}} \quad (1)$$

Where ε_{r1} represents relative dielectric constant of upper soil horizon and ε_{r2} represents the relative dielectric constant of lower soil horizon. Another research executed by Kofman et al. [8] that aims to examine the relationship between voids dimensions and wavelengths of various antennas, and the corresponding GPR responses, shows that there were strong reverberations generated by the inner surface of the void targets [9].

2.2. Geological Setting

The models, applied in forward modelling or numerical simulation, were constructed to be as similar as the geological setting had by the site of investigation. Japan Cave Taman Hutan Raya is in Dago Pakar, Bandung, Indonesia and geographically located in latitude 6° 51' 23.6844" S and longitude 107° 37' 56.5428"E.

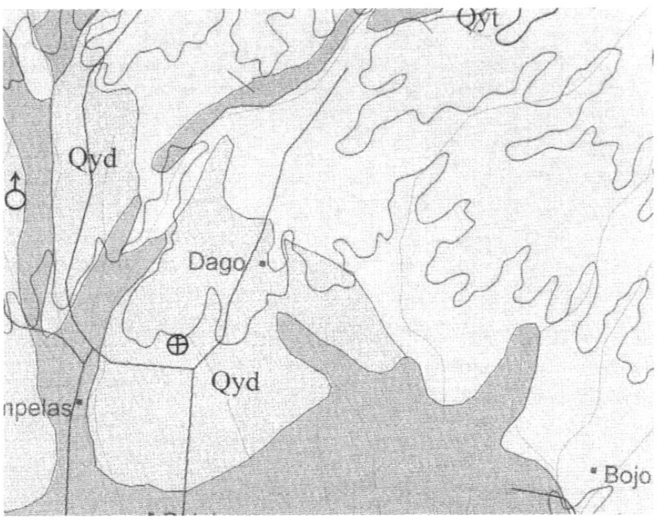

Figure 1. Geology map of Dago (Silitonga, 1979)

Type of the rocks dominated in Dago Pakar belongs to the group of sandy tuff originated in Dano mountain and Tangkuban Perahu volcano, according to regional geology map published by Silitonga in 1979 [10] (Figure 1). The map indicated that the type of rocks in Dago Pakar belongs to Qyd formation which consists of sandy tuff sourced by Dano and tangkuban Perahu volcano. The brown sandy tuff contains coarse and hollow hornblende mineral crystal and it has glassy luster and also resembles reddish lava, lapilli layers and breccia [11]. The material that forms the cave wall is merely hard rock without being layered by cement-concrete. Based on regional geology map, this hard rock is considered to be tuff that resembles the rock formed in Dago Pakar and it is likely to be emanated from the eruption of Tangkuban Perahu Volcano [12].

Figure 2. Pyroclastic tuff rock as the rock forming the cave wall

2.3. Numerical Simulation

To conduct a preliminary study of the signal characteristics of GPR reflections in Japan Cave Dago, the numerical simulation or forward modelling was done using MatGPR. In this stage, we constructed three synthetic models, and also one homogenous model which is used as a comparison to distinguish whether any cave or void exists. The simulation is done using finite different time domain (FDTD) method. The chosen profiles are shown in Figure 3 as red, yellow and purple lines.

Figure 3. The illustration of Japan Cave. The white cylinder represents cavity, and colored lines (red, yellow and purple) are the profiles chosen to be modeled.

The physical parameters used in forward modelling process is shown in Table 1.

Table 1. Rock electromagnetic properties in Japan Cave Dago

Rock Type	Resistivity	Relative dielectric constant	Relative magnetic permeability	Velocity (m/ns)
Tuff	10^5 [13]	2.6 [14]	1.021 [15]	0.14835
air	$1{,}3\times10^{16}$ [16]	1 [17]	1.00000037 [18]	0.2998
soil	1000 [19]	5 [19]	1.0006 [20]	0.13401

3. Result

The models used in this research and the synthetic radargrams resulted by those models are shown in the figures below:

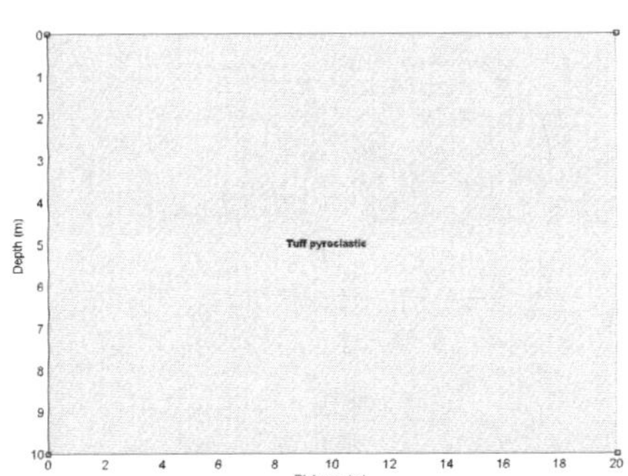

Figure 4. Homogenous model

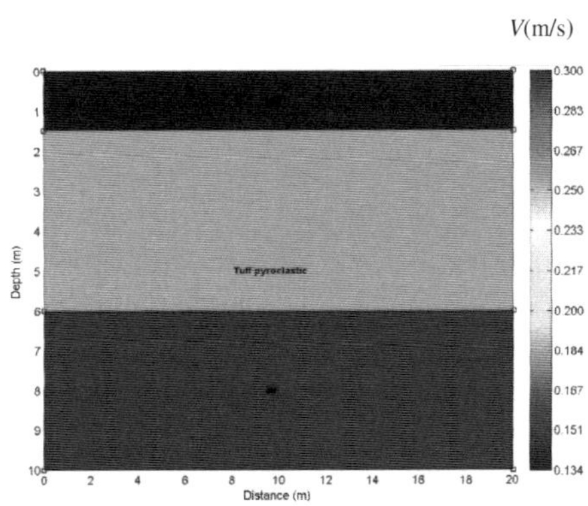

Figure 5. Line 1 geological model

IOP Conf. Series: Earth and Environmental Science **62** (2017) 012030 doi:10.1088/1755-1315/62/1/012030

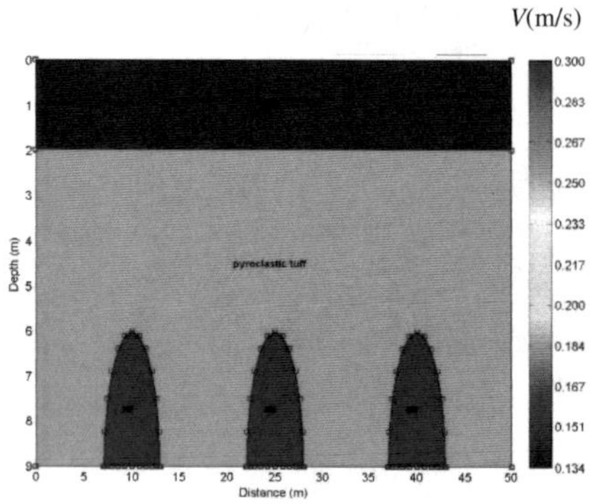

Figure 6. Line 2 geological model

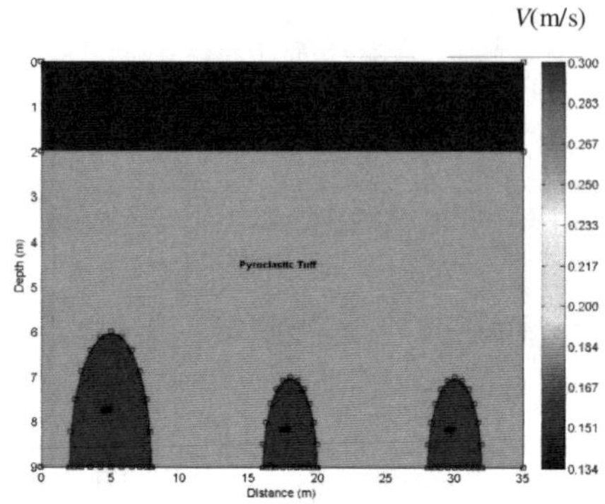

Figure 7. Line 3 geological model

Forward modelling in 100 MHz frequency:

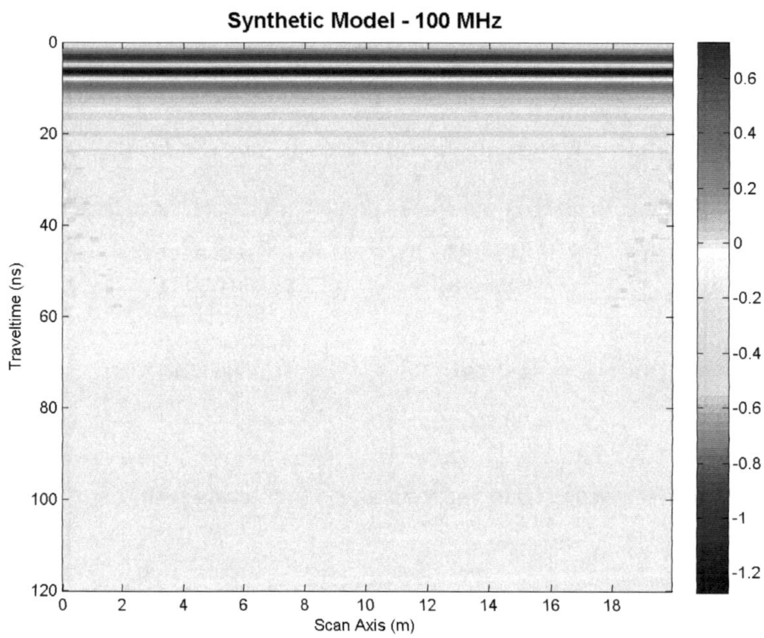

Figure 8. Homogenous synthetic radargram of 100MHz antenna

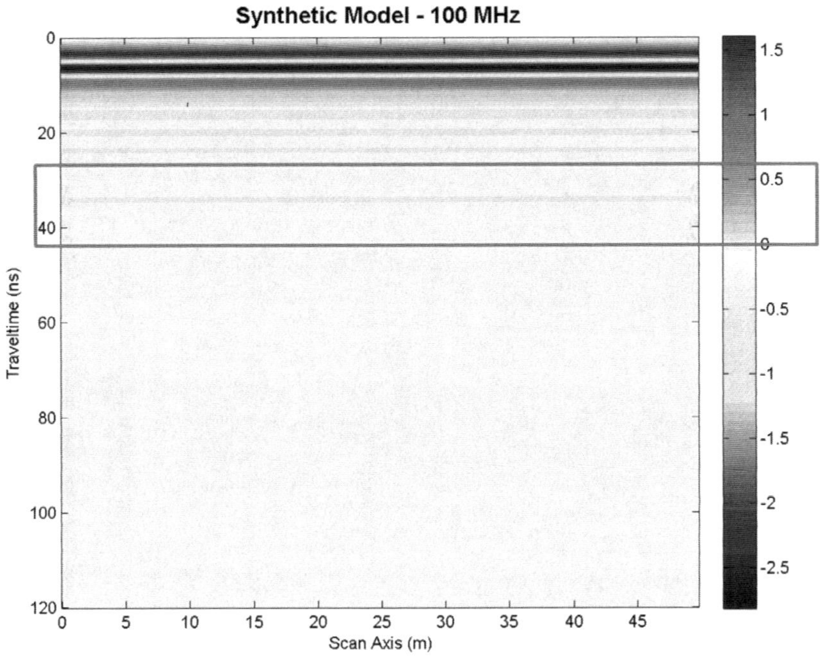

Figure 9. Line 1 synthetic radargram of 100MHz antenna

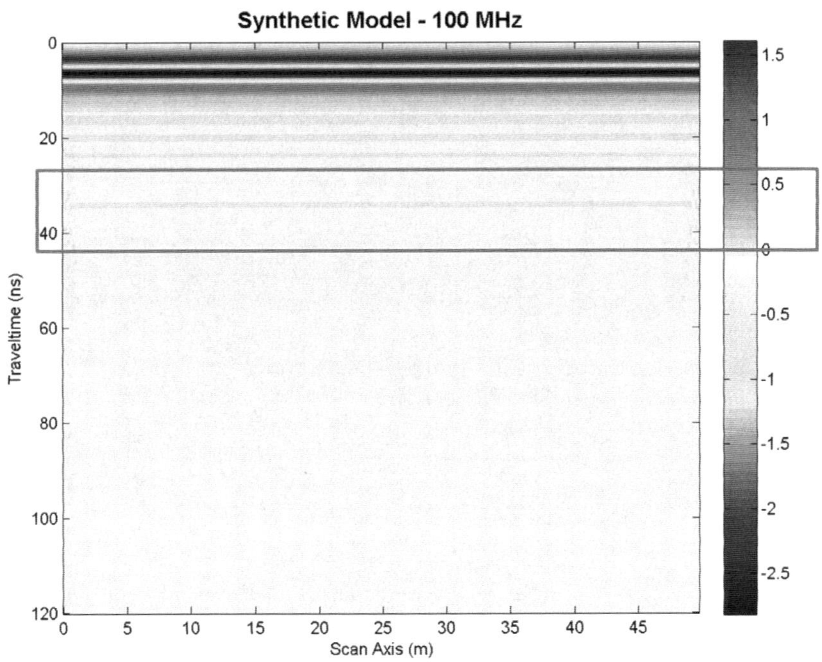

Figure 10. Line 2 synthetic radargram of 100MHz antenna

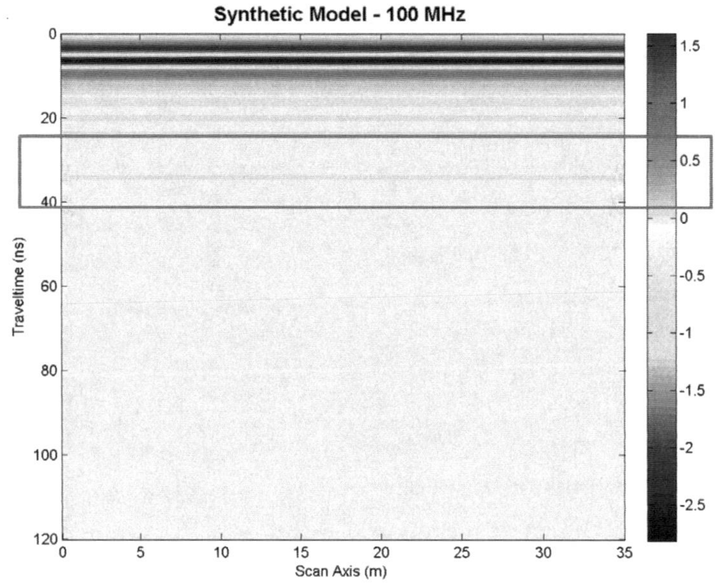

Figure 11. Line 3 synthetic radargram of 100MHz antenna

Forward modelling in 50 MHz frequency:

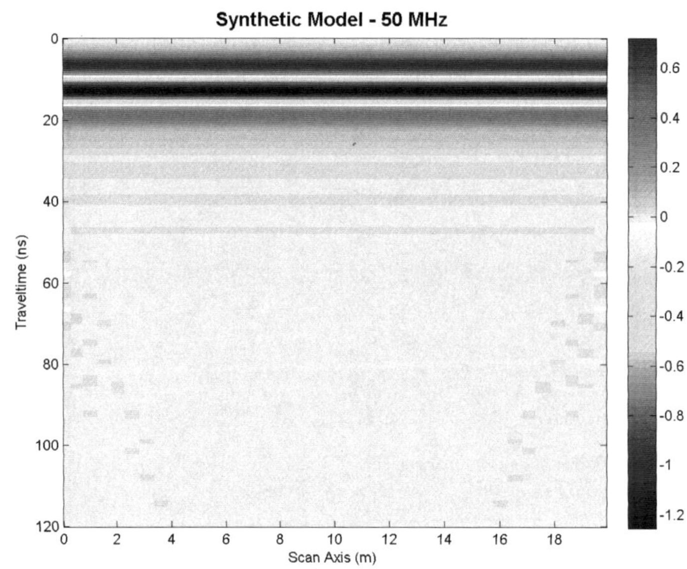

Figure 12. Homogenous synthetic radargram of 50MHz antenna

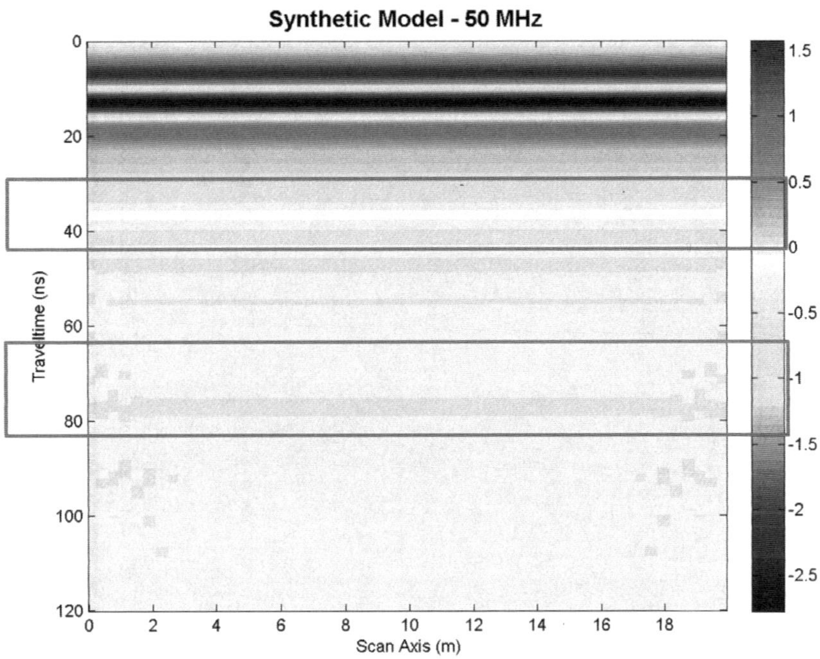

Figure 13. Line 1 synthetic radargram of 50MHz antenna

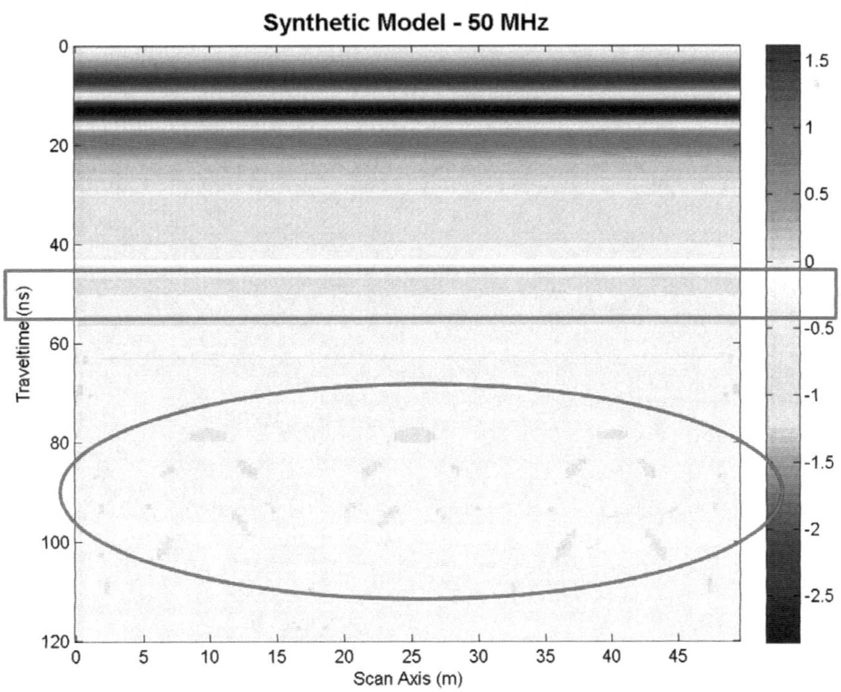

Figure 14. Line 2 synthetic radargram of 50MHz antenna

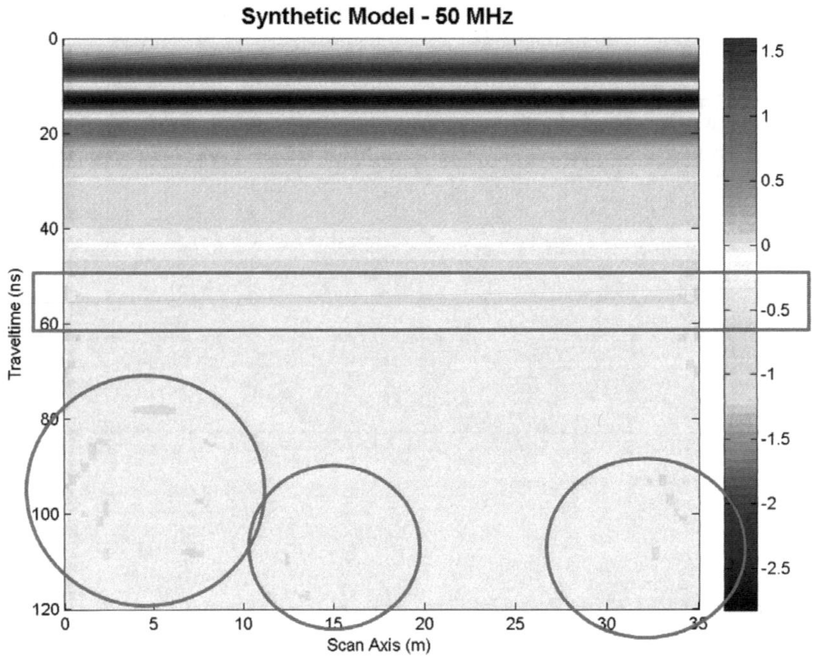

Figure 15. Line 3 synthetic radargram of 50MHz antenna

4. Conclusion

By assuming physical parameters used, modelling results indicate that forward modelling on a frequency of 100 MHz could not identify the interface between the tuff and air, but all the three synthetic models can still distinguish the boundary layer between the soil and tuff. However, forward modelling at a frequency of 50 MHz, with deeper penetration depth, all the three synthetic models can identify the boundary layer between the soil and tuff, and also the interface between the tuff and air. The comparison can be seen in Figure 16.

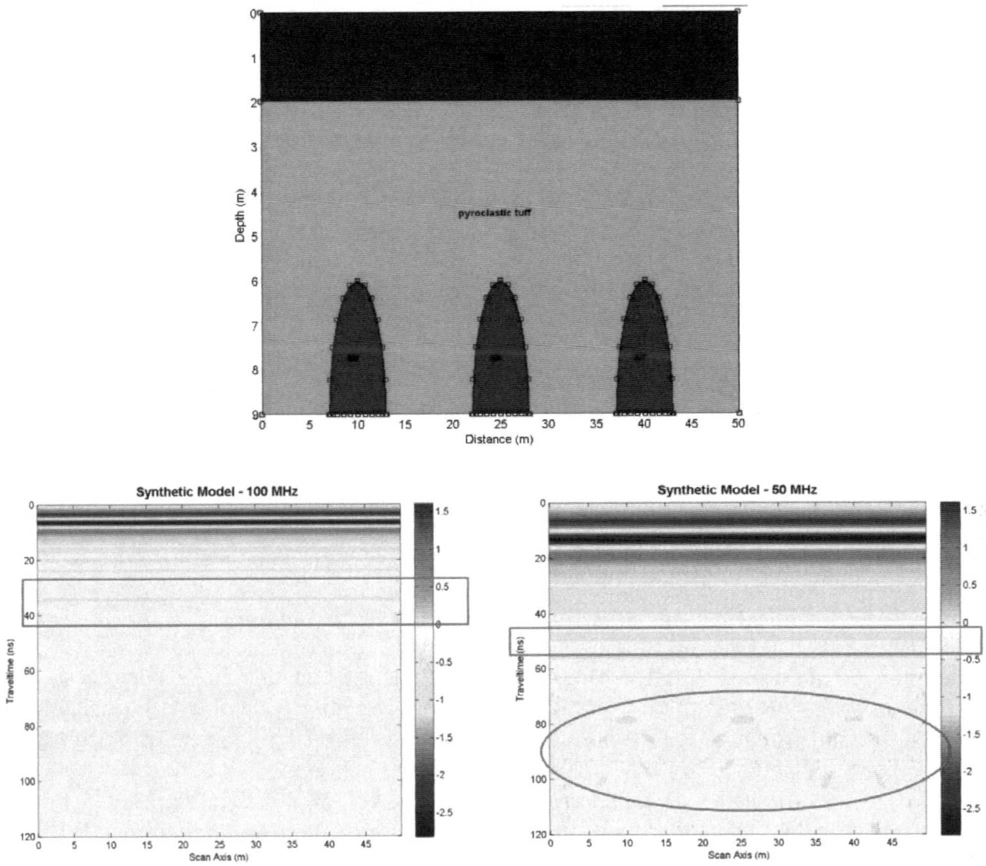

Figure 16. Comparison of forward modelling results of 50 MHz and 100 MHz frequency in line 2.

The differences between the forward modelling results happen because the energy assigned to send back wave reflection resulted by the interface of tuff-air toward the surface is insufficient at the test frequency of 100 MHz. Nevertheless, at a frequency of 50 MHz, with higher penetration depth, there is still enough energy left to deliver the reflection wave of tuff-air boundary layer toward the surface, and therefore the contrast can still be read via the receiver. However, this test should still be validated by using data from field acquisition; because the investigation depth of Japan cave (6-10 meters), should still be accommodated by a GPR 100 MHz.

References

[1] De Bruyn I A, Bell F G 2001 The occurrence of sinkholes and subsidence depressions in the far West Rand and Gauteng province, South Africa, and their engineering implications *J. Environ Eng. Geosci.* **7** 281-95

[2] Waltham T, Bell F 2005 *Sinkholes and Subsidence.Karst and Cavernous Rocks in Engineering and Construction* ed Culshaw M (Berlin: Springer) p 382

[3] Mareš S, Kelly, W E. & Mazač O 1997 *Applied Geophysics in Environmental Engineering and Science* (Prague: Universita Karlova) p 382

[4] Reynolds J M 1997 *An Introduction to Applied and Environmental Geophysics* (Chichester : Wiley)

[5] Davis J L and A P Annan 1989 Ground-penetrating radar for the high resolution mapping of soil and rock stratigraphy *J. Geophys. Prospec.* **37** 531–51
[6] Widodo, Aditama I F, Syaifullah K, Mahya M J, Hidayat M 2016 Detecting buried human bodies using ground-penetrating radar *J. Earth Sci. Res.* ISSN 1927-0542
[7] Annan A P 2001 *Ground penetrating radar workshop notes* (Mississauga: Sensors and Software Inc)
[8] Kofman L, Ronen A & Frydman S 2006 Detection of model voids by identifying reverberation phenomena in GPR records *J. Appl. Geophys.* **59** 284–99.
[9] Gosar A 2012 Analysis of the capabilities of low frequency ground penetrating radar for cavities detection in rough terrain conditions: the case of divača cave, Slovenia *Acta Carsologica* **1**(41), 77-88
[10] Silitonga P H 1973, *Peta Geologi Lembar Bandung* scale 1 : 100.000 (Bandung: PPPG)
[11] Silitonga P H 1979, *Peta Geologi Lembar Bandung* scale 1 : 100.000 (Bandung: PPPG)
[12] Ardi N D, Iryanti M 2009 Profil resistivitas 2D pada gua bawah tanah dengan metode geolistrik konfigurasi wenner-schlumberger (studi kasus Gua Dago Pakar, Bandung) *J. Pengajaran MIPA* **14** L2
[13] Telford W M, Geldart L P, Sheriff R E & Keys D A 1976 *Applied Geophysics* (London : Cambridge University Press)
[14] Campbell M J and Ulrichs J 1969 Electrical properties of rocks and their significance for lunar radar observations *J. Geophys. Res.* **74** 5867–81
[15] Scott J H *1971 Electrical And Magnetic Properties of Rock And Soils: Electromagnetic Pulse Theoretical Notes* AFWL EMP 2-1
[16] Pawar S D, Murugavel P, Lal D M 2009 Effect of relative humidity and sea level pressure on electrical conductivity of air over Indian Ocean *J. Geophys. Res.* **114** D02205
[17] Davis J L and Annan A P 1989 Ground-penetrating radar for high-resolution mapping of soil and rock stratigraphy *J. Geophys. Prospec.* **37** 531-51
[18] Cullity B D and Graham C D 2008 *Introduction to Magnetic Materials 2nd edition*
[19] Kapfer, Robert M 2005 *An advanced specular and diffuse Bidirectional Reflectance Distribution Function target model for a synthetic aperture ground penetrating radar* Thesis from Rochester Institute of Technology. Accessed from http://scholarworks.rit.edu/theses/8074
[20] Nettleton L.L 1940 *Geophysical Prospecting for Oil* (New York: McGraw Hill Book Co., Inc.) p 444

Identification of Non-Uniform Ground Comparing Responses Function of Incident SH Wave by DWM to Resonance Frequency by HVSR

Tun K. M.[1*], Pramumijoyo S.[2], Sismanto[3] and Kiyono J.[4]

[1] Ph.D. Candidate, Geological Engineering Department, Gadjah Mada University
[2] Professor, Geological Engineering Department, Gadjah Mada University
[3] Professor, Geophysics Department, Gadjah Mada University
[4] Professor, Department of Urban Management, Kyoto University

*Email:khinmyotun.geol@gmail.com

Abstract. In this paper, we simulated frequency responses of subsurface due to incident SH wave by using discrete wave number method (DWM: Aki-Larner Method). It is useful in constructing the models of the non-uniform subsurface. We estimated subsurface profile based on the resonance frequency of HVSR method comparing with frequency responses obtained from SH waves incident by DWM. The only information from HVSR is the resonance-frequency of the ground. Although what obtained from the numerical analysis by DWM is a response function, it coincides with the response of the surface ground if the input motion is a white noise and the ground behaviour is a liner. In this case, the peak of response function can be considered to be a resonance frequency of the ground. Therefore, we here compared the peak frequencies of HVSR curve and response functions. To estimate the resonance frequency of ground by HVSR, the single observations of microtremor at 161 sites were carried out by using the three-component accelerometer with a data logger, GPL-6A3P, in the study area of Ende Regency that is one of five regencies in East Nusa Tenggara Province on the island of Flores, Indonesia. The results show that site response in Ende area varies the resonance frequencies span 0.7-1.4 Hz. For DWM, we assume simple two layered media with irregular boundary aimed at making a simple ground profile of the study area. The two-layered model is composed of a 'soft basin' on engineering bedrock. We varied the depth and the shear wave velocity of the basin and calculated response functions by using DWM. Then we selected the best fit parameters by comparing the resonance frequency of H/V result. Finally, we proposed a possible non-uniform ground model in the study area.

1. Introduction

Many earthquakes occurred in Ende Regency in the past that is one of five regencies on the island of Flores at East Nusa Tenggara Province in Indonesia. Moreover, Flores Island is as a segment of the Volcanic Inner Banda Arc and has several active volcanoes. The region with respect to the earthquakes must also be considered a high seismic activity area. That's why we chose the study area as a site of interest, and highly susceptible of seismic hazard zone. Figure 1 shows the study area.

IOP Conf. Series: Earth and Environmental Science **62** (2017) 012031 doi:10.1088/1755-1315/62/1/012031

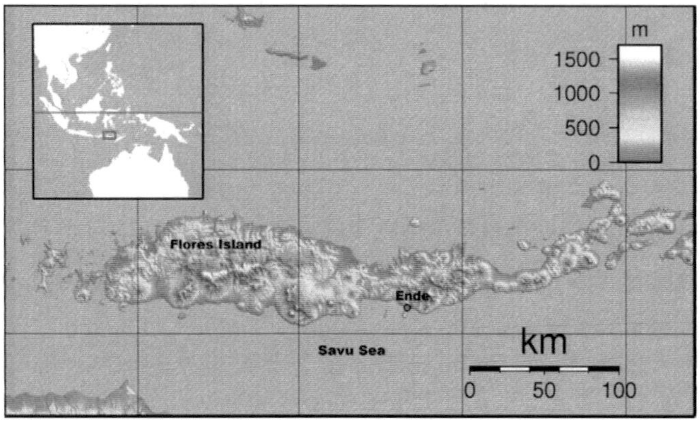

Figure 1. Location of the study area

Detailed subsurface explorations of such ground provide key factors for the evaluation of the seismic intensity at each site. In this study, we estimated subsurface profile based on the resonance frequency of Horizontal to Vertical Spectrum Ratio method [3] comparing frequency responses obtained from SH waves incident by discrete wave number method (DWM: Aki-Larner Method) [3].

2. Estimation of Resonance Frequency (*f*)

Single observations of microtremor at 161 sites with an interval of about 250 m in the study area were carried out by using the three-component accelerometer with a data logger, GPL-6A3P as shown in Figure2. This instrument is most suitable for detecting earthquake motions, ground vibration, and vibration characteristics (waveform) of buildings. The main object of this research is to perform the ground motion from especially the seismic microzonation.

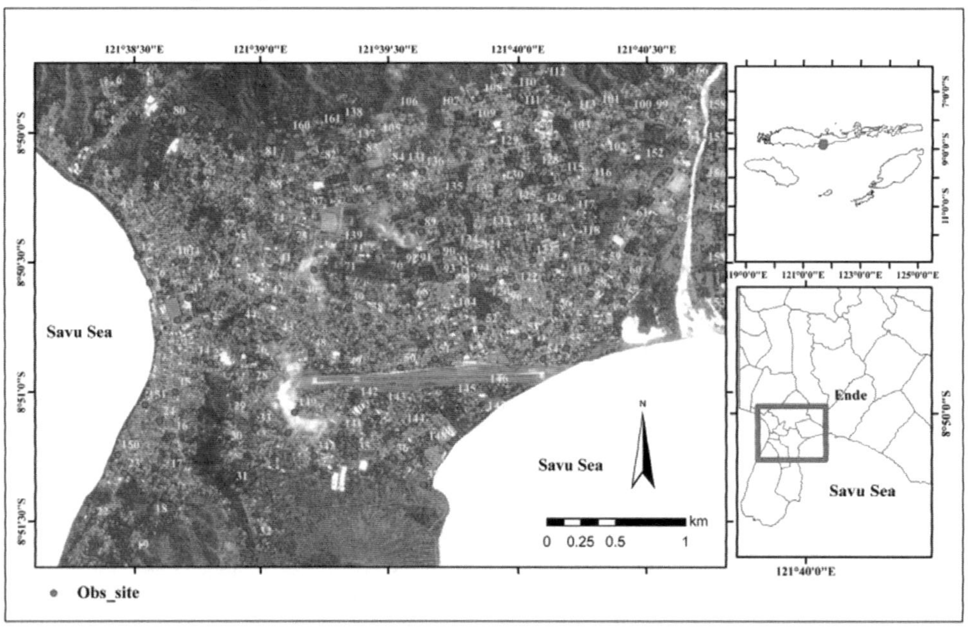

Figure 2. Location map of microtremor single station sites in Ende area

Microtremors, ground vibration, appear on the ground surface. Microtremor data analysis can be done using the method of Horizontal to Vertical Spectrum Ratio (HVSR). By looking at the peak frequency of the H/V spectral ratio, resonance frequency of ground can be estimated by [3].Equation (1) becomes the basis for calculating the ratio of the horizontal component of the spectrum of the vertical component as expressed below.

$$\frac{H_s}{V_s} = \frac{\sqrt{(S_{N-S})^2 + (S_{E-W})^2}}{S_{vertical}} \tag{1}$$

Where, H_S= Spectrum component horizontal layers of sediment
V_S=Spectrum component vertical layers of sediment
S_{N-S} = Fourier amplitude of the NS
S_{E-W}= Fourier amplitude of the EW
$S_{Vertical}$ = Fourier amplitude of the Vertical

The result of the analysis of the data using HVSR is resonance frequency (f_0) and the peak of the spectrum (A_g) of the H / V spectral peak that are the parameters from the reflection of dynamic characteristics of the soil layer as shown in Figure 3.

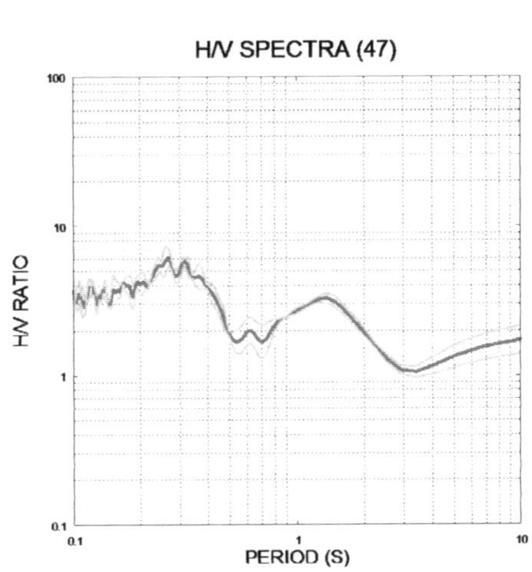

Figure 3. An example of the spectrum of H / V results analysis

Resonance frequency (f_0) values of the Horizontal to Vertical Spectrum Ratio (HVSR) in this study range between 0.7 and 1.4 Hz. Figure4 shows the distribution of resonance frequency estimated each measurement site. In the Figure 4 presents large (f_0) value on northern and southern part overall, and small (f_0) value on the middle part of the area having younger deposits, we could estimate that the thickness of the plane in the middle part was thicker than on the northern and southern part.

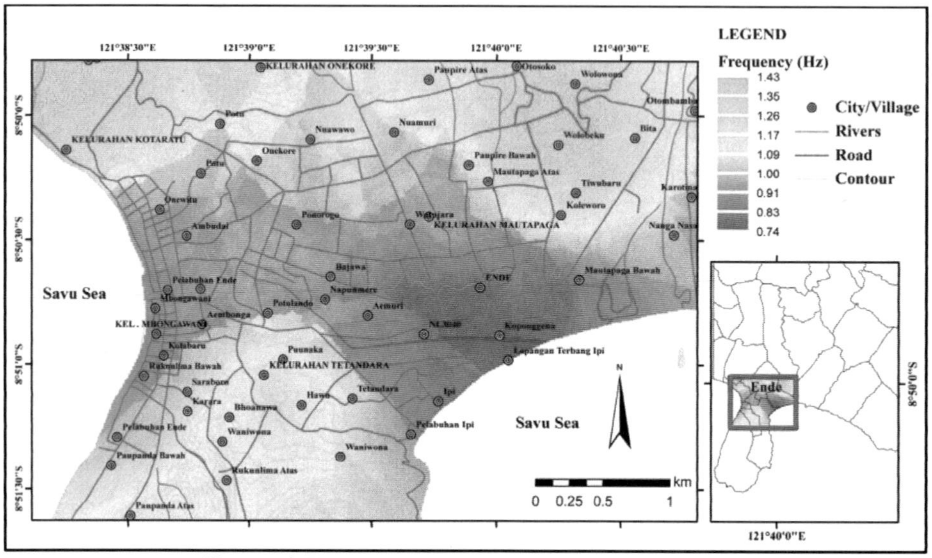

Figure 4. Resonance Frequency Map of Ende area

3. Identification of Simple Ground model

There is no information to estimate the ground profile by array observation or other ground exploration method in the study area. So, we here identify the ground model by using simple method [3]. In this article, we adopted discrete wavenumber Method. The method expresses the ground displacement as an infinity integral of wave number. If the irregularity of the interface is assumed to be periodic, the infinity summation of wave numbers replaced the place of the infinity integral. By truncating the infinity number and introducing continuity condition of displacement and stress at the interface, it determined the unknown coefficient of the up-going and down-going wave.

Now we consider the problem for SH-wave input to two layered media. The displacement of the first (upper) and second (basement) layer as expressed below.

$$u_1(x, z) = \int_{-\infty}^{\infty} \left\{ A_1(k) \exp(iv_1 z) + B_1(k) \exp(-iv_1 z \right\} \exp(ikx) dk \tag{2}$$

$$u_2(x, z) = \exp\left\{ i(k_0 x - v_0 z) \right\} + \int_{-\infty}^{\infty} A_2(k) \exp\left\{ i(kx + v_2 z \right\} dk \tag{3}$$

in which k and v are wave number for x- and z-direction, respectively, and A_1, B_1 and A_2 are unknown coefficients that should be solved.

The only information obtained from HVSR is a resonance frequency. Although what obtained from the numerical analysis by DWM is a response function, the response function coincides with the response of the surface ground if the input motion is a white noise and the ground behavior is a liner. In this case, the peak of response function can be considered to be a resonance frequency of the ground. Therefore, we here compared the peak frequencies of HVSR curve and response functions.

We, first, investigated the effects of incident angles on the response of the ground because the response amplitude is affected strongly by the incident angle of the wave. Figure5 shows frequency response curves for SH wave incident angles of 0°, 30°, 45°and 60°.

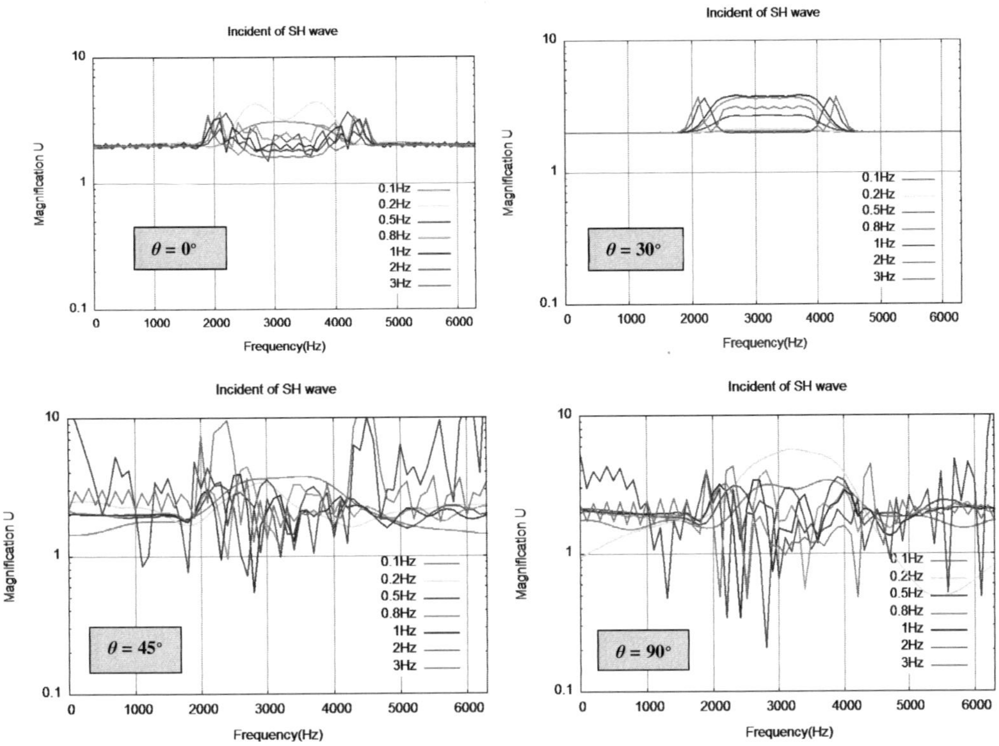

Figure 5. Magnification factors of some frequency for different incident wave angles (θ=0°, 30°, 45° and 60°) obtained by the discrete wave number method

For vertical incidence (θ = 0°) and almost vertical incidence (θ = 30°), the flat layer approximation made in calculations of the response amplitude is valid for certain frequency ranges. However, for larger incident angles (θ = 45° and 60°) the ground irregularity must be taken into account in evaluations of the seismic response of the ground; the simple multi-layered ground model gives very different and meaningless results as shown in Figure5. When the incident angles are 0° and 30°, the overall trend of the transfer functions have peaks coincide. Sharp fluctuations in the transfer function, however, take place at higher incident angles. This indicates that the incident angle is not a parameter that needs to be identified when the transfer function is used as an observation value. We use the value of the incident angle (θ) is 0° because we do not know exactly the direction of the source.

In this study, we considered many profiles depend on topography in the study area and frequency map from H/V analysis as shown in Figure6. We assumed a simple two-layered media (basin and bedrock) as in Figure 7. The ground structure of the section is roughly divided two layers; alluvial soil for the basin with an irregular interface and engineering bed rock. We fixed the shear wave velocity of the engineering bedrock as 400m/s. We changed randomly the value of the Shear wave velocity of the basin (V_1) between 100 m/s ~ 200m/s, the depth (D) between 50m ~ 500m and the length of the basin bottom (B) between 1km ~ 3km depend on each profile. Then we selected the best fit parameters comparing to resonance frequencies from H/V on each observation site. In Figure 8 showed an example of the comparison between the H/V ratios curves and response functions obtained results from SH wave input at an observation site.

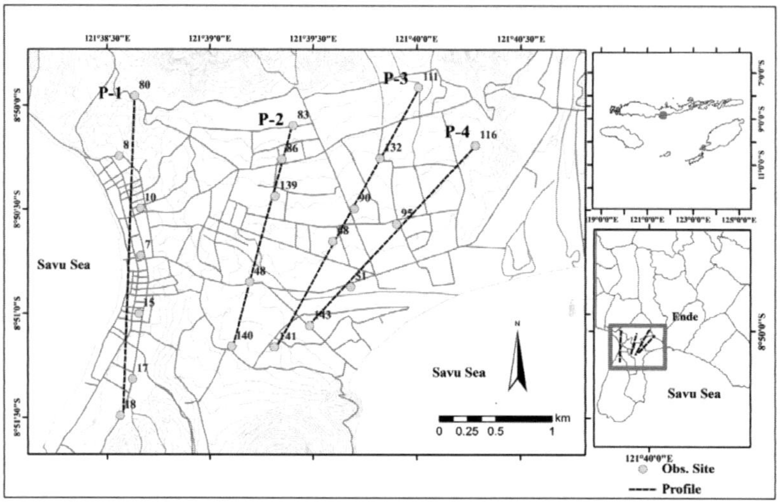

Figure 6. Four profiles map in Ende area.

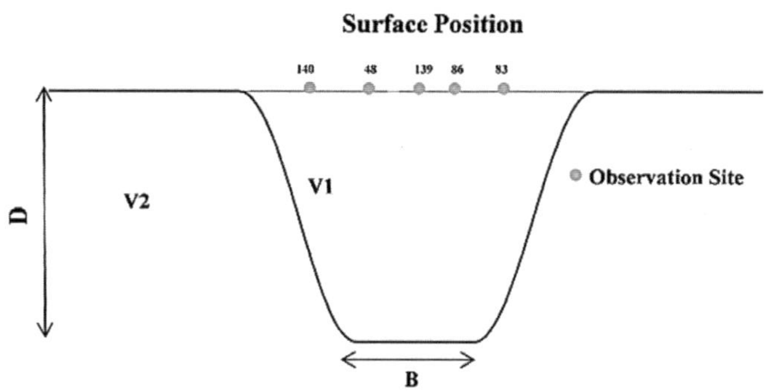

Figure 7. Schematic simple ground model for two layers media with Observation Points.

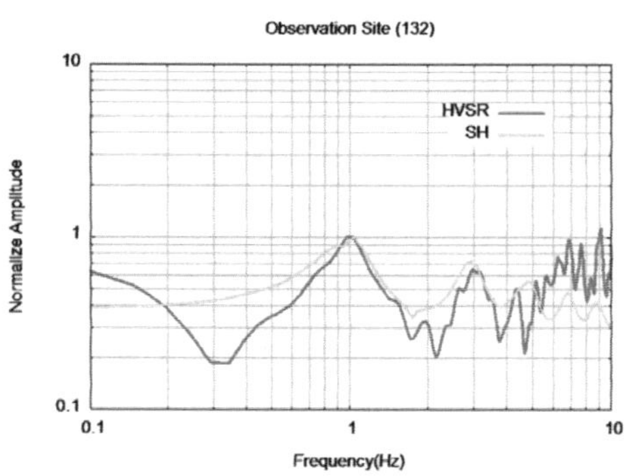

Figure 8. A comparisons between the H/V ratios curves and response functions obtained results from SH wave input at an observation site.

The results compared are listed in Table 1.1 and 1.2 for profile lines. This is the result from the best fit parameters. Figure 9 shows the ground models identified from each profile. Table 1.3 shows the shear wave velocity and depth of the basin (V_1) for each profile.

Table 1.1 Comparison the value of Frequency from SH by DWM method and HVSR for profile (1&2)

Obs. site	18	17	15	7	10	8	80	140	48	139	86	83
HVSR	1.02	0.82	0.73	0.82	0.92	1.22	1.36	2.34	0.78	0.78	0.87	1.80
SH	1.18	0.87	0.71	0.79	0.87	1.18	2.60	2.36	0.79	0.79	1.11	1.34

Table 1.2 Comparison the value of Frequency from SH by DWM method and HVSR for profile (3&4)

Obs. site	141	68	90	132	143	51	95	116
HVSR	1.02	0.82	0.73	0.82	0.92	1.22	1.36	2.34
SH	1.18	0.87	0.71	0.79	0.87	1.18	2.60	2.36

Table 1.3 Shear wave velocity and depth of soil layer from SH by DWM method for each profile

Profile	1	2	3	4
Shear wave velocity by SH (m/s)	150	169	149	168
Depth (m)	50	49	55	65

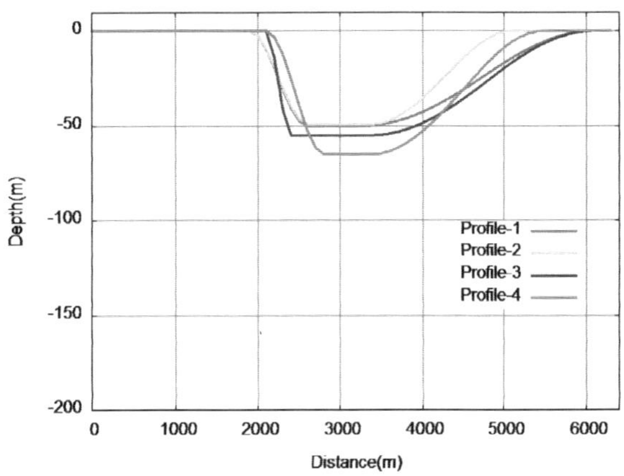

Figure 9. Identified ground models from SH obtained by DWM method.

4. Conclusion

In this paper, we conducted microtremor observations in Ende area, Indonesia, and numerical calculation to determine the ground profile. From the microtremor observation, we could find that the range of resonance frequency in Ende area is between 0.7 to 1.4 Hz. As the target area was between the mountains of its north and south, we could estimate that the sedimentary basin formed the plane in the middle part. Two-layered basin model was assumed. By comparing with the results of HVSR and DWM, we could propose the ground profile of a line. However, the actual ground would be more complicated.

Acknowledgements
This research was supported by JICA/AUN SEED-Net. I sincerely thank to the staffs at Gadjah Mada University and Kyoto University for their help and kind assistances during the analysis of this research.

References

[1] Nakamura, Y., Sato, T., and Nishinaga, M., 2000, Local Site Effect of Kobe Based on Microtremor Measurement, *Proceeding of the Sixth International Conference on Seismic Zonation EERI*, Palm Springs California.

[2] Aki, K. and Larner, K. L., 1970, Surface Motion of a layered Medium Having an Irregular Interface Due to Incident Plane SH Waves, *Journal. Of Geophys*. Res., Vol. **75**, No.5.

[3] Kiyono, J., Inoue, Y., and Shimizu, K., 2009, Earthquake Effects of layered Medium with Irregular Interface on Buried Pipeline, *Proc. Of EIT-JSCE Joint Symposium*, pp. 82-86, September 7-8, Bangkok, Thailand.

Development of 3D Soil Profiling System for A Damaged River Levee

Kazunori Murata[1], Takao Aizawa[1], Shunichiro Ito[1], and Toru Takahashi[2]

[1]Suncoh Consultants, Kameido, Koto-ku, Tokyo, 136-8522, Japan
[2]Fukada Geological Institute, Honkomagome, Bunkyo-ku, Tokyo, 113-0021, Japan

E-mail: k.murata@suncoh.co.jp

Abstract. A 3D soil profiling system for effective design of countermeasures for a damaged or potentially damaged river levee is developed. This system consists of a 3D integrated seismic and electric measurement tool and a 3D visualization system of its soil properties such as hydraulic and mechanical properties estimated from the geophysical data. For a quick survey, the 3D measurement tool is a sensor belt with both geophones and electrodes to be able to measure seismic and electric data simultaneously. S-wave velocity and resistivity data acquired by this tool are converted to soil properties such as hydraulic and mechanical properties and support a countermeasure design of levee. In this study, we introduce the experimental result of the 3D soil profiling by using the integrated seismic and electric measurement tool in the field.

1. Introduction

In Japan, recent localized torrential rainfalls have often caused serious damages to river levee as seen in the water floods in the Kinugawa River, 2015 in Ibaraki Pref. and in the Yabe River, 2012 in Fukuoka Pref. They have brought much public attention to the present state of river levee and to their investigation for safety evaluation and strengthening. In such investigations the internal structure of the earthen levee must be delineated for further analysis and design for countermeasures.

Geophysical methods can be one of the most effective tools for that purpose because of their quick and non-destructive subsurface profiling capability. Especially the seismic and electric methods with land streamer-type measurement systems have been aggressively employed for profiling the dike body and subsoil in the longitudinal direction along the levee.

On the other hand, geophysical methods have not been employed so often for profiling the cross-section of the levee, where mechanical and hydraulic analyses are performed for safety evaluation. The reason is said that the accuracy of the geophysical profile is not so enough for further analyses and geotechnical parameters such as permeability necessary for numerical simulations cannot be directly given. However recently 3D geophysical survey which is more accurate than conventional 2D survey is attempted to be applied to river levee and new interpretation techniques of multiple geophysical properties based on rock physics have been proposed to estimate geotechnical parameters from geophysical ones [1].

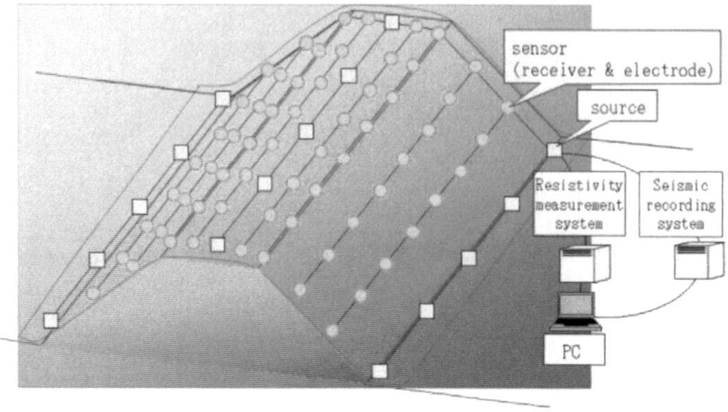

Figure 1. The schematic diagram of the 3D integrated seismic and electric measurement tool

In this study, therefore, we have started to develop a 3D geophysical soil profiling system consisting of a 3D integrated seismic and electric measurement tool and a 3D visualization system of soil profiles estimated from geophysical parameters. So far a prototype of the measurement system was manufactured and tested in a river levee and an interpretation technique of seismic velocity and resistivity was developed for estimating a soil type of a levee body. Figure 1 is the schematic diagram of the 3D integrated seismic and electric measurement tool.

2. 3D Integrated Seismic and Electric Measurement Tool

For safety evaluation of river levee, we have to know soil properties, permeability, strength of soil of river levee. These parameters can be estimated from seismic velocity, especially S-wave velocity, and resistivity because S-wave velocity strongly relates to strength of the soil and resistivity correlates with grain size for unsaturated soil. The measurement system we are developing, therefore, is designed for measuring S-wave and resistivity.

The prototype of the measurement tool is shown in Figure 2. The specification of the sensor belt is also shown Table 1. In the investigation of damaged and potentially damaged levee, the geophysical survey should be conducted as quickly and economically as possible. For that reason, the sensor belt of our measurement tool equips multiple geophones with a specially designed steel spikes which is used also as electrodes (Figure 2).

Figure 2. The prototype of the measurement tool

Table 1. specification of the sensor belt

Name of the part	Specification
Fabric belt	Width:70mm, Length:50m
Sensor interval	4m in basic
Geophone	28Hz, 12 string/belt
Spike electrode	Stainless steel with a base plate

This measurement tool is designed for measuring S-wave as well as P-wave by changing the geophone orientation to the ground surface. The fabric belt is very flexible so as to easily change the geophone interval. The steel spike electrode is designed and manufactured so as to be moved easily location by location.

An experiment for evaluation of the prototype of the measurement tool was carried out on a river levee. River front side of the levee body mainly consists of sand, and river back side mainly consists of cray. Its height and width are 5m and 30m, respectively. Eight sensor belts (96 channels) were set up on the ground surface. The geophone interval was 2m, meaning that the fabric belt was folded for adjusting the geophone interval to 2m from its original interval of 4m. S-wave seismic data acquisition was conducted by using a plank hammering as a seismic source. At the same time an electric survey with the pole-pole array (2m electrode interval) deploying the spike electrode of the sensor belt was carried out. The survey was completed within 90 min. for preparation, 90min. for the electric survey and 60 min. for the seismic survey. Our goal of this development is 180min. (3 hours) for all data acquisition procedures for a survey. In our final measurement system simultaneous seismic and electric data acquisition is designed so that the targeted goal is accomplished.

The result of S-wave velocity and resistivity are shown in Figure 3. River front side of the levee body (mainly sand) is relatively imaged in higher velocity and higher resistivity. And river back side (mainly clay) is relatively imaged in lower velocity and lower resistivity.

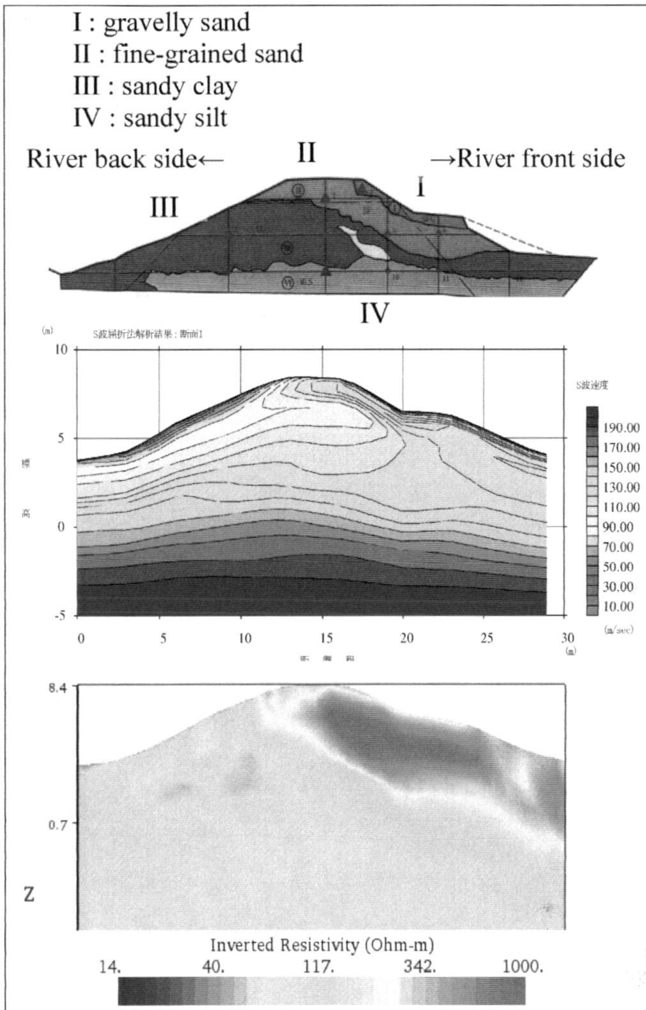

Figure 3. The result of S-wave velocity and resistivity

3. 3D Visualization System of Soil Properties

The system targeted in this project first builds a 3D geotechnical parameter model from a 3D geophysical model as output of the 3D geophysical measurement system. This model is of course built by interpreting the geophysical data in conjunction with the geotechnical data obtained in the boreholes and other investigations. The 3D visualization system helps the engineer to build the final model for the numerical analysis by visualizing the damaged interior of the levee in three dimensions.

Estimation of geotechnical parameters from geophysical parameters is conventionally performed by using a cross-plot and/or the empirical relation of both parameters. In this project to make estimation more accurate and reliable, rock (soil) physics is utilized. Figure 4 shows an example of prediction of S-wave velocity - porosity relation based on a rock physics model called as the unconsolidated sand model [2]. In this diagram actual data are also plotted, which is composed of clayey gravel, sand and clayey sand. This figure shows that the model predicted curves for different

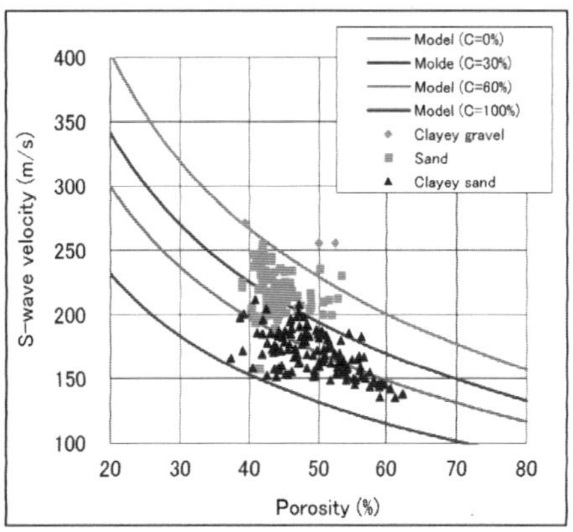

Figure 4. The prototype of the measurement tool

clay contents (0, 30, 60 and 100%) represent actual soil types very well.

We also estimate permeability by using kozeny-carman equation [3]. The equation is shown as follows.

$$k = \frac{1}{72\tau^2}\frac{\phi^3}{(1-\phi)^2}d^2$$

k: permeability
ϕ : porosity
d : particle size

Porosity can be estimated from S-wave velocity, and average particle size can be estimated from soil type. Therefore, permeability can be calculated from the two parameters. Soil type and permeability are shown in Figure 5. This figure shows good consistency between open-cut of the levee and the estimated soil profile. The estimated soil profile verified by the real data obtained in the open-cut of the levee body as shown in this figure proves that geophysical soil profiling can be feasible.

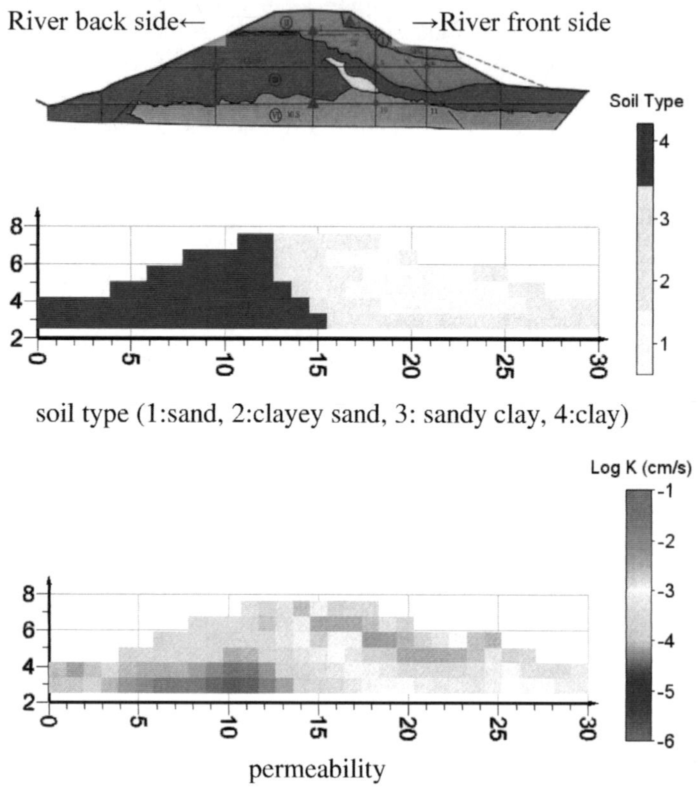

Figure 5. Soil type and coefficient of permeability estimated from geophysical data

4. Conclusion

For effective design of countermeasures for a damaged or potentially damaged river levee, we are developing a 3D geophysical soil profiling system, consisting of a 3D measurement tool for seismic and electric properties of the levee and a 3D visualization system of its soil properties such as hydraulic and mechanical properties estimated from the geophysical data. A prototype of the measurement system has been manufactures and tested on a real river levee to evaluate its performance. An estimation technique of geotechnical parameters from geophysical ones has been developed and verified through an application to real datasets of S-wave velocity and resistivity obtained seismic and electric surveys on a river levee.

References

[1] Takahashi and Yamamoto 2010 *Butsuri-Tansa* **63** (1) 102-108
[2] Avseth P et al 2005 *Cambridge University Press.*
[3] Carman P C 1956 *Academic Press, Butterworths Scientific Publications*

Acknowledgments

This study is funded by the research and development program in the MLIT (Ministry of Land, Infrastructure, Transport and Tourism).

Particle Swarm Optimization (PSO) for Magnetotelluric (MT) 1D Inversion Modeling

Hendra Grandis[1*] **and Yahya Maulana**[2]

[1]Applied and Exploration Geophysics Research Group, Faculty of Mining and Petroleum Engineering, Institut Teknologi Bandung, Jalan Ganesha 10 Bandung 40132, Indonesia

[2]Directorate General of New, Renewable Energy and Energy Conservation, Ministry of Energy and Mineral Resources, Jalan Pegangsaan Timur 1, Jakarta Pusat 10320, Indonesia

*corresponding author: grandis@itb.ac.id

Abstract. Particle Swarm Optimization (PSO) is one of nature-inspired optimization algorithms that adopts swarm (insects, school of fish, flock of birds etc.) behaviour in search for food or common target in a collaborative manner. The particles (or agents) in the swarm learn from their neighbours as well as themselves regarding the promising area in the search space. The information is then used to update their position in order to reach the target. The search algorithm of a particle is dictated by the best position of that particle during the process (individual learning term) and the best particle in its surroundings (social learning term) at a particular iteration. In terms of optimization, the particles are models defined by their parameters, while the promising area in the model space is characterized by a low misfit associated with optimum models. Being a global search approach, PSO is suitable for non-linear inverse problem resolution. The algorithm was applied to a simple minimization problem for illustration purpose. The application of PSO in geophysical inverse problem is demonstrated by inversion of synthetic magnetotelluric (MT) data associated with simple 1D models with satisfactory results in terms of model recovery as well as data misfit.

1. Introduction

In geophysics, the inverse modeling is performed to the observed (field) data to obtain the model representing the physical property (e.g. density, resistivity, magnetization, seismic velocity etc.) distribution of the sub-surface. The problem is a function minimization, where the function to be minimized is the misfit or difference between the observed data and the theoretical response of a model. For a linear inverse modeling, i.e. the model is linearly related to the data, the solution is relatively straightforward and well-known. For non-linear inverse problems, the solution is seek by linearization of the misfit function around a starting model and update that model iteratively until a minimum misfit is obtained. Such localized approach suffers from two main drawbacks, i.e. sensitivity to the starting model and possibility to converge to a local minimum rather than to a global one [1,2].

In the global approach, no linearization and gradient-free optimization is performed by intensive exploration of the model space in the search for solutions. This approach can effectively overcome difficulties of local search or linearized approach of strongly non-linear inverse problems. The Monte-Carlo based algorithms such as Simulated Annealing (SA), Genetic Algorithm (GA), Markov Chain

Monte Carlo (MCMC) are among the most popular non-linear inverse modeling algorithms that have been applied to geophysical inverse problems [3,4,5]. One of global approach algorithms which is gaining interest for geophysical inverse modeling is the Particle Swarm Optimization (PSO) proposed by Kennedy and Eberhart [6,7]. The algorithm is based on the behavior of animals in swarm to achieve a common goal with a collaborative manner. Individuals (models) occupying positions in the model space are modified iteratively by considering the best position of each individual (cognitive learning term) and the group's best position (social learning term) to reach the optimum position or solution. The position of a model in the model space is qualified with its misfit.

In electromagnetic (EM) geophysics (e.g. magnetotellurics, transient EM, etc.) the forward and inverse modeling for 1-D structure has been considered as solved problems. However, non-linearity and ill-posedness of the problem remain to be an interesting subject for research, especially for the application of global search approach e.g. [8,9,10]. The PSO algorithm has been applied to non-linear inversion of magnetotelluric (MT) and Vertical Electrical Sounding (VES) data using 1-D model with satisfactory results [11,12]. In this paper, we follow the same approach with the emphasis for MT 1-D inversion of synthetic data. It is also our purpose to illustrate the application of a global search approach to invert geophysical data with a relatively simple and yet non-linear forward problem such as MT.

2. Basic of Particle Swarm Optimization (PSO)

The Particle Swarm Optimization (PSO) is a stochastic optimization algorithm that belongs to the larger family of evolutionary computation techniques, e.g. GAs. Although it shares many similarities with GAs, PSO does not use a variety of selection, recombination and mutation operators to maintain and generate new population of potential solutions iteratively. The PSO adopts the collaborative behavior of a group of animals (insects, birds, fish) in search for a common target, e.g. food. Each member of the swarm (particle) in the search space adapts its trajectory according to the best experiences of the swarm to search for the target. The particle represents a potential solution, i.e. a model with associated fitness value, in the solution space. The target location is in fact the global optimum of the optimization problem.

In what follows, we describe only the basic PSO algorithm, while variations from the basic e.g. [13,14] are beyond the scope of this paper. The position and search trajectory of particles are defined in an N-dimensional model space. By adopting displacement formula in physics, the position $\mathbf{x} = [x_j]$; $j = 1, 2, \ldots N$ of the i-th particle at k-th iteration is expressed by,

$$\mathbf{x}_i(k+1) = \mathbf{x}_i(k) + \mathbf{v}_i(k) + \tfrac{1}{2}\mathbf{a}_i(k) \quad (1)$$

where $\mathbf{v}$ and $\mathbf{a}$ represent velocity and acceleration respectively each define in the N-dimensional space as for $\mathbf{x}$. With a unitary time step, the terms t and t^2 in velocity and acceleration respectively are eliminated. During each iteration, each particle learns from its own experience (personal influence) and from other members' experiences (social influence) as well. These two influences affect the particle's total acceleration such that,

$$\mathbf{a}_i(k) = c_1 R_{1i}\ (\mathbf{p}_i(k) - \mathbf{x}_i(k)) + c_2 R_{2i}\ (\mathbf{g}(k) - \mathbf{x}_i(k)) \quad (2)$$

where c_1 and c_2 denote the acceleration coefficients, R_{1i} and R_{2i} replacing the number ½ in equation (1) are random numbers introduced to include stochastic behavior of the process. The personal best $\mathbf{p}_i$ and the global best $\mathbf{g}$ in equation (2) are respectively: the best solution of the i-th particle and the best solution amongst all particles so far up to the k-th iteration.

The velocity update is the sum of the current velocity and the total acceleration expressed by equation (2), i.e.

$$\mathbf{v}_i(k+1) = \omega\ \mathbf{v}_i(k) + \mathbf{a}_i(k) \quad (3)$$

where the velocity update is expressed for the (k+1)-th iteration and ω set less than 1 is the inertia weight such that only a fraction of the velocity is carried over to the next iteration to prevent velocities from growing out of control. The position update of the i-th particle as in equation (1) becomes,

$$\mathbf{x}_i(k+1) = \mathbf{x}_i(k) + \mathbf{v}_i(k+1) \quad (4)$$

In practice, a particle's velocity is limited to a predefined range [$-\mathbf{v}_{max}$, $+\mathbf{v}_{max}$] to reduce the possibility of particles leaving the search space. The basic PSO algorithm is graphically illustrated in Figure 1. The term memory and cooperation denote personal and social influence respectively, while inertia is the previous velocity.

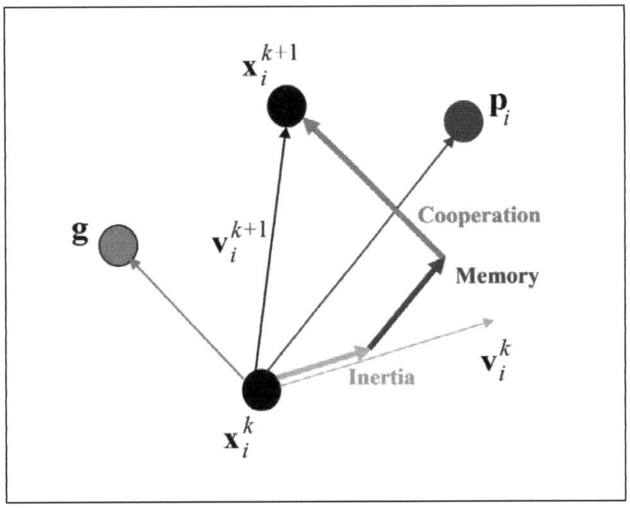

Figure 1. Graphical illustration of the basic PSO algorithm, in which particle $\mathbf{x}_i^k$ moves to $\mathbf{x}_i^{k+1}$ under the influence of its personal best $\mathbf{p}_i$ (memory) and the global best $\mathbf{g}$ (cooperation) while still having a part of velocity of the previous movement (inertia).

3. Magnetotelluric 1D Forward Modeling

The magnetotelluric (MT) response of a given conductivity model is obtained from the solution of Maxwell's differential equations. For one-dimensional (1D) model, the response function is well-known and usually expressed as a recursive formula relating the impedance at the surface of two successive layers (Z_j and Z_{j+1}) as follows [8],

$$Z_j = Z_{0j}\frac{1 - R_j \exp(-2(Z_{0j}/\rho_j)h_j)}{1 + R_j \exp(-2(Z_{0j}/\rho_j)h_j)} \; ; \quad R_j = \frac{Z_{0j} - Z_{j+1}}{Z_{0j} + Z_{j+1}} \; ; \quad Z_{0j} = \sqrt{i\,\omega\mu_0\,\rho_j} \quad (5)$$

where Z_{0j} denotes the impedance of a homogeneous medium with a resistivity ρ_j and h_j is the thickness of the j-th layer, while $\omega = 2\pi/T$ is the angular frequency for a period T and μ_0 is the magnetic permeability of a free space. All units are in SI. At surface (layer 1) of a model consisting of N-layer (Figure 2), the MT response is usually expressed as apparent resistivity (ρ_a) and phase (ϕ),

$$\rho_a = \frac{1}{\omega\mu_0}|Z_1|^2 \qquad \phi = \tan^{-1}\left(\frac{\mathrm{Im}\,Z_1}{\mathrm{Re}\,Z_1}\right) \quad (6)$$

From equations (5) and (6) it is obvious that the parameter models are non-linearly related to the data.

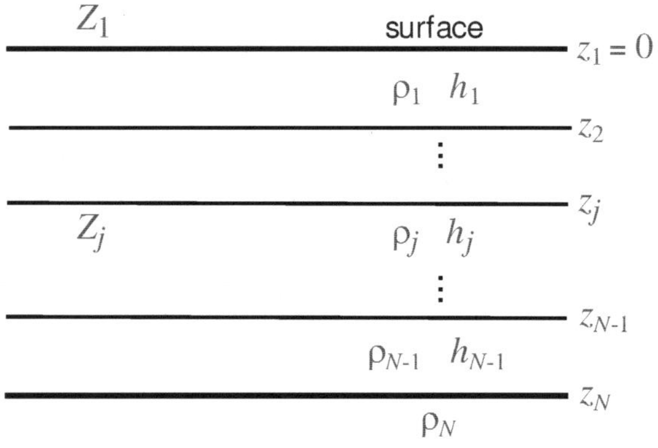

Figure 2. *N*-layered earth (1D) model, each layer with resistivity ρ_j, thickness h_j and impedance Z_j.

4. Inversion Results

The PSO algorithm was applied to invert synthetic MT data with 1D model. Four 3-layered synthetic models associated with H-, K-, Q-, and A-type sounding curves were used to generate the synthetic data (Table 1). Gaussian noise with 10% standard deviation was added to the theoretical response of the synthetic models. For the sake of brevity of the paper, only results with H- and K-type sounding curves (Model A and Model B) are presented. In general, models from both H- and K-type sounding curves are more difficult to resolve in the inversion.

Table 1. Synthetic model parameters used for generating the synthetic data.

	Model A (H-type)		Model B (K-type)		Model C (Q-type)		Model D (A-type)	
Layer	Resistivity (Ohm.m)	Thickness (m)	Resistivity (Ohm.m)	Thickness (m)	Resistivity (Ohm.m)	Thickness (m)	Resistivity (Ohm.m)	Thickness (m)
1	1000	500	100	500	1000	500	10	500
2	10	1000	1000	1000	100	1000	100	1000
3	100	-	10	-	10	-	1000	-

We performed inversions with the correct (3-layer) and un-correct (5-layer) number of layers to test the robustness of the algorithm faced to different *a priori* information on the number of layers. The PSO algorithm was also tested with different number of models (or particles), i.e. 50 and 200 particles. Inversions were done systematically up to 200 iterations, where in general, convergence was obtained since 100-th iteration. Therefore, the results presented in Table 2 (Model A) and Table 3 (Model B) are averaged models at 200-th iteration, with negligible uncertainties.

For the 3-layer model, the inverse models are practically similar and in very good agreement with the synthetic model, especially for the Model A (H-type) with both 50 and 200 particles. Similar observation is valid for Model B (K-type), except that the resistivity of the second layer is under-estimated. For the 5-layer model, the first 2 layers in the inverse models correspond relatively well with the same layers of the synthetic Model A. The last 3 layers of the inverse models are the equivalence layers of the third (last) layer of the Model A. This led to un-correct depth estimate of the last (moderately conductive) layer. The same difficulty in resolving the resistivity of the sandwiched resistive layer prevails for the Model B. However, the resistivity and depth of the last (conductive) layer were relatively well resolved. The comparisons of synthetic and inverse models along with the data fit for representative results are presented in Figure 3.

Table 2. Inverse model parameters from inversion of synthetic data (Model A).

	3-layer model				5-layer model			
	50 models (particles)		200 models (particles)		50 models (particles)		200 models (particles)	
Layer	Resistivity (Ohm.m)	Thickness (m)	Resistivity (Ohm.m)	Thickness (m)	Resistivity (Ohm.m)	Thickness (m)	Resistivity (Ohm.m)	Thickness (m)
1	1001.37	516.00	1001.37	516.00	1007.45	517.60	1051.01	511.31
2	10.03	973.69	10.03	973.69	10.07	876.06	11.05	1120.71
3	96.65	-	96.65	-	43.76	149.57	49.72	608.49
4	-	-	-	-	5.85	44.48	128.89	10345.62
5	-	-	-	-	97.63	-	81.73	-

Table 3. Inverse model parameters from inversion of synthetic data (Model B).

	3-layer model				5-layer model			
	50 models		200 models		50 models		200 models	
Layer	Resistivity (Ohm.m)	Thickness (m)	Resistivity (Ohm.m)	Thickness (m)	Resistivity (Ohm.m)	Thickness (m)	Resistivity (Ohm.m)	Thickness (m)
1	98.96	503.72	99.53	502.51	96.75	392.52	103.04	469.01
2	722.13	987.36	702.21	1004.68	286.19	457.89	411.41	738.15
3	9.48	-	9.75	-	248.94	582.21	200.46	310.16
4	-	-	-	-	22.16	198.02	37.94	53.86
5	-	-	-	-	9.78	-	9.91	-

5. Concluding Remarks

There are growing interests in computer intensive global optimization methods due to availability of computing power in recent years. The global and gradient-free approach avoids the use of linearization of highly non-linear inverse problems. Hence, it can overcome the fundamental limitations of linearized techniques. The Particle Swarm Optimization (PSO) which is one of nature-inspired global search algorithms has been presented with the application to invert MT data to obtain resistivity-depth variation of the subsurface represented by the 1D model. The results are encouraging in terms of model recovery as well as data misfit. The resolved inverse models are in accordance with the general characteristics of 1D MT modeling, i.e. difficulties in determining the resistivity of a resistive layer sandwiched between more conductive layers. It was also emphasized that a conductive layer at shallow depth may limit the ability of estimating parameters of the deeper layer, i.e. screening effects. The RMS misfits are in general around 18 to 19% which are consistent with the noise added in the synthetic data.

The PSO algorithm presented in this paper is relatively simple to implement with minimum inversion parameters to adjust, i.e. acceleration coefficients, inertia weight and number of particles or models. Those inversion paramaters have a relatively unsignificant influence on the overall inversion results. It is also possible to extend the basic PSO algorithm to solve more complex geophysical inverse problems. However, the application would be limited to a relatively small number of model parameters in order to be amenable with current computing capability. The possible applications include 1D inverse modeling of electromagnetic (EM) data, i.e. Transient EM (TEM), Vertical

Electrical Sounding (VES), Controlle-Source EM (CSEM), Nuclear Magnetic Resonance Sounding (NMRS) etc.

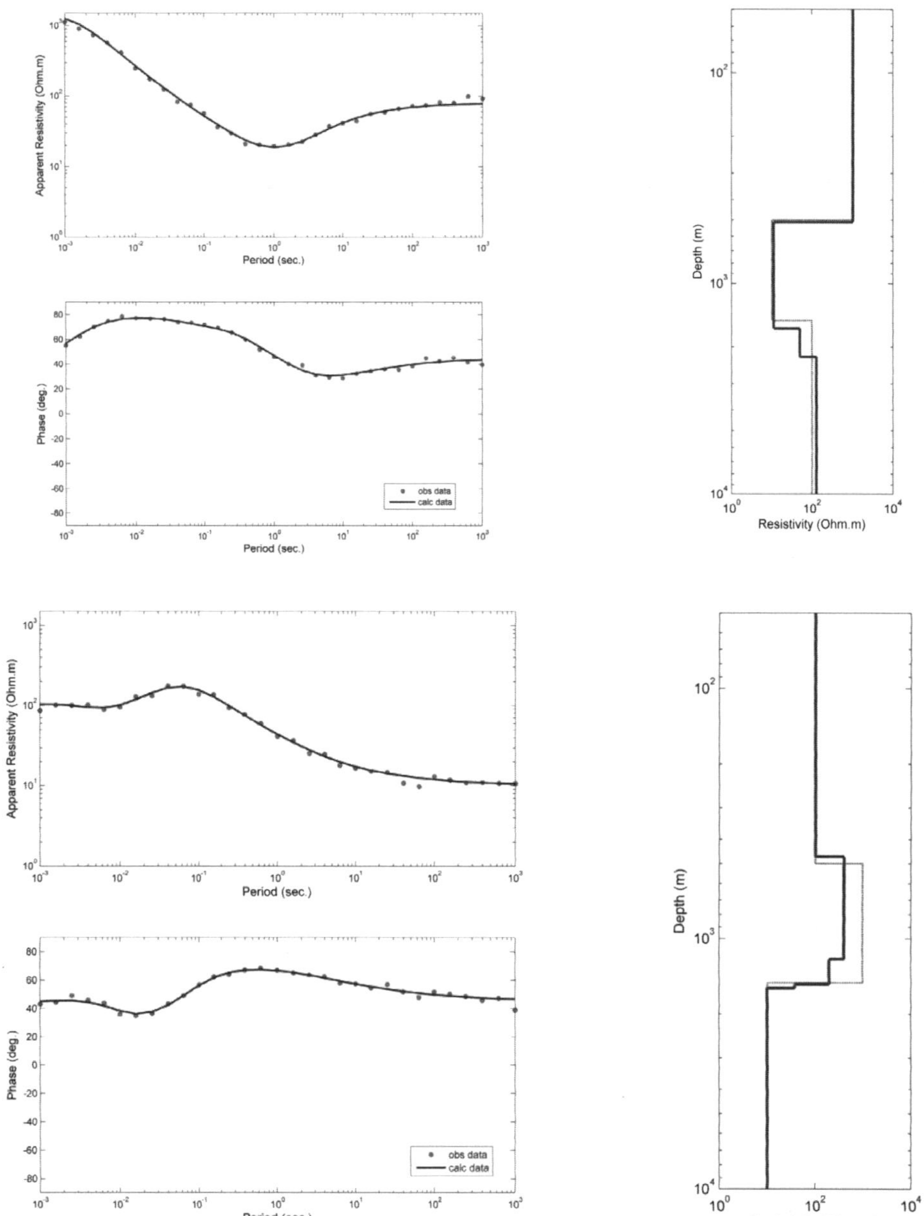

Figure 3. The fit between observed and calculated data along with the comparison of inverse (5 layers) model and synthetic model for Model A or H-type (top) and Model B or K-type sounding curve (bottom). In general, all results have RMS error around 18% to 19%.

References

[1] Tarantola A 1987 Inverse Problem Theory *Elsevier*
[2] Menke W 2012 Geophysical Data Analysis: Discrete Inverse Theory 3rd ed. *Elsevier*
[3] Sambridge M and Mosegaard K 2002 Montre Carlo methods for geophysical inverse problems *Review of Geophysics* **40**(3) 1-29.
[4] Sen K M and Stoffa P L 2013 Global Optimization Methods in Geophysical Inversion *2nd ed. Elsevier*
[5] Irawan D, Grandis H, and Sumintadireja P 2015 Quasi-2D resistivity model from inversion of vertical electrical sounding (VES) data using guided random search algorithm *Journal of Mathematical and Fundamental Sciences* **47**(3) 269-280.
[6] Kennedy J and Eberhart R C 1995 Particle Swarm Optimization *Proc. IEEE Int. Conf. on Neural Networks* **IV** 1942-1948.
[7] Clerc M and Kennedy J 2002 The Particle Swarm-Explosion, Stability, and Convergence in a Multidimensional Complex Space *IEEE Transaction on Evolutionary Comp* **6** 58-73.
[8] Grandis H 1999 An alternative algorithm for one-dimensional magnetotelluric response calculation *Computer & Geosciences* **25** 119-125.
[9] Auken E, Pellerin L, Christensen N B, and Sørensen K 2006 A survey of current trends in near-surface electrical and electromagnetic methods *Geophysics* **71** G249-G260.
[10] Ogaya X, Ledo J, Queralt P, Jones A G, and Marcuello A 2016 A layer stripping approach for monitoring resistivity variations using surface magnetotelluric response *Journal of Applied Geophysics* **132** 100-115.
[11] Shaw R and Srivastava S 2007 Particle swarm optimization: A new tool to invert geophysical data *Geophysics* **72** F75-F83.
[12] Cheng J, Li F, Penga S, Sun X, Zheng J, and Jia J 2015 Joint inversion of TEM and DC in roadway advanced detection based on particle swarm optimization *Journal of Applied Geophysics* **123** 30-35.

IOP Conf. Series: Earth and Environmental Science **62** (2017) 012034 doi:10.1088/1755-1315/62/1/012034

2-D Resistivity Structure of Cubadak Geothermal Area Revealed from Magnetotelluric Data

Hezliana Syahwanti[1], Wahyu Srigutomo[1], Muhammad Kholid[2]

[1]Physics of Earth and Complex System, Faculty of Mathematics and Natural Sciences, Institut Teknologi Bandung
[2]Subsurface Research Group, PSDG Bandung

E-mail: syahwantihezliana@gmail.com

Abstract. Cubadak area in Pasaman Regency, West Sumatra, Indonesia is one of many potential geothermal areas in Indonesia. This paper deals with resistivity structure beneath Cubadak area obtained from magnetotelluric data. The subsurface resistivity image of Cubadak reveals the components of Cubadak geothermal system clearly such as the altered clay having very low resistivity values, the possible reservoir zone having intermediate resistivity values underlying the altered clay as well as the indication of geological structures controlling the system. The top boundaries of the reservoir vary from elevation of 250 m to -500 m above sea level.

1. Introduction

Indonesia has abundant potential geothermal energy resources [1] due to its position as the meeting zone of 3 major tectonic plates (Indo-Australia, Eurasia and Pacific plates). One of them is located in Cubadak, Pasaman Regency, West Sumatera (Figure 1), situated at 597600 mE – 617700 mE and 21200 – 42500 mN [2]. The geothermal potential in Cubadak is indicated by surface manifestations such as Cubadak hot spring and Sawah Mudik hot spring. Therefore, it is important to provide information on the physical properties of Cubadak geothermal system in order to delineate the presence of its cap rock, reservoir and heat source as well as geological structures that may control the system. Resistivity method, like magnetotelluric method, is highly effective for such purpose due to its high sensitivity to detect resistivity contrasts that characterize the subsurface structure of a geothermal system.

Magnetotelluric (MT) method is a geophysical exploration method that uses natural electromagnetic (EM) wave as the source [3,4] and it is used for subsurface resistivity distribution mapping. In the data acquisition step, the EM responses are measured in time domain then they are converted into the frequency domain.

The EM sources in MT method come from the interaction between solar wind and magnetosphere for low frequency ($f<1$ Hz) and the thunder storms in ionosphere for the high frequency ($f>1$ Hz) [3]. The receiver records MT responses in terms of electric field (**E**) and magnetic field (**H**) at the surface [3]. The ratio between variation of electric field and magnetic field is called impedance (**Z**).

Z is a complex entity that can be express in amplitude and phase quantities. It can be utilized to obtain apparent resistivity value (ρ_a) and phase of impedance ($\emptyset$). Resistivity value was obtained by square absolute impedance divided by omega time vacuum permittivity. Phase value was obtained by arc tangential of comparison between imaginer and real of impedance.

Reservoir zones are found at a high-temperature geothermal system (Figure 2). The cooler zone at the upper layer is characterized by smectite alteration (formed at > 70^0 C) which is electrically conductive. In higher temperature, illite is formed under smectite layer. Increasing temperature will increase the composition of illite but decrease the composition of smectite [5]. If temperature increases, smectite will decrease and then the pure illite is formed at the temperature more than 220^0C. Clay and other alteration minerals (chlorite, epidote, etc.) are also formed at high temperature in propylitic zone [5].

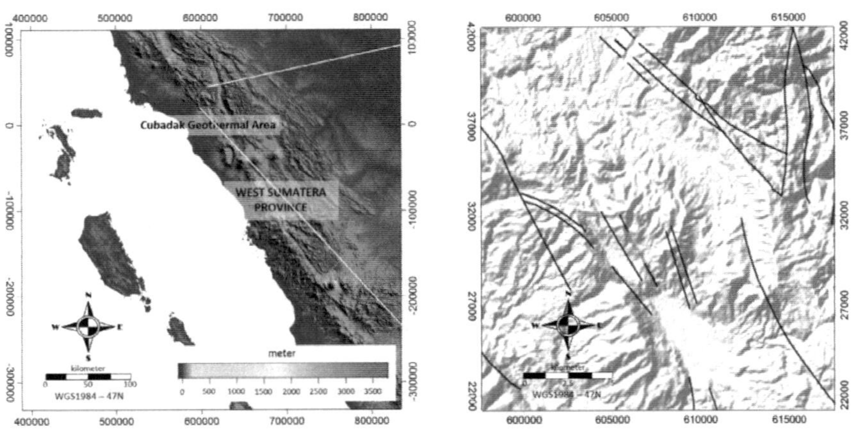

Figure 1. Location of Cubadak area, marked by red square (left) and it is zoomed in map (right) [2]

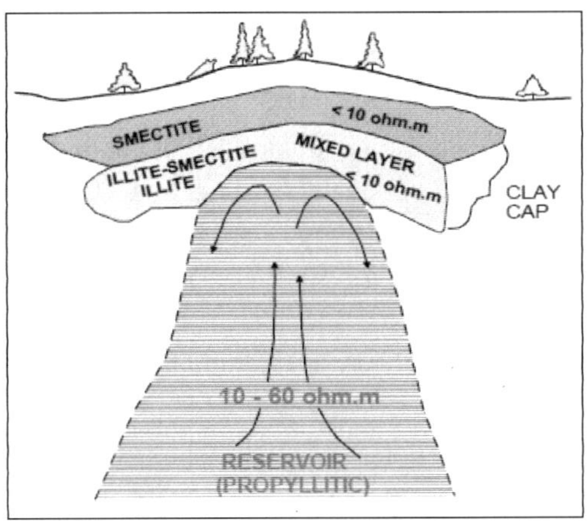

Figure 2. Scheme of the geothermal system [6]

In this study, the component of Cubadak geothermal system is estimated from 2-D resistivity cross section. The section was obtained from inversion of magnetotelluric data at 13 stations forming two parallel lines trending Southwest – Northeast (Figure 3).

2. Methodology

MT data acquisition was conducted in mid-2012 by Subsurface Research Group of Center of Geological Resources (PSDG), comprising 13 MT stations aligned in 2 parallel lines. At each stations, the total variation of electric in the magnetic field was recorded for about 18 hours. The above raw data were processed by using SSMT2000 software manufactured by Phoenix Geophysics to transform them into frequency domain expression in terms of apparent resistivities and phase of impedance as a function of frequency for each station. Further smoothing and editing of the apparent resistivity and phase of impedance were performed using MT Editor software develop by the same company. The final form of the data is manifested in EDI files ready for the inversion scheme.

The inversion of data into through resistivity model as a function of depth and horizontal distance (2-D Inversion) was performed by using WinGLink interpretation software. 2-D inversion uses nonlinear conjugate gradient inversion scheme [7]. The resulted 2-D resistivity sections were then interpreted as to identify the components of Cubadak geothermal system.

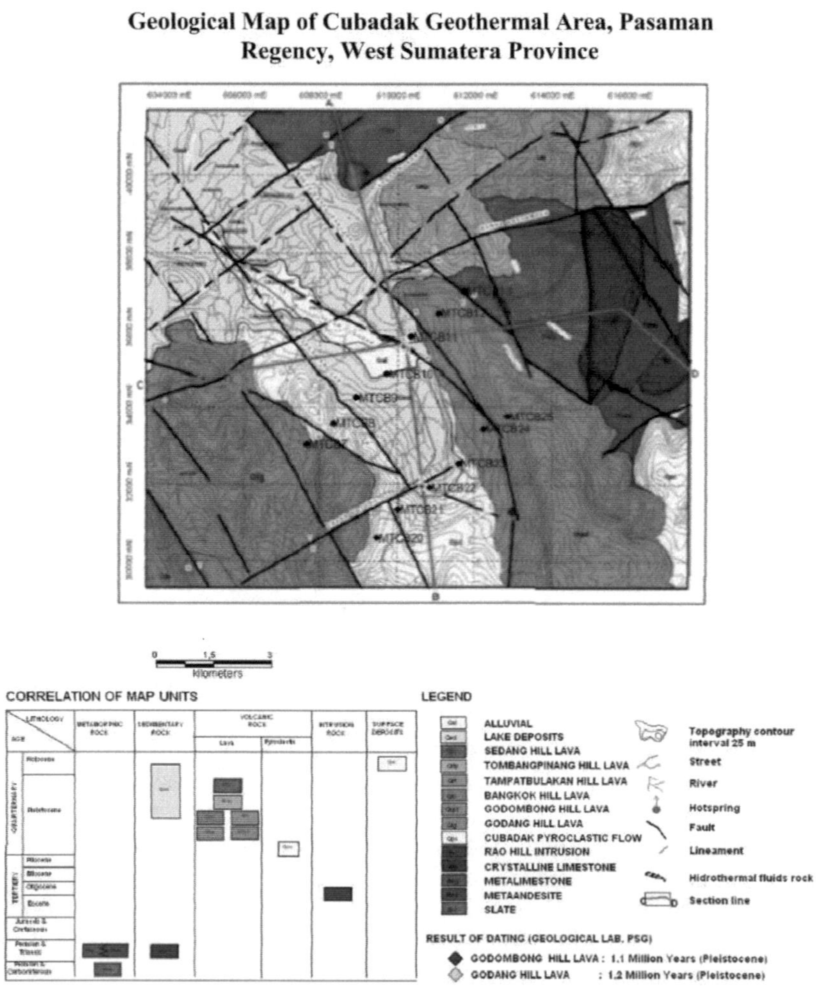

Figure 3. Regional Geology Map of Cubadak [8]

IOP Conf. Series: Earth and Environmental Science **62** (2017) 012034 doi:10.1088/1755-1315/62/1/012034

3. Results and Discussion

Figure 4 depicts two examples of data in terms of apparent resistivity and phase of impedance obtained in the field from MTCB 8 and MTCB 21 stations. The data generally can be categorized as having good or fair quantities. There are two sections obtained from the inversion scheme: Line 1 which comprises MTCB 7, MTCB 8, MTCB 9, MTCB 10, MTCB 11, MTCB 12, and MTCB 13 stations; and Line 2 which comprises MTCB 20, MTCB 21, MTCB 22, MTCB 23, MTCB 24, and MTCB 25 stations.

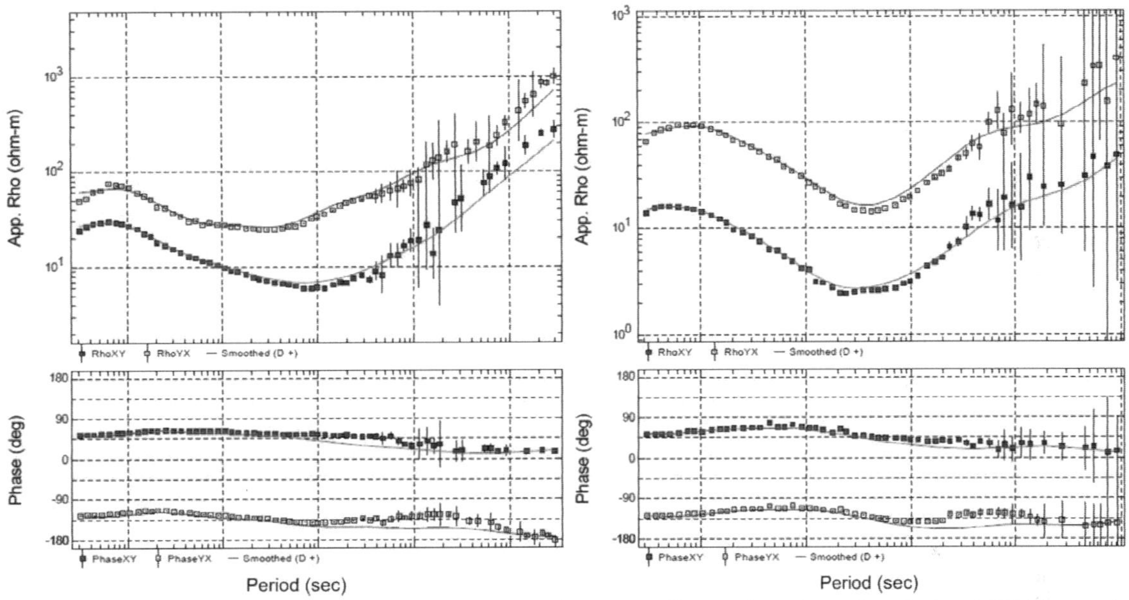

Figure 4. Apparent resistivity and phase of impedance profiles as function of period at MTCB 8 and MTCB 21

The resistivity section beneath Line 1 is shown in Figure 5 and for the Line 2 is shown in Figure 6. In general, the degree of acceptance of the modeled resistivity sections is judged by the fitness between the observed data and the calculated data from the inversion scheme. Figure 7 shows the comparison between the observed and calculated data for Line 1 and Line 2 in terms of pseudo sections of apparent resistivity and phase of impedance are relatively good as can be seen from the figure where both of pseudo sections are almost similar. However, the discrepancy between the observed and calculated phase of impedance is more obvious. Nevertheless, the modeled resistivity sections can be accepted to represent the resistivity structure beneath the two lines.

The root mean squares (RMS) misfit for both Line 1 and Line 2 inversions is shown in Figure 8. The RMS values converge towards a minimum value of about 2.3 at 30th iteration which is set as the maximum iteration number.

It can be seen from the modeled resistivity section beneath Line 1 that a low resistivity zone of about 10 Ωm or less spans from below MTCB 7 to MTCB 11. This low resistivity zone is associated with the presence of impermeable altered clay that serves as the cap rock of the geothermal system. The Cubadak hot spring is situated in the vicinity of MTCB 11 near from Cubadak fault. Underlying this low resistivity zone, a moderately resistive zone is found having resistivity values that vary from 20 to 70 Ωm which can be associated with the reservoir zone. The top boundaries of the reservoir vary from elevation of 250 m to -250 m above sea level suggesting shallow reservoir depths. The Cubadak fault near from MTCB 11 is interpreted as the pathway of thermal fluid that migrates to the surface and manifests as Cubadak hot spring. Cubadak hot spring is located at the NE end tip of the clay zone; this finding may be a strong evidence that the hot spring is situated within the outflow zone of the geothermal system. The high resistivity zone that is found below a moderately resistive overburden at the northeastern part of the resistivity section can be attributed to the presence of permo-carbon metamorphic rock as inferred from the geological map.

Beneath Line 2 a low resistivity zone below 10 Ωm spans from below MTCB 20 to an area between MTCB 23 and MTCB 24. This low resistivity zone is associated with the presence of impermeable altered clay. The Sawah Mudik hot spring is situated in the vicinity of MTCB 22. Underlying this low resistivity zone, a moderately resistive zone is found having resistivity values that vary from 20 to 70 Ωm which can be associated with the reservoir zone. The top boundaries of the reservoir vary from the elevation of 0 m to -1000 m above sea level suggesting shallow reservoir depths. Below the reservoir zone a high resistivity body exists indicating the presence of the hot intrusive rocks serving as the heat source of the Cubadak geothermal system. Judging from the shape of the altered clay and its related position from the interpreted heat source, the Sawah Mudik hot spring is located within the upflow zone of the Cubadak geothermal system.

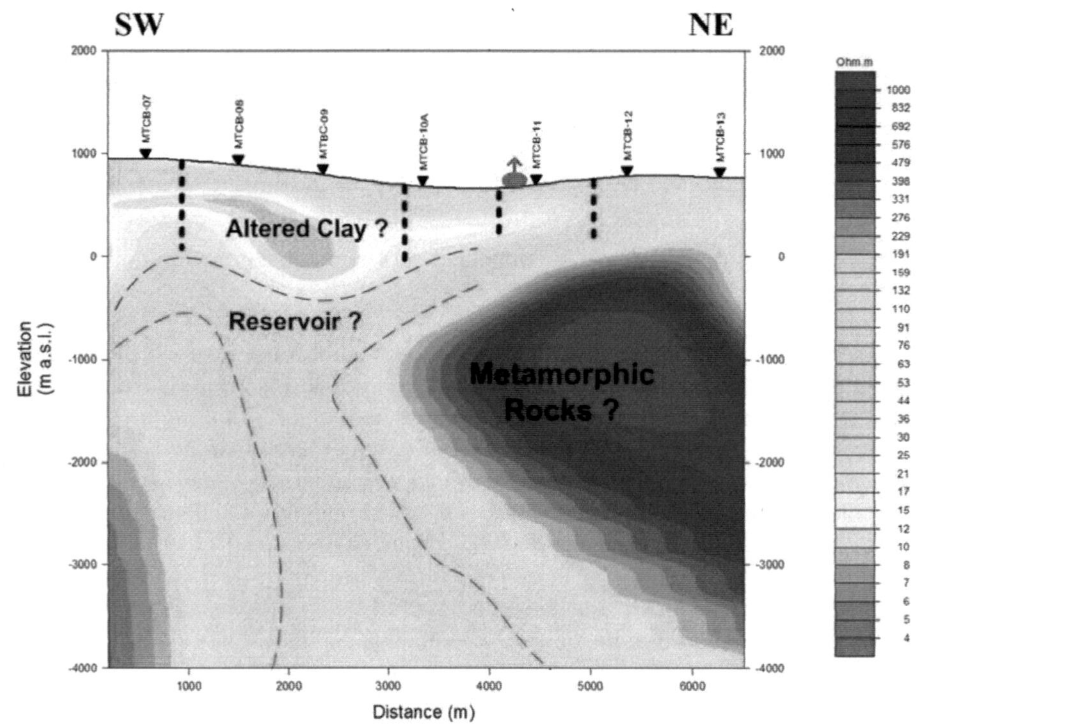

Figure 5. Resistivity section beneath Line 1. Vertical dashed lines represent faults taken from the geological map (Figure 3).

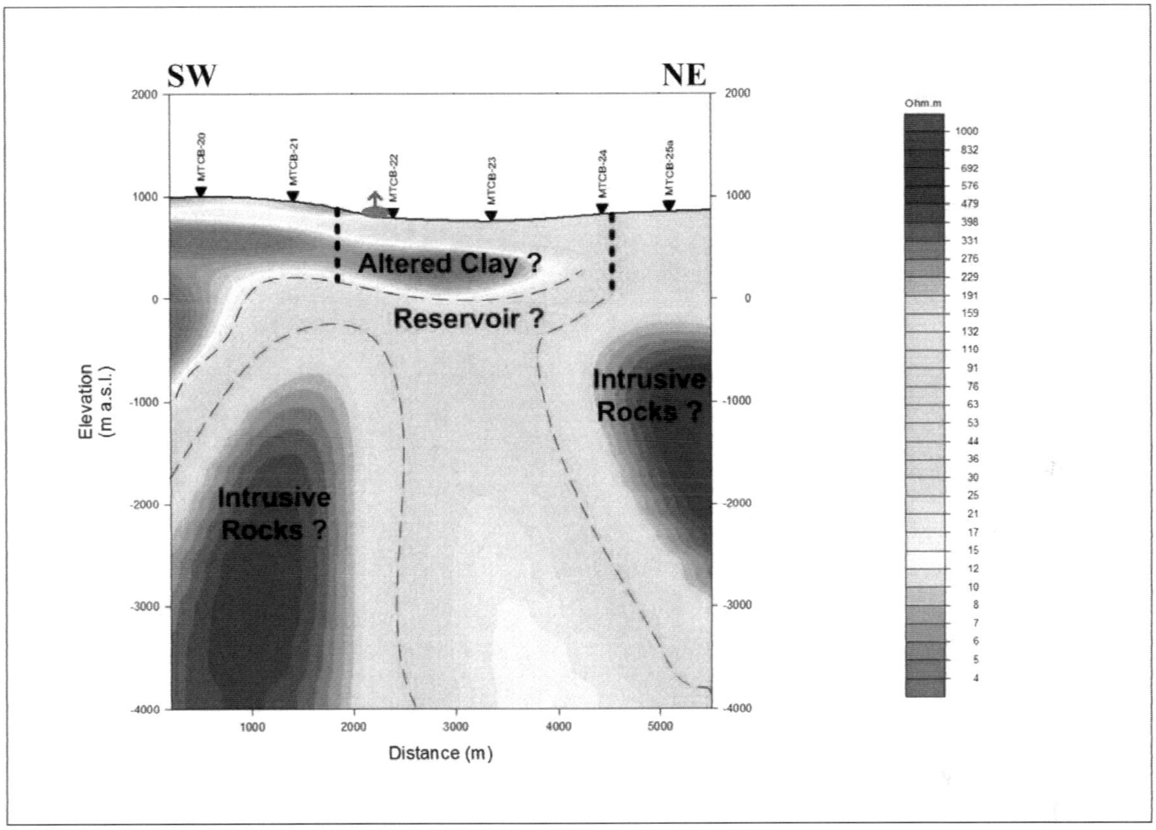

Figure 6. Resistivity section beneath Line 2. Vertical dashed lines represent faults taken from the geological map (Figure 3).

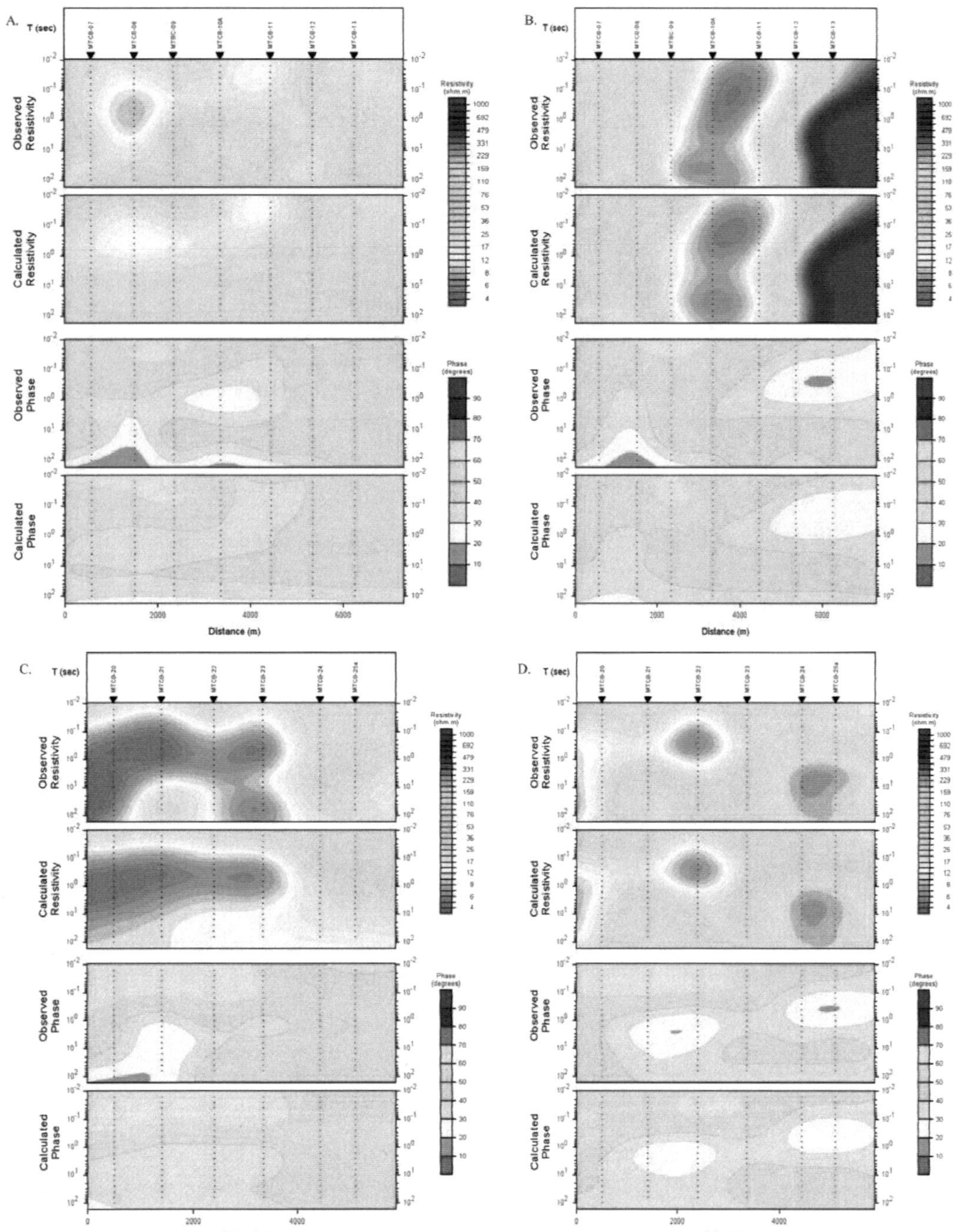

Figure 7. Pseudo sections of the observed and calculated data of: (A) Line 1 for TE mode; (B) Line 1 for TM mode; (C) Line 2 for TE mode; (D) Line 2 for TM mode.

IOP Conf. Series: Earth and Environmental Science **62** (2017) 012034 doi:10.1088/1755-1315/62/1/012034

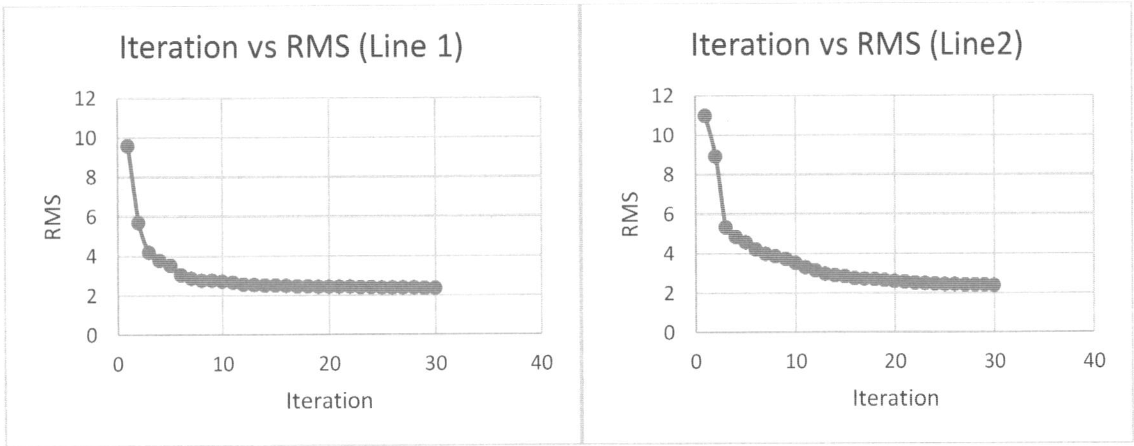

Figure 8. Graphic of Iteration vs. RMS

4. Conclusions

Based on the 2-D resistivity section obtained from the inversion of the MT data and the geological map of Cubadak area, the geothermal system of Cubadak area can be identified as altered clay zone is found between MTCB 7 to MTCB 11 trending SW-NE; between MTCB 20 to MTCB 24 trending SW-NE. The top boundaries of the reservoir beneath the altered clay vary from 250 m to -1000 m above sea level. Then Cubadak hot spring is located in the outflow zone whereas the Sawah Mudik hot spring is located in the upflow zone of Cubadak geothermal system.

References

[1] Hermawan D and Rezky Y 2010 The Role of Sumatera Fault Structures in Appearance of Geothermal Features at Cubadak Area, West Sumatera, Indonesia *Proceedings World Geothermal Congress 2010*

[2] Tim Survey Magnetotellurik 2012 Survey Magnetotellurik Daerah Panas Bumi Cubadak Kabupaten Pasaman, Sumatera Barat *Pusat Sumber Daya Geologi*

[3] Simpson F and Bahr K 2005 Practical Magnetotellurics *Cambridge University Press*

[4] Manoj C 2003 Magnetotelluric Data Analysis Using Advances In Signal Processing Techniques

[5] Anderson E, Crosby D and Ussher G 2000 Simple Resistivity Imaging To Reliably Locate The Geothermal Reservoir *Proceedings World Geothermal Congress 2000*

[6] Pellerin L, Johnston J M and Hohmann G W 1996 A Numerical Evaluation Of Electromagnetic Methods In Geothermal Exploration *Geophysics* **61** 121-130

[7] Rodi W L and Mackie R L 2001 Nonlinear Conjugate Gradients Algorithm For 2D Magnetotelluric Inversion Geophysics **66** 174-187

[8] Tim Survei Terpadu 2008 Survey Terpadu Geologi, Geokimia, dan Geofisika Daerah Panas Bumi Cubadak, Kabupaten Pasaman, Provinsi Sumatera Barat *Pusat Sumber Daya Geologi*

IOP Conf. Series: Earth and Environmental Science **62** (2017) 012035 doi:10.1088/1755-1315/62/1/012035

Improved Pseudo-section Representation for CSAMT Data in Geothermal Exploration

Hendra Grandis[1*], **Prihadi Sumintadireja**[2]

[1]Applied and Exploration Geophysics Research Group, Faculty of Mining and Petroleum Engineering, Institut Teknologi Bandung, Jalan Ganesha 10 Bandung 40132, Indonesia
[2]Applied Geology Group, Faculty of Earth Science and Technology, Institut Teknologi Bandung, Jalan Ganesha 10 Bandung 40132, Indonesia

*corresponding author: grandis@itb.ac.id

Abstract. Controlled-Source Audio-frequency Magnetotellurics (CSAMT) is a frequency domain sounding technique employing typically a grounded electric dipole as the primary electromagnetic (EM) source to infer the subsurface resistivity distribution. The use of an artificial source provides coherent signals with higher signal-to-noise ratio and overcomes the problems with randomness and fluctuation of the natural EM fields used in MT. However, being an extension of MT, the CSAMT data still uses apparent resistivity and phase for data representation. The finite transmitter-receiver distance in CSAMT leads to a somewhat "distorted" response of the subsurface compared to MT data. We propose a simple technique to present CSAMT data as an apparent resistivity pseudo-section with more meaningful information for qualitative interpretation. Tests with synthetic and field CSAMT data showed that the simple technique is valid only for sounding curves exhibiting a transition from high – low – high resistivity (i.e. H-type) prevailing in data from a geothermal prospect. For quantitative interpretation, we recommend the use of the full-solution of CSAMT modelling since our technique is not valid for more general cases.

1. Introduction

Controlled-Source Audio-frequency Magnetotellurics (CSAMT) typically uses a grounded electric dipole as an artificial electromagnetic (EM) signal source operating in the 0.1 Hz to 10 kHz frequency band. With higher amplitude signals, better quality CSAMT data can be achieved relative to those from MT sounding that uses the natural EM source [1,2]. In most cases, only one grounded electric dipole (1-2 km in length) is used and positioned along the x-axis. A multi-channel receiver records the electric field simultaneously at several dipoles (E_x) along a profile and the perpendicular magnetic field (H_y) at the center of the array. Therefore, only scalar impedance $Z_{xy} = E_x/H_y$ is obtained. The distance between CSAMT sounding sites along a profile is limited to around 100 m due to the restriction in cable transmission (see Figure 1). With such measurement configuration, the CSAMT technique tends to have a good lateral resolution.

With finite transmitter - receiver distance (R, typically 2 to 8 km), the EM field surrounding the transmitter behaves as far field and near field. In the far-field, the uniform plane-wave assumption prevailing in MT with distant source is still valid for CSAMT. In this case $R > 4\delta$, δ is the skin depth for a uniform equivalent material underlying the measurement area. The near-field is typically defined when $R < \delta$ and there is a transition between far field and near field [1,3]. In all cases, the apparent resistivity and phase derived from MT theory (i.e. Cagniard's equation) is used to convert the CSAMT scalar impedance into more meaningful information. The problem is that the apparent resistivity does not equal to the true resistivity of a homogeneous medium in the near field. For layered medium, the apparent resistivity curve does not exhibit the "correct" resistivity-depth variation of the subsurface. Therefore, qualitative interpretation of apparent resistivity sounding curves and pseudo-section from CSAMT data is less intuitive than its MT counterpart.

Several attempts were made to transform CSAMT data such that they represent adequately the resistivity variation with depth as in MT data. The term "near-field correction" emerged, although the possibility of such correction is debatable. The effort was driven mainly to be able to model CSAMT data using more commonly available MT modeling. Most of the techniques proposed are based on a generalization of the homogeneous medium response to non-homogeneous cases [4,5]. However, such generalization is not reliable especially for complex structures [e.g. 6,7]. This paper describes the recalculation of the apparent resistivity for CSAMT data along a profile such that they can be presented as a pseudo-section having better correlation with the resistivity structure. We start with nearly similar approach with the one proposed previously [4,5,6]. Then, we also introduce a semi-subjective method in the interpolation of apparent resistivity curve in the transition field. The resulting apparent resistivity curve will more likely represent the resistivity-depth variation of the subsurface.

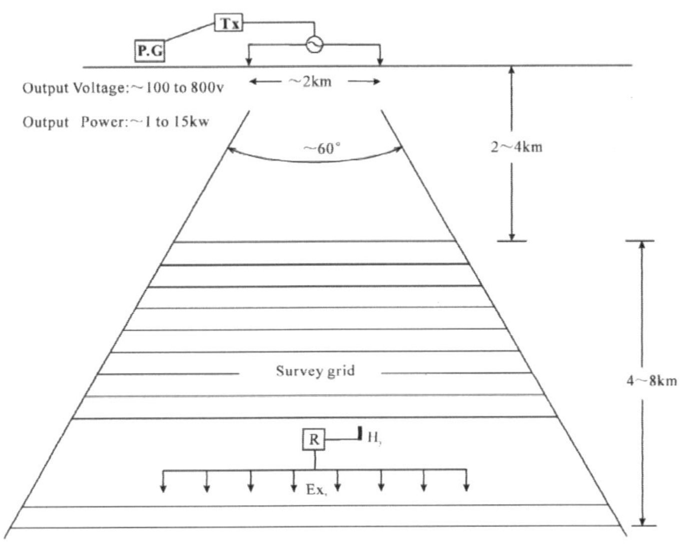

Figure 1. CSAMT field set-up with the electric dipole and transmitter (Tx) powered by an electrical power generator (PG). The receiver (R) records data from orthogonal components of electric and magnetic fields [1].

2. Far and near fields apparent resistivity

We performed MT and CSAMT 1D forward modeling using algorithms available in the literatures [8,9,10]. The so-called Cagniard's equation is commonly used to convert the MT complex scalar impedance Z_{xy} to apparent resistivity and phase. In a more generalized form with all quantities in SI units, the apparent resistivity equation used in both MT and CSAMT is,

$$\rho_a = \frac{1}{\omega\mu_0}\left|Z_{xy}\right|^2 \; ; \; \phi = \tan^{-1}\left(\frac{\mathrm{Im}\,Z_{xy}}{\mathrm{Re}\,Z_{xy}}\right) \tag{1}$$

where $\omega = 2\pi/T$, T is period (sec.) , μ_0 is the free-space magnetic permeability.

Figure 2 illustrates the differences between MT and CSAMT model response represented by the far field apparent resistivity and phase, i.e. equation (1) for simple layered models. We used the same 1D synthetic model with $R = 2$ km and $R = 8$ km. The model response exhibits an H-type sounding curves representing high - low - high resistivity variation with depth. In the far field, apparent resistivities for MT and CSAMT are identical up to $T = 0.05$ sec. for $R = 2$ km and to $T = 0.15$ sec. for $R = 8$ km. The near field prevails at longer periods where CSAMT apparent resistivity curves deviate and do not make an asymptote to the resistivity of the lowermost layer as for MT curves. The transition from far field to near field is usually identified by a "notch" represented by sharp resistivity change in a short interval of periods. Surprisingly, the "notch" appears only slightly with a shorter transmitter - receiver distance, i.e. $R = 2$ km. In all cases, a qualitative interpretation of apparent resistivity curves and pseudo-section from CSAMT data can be definitely misleading.

A more appropriate formula for the apparent resistivity in the near field can be used to obtain a sounding curve that might reflect the subsurface resistivity-depth variation. From the CSAMT response of a homogeneous medium [1], the the near field apparent resistivity can be calculated using,

$$\rho_a = 0.5\, R \left| Z_{xy} \right| \tag{2}$$

where R is the transmitter - receiver distance. Equation (2) is analogue to equation (1) in a sense that the apparent resistivity would be equal to the true resistivity of a homogeneous medium for near field and far field respectively. In the near field, the phase between electric and magnetic fields has no meaning since it falls almost strightly and makes an asymptote to 0° (see Figure 2 lower panels).

Figure 3 shows apparent resistivity curves from far field equation (1) and near field equation (2) for the same synthetic models presented in Figure 2. The intersection between far field and near field apparent resistivities can be identified as the center of the transition between far field and near field (or "transition field" for short). However, the extent of the transition field from this center towards shorter and longer periods is not obvious. Even for the same models, the position in the horizontal axis (i.e. period) of the transition field depends on the transmitter - receiver or R. Therefore, more objective criteria have to be set if the "correction" will be applied to the transition zone.

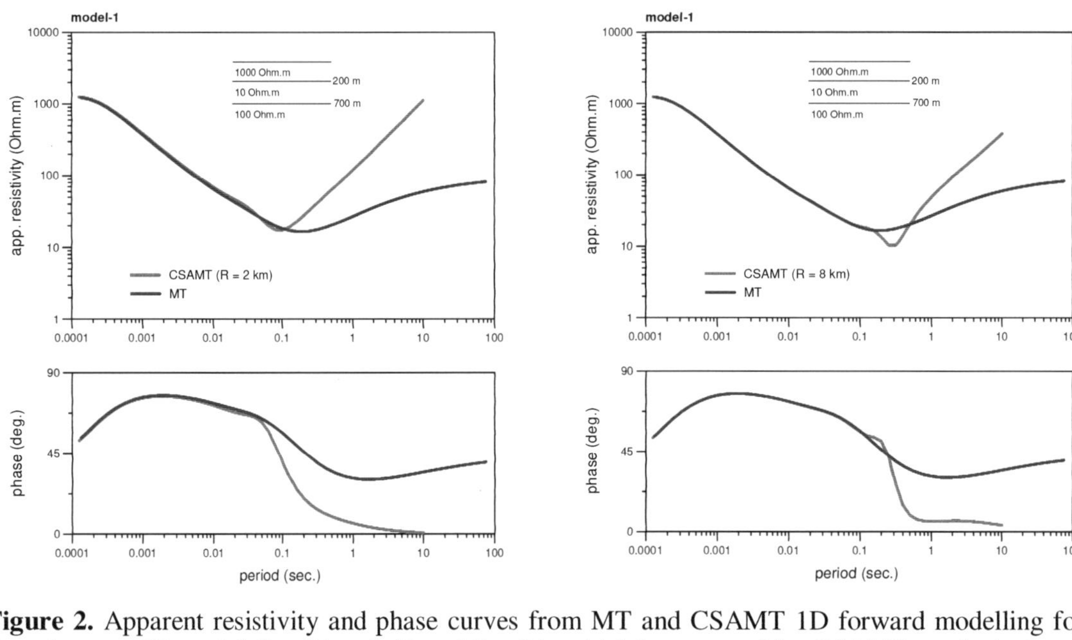

Figure 2. Apparent resistivity and phase curves from MT and CSAMT 1D forward modelling for a three-layered 1D model. $R = 2$ km (left) and $R = 8$ km (right) were used for CSAMT synthetic data.

IOP Conf. Series: Earth and Environmental Science **62** (2017) 012035 doi:10.1088/1755-1315/62/1/012035

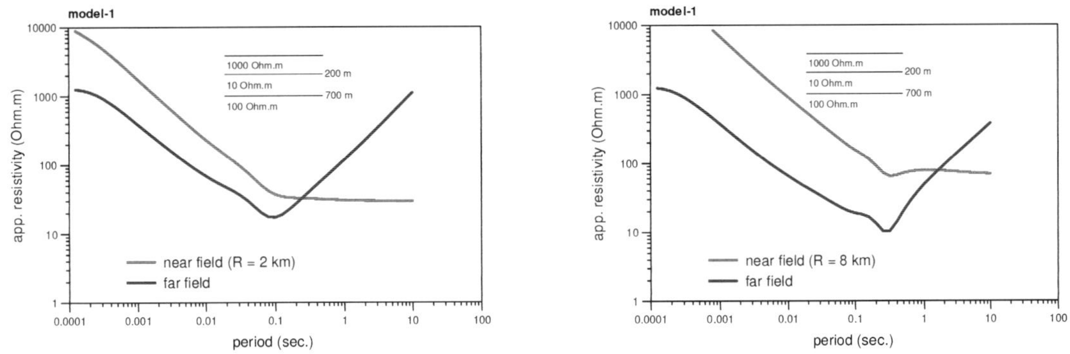

Figure 3. Apparent resistivity curves obtained from far field and near field equations for 1D synthetic models identical to those in Figure 3, with R = 2 km (left) and R = 8 km (right).

3. Transition field correction

In order to identify unambiguously the transition field, we study the characteristics of CSAMT response of a homogeneous medium and extend the results to more general cases, i.e. the non-homogeneous medium. The apparent resistivity curves for different homogeneous media with R= 2 km and R = 8 km are presented in Figure 4. Both transmitter - receiver distances represent the most commonly used field set-up in CSAMT. The curves are not plotted for all periods to avoid cluttering the figure. The transition from far field to near field is characterized by the deviation of apparent resistivity curves from the true resistivity of the homogeneous medium. The position of the transition field in the period axis depends both on the transmitter - receiver distance and the resistivity of the medium. In the resistive medium (e.g. 1000 Ohm.m) the far field extends only to short periods since the skin depth is large, and vice versa. In the same resistive medium, the far field extends towards longer periods for the larger transmitter- receiver distance (see Figure 4 right for R= 8 km).

To eliminate the dependence of the transition field on resistivity of the medium and transmitter - receiver distance, we propose to normalize the period as follows,

$$T^* = \frac{T}{R}\left|Z_{xy}\right| \tag{3}$$

where R is the transmitter - receiver distance in km, but we do not define the unit of T* for simplicity.

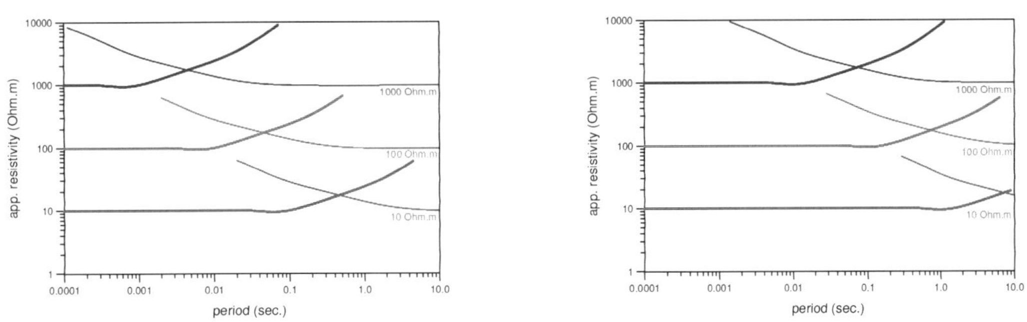

Figure 4. Apparent resistivity curves from far field equation (thick) and near field equation (thin) for different homogeneous media (10, 100 and 1000 Ohm.m) and transmitter - receiver distance, R = 2 km (left) and R = 8 km (right).

The apparent resistivity curves presented in Figure 4 are shown again in Figure 5 with the period transformed to the normalized period. It is obvious that the transition field position in the horizontal axis does not depend on the resistivity of the medium and transmitter - receiver distance. In all cases the intersection between far field and near field apparent resistivity curves always occurs at $T^* = 0.004$ unit. This fact will certainly help in identifying the transition zone and its correction using factors defined in the next section. The apparent resistivity curves are not plotted for the entire periods for the sake of simplicity of the figure. However, the near field apparent resistivity curves stop at different normalized periods. The results of the transformation from period to normalized period using equation (3) differ for different resistivity of the medium and transmitter - receiver distance.

For a homogeneous medium, we can calculate the correction factor to transform the apparent resistivity in the transition field to become equal to the resistivity of the medium. The correction factor is equal to one in far and near fields, and deviates from unity in the transition field. These factors are independent from the resistivity of the medium and the transmitter - receiver distance as long as they are plotted in the normalized period axis (see Figure 6). The curves in Figure 6 are segments with different colors showing that they are plotted from exactly the same data as in Figure 4 and Figure 5. In the transition field, K_f and K_n are used to correct apparent resistivity curves of a homogeneous medium in the far field and near field sides respectively. Previously, the value of K_f and K_n are interpolated from a predefined table which may not be valid for general cases [4].

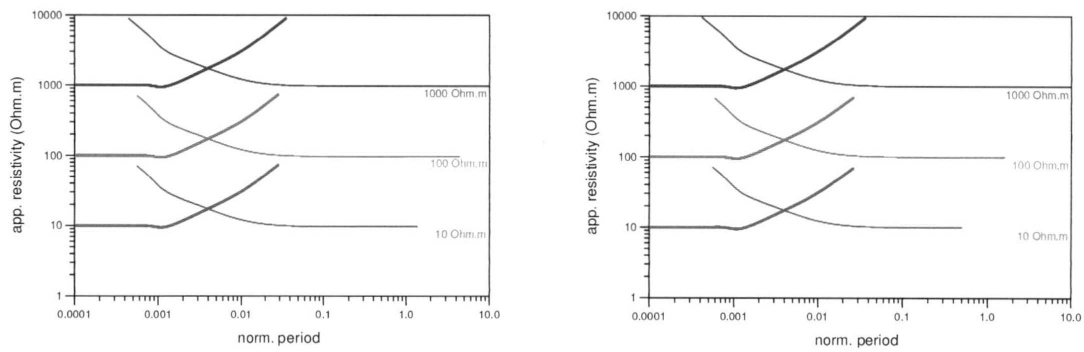

Figure 5. Apparent resistivity curves from far field equation (thin) and near field equation (thick) for different homogeneous media (10, 100 and 1000 Ohm.m) and transmitter - receiver distance R = 2 km (left) and R = 8 km (right). The normalized period is used in this figure.

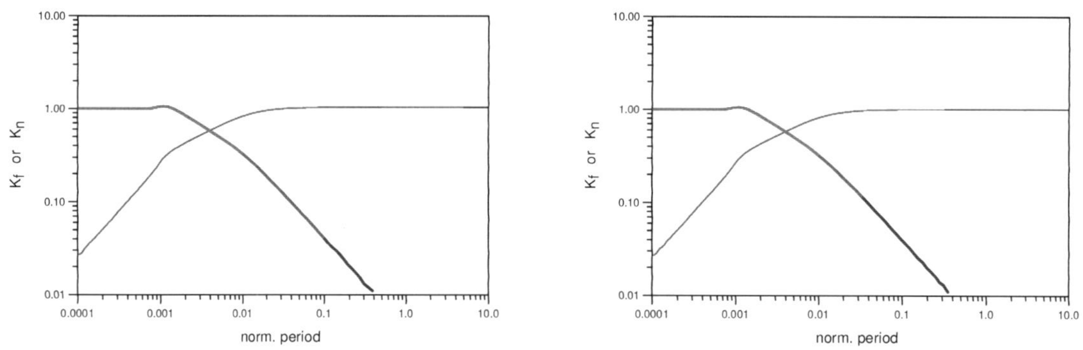

Figure 6. Correction factors for appparent resistivity curves from far field equation (thick) and near field equation (thin) to be equal to the resistivity of the medium for different homogeneous media (10, 100 and 1000 Ohm.m) and transmitter - receiver distance R = 2 km (left) and R = 8 km (right).

4. Results and discussion

The correction method for a homogeneous medium is considered valid and can be extended to CSAMT data associated with a non-homogeneous, i.e. layered medium. In the proposed technique, we calculate CSAMT response of an arbitrary homogeneous medium, R and period interval. The resistivity of the medium and the period interval are chosen such that the resulting response are in a sufficiently large interval of the normalized period. Although it is not critical, R is chosen equal to the actual transmitter - receiver distance of the CSAMT data to be corrected. Then, K_f and K_n are calculated as function of the normalized period, from which the correction factors at the transition field of the CSAMT data are interpolated.

Figure 7 illustrates the results for synthetic data presented in Figure 2 and Figure 3. By comparing them with the MT synthetic data, it is obvious that the correction works relatively well, although it is not perfect. For the model with R = 2 km, the lowermost layer's resistivity is under-estimated. For the same model with R = 8 km, the "notch" still exists in the corrected curve. The result for synthetic data with other types of sounding curve, i.e. ascending or A-type and bell-shaped or K-type sounding curves, are not reliable and therefore the method is considered not applicable [e.g. 11].

The ability of the method to correct CSAMT data that exhibit high - low - high resistivity variation, i.e. H-type sounding curve, is important in geothermal exploration using CSAMT. This resistivity-depth variation represent the conductive clay cap sandwiched by the upper resistive layer mostly associated with volcanic products and the moderately resistive reservoir [12,13]. The un-corrected CSAMT data with very high resistivity at longer periods might falsely indicate shallower reservoir or thinner cap rocks. The correction made to the CSAMT data leads to the correct trend of resistivity-depth variations and a qualitative interpretation can be adequately based on the corrected data.

The method was applied to real (field) CSAMT data distributed along a profile from Kamojang geothermal area in West Java, Indonesia. At most of the sounding sites, the data exhibits very high resistivity at longer periods, which is typical with CSAMT data. However, the transmitter - receiver distance is only 2.5 km such that the "notch" is absent in the data as previously indicated by the synthetic data with a relatively short R. The result for representative sounding sites is presented in Figure 8. It is obvious that the method has successfully corrected the high resistivity at longer periods to become more reasonable. The corrected sounding curve equals to the original only at a very limited interval in periods which shows that most of the data are in transition and near fields. For all data on the profile, the comparison of the apparent resistivity pseudo-sections before and after correction also shows that the corrected pseudo-section appears to be more realistic in terms of the subsurface resistivity distribution (Figure 9). The geological implications of the CSAMT data are beyond the scope of the paper. Nevertheless, the information gained after the correction of the CSAMT data is of significant value. The cap rocks thickness and the reservoir depth would otherwise be under-estimated.

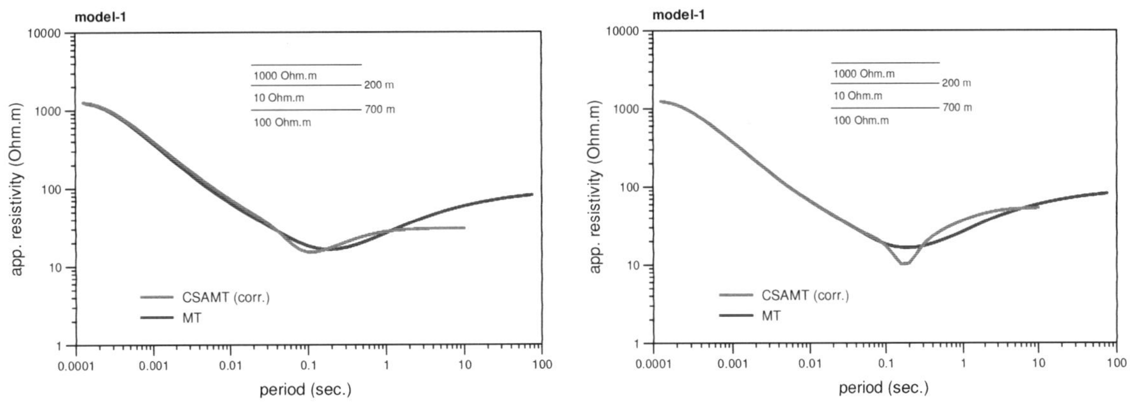

Figure 7. Synthetic CSAMT data with R = 2 km (left) and R = 8 km (right) after correction compared with the synthetic MT data of the same models (presented in Figure 2 and Figure 3).

IOP Conf. Series: Earth and Environmental Science **62** (2017) 012035 doi:10.1088/1755-1315/62/1/012035

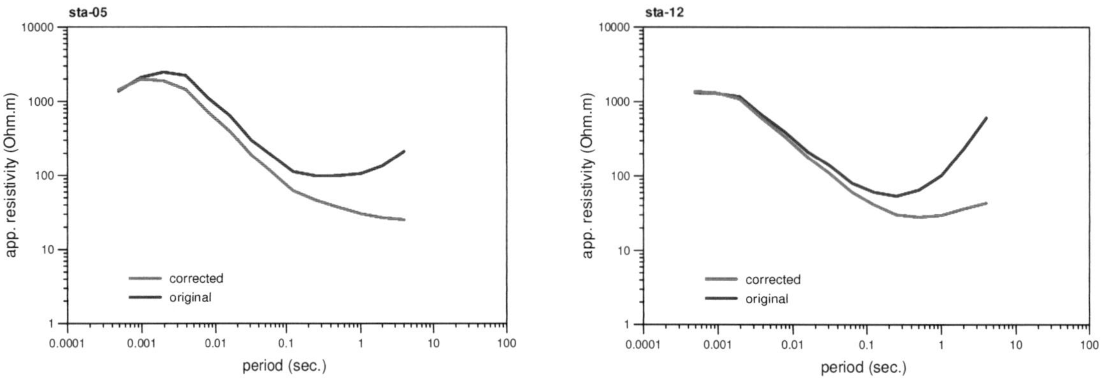

Figure 8. Result of CSAMT data correction from two sounding sites at Kamojang geothermal field, West Java, Indonesia.

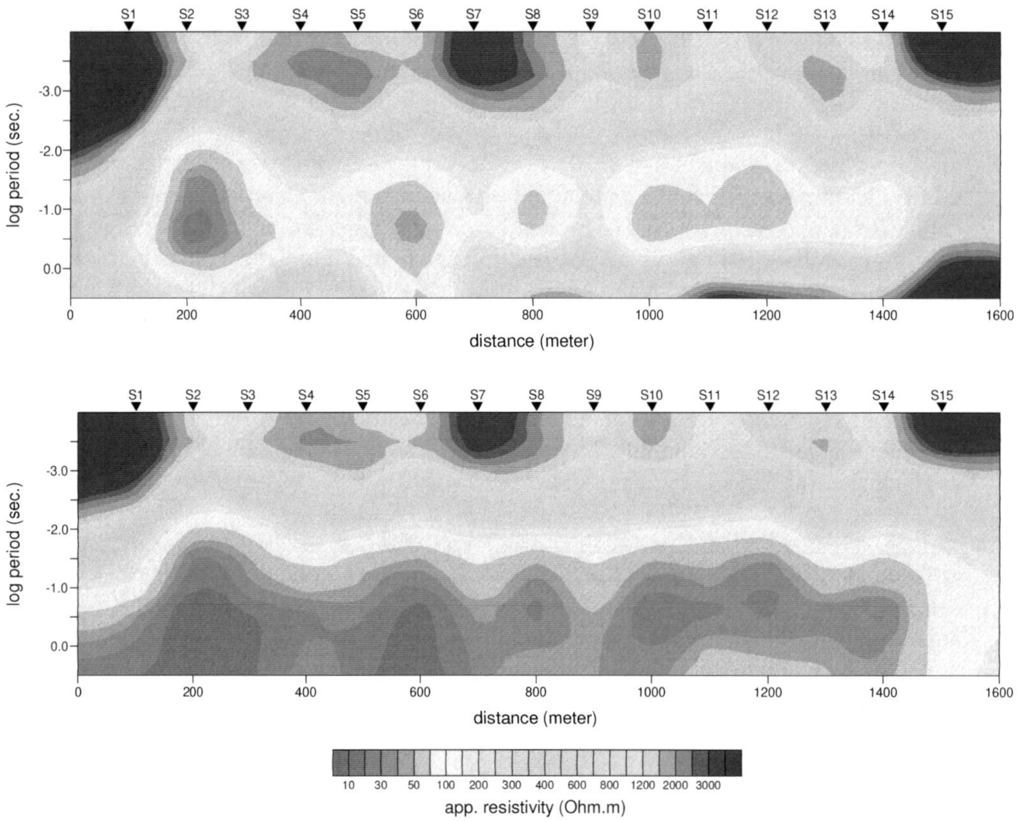

Figure 9. Apparent resistivity pseudo-sections of CSAMT data from Kamojang geothermal field, West Java, Indonesia, before (top panel) and after correction (lower panel) of the transition field. The apparent resistivity scale on the lower panel is for both pseudo-sections.

5. Conclusions

The CSAMT sounding technique appears to provide laterally high-resolution image of the subsurface thanks to the redundancy of the data along a profile line. However, CSAMT data and the pseudo-section representation suffer from distortion effects due to the finite transmitter - receiver separation. Therefore, direct qualitative interpretation from uncorrected CSAMT data might be erroneous. We have presented a method for correcting CSAMT data which leads to a more reasonable resistivity-depth variation. The method is valid only for an H-type (and most probably Q-type or descending) CSAMT sounding curves prevailing mostly in CSAMT data from geothermal prospects. The applicability of the method was tested using synthetic and field data from a geothermal field with satisfactory results. The "notch" still exists in the data after correction, especially for CSAMT data with a large transmitter - receiver distance. This is a relatively negligible impact on the overall quality of the results.

The proposed method is intended only to produce a better apparent resistivity pseudo-section of CSAMT data for qualitative interpretation. The result with field data shows: (i) improved lateral continuity of the conductive layer, (ii) thicker and more pronounced (lower resistivity) conductive clay cap. For more quantitative interpretation, it is necessary to perform the so-called full solution for CSAMT forward and inverse modelling [3,9,10,14]. This will exonerate the need for the transition field correction presented in this paper. In the use of full solution modelling apparent resistivity pseudo-sections would serve only to compare observed and calculated values or data misfit.

References

[1] Zonge K L and L J Hughes 1991 Controlled-source audio-frequency magnetotellurics, in Nabighian M N (Ed.) Electromagnetic Methods in Applied Geophysics Vol. 2 *Society of Exploration Geophysicists*
[2] Zhdanov M S and G V Keller 1994 The geolectrical methods in geophysical exploration *Methods in Geochemistry and Geophysics* **31** Elsevier
[3] L Da, W Xiaoping, D Qingyun, W Gang, L Xiangrong, W Ruo, Y Jun and Y Mingxin 2016 *Journal of Geophysics and Engineering* **13** 49-58
[4] Grandis H 2000 *Jurnal Teknologi Mineral* **VII/1** 43-50.
[5] Yamashita M, P G Hallof and W H Pelton 1985 CSAMT case histories with a multi-channel CSAMT system and discussion of near-field data correction *55th SEG Annual Meeting Expanded Abstracts*
[6] Bartel L C and R D Jacobson 1987 *Geophysics* **52** 665-677
[7] Szarka L 1988 On "Results of a controlled-source audio-frequency magnetotelluric survey at Puhimau thermal area, Kilauea volcano Hawaii, by L.C. Bartel and R.D. Jacobson (Geophysics, 52, 665-677) *Geophysics* **53** 726-727
[8] Grandis H 1999 An alternative algorithm for one-dimensional magnetotelluric response calculation *Computer and Geosciences* **25** 119-125.
[9] Li X and L B Pedersen 1991 *Geophysics* **56** 1456-1461.
[10] Li X, B Oskooi and L B Pedersen 2000 *Geophysics* **65** 452-464.
[11] Grandis H and P Sumintadireja 2012 A brief review for the proper application of magnetotellurics (MT) and Controlled-Source Audio-frequency Magnetotellurics (CSAMT) in geothermal exploration *Proceedings of 12th Indonesian Geothermal Association Annual Meeting and Conference*
[12] Munoz G 2014 Exploring for geothermal resources with electromagnetic methods *Surveys in Geophysics* **35** 101-122.
[13] Armadillo E, D Rizzello, M Verdoya, C Pasqua and P Pisani 2015 Cubic spline regularization applied to 1D magnetotelluric inverse modeling in geothermal areas *Proceedings of World Geothermal Congress*
[14] Routh P and D W Oldenburg 1999 *Geophysics* **64** 1689-1697

Synthetic Modeling of A Geothermal System Using Audio-magnetotelluric (AMT) and Magnetotelluric (MT)

Rifki Mega Saputra[1,*], **Widodo**[2]

[1]Undergraduate Program of Geophysical Engineering, Institut Teknologi Bandung Jalan Ganesha 10 Bandung 40132, Indonesia
[2]Applied and Exploration Geophysics Research Group, Institut Teknologi Bandung Jalan Ganesha 10 Bandung 40132, Indonesia

*E-mail: rigtra@students.itb.ac.id

Abstract. Indonesia has 40% of the world's potential geothermal resources with estimated capacity of 28,910 MW. Generally, the characteristic of the geothermal system in Indonesia is liquid-dominated systems, which driven by volcanic activities. In geothermal exploration, electromagnetic methods are used to map structures that could host potential reservoirs and source rocks. We want to know the responses of a geothermal system using synthetic data of Audio-magnetotelluric (AMT) and Magnetotelluric (MT). Due to frequency range, AMT and MT data can resolve the shallow and deeper structure, respectively. 1-D models have been performed using AMT and MT data. The results indicate that AMT and MT data give detailed conductivity distribution of geothermal structure.
Keywords: *magnetotelluric, audio-magnetotelluric, geothermal system, resistivity*

1. Introduction

Indonesia has 40% of the world's potential geothermal resources. There is 28,910 MW electricity that can be generated from geothermal [1]. Recently, only 1,343.5 MW geothermal power plant capacity is used.

Geothermal systems are indicated by the presence of heat source, reservoir, and a cap rock [2]. Mostly, geothermal systems in Indonesia are the result of volcano activity. The subduction of Indo-Australian plate – southern part of Indonesia- and Eurasian plate in the northwest is the main cause of volcanoes. The volcanoes lie along Indonesia region, from Sumatera Island in the west to Papua Island in the east.

There are several differences between the geothermal system between Sumatera and Java - including Nusa Tenggara and Sulawesi [3]. In Sumatera island, the geothermal systems relate with Andesitic-Rhyolitic volcanoes with its viscous and acid magma. The geothermal systems are controlled by a regional fault within Sumatera fault. The reservoir is on sediment rocks and can be discovered in the shallow zone. Whereas, the reservoir of geothermal systems in Java, Nusa Tenggara, and Sulawesi are in the deep zone. The systems are associated with andesitic-basaltic volcanism, where the magma is less viscous.

2. Method

2.1 Magnetotelluric theory

Magnetotelluric (MT) method measures natural electromagnetic field, which is used to investigate the electrical conductivity structure of the Earth [4]. The measurement on the surface is carried out to determine the resistivity of subsurface by investigating electrical and magnetic field. Basically, there is no significant difference between MT and AMT. It just depends on measured frequency.

The increased reception band of AMT method allows us to measure resistivity and improve the model in the shallow zone. Meanwhile, MT records frequency in range 0.001 Hz - 10 kHz that can be used to detect geological structures deeper than about 1,000 meters, where target depths are typically in the range of 1,000-3,000 meters.

High frequency of electromagnetic only gives a shallow depth of penetration, because of energy absorption approximately 1/e or ~37% from amplitude value on the surface [5]. This phenomenon is called skin depth (δ). Propagating of electromagnetics wave in the earth is diffusion event rather than wave propagation itself.

$$\delta = 503 \left(\frac{\rho}{f}\right)^{1/2} \text{ (meter)} \qquad (1)$$

Where, ρ is resistivity and *f* is frequency.

2.2 Apparent resistivity and phase

Generally, 1-D MT responses are expressed as apparent resistivity (ρ_a) and phase (ϕ) [6]. Those can be calculated using following equation, where Z_1 is impedance on the earth surface,

$$\rho_a = \frac{1}{\omega\mu_0} |Z_1|^2 \qquad (2)$$

$$\phi = \tan^{-1}\left(\frac{\text{Im } Z_1}{\text{Re } Z_1}\right) \qquad (3)$$

3. Modeling

Modeling and data processing were accomplished in the Geoelectric and Electromagnetic Laboratory, Department of Geophysical Engineering, Institut Teknologi Bandung. We used synthetic data to make geothermal system model.

3.1 Conceptual model

The conceptual model contains geological and geophysical profile about geothermal system in Indonesia. There are five layers i.e. overburden, clay cap, reservoir, bedrock, and heat source. The overburden layer has various in depth from 500-800 m, and the resistivity is 70 Ωm.

The clay cap lies under overburden layer between 700-1000 m depth. The clay cap has the lowest resistivity 8 Ωm. After that, the following layers are the reservoir, bedrock, and heat source with each depth is 1000-2200 m, 2200-2500 m, and 3100-4000 m. The resistivity is 60 Ωm, 250 Ωm, and 850 Ωm.

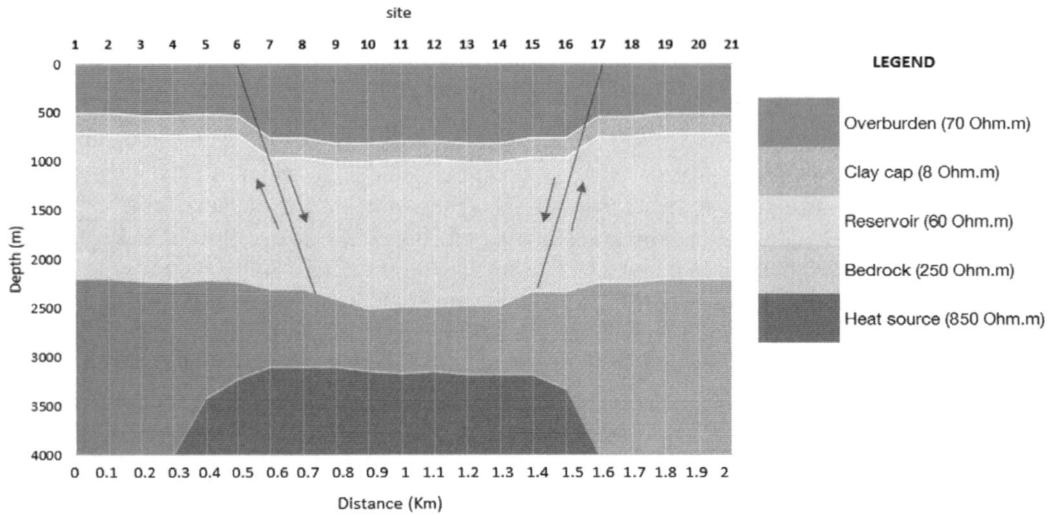

Fig. 1. A Conceptual model of the geothermal system in Indonesia with five layers, i.e. overburden, clay cap, reservoir, bedrock, and heat source. Each layer has resistivity value. The total depth is 4000 m. (Modified from [7])

3.2 Data processing

We obtained 21 synthetic data points from the conceptual model (see Fig. 1). Distance from a point to point is 100 meters. So, the total distance to be modeled is 2 km. We also need to input frequency range for each method, 0.001-1000 Hz for MT and 0.1 Hz-10 kHz for AMT. In the inversion processing, we used Levenberg-Marquardt algorithm.

In this paper, the model will be presented only in Site 1, Site 6, Site 10, and Site 18. In the Fig.2 shows the 1-D apparent resistivity models. The green line is resistivity model based on original information from resistivity value of forward modeling in each layer. The red line and blue line are the result of inversion of MT and AMT, respectively.

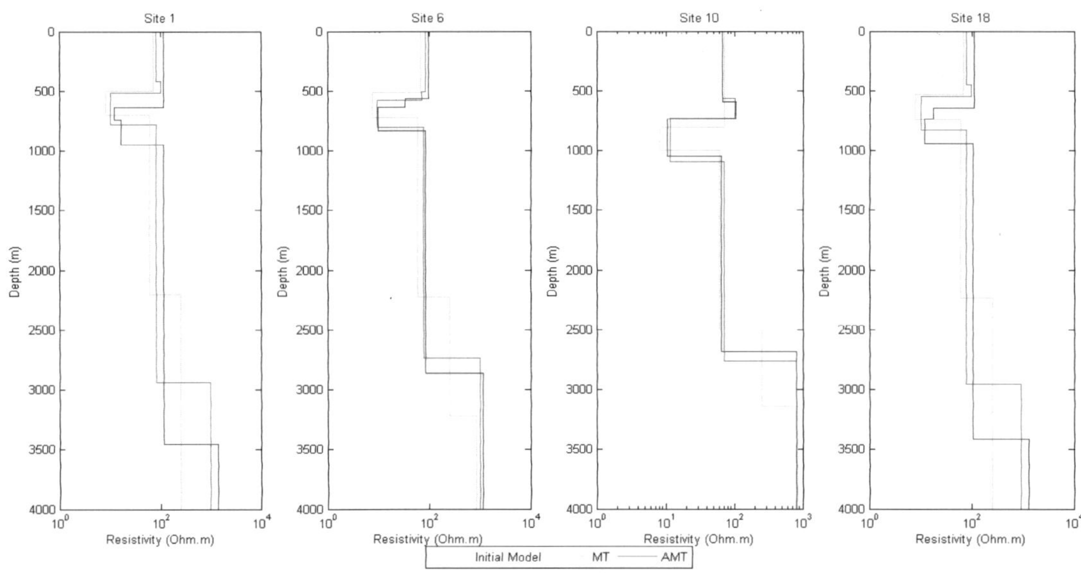

Fig. 2 1-D apparent resistivity models of MT and AMT in the Site-1, Site-6, Site-10, and Site-18.

4. Discussion

The 1-D resistivity models of some sites are plotted as in Fig.2. In this paper, we only discuss 1-D resistivity model of Site-1, Site-6, Site-10, and Site-18.

The curves (Fig. 3) can show the responses of MT and AMT data in Site-1 and Site-6. Responses for Site-10, and Site-18 in the (Fig.4). In the curve of Site-1, the graph increases when in the frequency 3-30 Hz. The curve between frequency and phase also increases at the same frequency. It shows phase change from less than 45 degrees become more than 45 degrees. It indicates that the electromagnetic wave passed through the conductive layer. From (Fig.2) the conductive layer lies between 500-1000 m depth.

Other curves for Site-6, Site-10, and Site-18 also give a similar result, that the conductive layer can be detected. The conductive layer is clay cap, and the reservoir lies underneath the clay cap, which becomes the main target of geothermal exploration.

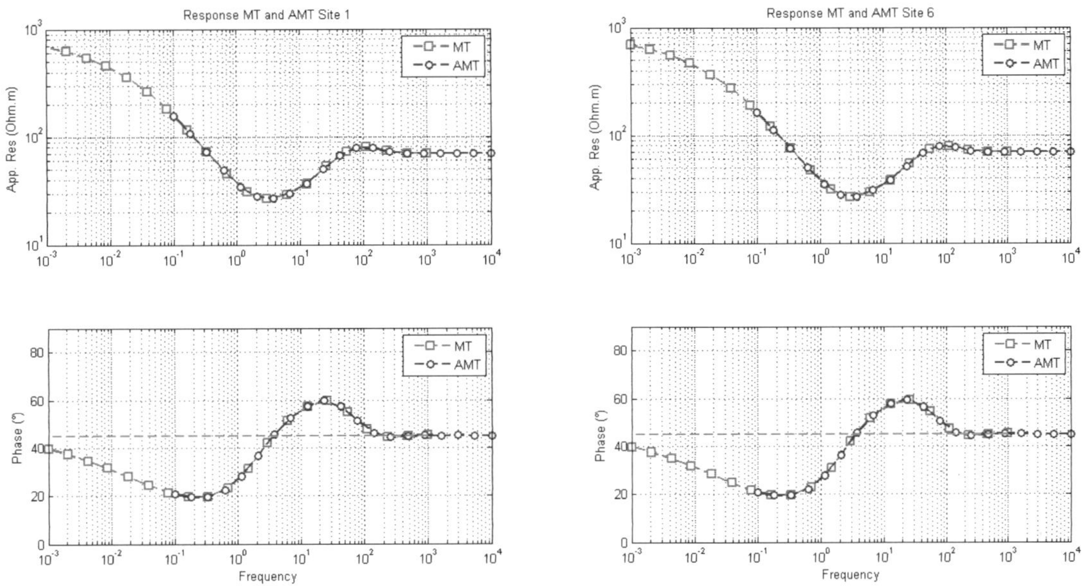

Fig.3 Response MT and AMT in Site-1 and Site-6. The curve between frequency and apparent resistivity (top) and curve between frequency and phase (bottom).

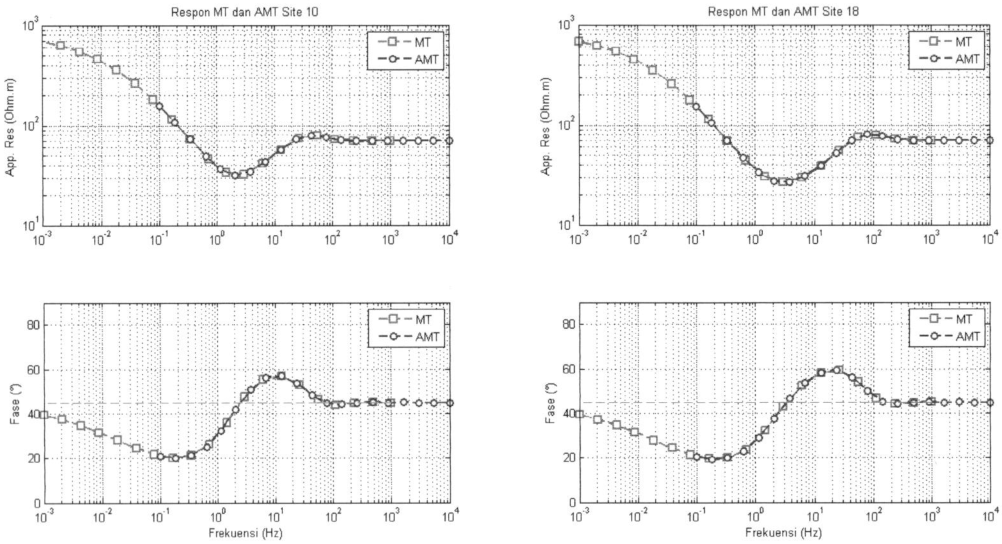

Fig.4 Response MT and AMT in Site-10, and Site-18. The curve between frequency and apparent resistivity (top) and curve between frequency and phase (bottom).

5. Conclusion

The result indicates that synthetic model of AMT and MT give detailed conductivity distribution of geothermal structure, from the surface to basement. AMT data can give an image of the structure in the shallow depth. While, from MT data we know better the structure in the deep zone where the heat source exists.

References

[1] Pusdatin ESDM 2014 Handbook of Energy & Economic Statistics of Indonesia *Ministry of Energy and Mineral Resources of The Republic of Indonesia*
[2] Saptadji N (n.d.) *Diktat Kuliah Teknik Panasbumi* (Bandung: Unpublished)
[3] Hochstein M, Oven S and Bromley C 1996 Thermal Springs at Hot Water Beach (Coromandel Peninsula, NZ) *Proceedings of the 18th NZ Geothermal Workshop New Zealand*
[4] Vozoff K 1991 The Magnetotelluric Method. In M. Nabighian Electromagnetic Methods in Applied Geophysics Volume 2, Application, Parts A and B *SEG* pp. 641-712
[5] Widodo 2012 Multidimensional Interpretation of Near Surface Electromagnetic Data Measured in Volvi Basin, Northen Greece *Koln: Universitat Zu Koln*
[6] Grandis H 2009 Pengantar Pemodelan Inversi Geofisika *HAGI*
[7] Meju M A 2002 Geoelectromagnetic Exploration For Natural Resources: Models, Case Studies, and Challenges *Surveys in Geophysics* **148**

Preliminary Study on Ground-Magnetic Data Near the Active Volcanoes in Konga Bay, East Flores Indonesia

Agus Laesanpura[1], Darharta Dahrin[1], Ivan Nurseptian[2]

[1]Applied and Exploration Geophysics Research Group, Faculty of Mining and Petroleum Engineering, Institut Teknologi Bandung, Jalan Ganesha 10 Bandung 40132, Indonesia
[2]Institut Teknologi Bandung, Jalan Ganesha 10 Bandung 40132, Indonesia

Email:laesanpura1@yahoo.com

Abstract. East Flores is part of Nusa Tenggara island belongs to volcanic arc zone, hence the active volcanoes surround the area about 60 x 50 square km. It is located at latitude south 8° 30', and longitude east 122° 45'. Geologically, the rock is mostly of volcanic material since Miocene age. The Intriguing question is where the volcanic feeder, pyroclastic, and how it vanish in subsurface. The magnetic data acquisitions were executed on land for 500 meter interval and denser through the bay surrounded by volcanoes. The combine reduction to pole and forward modelling is apply for serve interpretation using forward modelling technique. The two interpretation sections, show the body of magmatic may present at depth about 2 to 3 km. The observation show no significant decreasing or loosening of magnetic anomaly although near the active volcano. We suggest the thermal anomaly is just disturbing magnetic data in near surface but not in the depth one. Meanwhile the reduction to pole's section could distinguish the two group of rock. In assuming the layer is flat. The inferred peak of magmatic body near the existing volcano; and the active demagnetization associated through evidence of hot spring and inferred fault structure.

1. Introduction

East Flores is one of several chain of volcanic arc in Nusa Tenggara (South East islands) Indonesian, close to Bali and Lombok islands. It is on latitude south 8° 30', and longitude east 122° 45'. The island, tectonically is part of magmatic arc of Pacific's ring of fire, about 300 km from Indonesian subduction zone. The uniqueness, the small islands is encircled by volcanic product, and oceans: Indian Ocean in south, and Flores and Banda seas in North.

The magnetic data apply on this islands, for study of volcanic product, the magma feeder and the anomaly for its association using proton magnetometer. Because of dipolar magnetic and position in near equator, the qualitative interpretation will slightly hard. From here, the reduction to pole (RTP) idea [1][2][3] is apply for distinguishing anomaly and more distinct feature in subsurface. To interpret causative body use forward modelling technique. The coverage area is about 50 x 60 square km, and the magnetic data acquire along the existing road about 500 m interval and on the sea.

2. Geology of Konga bay

The study area is surround by volcano due to plate tectonic position in volcanic chain. Two main structure exist on this area. It is, the Miocene of strike slip fault, and the Pleistocene of normal fault in NW-SE and SW-NE.

The whole lithology [4][5], is mostly volcanic product, sometimes found marl, and the combined coral -beach sand. The oldest known rock crop out on this area is known as Kiro Formation (Tmk), Miocene age. It consist of product of lava, breccia, sometimes and sandy-tuff with composition of andesite until basalt.

The volcanic product of pyroclastic from Nangapada (Tmn) to Waihekang (Tmpw), is break by an intrusion of Wolowaru (Tmg) Formation of Middle Miocene. The most subtle and well-known is the old volcanic product (Pleistocene) and young volcanic product (quaternary). These two found in most volcanic Indonesian rock.

The quaternary-Pleistocene volcanic product consist two unit, the Quaternary Pleistocene Volcanic Product (QTv). It is consisting of Breccias, interbred sandy tuff with pumice or breccias tuff, and agglomerate. While the Holocene Volcanic Product (QHv) consisting lava of breccias, sandytuff. The QHv is found near the recent active volcano.

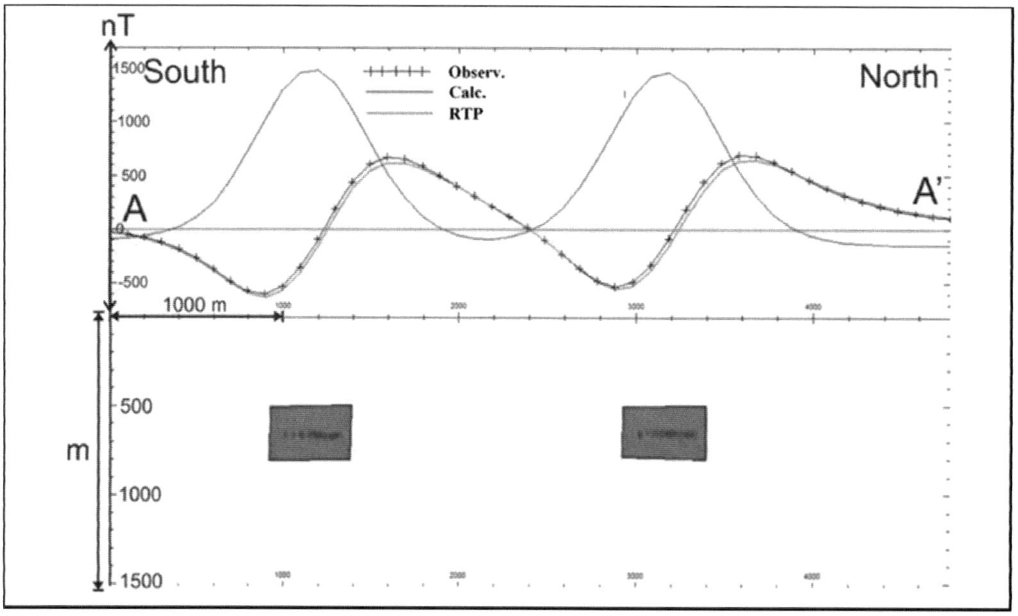

Figure 1. The reduction to pole to distinguishing anomaly, and the result is more distinct.

3. Reduction to pole and magnetic anomaly identification

The Reduction to pole (RTP)[1][2][3] is a method in magnetic interpretation to recognize the subsurface anomaly in case a complex or hardly identify the obscure subsurface object due to its dipolar responds. This formula apply for dipolar magnetic data near equator, transformed as it will be in polar position to easy interpretation in unique mono polar pattern. The original magnetic data T(O) in coordinate polar (ρ,ω) and inclination factor: i,μ, could be transformed to reduction to pole[g'] result, g', as:

$$g' = -\mu T(o) - \frac{1}{2\pi} \iint T(\rho,\omega)\,\Omega(\omega)\,\frac{d\rho}{\rho}\,d\omega \tag{1}$$

The following synthetic data show the effectiveness of reduction data in interpretation two closed object 2 km separation and 0.5 km depth. The curve show dipole pattern from the two object, and we are slightly speculation to decide. The reduction to pole pattern show the centre position of object, and it is on the flank of original data. We can simulate many possibilities for the object composition and the complexities of the pattern.

4. Interpretation two sections near the volcano

The magnetic data on this interpretation has deduced from unwanted effect of earth magnetic field and atmospheric disturb, applying IGRF formula and daily magnetic on base. Hence the result is magnetic represent the subsurface object only. The middle range of measurement of about 45600 nT, is reduced by IGRF, and the result is magnetic data ready to interpret on the range 1200 nT until 1400 nT. The interpretation use forward modelling technique software.

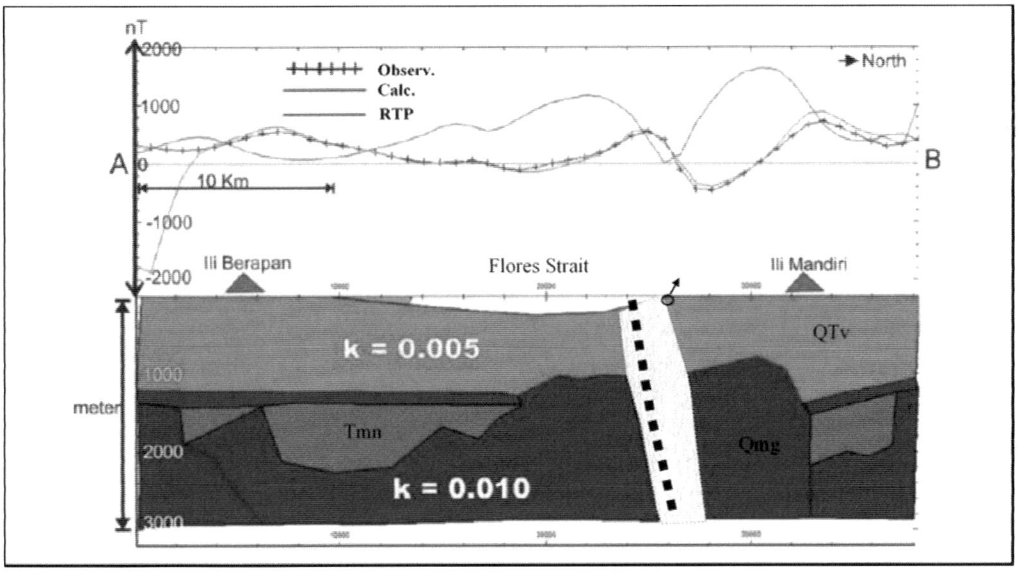

Figure 2. The AB section Interpretation. The slit of low RTP coincident with hot spring, and fault inferred.

4.1 AB Line

The AB cross section is a magnetic line in SW-NE where it is through almost volcano Ille Berapan, Flores strait, Oka bay, and the volcano Ille Mandiri. This interpretation line is also through the volcanic
deposit of quaternary (QTv), and the lineament's interpretation from Landsat near Oka Bay.

The anomaly show the depth magnetic effect (2500 meter), due to plutonic influence. The respond of reduction to pole is locally narrow low, near Oka bay. While in the adjacent curve and the magnetic data show the appearance double source of positive anomaly. The field evidence show near the center of anomaly in Oka bay found the hot spring, and on the Landsat map show the lineament. However, We guest the narrow low curve of reduction to pole is due to demagnetization by the existence of hot fluid, or mineralization among the (extinct) plutonic body (k=0.001). And, beneath the volcanic rock (K=0.005) consisting young andesit Formation.

4.2 EF Line

The EF section is crossing the profile AB. The two peak of anomaly as well as RTP show the possible evidence of a magmatic body. The SW vicinity is similar to AB section where we interpret the magmatic body beneath the Flores strait extend to this point. While in east vicinity, the Ili Boleng show up the volcano activities, but not present the strong demagnetization. This case is similar to other section where the effect fairly small representing by anomaly curve. But the plug as dyke maybe present strong magnetic such case Ili Boleng.

The RTP curve show the wide negative as well as the magnetic curve and it is related to volcanic rock of Kiro Formation. Hence these two curve could distinguish the volcanic rock of Kiro Formation (The oldest rock), from the others. The emergence the old Kiro Formation among the young in nearly flat sedimentary, means the throw is taken place. The body is either a horst or horizontal displacement. And now, Kiro Formation is evidence of Hiatus until recent years.

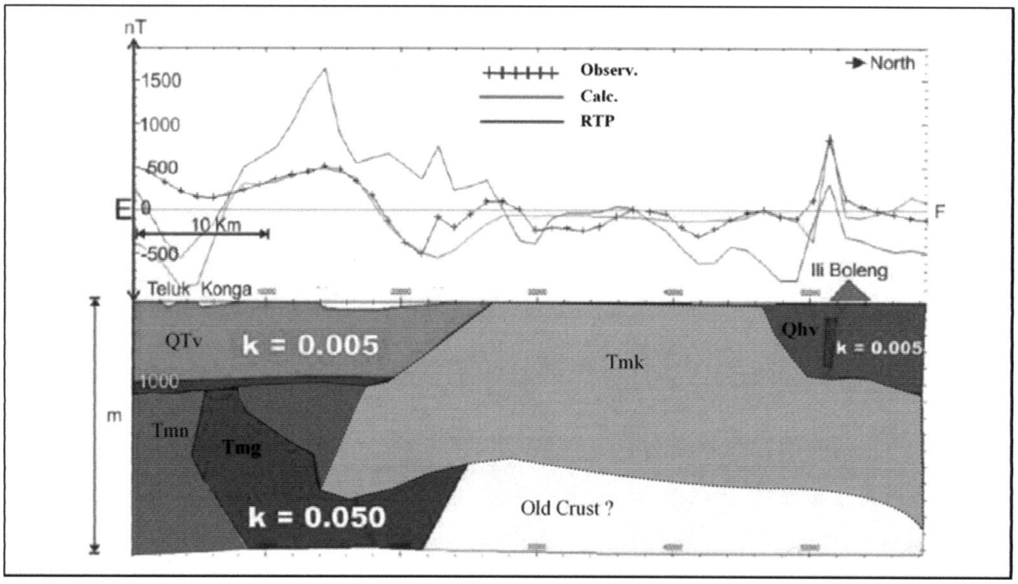

Figure 3. The EF section Interpretation. The old Kiro Formation show the combination of low magnetic and low RTP anomaly. Ili Boleng, an active Volcano show the peak anomaly.

5. Conclusion

At Konga bay, around volcanic product, the combination of Magnetic anomalies and the reduction to pole could distinguishing the body of plutonic by model of the plug, the oldest rock (Kori's Formation) from younger one. Meanwhile, the alteration is reveal by demagnetization around dominant high magnetic value.

Acknowledgments

Authors wishing to acknowledge the students and technician of Geophysical Engineering study program during the assessment, and the institution permitting to realize the research thanks to Maison du Monde Malais of Université de la Rochelle-French, and LPPM (Empowerment Community and research) ITB.

References

[1] Baranov V 1957 *Geophysics* **22** pp 359-383
[2] Baranov V, and Naudy H. 1964 *Geophysics* **29** pp 67-69
[3] Roest W R, Verhoef J, Pilkington M, 1992 *Geophysics* **57** pp 116-125
[4] Koesoemadinata S, and Noya N 1989 *Geological map of the Lomblen quadrangle*
[5] Suwarna N, Santosa S, Koeoemadinata S 1989 *Geological map of the Ende quadrangle, Nusatenggara*

IOP Conf. Series: Earth and Environmental Science **62** (2017) 012038 doi:10.1088/1755-1315/62/1/012038

Subsurface Structure Interpretation Beneath of Mt. Pandan Based on Gravity Data

D Santoso[1], E J Wahyudi[1], S Alawiyah[1], A D Nugraha[1], S Widiyantoro[1], W G A Kadir[1], P Supendi[2], S Wiyono[2], Zulkafriza[3]

[1]Faculty of Mining and Petroleum Engineering, Institut Teknologi Bandung, Indonesia
[2]Indonesian Agency for Meteorological, Climatological and Geophysical
[3]Postdoctoral researcher, GREAT, ITB

The corresponding author's e-mail: dsantoso78@yahoo.com

Abstract. Mt. Pandan is one of the volcano that state as dormant volcano. On the other hand, Smyth et al. (2008) defined that Mt. Pandan is an active volcano. This volcano is apart a volcanic chain in Java island which is trending east-west along the island. This volcanic chain known as present day volcanic arc. Mt. Wilis is located in the south and it relatively much bigger compare to Mt. Pandan. There were earthquakes activity experienced in the surrounding Mt. Pandan area in the past several years. This event is interesting, because Mt. Pandan is not classify as the active volcano according to the list of volcanoes in Indonesia. On the otherhand Smyth et. al. (2008) mentioned that G. Pandan as modern volcanic which is located in Kendeng Zone of East Java. Gravity measurement around Mt. Pandan area was done in order to understand subsurface structure of Mt. Pandan. Gravity interpretation results shows that there is a low density structure beneath Mt. Pandan. It could be interpreted as existing of magma body below the surface. Some indication of submagmatic activities were found as hot spring and warm ground. Therefore it could be concluded that there is a possibility of magmatic activity below the Mt. Pandan.

1. Introduction

Mt. Pandan is a volcano with 897 meters high located in the Eastern part of Java Island. It is approximately 40 km northeast of Madiun town. Mt. Pandan is not listed as one of active volcanoes in Indonesia. However, Java Island as part of Indonesia Island Arc become an active region. In the Island a recent active volcanic chain trending east-west is developed. Is is related to active margin as formed by Indian-Australian plate subducted underneath Eurasian plate in the southern part of Java island. Subduction process has occurred along the Java Trench that located at the south of Java Island begin from Middle Eocene to present day [2]. Then, in the volcanic island contains the active volcanism is developed in Java Island since Early Cenozoic. Some volcanoes of the modern Sunda Arc are distributed along the Java island. A second older arc of Eocene to Miocene volcanoes in the East Java region formed the Southern Mountain [6].

The modern Sunda Arc volcanoes in East Java are built mainly in the Kendeng Zone. Mt Wilis is one of the active volcanic centre in East Java. Mt. Pandan in the north of Mt Wilis is defined as one of Modern Volcanic Centres [7] (Figure 1).

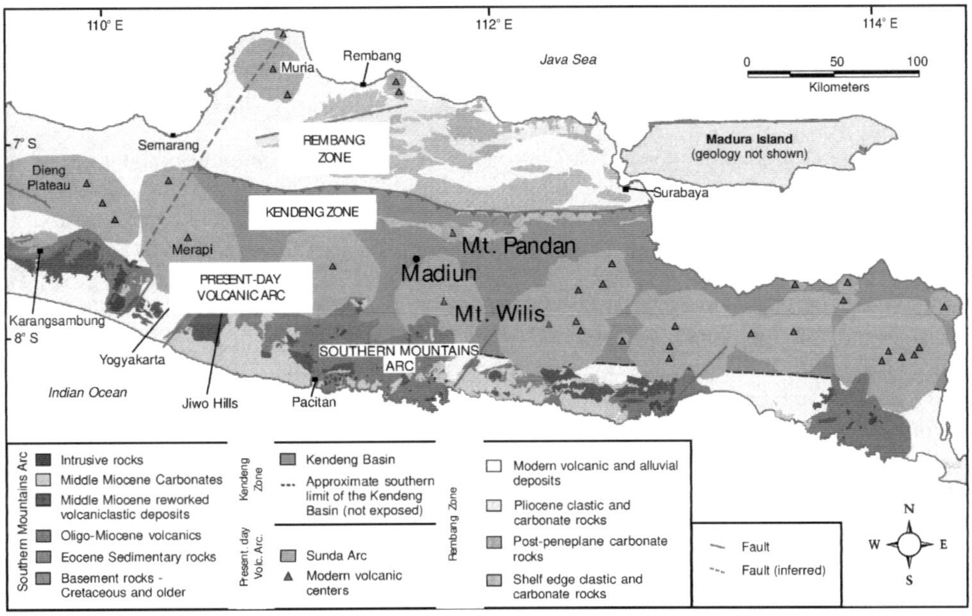

Figure 1. Simplified Geological Map East Java and its Tectonostratigraphy (modified from [6, 7])

Smyth [7] defined that Mt. Pandan with conical morphology in the north of Mt. Wilis which Sas a modern volcanic centre. On the other hand CVGHM [1] defined as a dormant volcano. Mt. Pandan has a relatively little conical shape with approximately 300 meters high compare to the surrounding area. Some recent earthquake events were recorded in the area of Mt. Pandan. It could indicate an activity that may be related to magmatic process. Structurally, in the north of Mt. Pandan a thrust fault is found.

Related to the occurrence of the earthquake there are two possibilities, first it was related to thrust fault activation and the second is related to submagmatic activation. There is also possibility that due to active fault movement may triggered upwelling magma.

Gravity measurement was conducted in the region in order to find an explanation relation to earthquake occurrence in the area of Mt. Pandan. By using Bouguer Anomaly Gravity data a subsurface structure beneath Mt. Pandan will be interpreted to explain the seismic activity in the area.

2. General Geology of East Java and important structural feature to Mt. Pandan

Physiographically, Java Island is divided into some region [8]. These zonations also reflect structural and stratigraphical features or tectonostratigraphy. By tectonostratigraphy, Smyth et al. [6] divided East Java into four zones, (Figure 1), they are (1) Southern Mountain Zone. The Eocene-Miocene volcanic arc built on Mesozoic basement. Deposits include siliciclastic, volcaniclastic, volcanic and carbonate rocks which generally dip uniformly to the south, (2) Present-day Volcanic Arc, which active from the Late Miocene (3) Kendeng Zone as the main Eocene-Miocene depocenter in East Java contains thick sequences of volcanogenic and pelagic sediments. It is now an east-west trending thrust belt, and (4) Rembang Zone, the Eocene-Pliocene sequence includes shelf-edge deposits such as shallow marine clastic sediments and extensive carbonates. This zone contains one major ENE-WSW fault-bounded high (Rembang High) and many east-west orientated folds.

Figure 1 also shows Simplified Geological Map of Mt. Pandan and surrounding area in East Java included tectonostratigraphical zonation. Thrust Fault that trending east west is the most important feature related to Mt. Pandan. It is become the boundary between Kendeng Zone and Rembang Zone.

3. Recent Seismic activity around Mt. Pandan

Some earthquakes happened in Madiun on June 25, 2015 in 7.73^0 S, 111.69^0 E with a depth of 10 km. The events were informed by Indonesian Agency for Meteorological, Climatological and Geophysical (MCG here in after we call BMKG). The earthquakes also caused some light to medium damages for several houses.

Nugraha et al. [4] studied hypocenter relocation of this earthquake by re-picking arrival of P wave and S wave phases, and then they done focal mechanism determination. The result showed that this earthquake was located in 7.6305^0, 111.7529^0 E with 14.81 km focus depth and a strike-slip mechanism with strike direction of 163^0 around Madiun and also close to Mt. Pandan.

Some small earthquakes of Magnitude less than 3 was happened in February 2016. They were observed by BMKG network as shown in Figure 2. In order to study the wave form and source mechanism of the earthquakes, the Team of Geophysical Engineering, Institut Teknologi Bandung from March to May 2016 was operated two broad band seismometer station around Mt. Pandan.

By studied the wave form, there were observed fairly clear of small earthquake that has very short time different the arrival of S and P phase wave about 0.14 second with frequency dominant of 20 Hz to 45 Hz. This event could be interpreted as related to an activity of local tectonic (fault). It is suggested that additional geophysical and geological investigation are needed in order to determine the source of mechanism of these earthquake activities in surrounding Mt. Pandan.

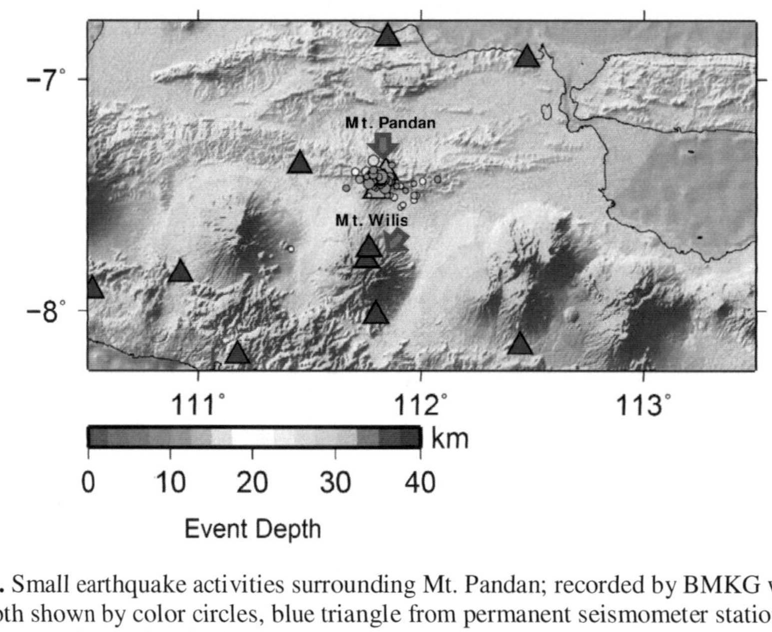

Figure 2. Small earthquake activities surrounding Mt. Pandan; recorded by BMKG with various focus depth shown by color circles, blue triangle from permanent seismometer station of BMKG in east Java, and gray tiangle from temporary seismometer station of ITB for March-May 2016 time periods (gray triangles) around Mt. Pandan. Beside ITB measurement, these events were occured in February 2016.

4. Gravity around Mt. Pandan

The Geophysical Engineering, Institute Teknologi Bandung Team conducted gravity measurement on June 2016. The purpose of the survey is to understand the subsurface structure beneath Mt. Pandan. By the data the Complete Bouguer Anamaly Map was constructed as shown on Figure 3 (1). Based on Complete Bouguer Anomaly Map of Mt. Pandan, the Residual Bouguer Anomaly Map was developed (figure 3 (2)). The gravity anomaly in the map shows interesting result as an existing of low density value beneath Mt. Pandan.

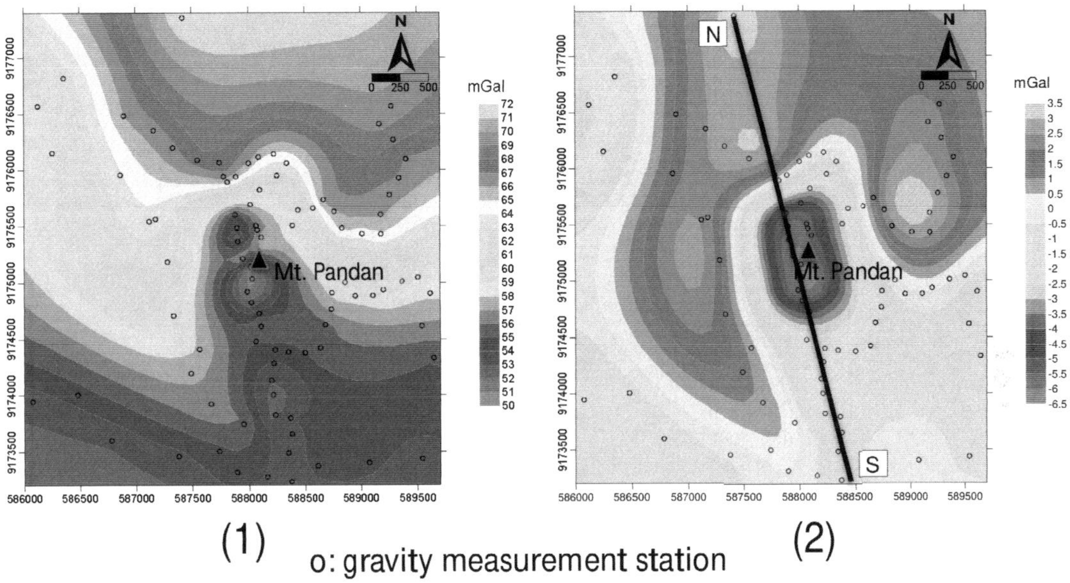

Figure 3. The result of gravity survey in Mt. Padang shown as Complete Bouguer Anomaly Map of Mt. Pandan (1) and Residual Anomaly Map of Mt. Pandan (2). Note the existing of low gravity anomaly in the area of Mt Pandan. N-S line shows location section for forward anomaly density model

Beside that it is found the trend less gravity anomaly value. The zone of less gravity anomaly is trending from north where Mt. Pandan located toward south where Mt. Wilis located. The density section was derived by Bouguer Residual Anomaly Map of Mt. Pandan can be seen in Figure 5. As the same an indication in the map low density anomaly is found underneath of Mt. Pandan. North south trending low density structure as shown in the map is identified also in the section. Anomaly density section derived from Residual Anomaly can be seen in Figure 5. According to the north-south section, it shows low density value in the area beneath Mt. Pandan (Figure 4).

5. Analysis

As mentioned before that the result of seismic wave data of earthquakes study in Mt. Pandan area, the events are occurred as earthquakes (Figure 2). Where as the focal mechanism of the earthquakes were as strike slip faults. They were indicating as shallow earthquakes. These findings indicated that they were not related to an upwelling of magma or movement of magma body.

On the contrary gravity data shows the indication the existing of low density value that could be interpreted related to existing of magma. The low density anomaly also formed a trend toward Mt. Wilis. There are also found a phenomena related to submagmatic activity such as hot springs and warm ground in the area. All the data could conclude that there is a possibility that both phenomena are related together. It could be explained that subduction process produced the fault movement and at the same time also triggered the upwelling magma to the surface.

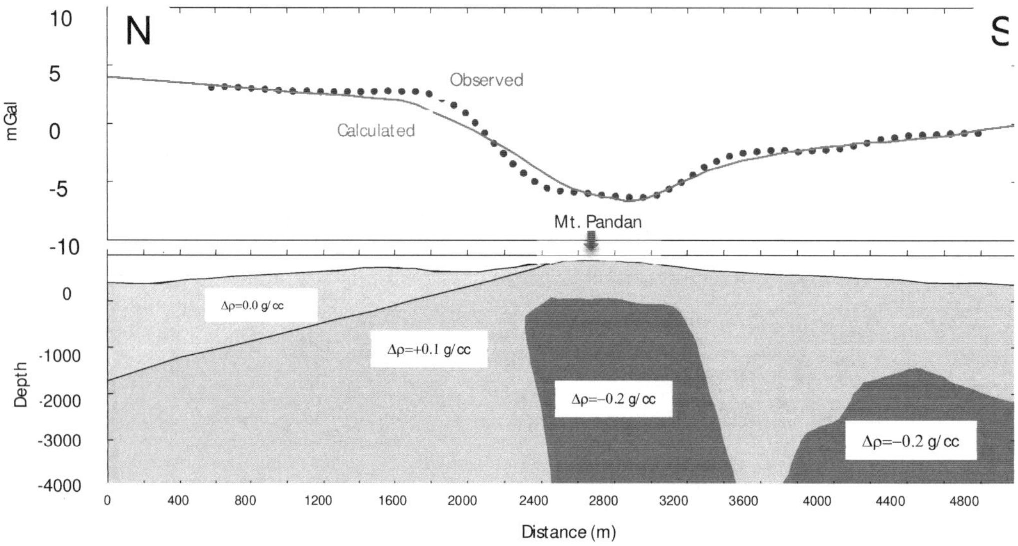

Figure 4. N-S density structure derive by forward from Bouguer Residual Anomaly section shows low density value underneath of Mt. Pandan which could be interpreted as magma body

6. Conclusions

According to available data on density structure beneath Mt. Pandan and supporting by earthquake data as well as submagmatic activity indication on the surface, Mt. Pandan could be determined as a modern volcano. Mt. Pandan volcano indicated has a relation of magma chamber to Wilis volcano. Both volcanoes are related to the same subduction process. The mechanism of tectonic process could be explained that the earthquakes may be triggered upwelling of the magma as they were happened due to subduction process.

As our experience before, the dynamic of subsurface condition could be studied by time lapse gravity study. Some examples of application the method was developed by [3, 5]. By using time lapse gravity method more clear understanding on subsurface change beneath of Mt. Pandan could be interpreted.

References

[1] Center for Volcanology and Geological Hazard Mitigation (CVGHM) 2015 *Brief report of Emergency Response Team of Madiun Earthquake 25 June 2015* 3 pp.
[2] Hall R 2002 *Journal of Asian Earth Sciences* **20** 353-434.
[3] Kadir W G A, Santoso D, and Alawiyah S 2007 Priciple and application of 4D microgravity survey for engineering purpose, case example: groundwater level lowering and subsidence in residential area of Jakarta *Proceedings of the Symposium on the Application of Geophyics to Engineering and Environmental Problems*, SAGEEP
[4] Nugraha A D, Supendi P, Shiddigqi H S, Widiyantoro S 2016 Uneexpected earthquake of June 25th, 2015 in Madiun, East Java *AIP Conf. Proc.* **1730** 020001
[5] Santoso D, Kadir W G A, Alawiyah S, Setianingsih, Sarkowi M, Minardi M. 2011 The time lapse microgravity response due to subsidence and groundwater level lowering, *The contribution of geosciences to human security engineering*, Institut Teknologi Bandung, Kyoto University, and Kyoto University GCOE Program of HSE, Eds.; Logos Verlag Berlin GmbH: Berlin, Germany, 2011; p. 27-48
[6] Smyth H, Hall R, Hamilton J P, and Kinny P 2005 East Java: Cenozoic basins, volcanoes and

ancient basement: Jakarta *Proceedings, Indonesian Petroleum Association Annual Convention 30th* p.251–266

[7] Smyth H, Hall R, and Nichols G J 2008 Cenozoic volcanic arc history of East Java, Indonesia: The stratigraphic record of eruptions on an active continental margin, in A.E. Draut, P.D. Clift, and D.W. Scholl, SPE436: Formation and Application of the Sedimentary Record in Arc Collision Zones, *Geological Society of America*, pp. 199-222

[8] van Bemmelen, R W 1949 The Geology of Indonesia *Govt. Printing Office, Nijhoff, The Hague* 732 pp

Seismic Travel-time Tomography beneath Merapi Volcano and its Surroundings: A Preliminary Result from DOMERAPI Project

Mohamad Ramdhan[1,2], Sri Widiyantoro[3], Andri Dian Nugraha[3], Asep Saepuloh[4], Jean-Philippe Métaxian[5], Said Kristyawan[1,2], Andry Syaly Sembiring[1,2], Agus Budi Santoso[6]

[1]Graduate Program of Earth Sciences, Faculty of Earth Sciences and Technology, Institut Teknologi Bandung, Indonesia
[2]Agency for Meteorology, Climatology, and Geophysics (BMKG), Indonesia
[3]Global Geophysical Research Group, Faculty of Mining and Petroleum Engineering, Institut Teknologi Bandung
[4]Faculty of Earth Sciences and Technology, Institut Teknologi Bandung
[5]ISTerre, IRD R219, CNRS, Universit'e de Savoie, Le Bourget-du-Lac, France; [6]Center for Research and Technology Development of Geological Disaster (BPPTKG), Indonesia

E-mail: *mramdhan123@gmail.com*

Abstract. Many geophysical methods at various scales have been applied to understand the internal structure beneath Merapi volcano and its magmatic process. As part of the DOMERAPI project, a seismic experiment was conducted from October 2013 to mid April 2015 in order to determine the deep magma source beneath the volcano through seismic travel-time tomography. The earthquake events were identified and picked manually and carefully to determine the hypocenters. They were then relocated to get precise hypocenter locations before running the seismic tomographic imaging. The data from the BMKG network from the same period of time as mentioned above were also incorporated to minimize azimuthal gap, because the majority of events occurred outside the DOMERAPI network. The checkerboard resolution test result depicts that the area around the network can be well resolved. Compared to previous studies, our result shows a higher resolution at shallow depths, i.e., less than 35 km and a low velocity material imaged to ascend diagonally from the deeper area.
Keywords: Merapi, DOMERAPI, BMKG seismic tomography, low velocity.
PACS: 91.30.Jk

1. Introduction

The Merapi volcano is one of the most active volcanos in the world. The eruption period occurred from two to six years [1]. The Merapi eruption type is dominated by pyroclastic flows caused by lava dome collapse [1,2]. A different eruption type, which is more explosive and of higher magnitude, such as the 2010 eruption, occurs less frequently, i.e. every 50 to 100 years [3,4]. Many earth scientists have studied various methods to determine the Merapi internal structure from local to regional scale. Two magmatic systems have been identified beneath Merapi down to 5 km depth beneath the summit by seismological studies [1,2]. A low velocity material ascending from the partial melting zone to volcanic arc was detected by travel-time seismic tomography by using MERAMEX network [5,6]. The result was updated by incorporating data from the BMKG network [7]. The seismic ambient noise tomography using MERAMEX network showed the low group velocities beneath Merapi volcano could be deleniated from 2 to 12 s periods [8]. The same method had been applied using DOMERAPI network [9]. The result showed the higher resolution than the previous study. It showed not only low group velocites beneath Merapi and Merbabu but also high group velocites beneath them [9]. The Merapi eruption system is controlled by a deep magma source [10]. It could not be detected clearly by seismic studies as mentioned above because of the limitation in data resolution. The DOMERAPI seismic experiment, which was conducted from October 2013 to mid April 2015, has a target, i.e. to determine the deep magma source and its magmatic process beneath the volcano through travel-time seismic tomography.

2. Data and Method

The DOMERAPI seismic network covers Merapi and Merbabu volcanoes. The average distance between seismometers is about 4 km. Opak fault is also covered by the network with a fewer density of seismometers compared to the two volcanoes (see figure 1.a). The majority of earthquakes occurred outside of the network. Data from the BMKG network were included in order to minimize the azimuthal gap [11]. The contribution of the earthquakes located outside the DOMERAPI network is essential to delineate the deep structure beneath Merapi volcano given that the volcano-tectonic earthquakes (VT) occur at maximum 5 km below the summit [1,2]. The total number of earthquakes used in this study is 372. These events were recorded by 70 seismic stations (53 from DOMERAPI network and 17 from BMKG network) and were used to compute travel-time seismic tomography. The events and the network are restricted in the area: 108°-112°E and 6°-11°S. Computation of hypocenters and travel-time seismic tomography for P and S phases or P phase only was performed for events recorded at least by six stations. The minimum of six stations is used to obtain over-determined criteria. The Geiger method was applied to compute hypocenter determination with Hypoellipse software [12,13]. The hypocenters were relocated firstly before running travel-time tomography using a double-difference method [14,15]. The SIMULPS12 was used to compute travel-time seismic tomography [16,17]. The method has been successfully applied widely in various places to delineate the subsurface structure. The subducting slab in the Bungo chanel and Shikoku area had been detected successfully by the method [18]. The method also could detect clearly subduction slab beneath Bali island [19]. The method is also reliable to delineate subsurface structure beneath volcano. The magma source beneath Lokon and Guntur volcanoes could be delineated clearly by the method [20,21].

Figure 1.b shows epicenter distribution and seismic networks in the study area. The initial model used for travel-time seismic tomography computation is a 1-D P velocity model obtained with a previous regional study of central Java [5]. It used a Vp/Vs ratio of 1.73 [22]. The smallest horizontal and vertical grid is 10 km and 5 km, respectively. This grid size is smaller than those used in the previous studies [5,6,7].

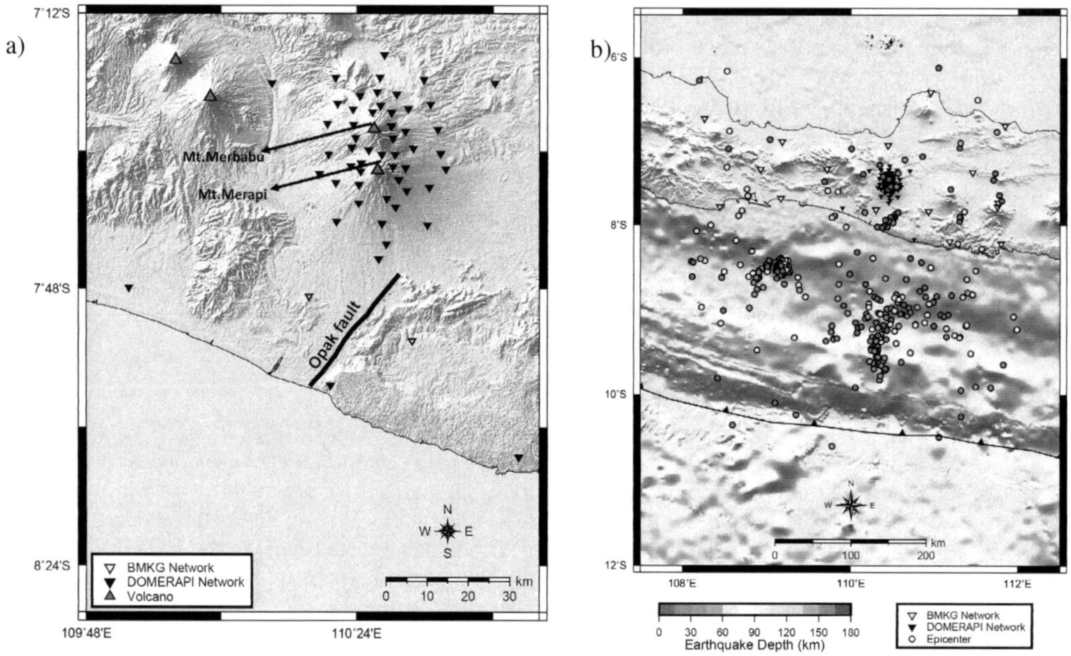

Figure 1. **a.** The stations distribution of DOMERAPI (black inverted triangles) and BMKG (yellow inverted triangles) networks in the study area.
b. The seismometer and epicenter distribution used in the travel-time seismic tomographic imaging.

3. Results and Discussion

The total number of ray paths we used is 16,101 (11,074 for P wave and 5,054 for S wave). The inversion was performed for Vp and Vp/Vs ratio. For the Vp/Vs ratio inversion we used an equal ray path number of P and S. Ideally, the total number of of P and S wave data is equal, but in the real condition the number of S is less than P wave because in general the signal to noise (S/N) ratio is lower for S wave than for P wave. The low S/N ratio produces an un-clear onset. The Checkerboard Resolution Test (CRT) that we conducted shows that the area beneath the study area could be recovered well, especiallly for depths lower than 35 km. The recovery of the CRT for the depth of more than 60 km is still reasonably good although the anomalies are weaker than the upper anomalies. In this paper we only shows P wave tomogram as a premilinary result of DOMERAPI project. Figure 2 shows the CRT result and Vp tomogram beneath Merapi volcano.

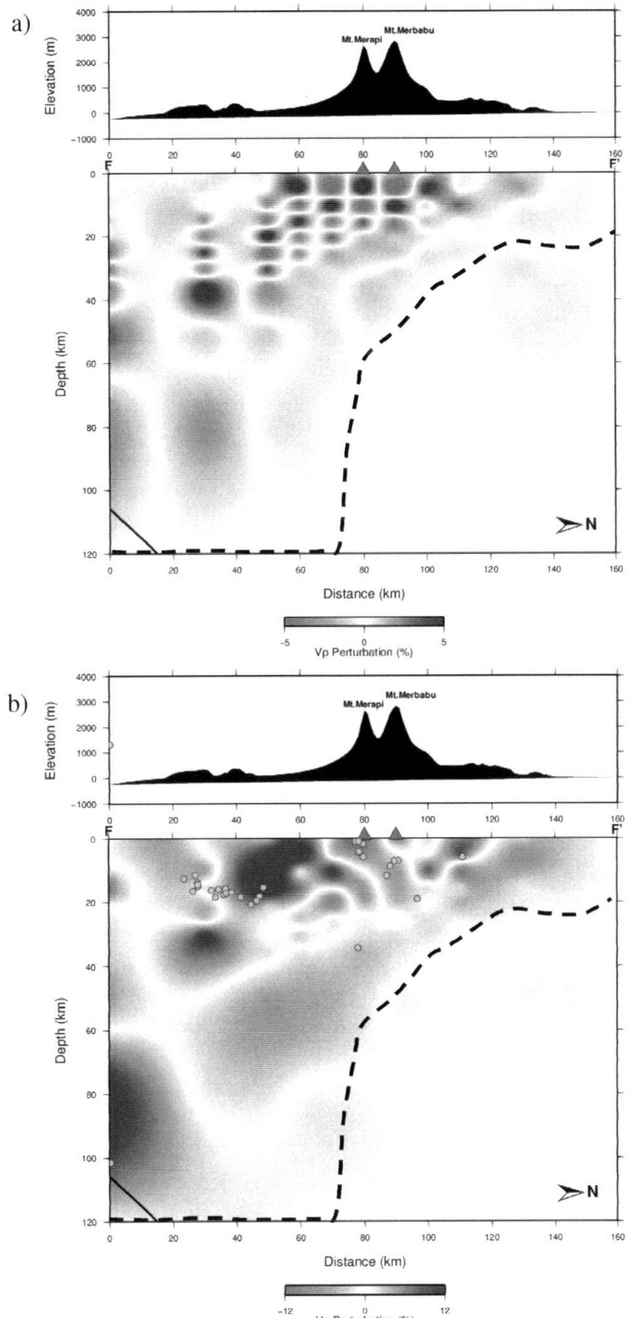

Figure 2. **a.** The CRT result beneath Merapi volcano and its surroundings. Black line depicts the slab 1.0 model [23]. The Dashed line indicates the area with good resolution.
b. Tomogram Vp beneath Merapi volcano and its surroundings. Black and dashed lines are the same with those in (a).

The previous results had not showed a low velocity material ascends diagonally from depth of about 100 km [5,6,7]. The denser station distribution of this study is likely to enhance the resolution beneath study area. The Vp tomogram shows that the low velocity beneath Merapi volcano is dipping southward. It is consistent with the result from a previous study [5]. The other study with D-InSAR method stated that the conduit is dipping to the north-west direction [24]. An impermeabel zone with high velocity was detected beneath Merapi volcano from the depths between 12 and 27 km. The zone could not be detected clearly in the previous studies due to limitation of data. The low and high velocity zones at 100 km and 12-27 km depths, respectively are consistent with the two main magmatic systems characterized by petrological studies beneath Merapi volcano, known as deep and intermediate magma reservoirs [10]. Shallow magma reservoir could not be detected by this study because of the limited data coverage [1,2,10]. The high velocity materials beneath Merbabu volcano is interpreted as a large solidified magma body/intrusion/conduit. The large body serves as a barrier of melting rock from the deep reservoir to the surface. It may infer that Merbabu volcano is less active than Merapi volcano. The fact shows that the last eruption of the Merbabu volcano was in 1797 [25]. We also interpreted that the boundary between low and high velocities at the southern part of the study area represents the Opak fault. The boundary of these low and high velocities originated from the difference of rock physical properties [5]. The existence of low velocity around the seismogenic zone may change the long-term stress concentration [26, 27].

4. Concluding Remarks

The data taken from both DOMERAPI and BMKG networks are effective to improve the resolution beneath the Merapi volcano, especially for depths less than 35 km. The existence of deep and intermediate reservoirs could be detected clearly at depths of around 27 km and 12 km, respectively. The dipping low velocity material ascends from the depth about 100 km. The results also imply that Merbabu volcano is less active than Merapi volcano. To the south of Merapi volcano there is the contact zone between high and low velocity anomalies interpreted as the Opak fault zone.

5. Acknowledgements

We are grateful to Institut de Recherche pour le Développement (IRD), France for funding DOMERAPI network, Center for Volcanology and Geohazard Mitigation as main counterpart of the DOMERAPI Project in Indonesia, and the Agency for Meteorology, Climatology and Geophysics (BMKG), Indonesia for providing waveform data used in this study. The gratefulness is extended also to the Indonesia Endowment Fund for Education (LPDP) for granting Doctoral scholarship to the first author to study at Institute of Technology Bandung (ITB), the Ministry of Research, Technology, and Higher Education of Indonesia (KEMENRISTEK DIKTI) for the research funding, and Near Surface Geophysics Laboratory, ITB for providing the facility to process the earthquake data and travel-time seismic tomography.

References

[1] Ratdomopurbo A and Poupinet G 2000 *J. Volcanol. Geotherm. Res.* **100** 193-214.
[2] Hidayati S Ishihara K Iguchi M and Ratdomopurbo A 2008 *Indonesian Journal of Physics.* **9**(3) 75-82.
[3] Surono Jousset P Pallister J Boichu M Buongiorno M F Budisantoso A Costa F Andreastuti A Prata F Schneider D Clarisse L Humaida H Sumarti S Bignami C Griswold J Carn S Oppenheimer C and Lavigne F 2012 *J. Volcanol. Geotherm. Res.* **241-242** 121-135.
[4] Luehr B-G Koulakov I Rabbel W Zschau J Ratdomopurbo A Brotopuspito K S Fauzi and Sahara D P 2013 *J. Volcanol. Geotherm. Res.* **261** 7-19.
[5] Koulakov I Bohm M Asch G Luhr B-G Manzanares A Brotopuspito K S Fauzi Purbawinata M

A Puspito N T Ratdomopurbo A Kopp H Rabbel W and Shevkunova E 2007 *J. Geophys. Res.* **112** B08310 doi:10.1029/2006JB004712.
[6] Wagner D Koulakov I Rabbel W Lüehr B-G Wittwer A Kopp H Bohm M Asch G and MERAMEX Scientist 2007 *Geophys J Int.* **170(2)** 923-932 doi:101111/j1365-46X200703435x.
[7] Rohadi S Widiyantoro S Nugraha A D and Masturyono 2013 International Journal of Tomography and Simulation. **24(3)** 1-16.
[8] Zulfakriza Z Saygin E Cummins P R Widiyantoro S Nugraha A D Lühr B-G and Bodin T 2014 *Geophys J Int.* **197** 630-635 doi: 10.1093/gji/ggu016.
[9] Trichandi R Yudistira T Nugraha A D Zulfakriza Z and Saygin E 2015 *AIP Conference Proceedings.* **1658** 050010 doi: 10.1063/1.4915049.
[10] Costa F Andreastuti S Bouvet de Maisonneuve C and Pallister J S 2013 *J. Volcanol. Geotherm. Res.* **261** 209-235.
[11] Ramdhan M Nugraha A D Widiyantoro S Métaxian J P and Valencia A A 2015 *AIP Conference Proceedings.* **1658** 030007; doi: 10.1063/1.4915015.
[12] Geiger L 1912 *Bulletin of St. Louis University.* **8(1)** 56-71.
[13] Lahr J C 1989 HYPOELLIPSE/Version 2.0: A computer program for determining local earthquakes hypocentral parameters magnitude and first-motion pattern. *U.S Geol. Surv. Open File Rep.* 89-116. http://pubs.usgs.gov/of/1989/0116/report.pdf.
[14] Waldhauser F and Ellsworth W L 2000 *Bull. Seismol. Soc. Am.* **90** 1353-1368.
[15] Waldhauser F 2001 HypoDD A Program to Compute Double-Difference Hypocenter Locations *U.S Geol. Surv. Open File Rep.* 01-113.
[16] Eberhart-Phillips D 1993 Local earthquake tomography: Earthquake source regions, in Seismic Tomography: Theory and Practice, edited by H. M. Iyer and K. Hirahara. *CRC Press, Boca Raton, Fla.* 613-643.
[17] Evans J Eberhart-Phillips D and Thurber C H 1994 User's manual for SIMULPS12 for imaging *Vp* and *Vp/Vs*: A derivative of the 'Thurber' tomographic inversion SIMUL3 for local earthquakes. *U.S Geol. Surv. Open File Rep.* 94-431.
[18] Nugraha A D and Mori J 2006 *Geophys. Res. Lett.* **33 1-5** doi:10.1029/2006GL028479.
[19] Nugraha A D Kusnandar R Puspito N T Sakti A P and Yudistira T 2015 *AIP Conference Proceedings.* **1658** 030019 doi: 10.1063/1.4915027.
[20] Firmansyah R Nugraha A D and Kristianto 2015 *AIP Conference Proceedings.* **1658** 050012 doi: 10.1063/1.4915051.
[21] Nugraha A D Widiyantoro S Gunawan A and Suantika G 2013 *J. Math. Fund. Sci.* **45(1)** 17-28.
[22] Wadati K 1933 *Geophys. Mag.* **7** 101-111.
[23] http://earthquake.usgs.gov/research/data/slab/
[24] Saepuloh A Urai M Aisyah N Sunarta Widiwijayanti C Subandriyo and Jousset P 2013 *J. Volcanol. Geotherm. Res.* **261** 130-143.
[25] van Hinlopen Labberton D 1921 Oud Javaansche gegevens omtrent de vulkanologie van Java *Natuurk.* Tijdschr. *Nederl. Ind. v.81.* 154-158.
[26] Nugraha A D Ohmi S Mori J and Shibutani T *Earth Planets Space.* **65** 871-881.
[27] Ry R V, Priyono A, Nugraha A D, Basuki A 2015 Seismicity Study of Volcano-Tectonic In and Around Tangkuban Parahu Active Volcano in West Java Region, Indonesia *AIP Conf. Proc.* **1730** 020004.

Ambient Noise Tomography of Merapi Complex, Central Java, Indonesia: A Preliminary Result

Massita Putriastuti [1], Tedi Yudistira[2], Andri Dian Nugraha[2], Sri Widiantoro[2] and Jean-Philippe Metaxian[3]

[1]Master Program of Geophysical Engineering, Faculty of Mining and Petroleum Engineering, Institut Teknologi Bandung, 40132, Bandung, Indonesia.
[2]Global Geophysics Research Group, Faculty of Mining and Petroleum Engineering, Institut Teknologi Bandung, Bandung, 40132, Indonesia
[3]ISTerre, IRD R219, CNRS, Université de Savoie, Le Bourget-du-Lac, France

Email: massitacindy@gmail.com

Abstract. Mt. Merapi is one of the most active and hazardous volcanoes not only in Indonesia but also in the world. Having a height of about 2968 meter above sea level it contains an active lava dome, which regularly produces pyroclastic flows and is categorized as a stratovolcano. It erupts on average every 2-5 years, in which thousands of people live on the flanks of the volcano. The last eruption occurred on 26 October 2010 and after the large eruption in 2010 the characteristic of Mt. Merapi was changed. Due to its uniqueness, Merapi is closely monitored by the many geoscientists, particularly through volcanological surveys.This study is concerned with the application of ambient noise tomography (ANT) to create Rayleigh wave group velocity maps around Mt. Merapi. The continuous data set is taken from the DOMERAPI project, which consists of temporary seismic array of 40 broadband seismometers for approximately ten months. The resulting group velocity maps of Rayleigh wave show an anomaly pattern that agrees with previous geological and geophysical study results. A pronounced, positive anomaly is clearly imaged with direction about 152°N beneath Mt. Merapi through to Mt. Merbabu. In addition, negative anomalies are observed in its east and west flanks.

Keywords: Volcano, Ambient Noise, Tomography, Rayleigh Wave

1. Introduction

In the last decade, a number of researches have been developed for creating more effective methodologies to be applied in seismological studies to cope with the recent regulations which addressing the assessment of seismic hazard. Since the distribution of earthquakes is poor and almost impossible to controlled, seismologist need another tools to overcome the lack of earthquake data in the recording so that we still can produce a high resolution images in insufficient path coverage seismically inactive regions. On the other hand, seismometer will always contain noise in the recording with or without the presence of earthquake signal. Noise is conventionally described as disturbance in the seismogram which usually is removed by seismologist for not contain any useful information. It can comes from the instrumentation itself or mostly from ambient noise vibration. Ambient noise can take a wide variety of forms, for example, it might come from urban dwellers, rural regions, oceanic activities or even atmospheric activities. This ambient noise will be recorded by seismometer as a surface waves. There are two kinds of surface waves; the first ones is Rayleigh waves, the wave whose induces an elliptical motion with the amplitude exponentially decreases by increasing in depth. Such elliptical motion is the result of the superposition of the vertical and radial components. The other surface waves called Love waves are somehow simpler because they move only on the horizontal plane, transversally with respect to the direction of propagation.

Using basic principle of seismic interferometry (SI), Green's function from surface waves can be extracted between two stations in different location by long term correlations of ambient seismic noise recording. Now days, surface wave tomography based on cross-correlation of ambient seismic noise has been applied successfully at local and regional scales. This new technique promises significant improvements in seismology especially in obtaining high resolution and accuracy of crustal and upper mantel images without needing the presence of any earthquake signals. This paper will merely describe some preliminary results and analysis on applying tomography using seismic ambient noise data in Merapi volcano as one of the most active volcano in Indonesia. Based on authorized record, this volcano has already erupted 33 times with a short rest period from about 1 to 6 year.

2. Geology Regional

Merapi volcano is an almost persistently active basalt to basaltic andesite volcano in Central Java, Indonesia. It is one of the world's most active and dangerous volcanoes in Indonesia. According to Van Bemmelen [21], the area around Merapi volcano consist of two different materials; the basic material is low density alluvial, whereas the volcanoes themselves consist of denser material from Holocene volcanic material.

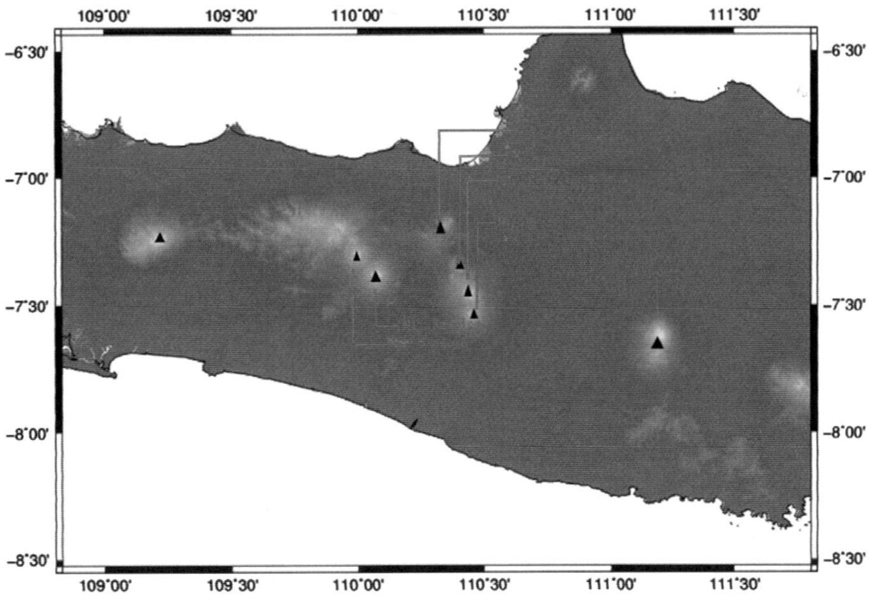

Figure 1. Map of volcanoes surrounding Merapi complex. This picture shows that Merapi volcano is bounded by two volcanic alignments intersections.

Merapi volcano formed on Late Pleistocene – Early Holocene period, above the Eurasia – Indo-Australia tectonic plates subduction zone and is bounded by two main volcanic alignments intersections, Ungaran – Telomoyo – Merbabu - Merapi (MMTU) and Lawu – Merapi – Sumbing –

Sindoro – Slamet (SSSML) (figure 1) [1]. MMTU volcanoes is an example of the double-chain volcanic arc, phenomenon of the existence of two or more volcanoes across the arc and its volcanoes can be divided into the trench-side and back-arc side volcanoes. The back-arc side volcanoes of Slamet (still active) and Ungaran (dormant) are located about 210 – 225 km above the slab while the vertical distance of Merapi volcano from the slab is about 175 km. It is clearly demonstrated that the vertical distance between each volcano and subducted slab increases constantly from the trench side or in this case Merapi volcano in the south towards back-arc side in the north. Based on radiometric data from Kohno *et al.* [6], the volcanism along MMTU double volcanic chain seems to occur during the Quarternary period up to now with Ungaran volcano is the oldest followed by the Telomoyo, Merbabu, and Merapi as the youngest and also the only active volcano in MMTU section. While in MMTU only have Merapi as the active volcano, all the volcanoes in SSSML is still classified as active volcanoes.

3. Data Sets and Processing

We used DOMERAPI array which continuously recording for about ten months from October 2013 to July 2014. There are 42 three component broadband seismometers, placed surrounding Merapi and Merbabu volcanoes (figure 2). For this preliminary result, we obtain the Green's function from vertical component only which representing Rayleigh wave. Also this study used ambient noise from microseism activity generated by oceanic waves with frequency ranges about 0.03 – 0.3 Hz consist of Single Frequency Microseism (0.03 – 0.1 Hz) and Double Frequency Microseism (0.1 – 0.3 Hz).

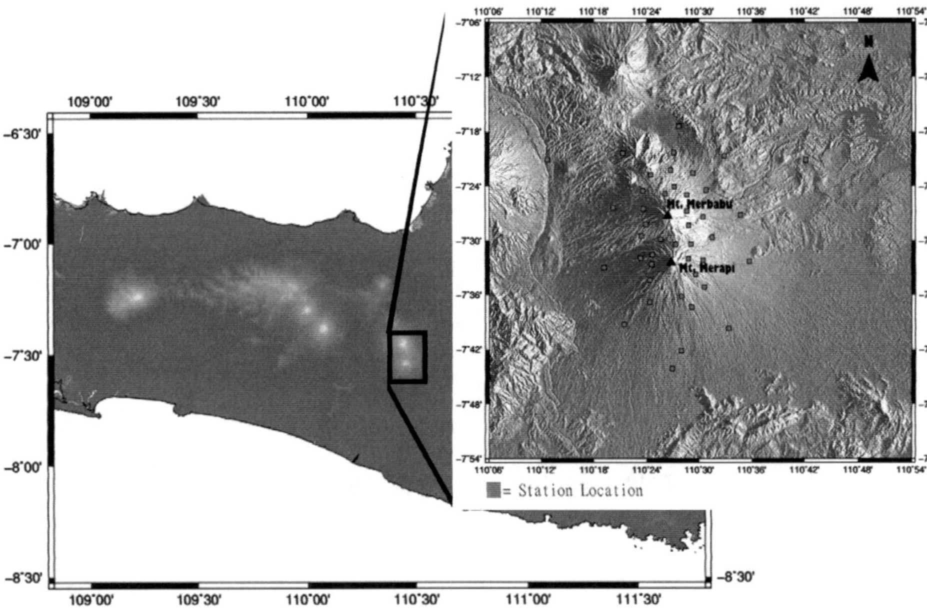

Figure 2. The area inside the black square is our study area with stations distribution location in the inset.

This study followed the method published by Bensen *et al.* [2] to produce the Green's function from cross correlating data between two different stations. There are some routine data processing which aim to improve the data quality. It basically contains some signal analysis and processing to obscure the earthquake signal and enhance the noise. The first step is removing mean, followed by removing trend, and some normalization technique. The most important part in routine processing is

IOP Conf. Series: Earth and Environmental Science **62** (2017) 012040 doi:10.1088/1755-1315/62/1/012040

doing the time domain normalization. We performed one bit normalization as tools to accentuate the noise. We also applied spectral whitening as additional method. The next step is the main processing of seismic ambient noise, it is cross correlation and stacking. We performed daily cross correlation and then stacked it over ten months. There are 770 Green's function as a results with two sides time function consist of positive (causal) and negative (acausal) time lags.

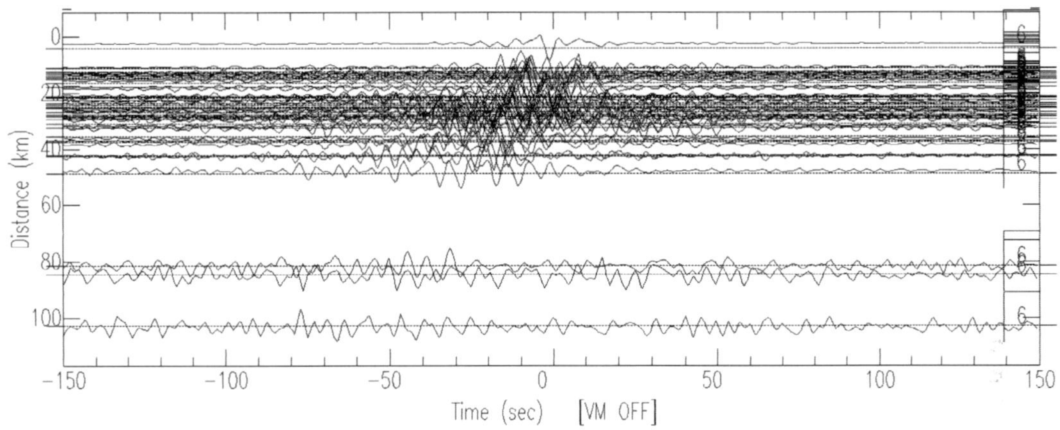

Figure 3. The result of waveform cross-correlated from ME01 to all other stations. Two sided of GF (causal and acausal) filtered between 0.03 and 0.1 Hz frequencies.

Through cross correlated the data from relatively south seismometers to relatively north ones, we obtained unique Green's functions in which acausal part mostly being the dominant part in Green's function results (figure 3). It indicating that Rayleigh waves are dominantly comes from the north of the Merapi volcano. Then we measured the dispersion curves from acausal Green's function in all station pairs using multiple filtering technique which is introduced by Dziewonski *et al.* [4] to produce the group velocity of various frequencies in every station pairs. According to Bensen *et al.* [2], we found that a reliable dispersion measurement at a certain period correspond to its interstation distance (Δ in km) it should at least 3 wavelength (λ) or the maximum cut off period is in about 11.1 second with 1.55 kms-1 as phase speed estimation. For the tomographic inversion we integrated fast marching method and subspace inversion provided by Rawlinson (2005). The detailed is given in Rawlinson & Sambridge [12].

4. Results and Discussion

Tomography map from group velocity shown two interesting structures as seen in figure 4. The first part is the relatively high group velocity anomaly right beneath Merapi and Merbabu volcano, this structure linearly connecting Merbabu volcano to Merapi volcano with direction about 152°N. The second part comes from the wide relatively low group velocity anomaly in about southwest of Merapi volcano. Based on previous study, we then compared and analyzed these two anomalies corresponding with the other Merapi volcano studies. Compared with studies from Koulakov et al. [7] and Zulfakriza et al. [19], the high group velocity anomaly is not clearly emerge since they study regional zone. However, Koulakov have noted that even Merapi volcano lies in strong low velocity anomaly, MLA (Merapi Lawu Anomaly) but the volcanoes themselves appear to be located above relatively higher-velocity patterns [7].

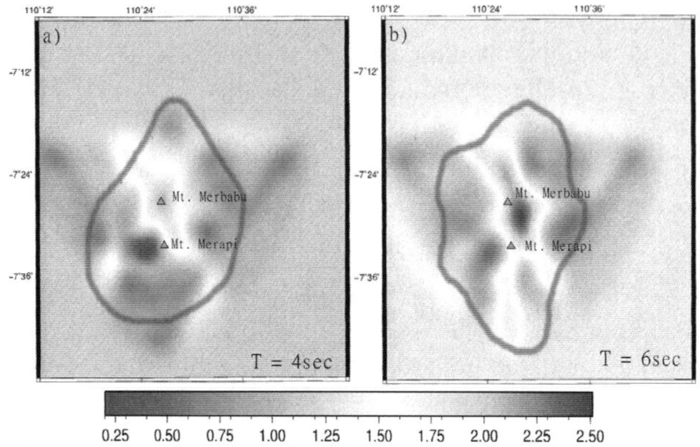

Figure 4. Group velocity tomography result for 4 (a) and 6 (b) period. Interpretation only focusing inside the red lines which is having the best resolution image from checkerboard test. It consist of two main bodies representing high velocity anomaly and low velocity anomaly.

More similar results came from Bouguer gravity map by Tiede *et al.* [16]. Apart from Bouguer anomaly gravity maps, Tiede *et al.* [16] also provides the 3 dimensional modelling for high and low gravity anomaly separately. From their results, we can observe a number of similarities in both interesting structures of this study. The high density structures with northwest – southeast direction which lies along Telomoyo – Merbabu – Merapi volcanoes and 2 – 7 km depth surrounded by lower gravity is directly proportional with the group velocity anomaly from this study. This arc structure with high density and high group velocity anomaly showing the NS striking transverse fault zone that is mentioned by Van Bemmelen [17]. It is underlines the history of volcanic origin in which shows the older basaltic lavas derived from subduction's partial melts passing the fault zones, and through the cooling and solidification so that it's becoming intrusive complex. Moreover the low group velocity anomaly in the southwest of Merapi volcano not only related with the low density structure from Tiede *et al*, [16] but it also coincidence with low resistivity structure from magnetotelluric method provided by Müller and Haak [9]. This anomaly structure indicates a potential location of Merapi volcano's magma chamber.

5. Conclusions

We used ten months of seismic noise data recorded by DOMERAPI seismic array to extract the Rayleigh wave Green's function. This Green's function is used to perform a Rayleigh waves tomography around Merapi complex. This tomography ambient noise method successfully improved the resolution of the the shallow depth right beneath Merapi complex in which have less seismicity activity compared to subduction's activity in the southern part of Java.

Tomographic inversion shows a coincidence results with some previous studies. There is an high group velocity anomaly and high density anomaly linearly running direction about 152°N beneath Merapi complex which is indicated as old basaltic lava, intrusively starting from higher anomaly in NW to the SE alongside the transverse faults. Furthermore there is an indication of magma chamber in the southwest flank of Merapi volcano according to its properties such as low group velocity, low density and also low resistivity anomalies.

Acknowledgment
The authors are grateful to the funding of The French National Research Agency and the Institut de recherche pour le développemen to the DOMERAPI project and all of the DOMERAPI project partner organization. We also thank the Near Surface Geophysics laboratory FTTM-ITB to facilitate the author during the research project.

References

[1] Beauducel F 1998 Structures and Mechanical Behaviour of Merapi Volcano, Java: A Methodological Approach of The Deformation Field *Thèse de Doctorat, Universiteit Denis Diderot Paris VII* 260
[2] Bensen G D, Ritzwoller M H, Barmin M P, Levshin A L, Lin F, Moschetti M P, Shapiro N M, Yang Y 2007 *Geophys J.Int.* **169** 1239-1260
[3] Claerbout J F 1968 *Geophysics* **33** 264–269.
[4] Dziewonski A, Bloch S, and Landisman M 1969 *Bulletin of the Seismological Society of America* **59**(1) 427-444.
[5] Firmansyah R, Nugraha A D, Kristianto 2014 The Preliminary Results : Internal Seismic Velocity Structure Imaging Beneath Mt. Lokon *AIP Conf. Proc.* **1658** 050012.
[6] Herrmann R B 1987 Computer Programs in Seismology *User's manual Vol. II-IV St. Louis University, Missouri*
[7] Kohno Y, Setijadji L D, Zoltan P 2006 Geochronology and Petrogenetic Aspects of Quaternary Across Arc Magmatism on Merapi, Merbabu, Telomoyo, Ungaran Volcanoes, Central Java, Indonesia *Proc. of the 3rd Int. Symp. Earth Resources and Geological Engineering Education* 194 – 201
[8] Koulakov I, Bohm M, Asch G, Luhr B G, Manzanares A, Brotopuspito K S, Fauzi, Purbawinata M A, Puspito N T, Ratdomopurbo A, Kopp H, Rabbel W, Sheykunova E 2007 *J. of Geophysical Research* **112** B08310.
[9] Nugraha A D, Widiyantoro S, Gunawan A, Suantika G 2013. Seismic Velocity Structures Beneath the Guntur Volcano Complex, West Java, Derived from Simultaneous Tomography Inversion and Hypocenter Relocation *J. of Mathematical and Fundamental Sciences* **45**
[10] Moro D G 2015 Surface Wave Analysis for Near Surface Applications *Elsevier* ISBN:978-0-12-800770-9.
[11] Müller A and Haak V 2004 *J. Volcanol. Geotherm. Res.* **138** (3–4) 205–222
[12] Neuman Van Padang M 1951 Catalogue of the Active Volcanoes of Indonesia *Catalogue of the Active Volcanoes of the World. Int. Volcanological Associatio. Naples* 271 pp.
[13] Rawlinson N, Reading A M, Kennet B L N 2006 *J. Geophys. Res.* **111** B02301. doi:10.1029/2005JB003803.
[14] Rawlinson N, Sambridge M 2003 Seismic Traveltime Tomography of the Crust and Lithosphere *Advances In Geophysics* **46** 81-197.
[15] Ry R V, Nugraha A D 2014 Improve Earthquake Hypocenter Using Adaptive Simulated Annealing Inversion in Regional Tectonic, Volcano Tectonic, and Geothermal Observation *AIP Conf. Proc.* **1658** 030004.
[16] Ry R V, Priyono A, Nugraha A D, Basuki A 2015 Seismicity Study of Volcano-Tectonic In and Around Tangkuban Parahu Active Volcano in West Java Region, Indonesia *AIP Conf. Proc.* **1730** 020004.
[17] Sethian J C 1999 Advancing Interfaces: Level Set and Fast Marching Method *Dept. of Mathematics, University of California, Berkeley*
[18] Setijadji L D 2010 Segmented Volcanic Arc and Its Association with Geothermal Field in Java Island, Indonesia *World Geothermal Cong. 2010*
[19] Tatsumi Y and Eggins S 1995 Subduction Zone Magmatism: Boston *Blackwell Science* 211 p.
[20] Tiede C, Camacho A G, and Gerstenecker C 2005 *Geochemistry, Geophysics, Geosystems* **6** (9) 679 pp 1-13
[21] Van Bemmelen R W 1949 The Geology of Indonesia *Government Printing Office, Nijhoff*
[22] Villaseñor A and Engdahl E R 2007 *Bull. Seism. Soc. Am.* **97** 1820-1832.
[23] Zulfakriza Z, Saygin E, Cummins P R, Widiyantoro S, Nugraha A D 2014 *Geophys. J. Int.* **197** (1) 630-635

Aquifer Distribution Based on 1D Resistivity Method at Jatinangor Educational Area, Sumedang Regency, West Java Province

Andreas Franzona Pangaribuan[1], Febriwan Mohammad[1], Muhammad Fadly [1], Deden Zaenuddin Muttaqin[1]

[1]Undergraduate Program of Geological Sciences, Geological Faculty, Universitas Padjadjaran, Jalan Raya Bandung-Sumedang km.21, Sumedang 45363, Indonesia

Abstract. Continuous development in the area of Jatinangor campus is becoming one of the problems threatening the groundwater supply. To support the availability of groundwater in the area of Jatinangor campus, a geophysical investigation with the geo-electric method is conducted to determine the condition of the subsurface based on the value of resistivity of rock. Based on Bandung's regional geological map of Silitonga in 2003, rocks in the Jatinangor area consist of volcanic rock breccia, tuffs, and lapilli that makes it possible to contain the groundwater. 32 stations of 1-Dimensional (DC sounding) geo-electric measurement using Schlumberger configuration are performed in Jatinangor area. We integrated the results of measurement with geological and hydrogeological observation information with the aim of producing images of subsurface rocks and distribution models. Based on the model, the type of aquifer contained in the study area as well as its potential reserve can be determined. This research aims to know the potential groundwater zone region to support the availability of groundwater for Jatinangor education region. Furthermore, the results are expected to provide insights in implementing conservation strategies for Jatinangor educational area, Sumedang Regency.

1. Intoduction

Jatinangor Education Area is located in the 6°52'30"- 7°00'00" of south latitude and 107°45'00"- 107°52'30" of east latitude. Administratively, This area lies on the border between The Sumedang Regency and Bandung City. High population growth in this region resulted in a high population density which is around 4,270 people / km^2 [1]. This research aims to determine the potential groundwater zone region to support the availability of groundwater for Jatinangor education region. Furthermore, the results of this research are expected to provide insights in implementing conservation strategies for Jatinangor educational area, Sumedang Regency.

The resistivity method is a geophysical method that is very popular and frequently used both in geological survey and exploration. The method is better to be used for knowing the geological condition or structure of the subsurface based on the resistivity variety of the rocks. The value depends on several factors, including the age of rocks, electrolyte contents, rock density, a number of minerals it contains, porosity, permeability and so forth. One of the applications of this method is groundwater exploration. When the direct current flows into the earth through two current electrodes A and B, and the flow of the stream causing the potential difference in two potential electrode M and N is measured, we can obtain the apparent resistivity [2].

2. Geological Condition

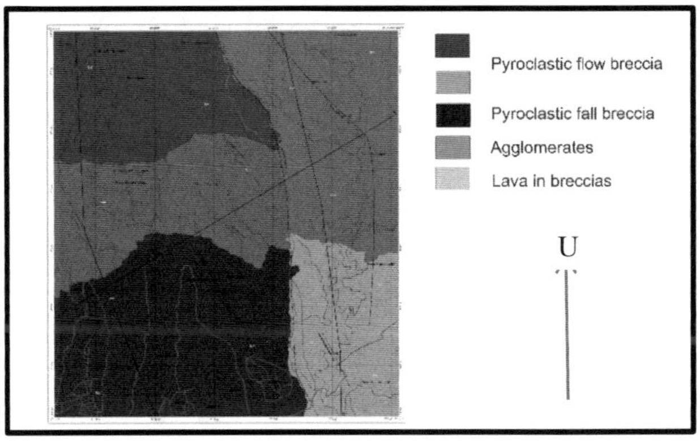

Figure 1. Geological Map of Study Area, modification of Silitonga [3]

The exposed rocks in study area consists of five main units which all of them are young volcanic products, as can be seen in the Figure 1 [3]. they consist of irreducibly young volcanic deposits which are tuffaceous sand, lapilli, breccia, lava, and agglomerate. These rocks came from The Tangkubanparahu Mountains and also from some parts of The Tampomas Mount, which spread to many places within Sumedang-Bandung Border. These Quarter-age rocks formed a small and flat plains. The rocks also formed lowlands covered by yellowish-grey and reddish-grey soils [4].

3. Hydrogeological Condition

There are two regions of the groundwater basin in this area, namely the area of groundwater with the discharge rate of less than 5 l/second and having deep, widespread moderate productivity aquifer, as can be seen in the Figure 2. This region occupies 60% of the study area especially in the southern part. The second region is the area of the shallow, productive aquifer with a deep water table. There are several springs which were encountered in the area. This region occupies 40% of the study area especially in the northern part [5].

4. Method

The research was conducted on Universitas Padjadjaran Campus by using a single Geoelectric Tool Unit ABEM DC Terrameter - SAZ 2000. We used The Schlumberger configuration of Resistivity method. We placed the potential electrodes M and N between the two current electrodes A and B. We carried out the measurement at 32 points with a range as long as 300 meters between electrodes (AB / 2). The estimation of the depth of each point was expected to reach 175 meters.

Every point of measurement provides an estimation model of rock layers in an upright or vertical direction (stratigraphy) based on the resistance value (ρ). Each point has a coordinates obtained from measurements with GPS, so we can also make a contour map based on the depths.

We also make the cross section model that made of few measurement points with the direction of N-S and NE-SW. The cross section model was made to understand the spread of rock layers in the lateral direction. The Data readings from each measurement tools were processed and correlated with geological and hydrogeological data from the field observations of previous studies that have been done by Geophysics Laboratory, to identify the composition of rock layers in the subsurface.

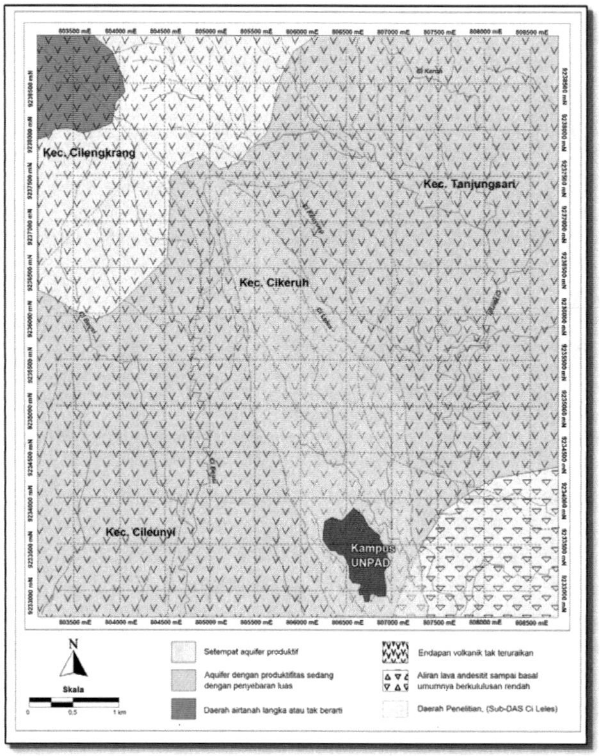

Figure 2. Hydrogeology regional map, modification of Bandung Quadrangle Hydrogeological map [5]

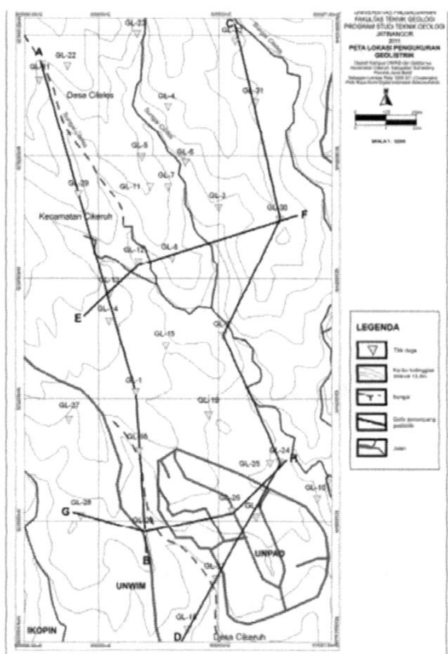

Figure 3. The measurement points

5. Result and Discussion

After that, we obtained the variation of the apparent resistivity values and the current electrode spacing stretch (AB / 2). Then, the data were processed by the software Progress ver.3.0 to obtain the 'true' resistivity value and the actual thickness of the layer. Each resistivity value has a picture prediction of a rock at a certain depth because every rock has a resistivity value of which varies depending on rock hardness, porosity, permeability, the age of rocks, the content of electrolytes, minerals, rock structures and geological phenomena otherwise influence the value resistivity in these rocks.

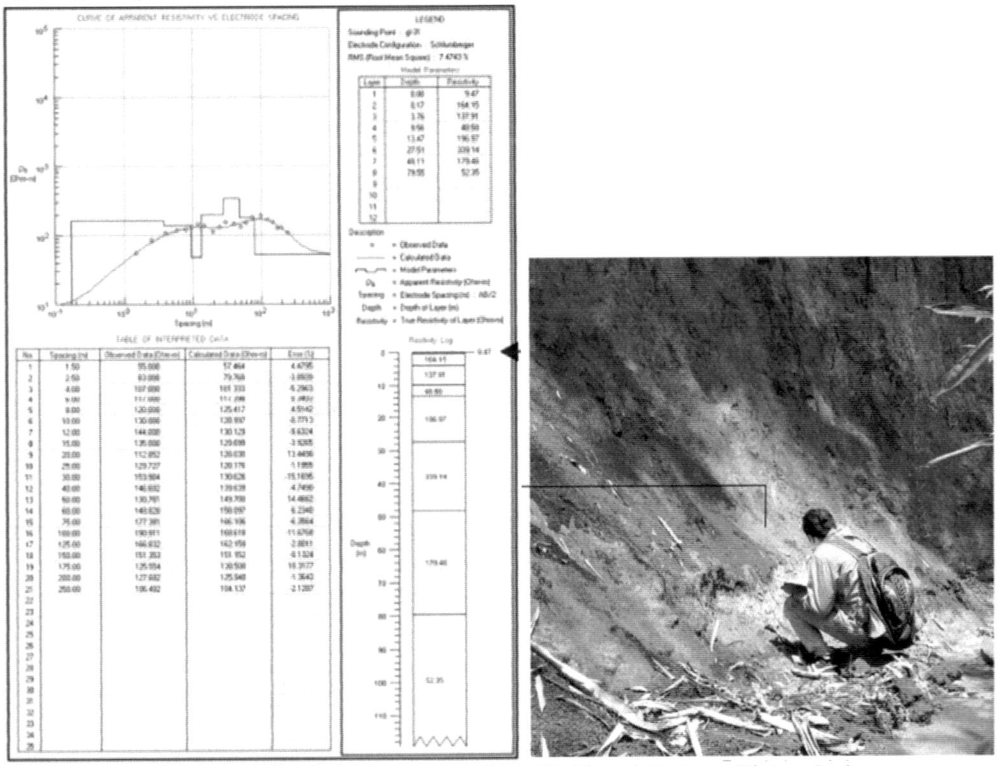

Figure 4. Resistivity cross section GL-31 and Outcrop Picture L-4

Based on the results of the geoelectrical data processing correlated with the geological conditions around the study site, the study area is mostly composed of volcanic activity deposits. Several variations of the resistivity values can be divided into several groups to indicate the type of rock below the surface. Here is the range of resistivity values for each rock:

Table 1. Resistivity variations with their estimated lithology and hydrogeological properties

ρ (Ω.m)	Estimated Lithology	Estimated Hydrogeology
< 30	Fine tuff layer	Impermeable Layer
31 - 60	Sandy Tuff	Aquifer
61 - 100	Tuffaceous sandstones	Aquifer
101 - 200	Matrix supported breccia (laharic breccia)	Aquifer
201 - 350	Grain supported breccia (volcanic breccia)	Impermeable Layer
> 350	Andesitic Lava	Impermeable Layer

The Isoresistivity maps were created to determine the distribution of resistivity value at each depth. The resistivity map can be used to delineate the distribution of 1D resistivity laterally (Figure 5). The resistivity value distributions on the iso resistivity map can provide a picture of the rock layers below the surface by correlating the existing geological data into it. The table 1 was being used as the reference for the map. The table was made according the existing geological condition in this area based on the previous geological studies of the department of Geology. The sameness resistivity values can be considered as the same layer of a rock while the contrast value is expected to be a striking contact between the different rock layers or it may be the emergence of geological structure. In this study, the authors created a map iso resistivity at a depth of 0m, 1m, 5m, 10m, 15m, 20m, 30m, 40m, 50m, 70m, 100m, to 120m.

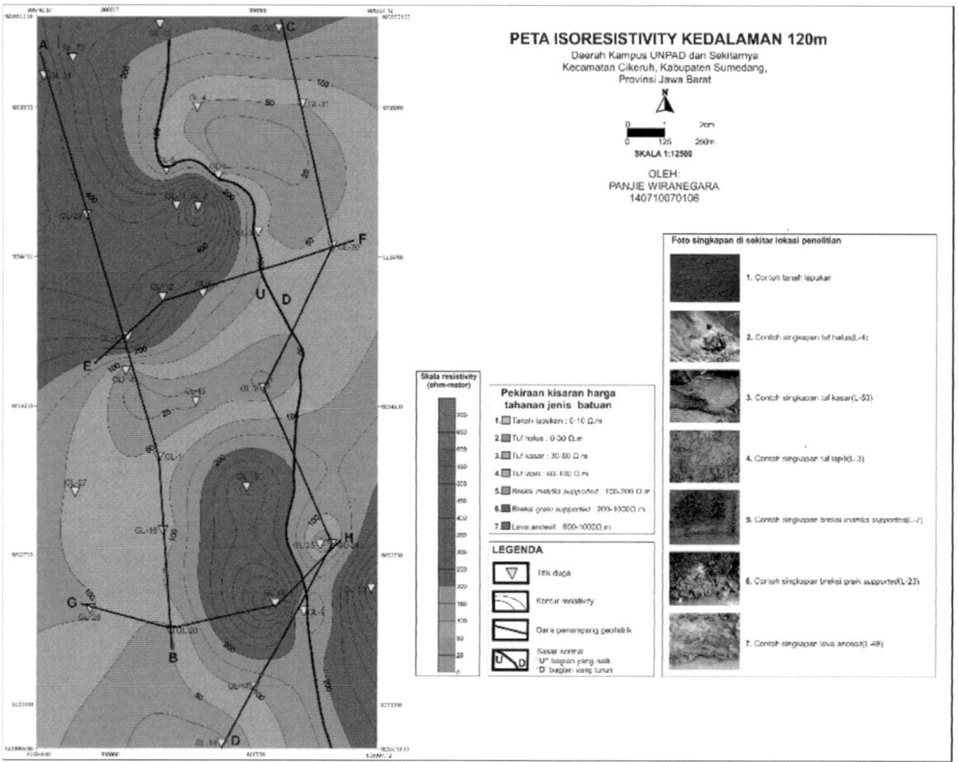

Figure 5. The Iso resistivity Map at the depth of 120 m

The Geoelectric cross-section can give us a presence, distribution and thickness of the rock layers below the surface concerning the geological data around the study site. It can also provide an overview of the aquifer characteristics beneath the surface. Therefore, the authors made the cross-section line passing through the several measurement points (Figure 6). The authors made at least four cross-section lines through the study area (Figure 3).

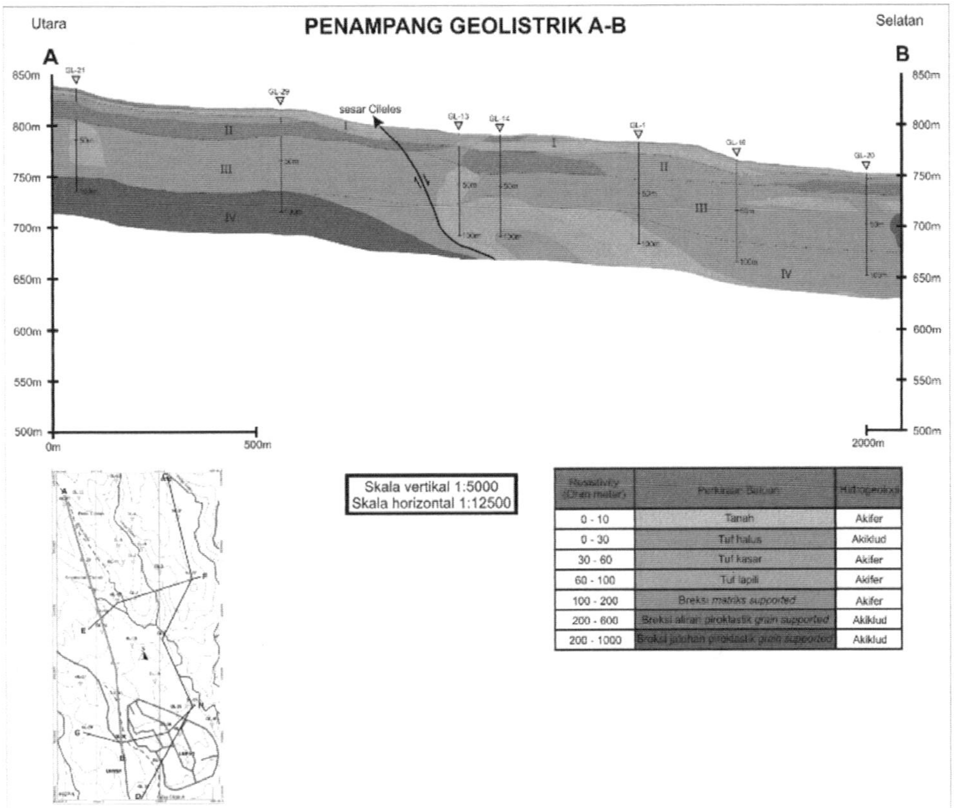

Figure 6. The Cross section A-B and its lithology and Hydrogeological properties

The Geoelectric block diagram (Figure 7) also was made to obtain the distribution picture of the rock layers in 3D. The block was created by linking the spread of the rock layers based on the cross-section line reconstruction and, using Rock Work 14 application, producing the distribution of the rock layers in 3D. It is easier to interpret the rock layer distribution in 3D.

The southwestern and northeastern parts of the study area are expected to have the considerable groundwater resources. The rock layer that serves as an aquifer is dominating these two areas including coarse tuff, lapilli tuff, and matrix supported breccia. While the layer of fine tuffs, grain supported pyroclastic flow breccias and fall breccias that serve as aquiclude only exist in a small part of the entire layers that make up this region.

On the contrary, the northwestern and southeastern parts of the study area are expected to have fewer groundwater resources. The grain supported pyroclastic flow breccias and pyroclastic breccias which serve as aquiclude are dominating these two areas. The Distribution of these rock layers is very dominating and quite thick. So that the space for groundwater flow in these areas is very limited. Besides, the distribution of the aquifer layers such as coarse tuff, lapilli tuff, and matrix supported breccias and also andesitic lava that serves as aquicrack spread only in a few places and depths. The groundwater flow in this area is very narrow and are much less likely to be found with groundwater resources.

There are Cileles Fault and Cikeuyeup Fault found in the study area. The Cileles fault is a normal fault trending from the northwest to the southeast with lowering depth at the east side. While Cikeuyeup fault is a similar fault with lowering depth at the west side. So the situation under the surface in the middle between these faults form a basin. We consider that the groundwater basin in this area is as a good groundwater. Aquifer layers (coarse tuff, lapilli tuff, and matrix supported breccia) is dominating the basin below the surface. While aquiclude layers (grain supported breccia and fine tuff) are rare in some places and their thickness does not dominate those areas. Besides, both of these faults may act as an exit-entry barrier of the groundwater flow. So that the groundwater cannot flow out properly from the basin.

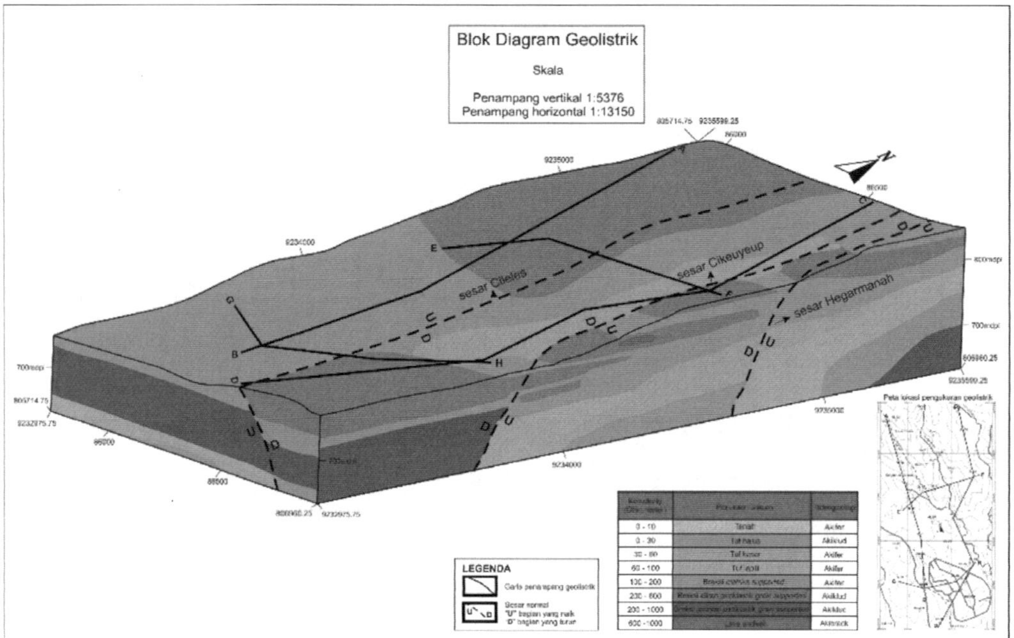

Figure 7. The geoelectric block diagram

6. Conclusion

As explained in the discussion, we can see that at least there are two locations that are expected to be aquifers, which are The southwestern and northeastern parts of the study area. The subsurface rock layers in these two areas are dominated by coarse tuff, lapilli tuff, and matrix supported breccia which are considered a good aquifer. While the layer of fine tuffs, grain supported pyroclastic flow breccias and fall breccias that serve as aquiclude only exist in a small part of the entire layers that make up this region.

References

[1] Fadly M *et al* 2016 *Proc.Sustainable Development Conference 2016 ed.* (Kuching: Malaysia) p 46-62

[2] Telford W M, Geldart L P, and Sheriff R E 1990 *Applied Geophysics* **1** Cambridge University press

[3] Hendarmawan *et al* 2014 Pola Aliran Air Tanah Berdasarkan Kandungan Isotop Stabil Deuterium-Oksigen Di Tiga Cekungan Pada Lereng Gunungapi Manglayang, Kawasan Kampus Unpad, Jatinangor.Bandung: Universitas Padjadjaran

[4] Silitonga 2003 *Geological Map of The Bandung Quadrangle, Java.* Scale :100,000

[5] Soetrisno 1983 *Hydrogeological Map of The Bandung Quadrangle, Java.* Scale 1:250,000

The Implementation of 2-D Resistivity Method in Verifying Paleozoic Aquifer Properties at Bukit Chondong, Perlis (Malaysia)

Umi Maslinda [1], M.M. Nordiana [1] , A.A. Bery [1], Muhamad Afiq Saharudin [1], Hazrul Hisham [1], Z.M. Taqiuddin [1], Nabila Sulaiman [1], M.K.A. Nur Amalina [1] and A.N. Nordiana [1]

[1] Geophysics Section, School of Physics, 11800 Universiti Sains Malaysia, Penang, Malaysia.

E-mail: umimaslinda@gmail.com

Abstract. The research was conducted using 2-D resistivity in verifying Paleozoic aquifer. Since most geologic materials behave as electrical insulators, surface measurements of earth resistivity are controlled by the electrolytic ability of interstitial water. The subsurface distribution of water is controlled by the porosity of the formations. The study area is at Bukit Chondong, Beseri, Perlis. Bukit Chondong is made of sedimentary rock which mostly is sandstone. Bukit Chondong is from uppermost of the Kubang Pasu Formation that represented by a thick unit of grey mudstone interbedded with sandstone. The Kubang Pasu Formation was influenced by shallow marine during the early age. Paleocurrent and fossils traces were found on the mudstone at the study area. The area is suspected to be a Paleozoic aquifer because the sandstone can be a productive aquifer with diffuse flow. The water movement in sandstone is through the fractures and joints. Most of the water stores and transmits in sandstone. The interbedded sandstone and mudstone is one of the aquifer characteristic. Sandstone and mudstone are water-bearing rocks and low-permeable rocks respectively. The data was processed according to the geological information of the study area since there was an outcrop. The study area have low resistivity value which both sandstone and mudstone give less than 800 Ohm-m due to the water content (Sulphide and clay).

1. Introduction

The aquifer is an underground layer of water bearing permeable rock, rock fractures or unconsolidated materials such as gravel, sand, or silt from which the groundwater can be exploited using water well. Bukit Chondong is mostly consists of sedimentary rocks where mostly is sandstone. The outcrops in Bukit Chondong have been studied by geologist until today. Bukit Chondong is from uppermost Kubang Pasu Formation represented by a thick unit of grey mudstone interbedded with sandstone. The Kubang Pasu Formation consists of coarsening upward cycles of clastic during Kungurian age which representing shallow marine, wave- and storm-influenced shoreline [1]. Thus, geophysics plays important roles to identify the outcrops and to prove the area was the shallow marine influenced.

The 2-D resistivity method is related to different geological parameters such as the mineral and fluid content, porosity and degree of water saturation in the rock. The method also referred to as galvanic electrical method which sometimes useful for determining shallow and deep geologic and hydrogeologic conditions [2]. Electrical resistivity is measure the electrical resistance to a direct current applied at the surface. Resistivity method is a non-destructive ground survey method and its main advantage is produced the subsurface resistivity distribution. Resistivity method has many applications in a number of fields. It provides accurate depth information of uncertain target through visualization. The depth information is important to detect the aquifer. The sedimentary rocks usually are more porous and have a higher water content, which result in lower resistivity values. Wet soils and fresh ground water have even lower resistivity values [3].

2. Study Area

The study was carried out at Bukit Chondong, Beseri, Perlis, Malaysia. Bukit Chondong is located at east of the Maktab Rendah Sains Mara (MRSM) and the line is facing MRSM as shown in Figure 1. The line starts with coordinate of N 6.55299°, E 100.23647° and end at N 6.55265°, E 100.23638°. Chondong is a small steep hill forming part of the N-S trending Chuping Hills in central Perlis. The study area was a quarry and the quarry exposes the uppermost of the Kubang Pasu Formation which mostly consist of mudstone that interbedded with sandstone.

Figure 1. The survey line was conducted on top of the outcrop.

3. Methodology

The 2-D resistivity survey was conducted using Pole-Dipole and Wenner-Schlumberger array with one meter electrode spacing. The total length of the survey is 40 meter with 41 electrodes. The survey used two cables, ABEM SAS4000 System and a remote cable for Pole-Dipole array. The processing was performed using Res2Dinv and Surfer10 software. The results are then interpreted by referring to the geological information of the study area.

4. Results and Discussion

The survey area is mostly made of brownish red colour sandstone and dark grey colour mudstone. The outcrop shows both sandstone and mudstone are interbedded to each other as shown in Figure 2. The depth of the top sandstone is about 5.8 m and followed by mudstone with depth about 7.5 m. The third and fourth layer is sandstone and mudstone with depth about 8.8 m and 10.5 m respectively. The thickness of the layer of mudstone is same with 1.7 m. A productive aquifer can be comprised of sand and gravel, sandstone and limestone. Thus, this study area that composed of sandstone has potential to be an aquifer.

Figure 2. The outcrop of mudstone interbedded with sandstone.

Moreover, there are also presence of sulphur (Figure 3) and black shales in this study area. The presence of sulphur provides information that the area was influenced by marine because the sulphur ion is stable in aqueous solution, where the polar water reduces the energy of the charge [4]. The sulphur is made from the oxidation of sulfide. The main pathways of sulfide oxidation in marine sediments associate with complex interactions of chemical reaction and microbial metabolism. The black shale is a fine-grained, indurated detrital sedimentary rock formed by the consolidation of clay, silt or mud.

Figure 3. Presence of sulphur at the survey area.

IOP Conf. Series: Earth and Environmental Science **62** (2017) 012042 doi:10.1088/1755-1315/62/1/012042

Around the survey area also found mudstone with ripples mark and fossils traces on the mudstone surface as shown in Figure 4. It is also known as paleocurrent. Paleocurrent is a geological feature that helps determine the direction of flowing water in the geologic part which typically a sedimentary structure [5]. This is the other evidence that prove the area was influenced by marine. The water current cause ripples to move by erosion and deposition. Ripples grow up in either one direction or oscillatory currents. Oscillation ripples form by wave movement and tend to be symmetric in shape with straight crests. Besides that, fossils traces is found on mudstone surface at the survey area or known as bioturbation. The bioturbation process is from organism activities that disturbed the formation of the rock. Organism may disturb soft sediment after deposition such as mudstone. The bioturbation of the mudstone happened after deposition. Bioturbation features can be seen as trace fossils and include the tracks, trails and burrow of animals, as well as root casts, bark impressions that left by plants [6].

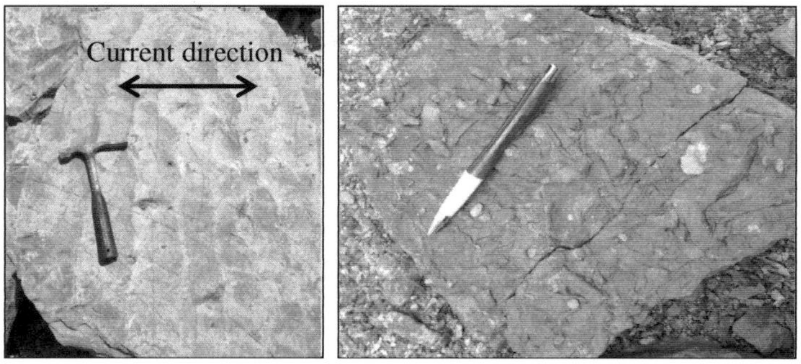

Figure 4. Paleocurrent and fossils traces on mudstone surface.

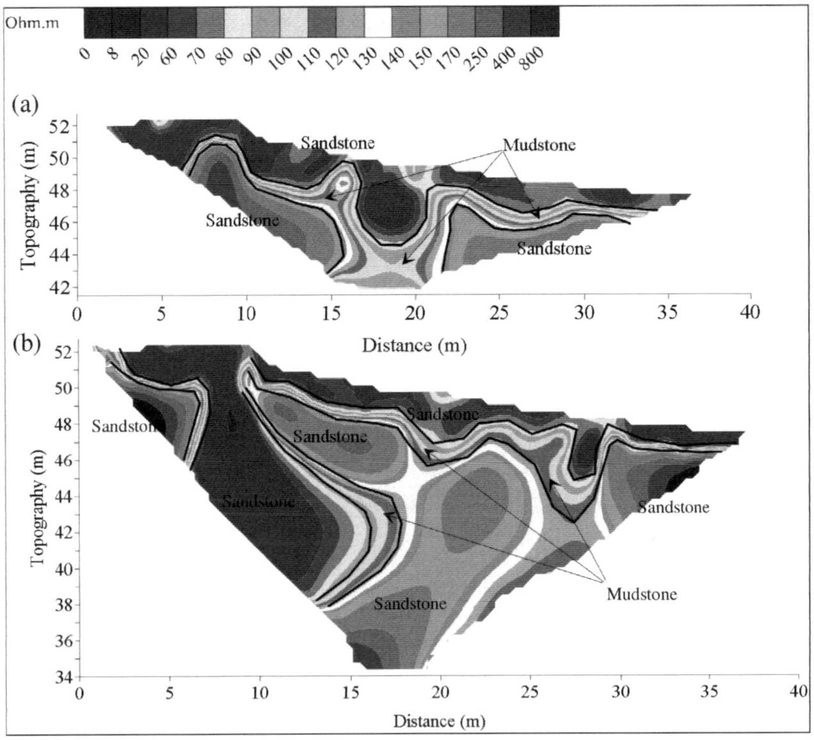

Figure 5. 2-D resistivity results from (a) Wenner-Schlumberger and (b) Pole-Dipole arrays.

Figure 5 is the interpreted results from both Wenner-Schlumberger and Pole-Dipole arrays that show the layering of the subsurface. The different of the arrays are only the depth of penetration. Pole-Dipole array can penetrate deeper than Wenner-Schlumberger array. The different in depth is not significant in this study, but both results produced the same contour maps that show the layering of the different rocks. A standard value of resistivity range is used for both arrays according to the type of rock that found at the study area, sandstone and mudstone. The range of resistivity value for sedimentary rock is 5 to 1000 Ohm-m [7]. This is due to sedimentary rocks are usually porous and have pores that filled with fluids which mainly water causing the rocks become electrolytic conductors. Electrical current is passing through a rock mainly by the passage of ions in pore of waters. The sedimentary rocks tend to be conductive due to their high fluid content [8]. The standard value used is in the range of aquifer resistivity value from 50 to 2000 Ohm-m [9]. Therefore, the study area has potential to be an aquifer.

Both outcrop and the 2-D resistivity results show a better correlation as can be seen from the outcrop, the top layer is sandstone with depth about 5.8 m while from the 2-D resistivity result, the top layer has interpreted as about 4-6 m thick from the surface with the resistivity value range of 8.00 to 80.0 Ohm-m. The second layer is mudstone with resistivity value between 100 and 130 Ohm-m which about 1 m thick. It is almost same depth as the measurement from the outcrop which is 1.7 m thick. The third layer of sandstone was measured at depth of 8.8 m from the outcrop. Similarly, the third layer sandstone that interpreted from the 2-D resistivity results lies between 5-8 m depth with resistivity value of less than 150 Ohm-m which has different resistivity value with the top layer of sandstone. The value for this layer is higher than both two layers above it though the resistivity value for mudstone usually is lower than sandstone. It is because the sandstone at the top layer was affected by weathering process due to the physical, chemical or biological activity. Then at depth about 8 m, it is interpreted as mudstone with resistivity value range same as the second layer of mudstone.

Aquifer occurs in porous or fractured rocks and moves within them when there is a pressure difference. Fractured porous rocks are characterized by presence of fissure and weakly cemented sandstone. According to their ability to transfer water, rocks are classified as either water bearing or well-permeable, low-permeable and practically impermeable or water resistant. The water-bearing and low-permeable rocks usually occur as alternating layers of deposits or zones with different thickness [10]. The outcrop shows that the sandstone layer is alternate with mudstone layer. Both sandstone and mudstone are high porosity but only sandstone is high permeability. Between the grains of sand in sandstone, there is space that can store the fluids. Thus, the sandstone may acts as water-bearing rocks while mudstone acts as low-permeable rocks. The lower permeability of mudstone can be an effective seals for an aquifer.

5. Conclusion

The 2-D resistivity is a very effective method to estimate the hydraulic properties of aquifers. This method successfully differentiate sandstone and mudstone. The study area have low resistivity value and expected as aquifer. The resistivity value of both sandstone and mudstone less than 800 Ohm.m. Sandstone and mudstone are acting as water-bearing rocks and low-permeable rocks respectively and the layering of sandstone and mudstone can be an effective aquifer. The outcrop and the 2-D resistivity results show a good correlation.

Acknowledgements

The authors thanked all the technical staffs of the Geophysics laboratory and postgraduate students, School of Physics, Universiti Sains Malaysia for their assistance during the data acquisition. A sincere gratitude and appreciation is given to Institute of Postgraduate Studies, Universiti Sains Malaysia for their sponsorship (IPS Conference Fund).

References

[1] Meor H A H, Aye-Ko A and Becker RT 2014 Stratigraphy and Palaeoenvironmental Evolution of the mid- to upper Palaezoic Succession in Northwest Peninsular Malaysia *Journal of Asian Earth Sciences* **83** 60-79.

[2] Sultan A S A 2012 Groundwater Management by Using Hydro-Geophysical Investigation: Case Study: An Area Located at North Abu Zabal City *National Research Institute of Astronomy and Geophysics, Helwan, Cairo* 183

[3] Loke M H 1999 Electrical Imaging Surveys for Environmental and Engineering Studies **6574525** 4

[4] Mohamed, A M O and El Gamel M M 2010 Sulphur Concrete for the Construction Industry: A Sustainable Development Approach *J Ross Publishing USA* 89-102

[5] Plummer C C 2014 Physical Geology 12th Edition *Content Tecnologies Inc.*

[6] Gingras M K, Pemberton S G and Smith M 2015 Bioturbation: Reworking sediments for better or worse *Oilfield Review Winter 2014/2015* **26** 46-51

[7] Keller G V and Frischknecht F C 1996 Electrical Methods in Geophysical Prospecting *Pergamon Press Inc.* Oxford

[8] Nagaratnam S, Sanjay K S, Braja M D 2013 Rock Mechanics: An Introduction *CRC Press, Taylor and Francis Group*

[9] Bernard J 2003 The Principles of Geophysical Methods for Groundwater Investigations Terraplus 2

[10] Shestopalov V M 2003 Types and Properties of Water: Typification of Groundwater Characteristic *Encyclopedia of Life Support Systems (EOLSS)* **1** 95

IOP Conf. Series: Earth and Environmental Science **62** (2017) 012043 doi:10.1088/1755-1315/62/1/012043

Three-Dimensional Modeling of TDEM Study for Carbon Capture Storage

Dias Nurazna Pramukusuma[1], Diniasa Manoti[1], Warsa[2]

[1]Undergraduate Program of Geophysical Engineering, Institut Teknologi Bandung
[1]Applied and Exploration Geophysics Research Group, Faculty of Mining and Petroleum Engineering, Institut Teknologi Bandung

Email: nuraznadias@gmail.com, dinibedew@gmail.com, warsa@gf.itb.ac.id

Abstract. TDEM (Time Domain Electromagnetic) method is a sounding geophysical method using electromagnetic wave generated by the electrical current source to image cross section of earth's subsurface. TDEM using galvanic source (grounded wire) is one of TDEM acquisition technique which has flexible properties for the complex topography of measurement area. Three-dimensional forward and inversion modeling technique of TDEM with galvanic source still develops until now. Three-dimensional modeling is required because three-dimentional structure can influence TDEM signal which has a dimension of z-axis direction only. In this paper, We have been developing a 3D forward modeling program applying finite difference time domain method for 3D subsurface models. We can assess a distribution of TDEM synthetic data affected by 3D model. Using this 3D TDEM modeling program synthetic data has been generated as the function of transmitter and receiver position in 3D model space. Furthermore, this program can be used to simulate a conductivity change of reservoir model as consideration of reservoir monitoring implementation.
Keyword: finite difference, electromagnetic, galvanic, three-dimensional

1. Introduction

TDEM is a sounding geophysical method using electromagnetic wave propagation with galvanic or loop source to image subsurface cross section. For a complex topography field, a galvanic source is required to use because this configuration can be placed anywhere. Three-dimensional forward and inverse modeling of galvanic source is still improved until now. In some discontinuities structure case, we have to model with 3D modeling because TDEM data which generate 1D sounding data but it affected by 3D subsurface model. For example is sand lenses, receiver outside the x and y coordinate of lenses concerning homogeneous structure theoretically. But, a real data measurement, the receiver generates a response from sand lenses. This is an attractive thing to build 3D forward modeling program for TDEM.

Electromagnetic problem formulated by Maxwell equation which integral, finite difference, or analytical solution can solves that. In this paper, we want to solve with finite difference time domain in 3D model space. Because of TDEM sensitive for conductivity distribution, we want to imply 3D forward modeling to detect conductivity change in Carbon Capture Storage (CCS) monitoring project, like Gundih field. Carbon Capture Storage (CCS) pilot project Gundih is project to improve new technique for CO_2 injection monitoring technology which would be better enabled in a shallow reservoir like Ngrayong formation. Resistivity change that is attributed to CO_2 injection may vary from 0.2 to 10 times its initial value [1]. We want to pre-observe CO_2 injection and TDEM data relation, so that, we can understand about TDEM data characteristic and plan acquisition parameter in monitoring CCS project case. This research can become preliminary study about 3D TDEM data processing and this is an important thing that best monitoring technique can get the best CCS reservoir which has high CO_2 concentration and low leakage surrounding that reservoir after injection.

2. Basic Concept

2.1 Electromagnetic Concept

Maxwell equation consisted of Gauss law in electricity, Gauss law in magnetism, Faraday law, and Ampere-Maxwell law [2] which formulated by

$$\nabla \times \boldsymbol{H} = \boldsymbol{j} + \frac{\partial \boldsymbol{D}}{\partial t} = \boldsymbol{c} \tag{1}$$

$$\nabla \times \boldsymbol{E} = -\frac{\partial \boldsymbol{B}}{\partial t} \tag{2}$$

$$\nabla \cdot \boldsymbol{B} = 0 \tag{3}$$

$$\nabla \cdot \boldsymbol{D} = q \tag{4}$$

where H and B are magnet field vector (A/m) and magnet induction field vector (T), D, q, and c are electric field displacement (C/m^2), electric charge (C), and current density (A/m^2). Whereas, material characteristic abbreviated to

$$\boldsymbol{D} = \epsilon \boldsymbol{E} \tag{5}$$

$$\boldsymbol{B} = \mu \boldsymbol{H} \tag{6}$$

where ϵ and μ are permittivity (F/m) and permeability (N/A^2). The geophysical principle for the electromagnetic survey is the transmitter (active or passive) generate induced magnetic field to earth. Induced magnetic field exceeds conductive medium and then galvanize Eddy current inside the medium. Eddy current generate secondary induced magnetic field through to receivers at surface or borehole [3].

2.2 TDEM Measurement Method

Current source type of TDEM survey consisting loop current and dipole current. In this paper, we discuss about dipole source (see Figure 1) because we imply it in CCS which has the deep target and complex topography. Dipole source is more flexible and faster acquisition than loop source. One period of TDEM measurement data called like 50% duty cycle (¼ positive current on, ¼ current off, ¼ negative current on, ¼ current off). The receiver record induced magnetic field when current off. Interest thing caused by transmitter and receiver offset is we obtain TDEM data has two poles in one decay time. This is one of reason, we must generate negative current when TDEM was measuring.

b)

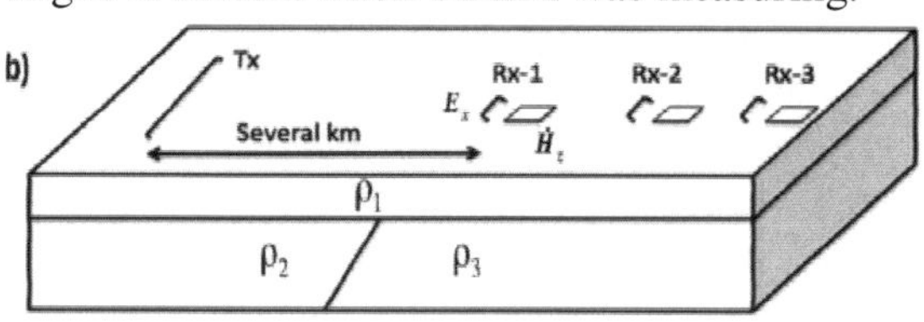

Figure 1. TDEM acquisition with galvanic source [4]

3. Methodology

TDEM dipole source as same as the basic principle of geoelectric when current injection is working. So that, TDEM forward modeling must concern geoelectric forward modeling. In this paper, we apply geoelectric forward modeling equation [5] and electromagnetic [6], it can be written be

$$\vec{\nabla}\cdot\left(-\sigma(x,y,z)\vec{\nabla}\varphi\right) = I\delta(x - x_s)\delta(y - y_s)\delta(z - z_s) \tag{7}$$

$$\nabla \times \boldsymbol{H} - \sigma\boldsymbol{E} - \varepsilon\frac{\partial}{\partial t}\boldsymbol{E} = \boldsymbol{j_s} \tag{8}$$

where φ is a electric potential (V), δ is Dirac function, (x_s, y_s, z_s) is source coordinate (m). Geoelectric forward modeling generates electric potential distribution and we determine electric field from a gradient of the potential. Even though, electromagnetic equation use to simulate electro induced magnetic field decay. Matrix system of geoelectric forward modeling [7] formulated by

$$(\boldsymbol{D}S(\sigma)\boldsymbol{G})\boldsymbol{u} = \boldsymbol{A}(\sigma)\boldsymbol{u} = \boldsymbol{q} \tag{9}$$

D, G, S, u, and q are divergence, operator gradient, conductivity, potential, and current source position. In this paper, delta Dirac function (q matrix) approached by Gaussian function. We have applied conjugate-gradient solver (CG solver) because of large size matrix dimension.

Electromagnetic mechanism formulated to simulate electro induced magnetic field decay in 3D space model. Electromagnetic wave propagation [8] can be written as

$$B^n = -\nabla \times E^n \tag{10}$$

$$H^n = \frac{B^n}{\mu} \tag{11}$$

$$E^{n+1} = \frac{2\gamma - \sigma\Delta t_n}{2\gamma + \sigma\Delta t_n}E^n + \frac{2\Delta t_n}{2\gamma + \sigma\Delta t_n}\left(\nabla \times H^{n+\frac{1}{2}}\right) \tag{12}$$

where γ and σ are permittivity (F/m) and conductivity of medium (S/m). Solving the electromagnetic equation, we use FDTD (finite difference time domain) method [8]. Explicitly, this method can determine unknown parameter from known parameter evaluated before.

4. Result

3D grid model formed by heterogenous grid which expands together with increasing source and transmitter distance. Acquisition design of 3D forward modeling in Figure 2 adjusted to acquisition design of Gundih field which set to an equatorial array and inline array [9]. We do that to consider synthetic data and field data to analysis program quality.

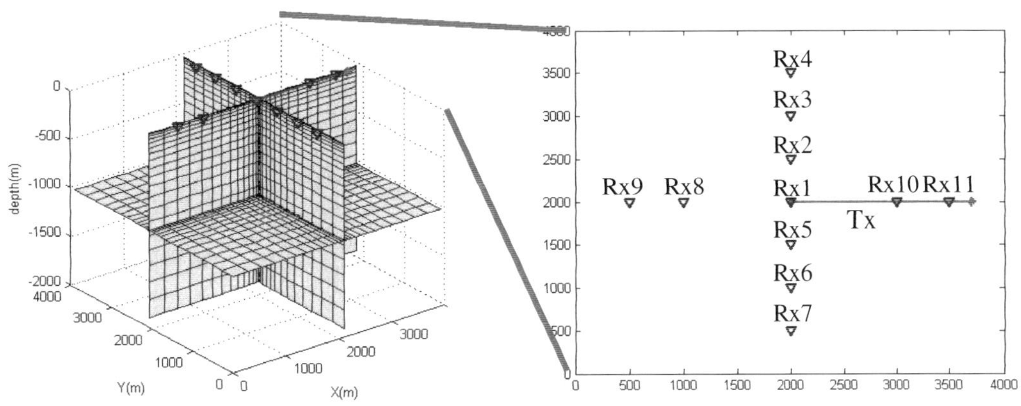

Figure 2. Transmitter and receiver set up of forward modeling

4.1. Homogeneous Synthetic Model

In homogeneous synthetic model case, we have analysed some parameter such us electromagnetic wave propagation and sensitivity at some depths. This matter becomes information to understand signal/ synthetic data generated by the program. Because of computation time process equivalent to conductivity medium, therefore we use 100 S/m of conductivity. When t=0 second, induced magnetic field distribution expand and has lower value with increasing of depths. It appropriates to smoking ring principle which TDEM was measuring can forms that. When t=10 second, induced magnetic field distribution spread at a whole of depths because decay mechanism has happened. At sensitivity analysis indicate that increasing depth cause decreasing sensitivity block value. High sensitivity located at source electrode coordinate. So that, when we plan TDEM survey design, we locate the source to target coordinate, although source coordinate can be located outside target with receivers are located to target coordinate, methodically.

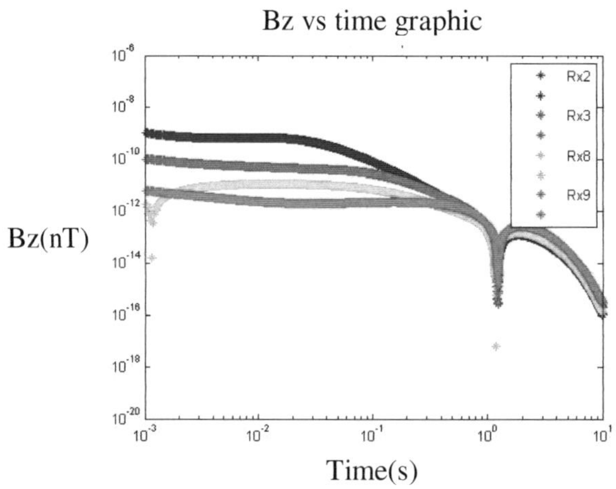

Figure 3. Synthetic data of induced magnetic field z-direction at receiver Rx2, Rx3, Rx8, and Rx9

Synthetic data generated by program at Figure 3 indicate similar pattern with field data. In one decay cycle, signal polarization changes was happened. This polarization changes caused by electromagnetic wave propagation at the surface. So that, to obtain complete TDEM synthetic data, we have to do forward modeling in one period of TDEM measurement.

4.2. Reservoir Block Model

Reservoir model adjusted with field condition of CCS at Gundih, it is sandy lenses structure with clay matrix. Some assumption used to this model are gas conductivity 0.01 S/m, CO_2 gas conductivity 0.2 S/m with 21% fraction, 28-40% of porosity, and geometrical function m=2.15 [10]. To calculate average conductivity of reservoir, we apply Archie law calculation for a heterogenous phase case with a parallel model approach. We get reservoir conductivity is 0.098 S/m (Figure 4).

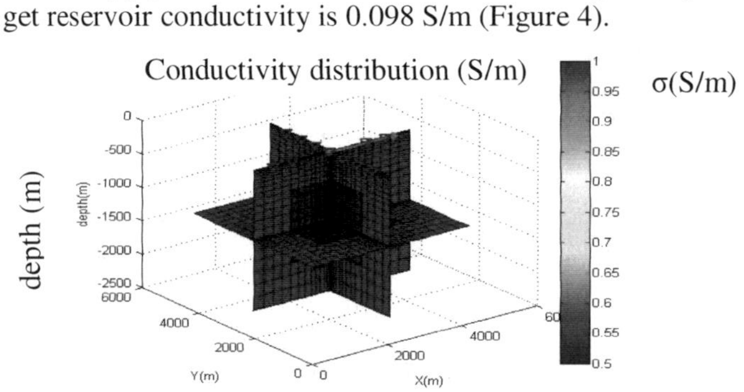

Figure 4. Reservoir blocks model

In 3D FDTD simulation at reservoir model, we obtain synthetic data which when t=0.2 second are signal because the reservoir effect (see Figure 5). That can be analyzed by 3D electromagnetic wave propagation. 3D induced magnetic field indicate that when t=0.2 second, induced magnetic field propagate at reservoir coordinate and received by receivers at a surface (see Figure 6). Because of the electromagnetic wave velocity is very high, the different time between time of electromagnetic wave propagated in reservoir and received by receivers is very fast. So that, we said that electromagnetic wave was received by receiver represents wave propagation in subsurface model. The generated signal from this program have been compared with signal pattern from analythical formulation. There are gradient change and form curved at detected anomaly time. Stronger signal equivalent with longer distance from transmitter. At long distance, wave propagation can separate signal from different medium when we apply the ray tracing illustration. It is the advantage of 3D modeling which we can analyze synthetic data anomaly from electromagnetic wave propagation. And we know that signal pattern which is generated by subsurface model.

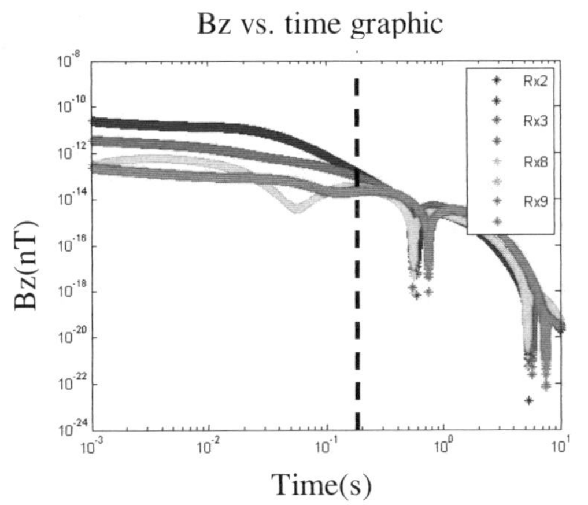

Figure 5. Synthetic data of induced magnetic field z-direction at receiver Rx2, Rx3, Rx8, and Rx9

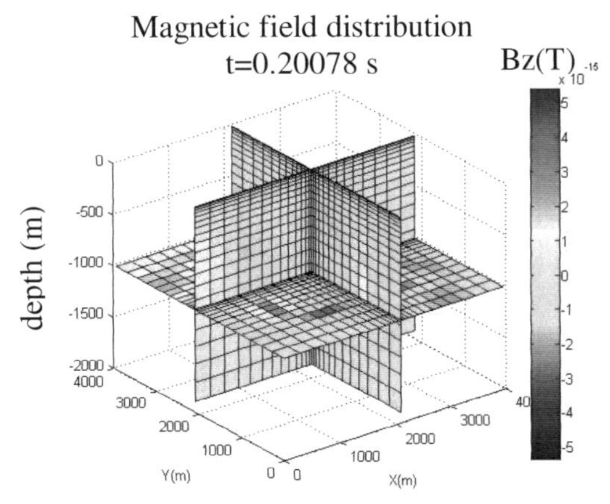

Figure 6. 3D Induced magnetic field modeling at t=0.2 second

TDEM method responsive to conductivity contrast, including to conductivity changes. In CCS case which CO_2 injection can affect the conductivity of reservoir. Theoretically, resistivity changes caused by CO_2 injection around 0.2 to 10 times from initial value [1]. So that, we consider conductivity of reservoir between 0.1 S/m (initial conductivity) and 0.5 S/m (0.2 times of initial resistivity or 5 times of initial conductivity). The result, conductivity changes can be detected by TDEM method based on FDTD simulation (Figure 7). There are different between blue line and red line, higher conductivity (red line) has longer decay time because the charge properties. Both lines, can indicate the reservoir signal at t=0.2 seconds like the explanation before about reservoir signal. At the higher conductivity, reservoir impact to signal is stronger than lower conductivity. After t=0.2 seconds, the signal at 0.5 S/m increases gradually but at 0.1 S/m the signal ... e.

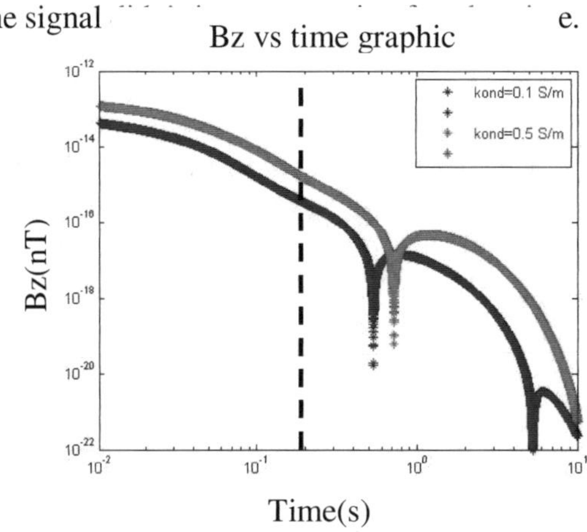

Figure 7. Synthetic data of induced magnetic field z-direction at receiver Rx5

5. Conclusion

A 3D grounded wire (galvanic source) TDEM forward modeling using FDTD simulate electric field modeling caused by current injection and done in one period TDEM measurement. Synthetic data generated by 3D TDEM forward modeling with FDTD method can indicate anomaly respond which observed from electromagnetic wave propagation in 3D space model. Based on 3D TDEM with FDTD, TDEM method can give conductivity changes respond in reservoir model and also prove that one-dimensional data of TDEM affected by three-dimensional structure.

6. Acknowledgement

The authors thank to JICA-JST, Pertamina EP &UTC, and DGHE Indonesia for supporting this study.

References

[1] Newmark R L, Ramirez A L, Daily W D 2001 Monitoring Carbon Oxide Sequestration Using Electrical Resistance Tomography (ERT) *Sensitivity Studies. Natl Energy Technol Lab*

[2] Zhdanov M S 2009 Geophysical Electromagnetic Theory and Method *Elsevier*

[3] Meunier E 2015 Exploring for Graphite Using A New Ground-Based Time-Domain *Carleton University: Master Thesis Department of Earth Science*

[4] Haroon A, Adrian J, Bergers R, Gurk M, Tezkan B and Mammadov A L 2014 Joint Inversion of Long-Offset and Central-Loop Transient Electromagnetic Data: Application to A Mud Vulcano Exploration in Perekishkul, Azerbaijan. *Geophysics Prospecting*

[5] Telford W M 1990 Applied Geophysics Second Edition (Cambridge: Cambridge University Press)

[6] Ward S H and Hohmann G W 1991 Electromagnetic Theory for Geophysical Exploration in Electromagnetic methods in applied geophysics. Nabighian, M N (Ed.) *Society of Exploration Geophysicists* **1**

[7] Pidlisecky A, Haber E and Knight R 2007 *Geophysics* **72** No 2

[8] Commer M 2003 Three-Dimensional Inversion of Transient Electromagnetic Data: A Comparative Study *Mathematisch-Naturwissenschaftlichen Fakultat der Universitat Zu Koln*

[9] Srigutomo W, Warsa, Sule R, Saito A, Nakayama K, Hato M, Trimadona, Prasetyo D, dan Widarto S 2014 Time Domain Electromagnetic (TDEM) Baseline Survey for CCS in Gundih Area, Central Java, Indonesia *Proceeding of the 12th SEGJ International Symposium*
[10] Glover P 2010 *Geophysics* **75**

Modeling of Floating Time Domain Electromagnetic Method to Detect Dissolved Sediment

Siti Nurjanah[1], Widodo[2]

[1]Undergraduate Program of Geophysical Engineering, Institut Teknologi Bandung
[2]Applied Geophysics and Exploration Research Group, Institut Teknologi Bandung

Email: nana.anhan@gmail.com

Abstract. In hydrology context, sediment can be interpreted as inorganic and organic material that is transported by, suspended in, or deposited by streams. It is important to know the function of soil, stream discharge, land-cover features, weather conditions and land-use activities. Sediment load carried by streams and rivers can be composed either of fine materials, mostly silts, and clays, or coarse materials such as sand. One product of sediment is dissolved load consists of indistinct material in solution moving downstream. It is produced by chemical weathering processes and does not move out of the water. To investigate the dissolved sediment, we have applied the floating of Time Domain Electromagnetic (TDEM) method. The acquisition of TDEM data has been performed use tires and small ship as innovation measurements. The calculated data model using Occam and Marquardt Algorithms. The responses of data show the sedimentation has less resistive compare the surrounding structures. This innovation is very helpful to know the environmental condition, especially in the water.
Keywords: Dissolved sediment, floating loop, Time Domain Electromagnetic

1. Introduction

Until recently there has not been a lot of discussion about the use of geophysical techniques to detect dissolved sediment. The sedimentary environment can be occurred on the land or under the water based on the physical and chemical processes that are active and the organisms that live under those conditions [1]. Seismic and gravity methods were used commonly for dissolved sediment mapping purposes. Those methods involve measurement of specific parameters generated by the anomalies. Dissolved sediments determined by conductivity contrast from others. The conductivity anomaly was generally accepted as the best indicator of the sediment. But it became difficult to give good results with just conductance parameter.

Time domain electromagnetic (TDEM) method also can be used to “see” dissolved sediments because of its sensitivity in conductivity contrast. It was more effective to use because the acquisition technique is simple and the instrumentation costs relatively cheaper, rather than seismic method. This method relies on detection of induced voltage generated by the variance of the earth secondary magnetic field.

Unfortunately, the potential of TDEM method was not fully realized for aquatic environments and sediment mapping beneath it because this method still focused on finding conductivity anomalies on the land. TDEM method particularly has not been applied in deposit exposes in which located under the water. It requires instrument modification, and forward modeling processes to assure the effectiveness of the method for beneath the water sediments detecting.

This paper illustrates the configuration that used to modify the TDEM instruments so that it stays above the water when acquisition processes. Also, it presents some synthetic models of layers that calculated by Occam [2] and Levenberg-Marquardt [3,4] algorithms. The algorithms have allowed us evaluating the similarity of forward modeling and calculated data model. The resemblance indicates the applicability of TDEM method to detect dissolved sediments in which lied underneath the water.

2. Method

2.1. Time Domain Electromagnetic

TDEM methods need a transmitter to induce time-varying current into a transmitter loop. The current of transmitter loop generates a primary magnetic field that forms a smoke ring [5,6] into the subsurface. When the current finds different subsurface materials, it induces eddy currents that generate secondary magnetic field based on Maxwell equations. This secondary magnetic field is recorded in receiver at specific time periods.

The skin depth is denoted as a depth when the amplitude is dissipated by 1/e or about 37% of the initial rate. It is used as an estimation of the exploration depth of the electromagnetic schemes [7]. The skin depth of time domain (δ_{TD}) is expressed by

$$\delta_{TD} = \sqrt{\frac{2t}{\mu\sigma}}$$

The depth is illustrated as the maximal depth in which a specified object can be observed during the investigation.

2.2. Modification Design

Design modification made by using pipes, tires, and straps that are formed with a central loop [8,9] configuration of transmitter and receiver. We used pipes with its size and length of 2 inches and 1 m, respectively. These pipes can be combined with other pipes and are easily be adjusted. Straps used to strengthen the connection of pipes, so that are not easily separated. Once the pipeline is ready, then the tire is installed on each side of the circuit pipes as necessary. The tires will support the pipeline over the water. Figure 1 shows TDEM modification design on the water with combined pipes with a size of transmitter and receiver loop is 6 x 6 m^2 and 5 x 5 m^2, respectively.

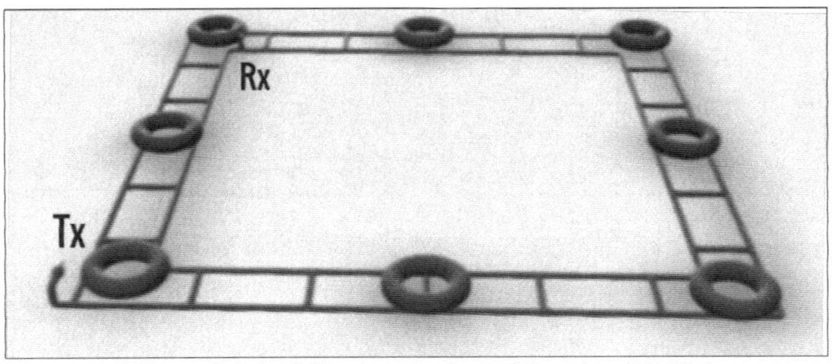

Figure 1. Modification design of floating TDEM.

The assembled pipeline used to form a loop of transmitter and receiver and insert the cables into the pipeline. In each pipe used such of rubber to assure that the cables keep straight inside the pipeline. Detail of the pipe shown in figure 2. The tool of TDEM will remain in dry place (or small ship) during the measurement.

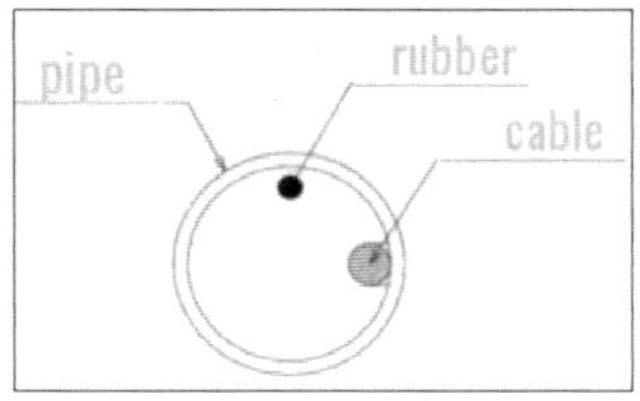

Figure 2. Detail of cable clamp inside the pipe.

2.3. Numerical Simulation

To find out the applicability of the proposed design, a forward modeling and data inversion is conducted. In this case, a generalized stratigraphic sequence [10] off the west coast of Peninsular Malaysia is chosen to become the synthetic model. The model consists of 6 layers with five different geological layers: soft marine mud, sandy clay, clay, gravels or pebbles, and granite bedrock. We assume there is a presence of secondary tin deposits within gravels. The chosen lines of the synthetic model and the physical parameters [11] (estimation value) used in forward modeling process are shown in figure 3 and table 1, respectively.

3. Results and Discussions

The resistivity models (Line 1, 2, and 3) from Occam inversion can be viewed in figure 4. The resistivity model is taken from two order smoothness derivatives at three different lines using starting model with homogeneous resistivity and thickness of 5 Ωm and 10 m, respectively, computed for times of 1 μs and 1000 μs. The first and second order smoothness derivatives of data mostly fitted from 5 - 90 m depth.

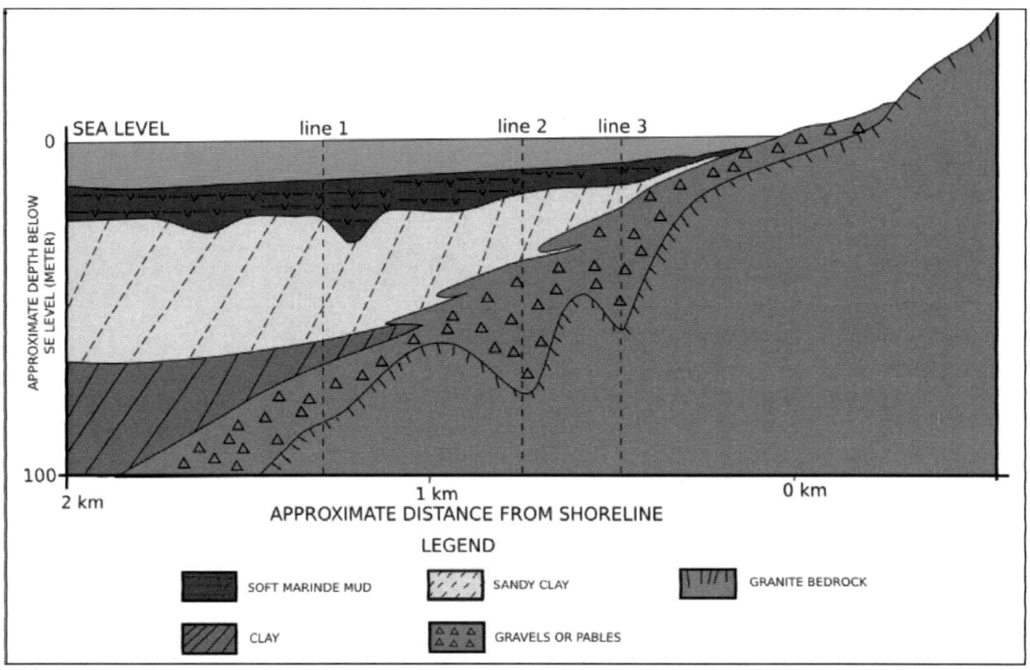

Figure 3. Generalized stratigraphic sequences off west coast Peninsular Malaysia (modified [10]).

Table 1. Physical properties of synthetic model

Material	Resistivity (Ωm)	Thickness (m)		
		Line 1	Line 2	Line 3
water	5	10	10	8
Mud	50	10	10	10
Sandy clay	150	35	20	7
Clay	100	10	-	-
Tin deposits	20	15	35	35
Granite bedrock	1000	∞	∞	∞

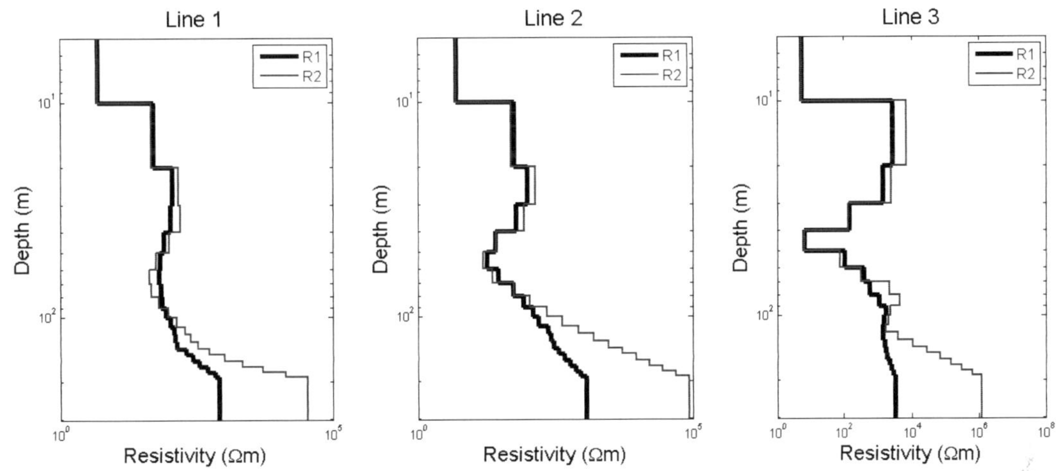

Figure 4. The first (R1) and the second (R2) order of smoothness restrictions of Occam inversion at line 1, 2, and 3.

In the case of Marquardt inversion, the data inverted by using four layers of starting models with homogeneous resistivity and thickness of 10Ωm and 10m, respectively. In this stage, the correlation of measured and calculated data has a good fitting and RMS values are varying from 0.3 – 2%. Figure 5 shows the resistivity models of Occam and Marquardt inversion. It shows that the Marquardt and Occam models are mostly suitable with each other. The most important thing is the correlation models can show the presence of secondary tin layer.

4. Conclusion

The proposed design can be used to implement geophysical measurements of TDEM method on the water. The method has good resolution, particularly for mapping conductive layers in resistive sections. The inversion results of the synthetic models proved that the numerical simulation of the design is capable of detecting dissolved sediments in which occurred beneath the water. The dissolved sediments (secondary tin deposits) have less resistive layer comparing to the other layers. The presence of deposits with a conductive layer (20 Ωm) can be seen at a depth of ± 60 meters, 40 meters, and 30 meters for line 1, line 2, and line 3, respectively. The result of this study concluded this innovation is very helpful to know the environmental condition, especially in the water.

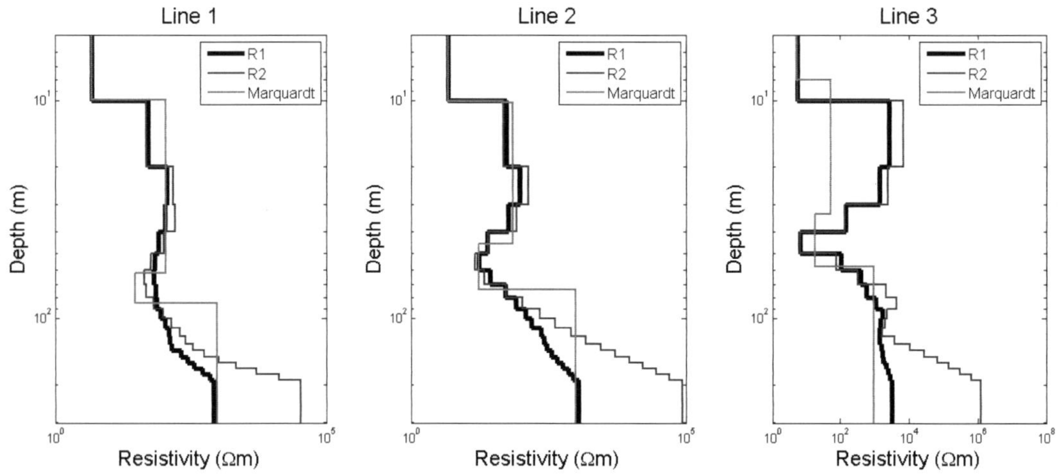

Figure 5. Marquardt and Occam's (first (R1) and second (R2) order of smoothness constraints) inversion models for line 1, 2, and 3.

Figure 5 shows the resistivity models of Occam and Marquardt inversion. It shows that the resulting Marquardt and Occam models are mostly suitable with each other. The most important thing is the correlation models can show the presence of secondary tin layer.

References

[1] Nichols G 2009 Sedimentology and Stratigraphy *United Kingdom: Blackwell*
[2] Constable S C, Parker R L and Constable C G 1987 *Geophysics* **52** 289-300
[3] Levenberg K 1944 A Method for the Solution of Certain Problems in Least Squares *Quarterly of Appl. Mathematics* **2** 164-168
[4] Marquardt D 1963 Algorithm for Least-Squares Estimation of Nonlinear Parameters *SIAM Journal of Appl. Mathematics* **11** 431-441
[5] Nabighian M N 1979 *Geophysics* **44** 1700-05
[6] Wameyo P M 2005 Magnetotelluric And Transient Electromagnetic Methods In Geothermal Prospecting, With Examples From Menengai, Kenya *The United Nation University Geothermal Training Programme Report* **21** 416
[7] Spies B R and Frischknecht F C 1989 Depth of investigation in electromagnetic sounding methods *Geophysics* 872-888
[8] Widodo 2012 Multidimensional Interpretation of Near Surface Electromagnetic Data Measured in Volvi Basin *Journal of Environmental and Engineering Geophysics* 121-133
[9] Scholl C 2005 The influence of multidimensional structures on the interpretation of LOTEM data with one-dimensional models and the application to Israel *Koln: Universität Zu Köln*
[10] Ringis J 1979 *Geophysics* **10** 3 213-215
[11] Telford W M, Geldart L P and Sheriff R E 1990 Applied Geophysics 2nd Ed *Cambridge: Cambridge University Press*

IOP Conf. Series: Earth and Environmental Science **62** (2017) 012045 doi:10.1088/1755-1315/62/1/012045

Decomposing Worldwide Complete Spherical Bouguer Gravity Anomaly Using 2-D Empirical Method

Ruhul Firdaus[1] and Gestin Mey Ekawati[1]

[1]Department of Geophysical Engineering, Faculty of Science, Institut Teknologi Sumatera, Jl. Terusan Ryacudu, Desa Way Hui, Kecamatan Jati Agung, Lampung Selatan, 35365, Indonesia

E-mail : ruhulhajjaj@gmail.com

Abstract. Currently available worldwide gravity anomaly data provides a high-resolution (2'x2') of Complete Spherical Bouguer Anomaly (CSBA) based on the available information of the Earth gravity field from surface and satellite measurements. The data has not only been provided and processed thoroughly but it also has been claimed to be appropriate for various geophysical applications. Therefore, the analysis of gravity anomaly is becoming increasingly significant for the earth sciences as a whole and assisting both shallow and deep geological problems. Earth gravity anomaly has to be analyzed carefully as it has very complex data due to anomaly mixing of the density masses spread over the Earth horizontally and vertically. The bigger the spatial coverage of data (e.g. global scale data), the more severe the data from anomaly mixing due to various wavelength. BEMD is an empirical method supposedly suitable with highly oscillation-mixing data. It can effectively isolate each local anomaly in details and is analogized as successively reverse moving average with local windowing. BEMD is designed to reduce multi-component, non-linear gravity field data to a series of single local anomaly contributions. Anomaly from a single body was assumed as a mono-component signal. The main advantage of BEMD processing techniques is to present the subtle details in the data which are not clearly identified in anomaly maps, without specifying any prior information about the nature of the source bodies. As the result, we have identified regional anomalies due to the drift of continental and oceanic masses considered as *crust-regional anomaly* (CRA). We remove the CRA from the CBA to provide *surface-residual anomaly* (SRA) where shallow geologic bodies reveal. Meanwhile, the CRA itself can be used as reference to reduce this high magnitude anomaly from any measurement data to exhibit only shallow body anomaly. Further analysis can be carried out to build a general understanding of the details and parameters of the shallower or deeper causative body distributions.

1. Introduction

The gravity method involves measurement of very small variations in the Earth's gravitational field, of the order of a few parts per million or lower, caused by lateral variation in density. The gravity anomaly is the contrast of gravitational accelaration between Earth's masses and that of a reference mass distribution. It has a wide variety of applications in the Earth sciences, espescially in geophysics. Gravity variations over thousands of kilometres can be used for studies of mantle convection, variations over hundreds and tens of kilometres are relevant for studies such as lithospheric flexure, plate tectonics, crustal structure, sedimentary basin development, hydrocarbon and mineral exploration, while gravity variations over tens of metres can be used in civil engineering applications

[1]. The potentially broad range of contrasting densities in the near-surface, in the crust, and in subcrustal rock materials leads to the wide range of applications of the gravity method [2]. Depth of the source anomaly relates to its wavelength in the surface due to potential field expansion. While the amplitude is proportional to the lateral density contrast, it follows the rule of Newton's inverse square of mass distance. Therefore, measurement of the Earth gravity field is a useful tool for investigating the sub-surface because mass variations below the surface cause variations in the gravity field.

Separation of gravity anomaly is considered as an important step for interpreting and modeling subsurface condition related to density contrast. Observed gravity data is a resultant from entire sub-surface bodies which one type of them might be considered as target anomaly (the *residual*). The idea of such separation process relies on different scale of the wavelength between shallow-small bodies and deep-regional bodies. Several techniques have been developed and applied to do anomaly separation graphically and matematically. The graphical method is not used these days anymore but it has historical importance [3]. Among the distinguished analytical methods are polynomial fitting, moving average, upward and downward continuation and wavelength filtering [3],[5]. Complexity of geological structure and mass body distribution derives multi-component, non-stationary, and non-linear gravity anomaly [4][5],[6]. These facts lead to some difficulties in identification, localizing and examination the anomalous suspected body using the gravity field anomaly. The drawback of all of the existing methods is, when the gravity bodies extracted by averaging the anomalous target zone, some residual still remains mixed with the regional [2]. BEMD of gravity field can get more information and richer physical geographical (geological) meaning than the previous analytic continuation and other traditional field decomposition methods [6].

There are now many ways to acquire gravity data on land, sea, air, and from space. These data play important role in gravity studies in many scales. The World Gravity Map (WGM) is a set of high-resolution (2'x2') gravity anomaly maps at global scale. This is the first dataset that take into account a realistic Earth model and that consider the contribution of most surface masses (atmosphere, land, oceans, inland seas, lakes, ice caps, and ice shelves) [7]. One of the data contained in WGM is Complete Spherical Bouguer Anomaly (CSBA) which is provided to meet the definition of gravity anomaly in geophysics. The source of data is taken from surface gravity measurements (from land, marine or airborne surveys), satellite altimetry and satellite gravimetry (GRACE mission) measurements. Availibility of this data has increased the interest to study the Earth's gravitational field on local and global scale.

2. Methodology

The geophysical potential data are nonlinear and nonstationary. A relatively new technique named *empirical mode decomposition* (EMD) has been developed to deal with such type of data. EMD is developed by Huang et al. (1998) to establish a multi-resolution decomposition technique[8]. EMD is an adaptive decomposition of one-dimensional data and widely used to analyze different variety of signals. In geophysical field, EMD has been applied to seismic data [9]9,[10] and log evaluation [11]. EMD gives the decomposition result in small number of what so called *intrinsic mode functions* (IMFs). EMD works efficiently and it decomposes the input signal completely. There is no information loss, i.e., the original signal is reconstructed by summing all IMFs. The result is excellent in resolution and separation between details so we can get multi-resolution signal separated each other entirely. It is a quasi-orthogonal decomposition in that the cross-correlation coefficients between the different IMFs are always close to zero. This minimizes energy leakage between the IMFs[12]. The residual mixing within regional can be handled using this methodology.

EMD is one-dimensional data analysis while WGM is two-dimensional. The concept of EMD was implemented to analyze two-dimensional signal (image or map) that is called *Bidimensional Empirical Mode Decomposition* (BEMD) [[13]3,[14]4]. This method has been successfully applied to examine gravity map and shows good conformity with geological environment by Hasan (2008) [9], Huang et al. (2010) [16], Morris et al. (2010) [17], Firdaus et al. (2012) [5], Rahim (2015) [4]. The fundamental principle of BEMD is to extract mono-component image from multi-component image by removing local trends. The BEMD results are Bidimensional IMF (BIMF) which are a set of mono-component multi-resolution image arranged by the order of detail component. It is designed to reduce a complex gravity field data to a series of single local anomaly contributions. The highest detail appears as the first BIMF and than come BIMFs with less and less detail until the last one called *residual component* coming out bringing global trend for the entire map. This methodology somewhat is in kind with successively reverse moving average with local windowing. We use this kind of image analysis to isolate every single body anomaly contained in WGM data.

3. Data

CSBA has a very highly non-linear and non-stationary data as it represent global and local anomalies. It contains very wide information about earth gravitational field. Figure 1 presents several features brought by the data. Negative anomalies (greenish to purple) occure on regions ascociated with very thick continental crust where the dominant part of continental crust is reflected by relatively low anomaly ($\leq$ 250 mGal, yellow to greenish). Oceanic crust has higher anomaly ($\geq$ 400 mGal, shining red) which can be divided into two parts. The first part has lower anomaly than the second part. The first part is more likely ascociated with younger oceanic crust while the second part represents more mature oceanic crust. Plate boundaries also can be observed directly by their distinguished anomalous value between 250 mGal to 400 mGal (darker red). Such kind of information is very valuable to understand the Earth through gravity anomaly. Higher resolution data contain much more detail informations thus more local which permit us to interpret and model a very specific part of the Earth using gravity anomaly. However, these details are hardly to be observed in the original map. Hence the anomalies appear in different level of details, BEMD technique can be used to automatically gives us multi-resolution results separated by order of detail.

Figure 1. CSBA is one of the data provided within World Gravity Map (WGM) digital dataset. CSBA is very non-stationary and non-linear data. It exhibits broad variation of periodicity ($1' - 180^{o}$) and magnitude (0 – 600 mGal). Colormap has been adjust to give the best representation of features contained in the data.

4. Results

Applying BEMD to CSBA data generates nine BIMF extractions which are each of them gather all local oscillations. Assuming the CSBA generated by a superposition of their local oscillations, BIMF could be considered as a bidimensional monocomponent representation of gravity anomalies resulted from removing the local trend. Figure 2 shows different scale of detail and complete separation of anomaly mixing into each single local body anomaly without specifying any prior information about the nature of the source bodies. It could be considered as an anomaly from a single source body under assumption that a body anomaly like point mass approximately contribute only one single oscillation to the data. Each BIMF carries intense physical meaning from the original data. It brings out of each anomaly and places them in the order of local frequency. A wide range of densities occurs within the crust, from the essentially zero density of air-filled voids in near-surface formations, to density of unconsolidated sediments with their interstitial openings filled with either air or water, to the highest densities related to Fe/Mg-rich crystalline rocks and metallic ores. Even higher densities are associated with the radial shells that make up the mantle and the core of the Earth [2].

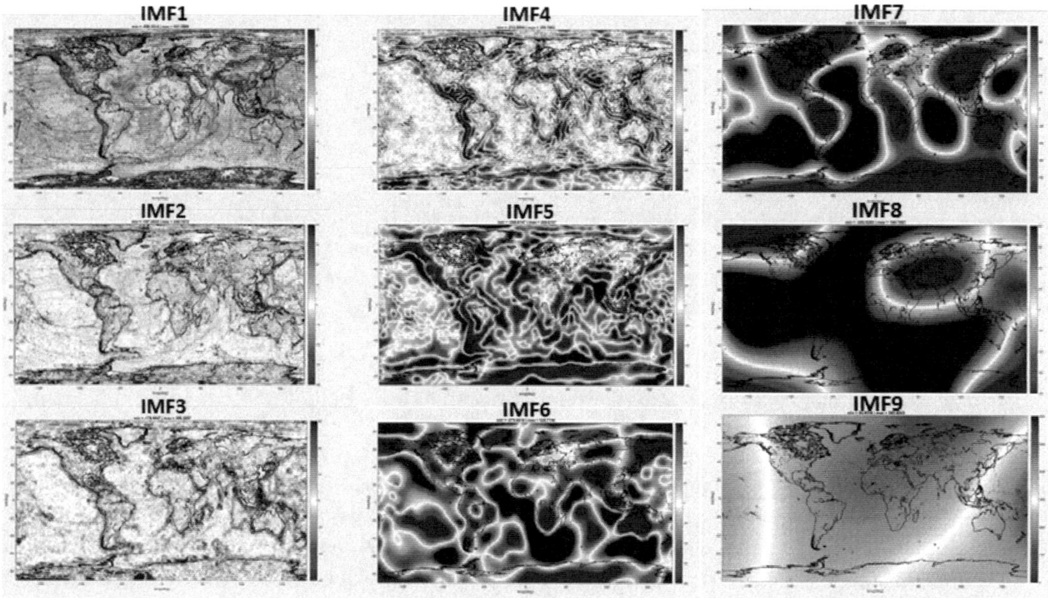

Figure 2. Bidimensional IMFs (BIMFs) expose the relative local gravity anomaly in the order of detail.The higher to lower order of detail anomalies is shown as BIMF1 to BIMF9 successively. Relative residual anomalies appear in BIMF1 to BIMF3, whereas the relative regional anomalies appear in BIMF4 to BIMF9. BIMF9 could also be considered as an absolute regional of the Earth gravity field.

The first three BIMFs contain the majority of body anomaly yet less dominant in the original data (CSBA). They reflect a great detail of gravity anomalies around the earth. Hydrocarbon and mineral exploration are strongly related to these scale of anomaly, espescially BIMF1. The next BIMFs exhibit less and less detail anomaly. BIMF4 to BIMF5 might be suited for basin and tectonic analysis. They have wider wavelength and the amplitude is relatively increasing. The longer-wavelength anomalies, related to deeper and larger sources, typically have greater amplitudes than the local anomalies of interest thus distort the target anomalies [18]. This explains why it is so hard to notice very small anomalous magnitude in the order of less than 50 mGal in the original data which has amplitude range from -600 to +600 mGal. BEMD analogous to microscope instrument in physical optics. BIMF1 to BIMF3 might be considered as the type of residual anomaly. These residuals often mixed together into a single map so we never have noticed them in such detail and fully understand what the physical and the geological meaning of them directly without further interpretation. The next BIMFs would be the regional anomalies for them all. The last BIMF is global trend of entire earth gravity field in the shape of monotonic signal. It is the most regional of all regional anomalies. Looking at the amplitude and the size of the anomaly, it is truly the most dominant contributor of observable Earth gravity field magnitude. It is tough to explain what physical earth yields such anomaly however all geophysicist believe that what we measure on the surface must have some relation to that of the sub-surface.

The edge of body anomaly is can also be determined sharply as it falls on zero crossings of BIMF. Therefore, one can make geometrical and structural interpretation using these results. BIMF1, for instance, can be used for near-surface structure mapping espescially at the location where surface covered by alluvial deposits at land or deep water at sea, on such condition where we can not get direct observation either on the ground or up from the space. Also, these set of anomalies particularly give insight about gravity field stratification from surface to the deeper part of the Earth. This anomaly stratification represents structure, dimension and mass body distribution from the surface to the mantle or even deeper.

The ambiguity of gravity interpretation can be supressed by using auxiliary geophysical and geological data [18]. Although non-uniqueness is an overriding concern in all gravity interpretation, the availibility of appropriate constrains surely can limit these ambiguities [2]. BEMD reduces some amounts of ambiguity inherited in gravity method by separating oscillation redundancy at local points, giving monocomponent which equal to single body anomaly respon of lateral density contrast. The relevant and appropriate constrains expressed by BIMFs could be infered by integrating them with geological and other geophysical constrains. We would like to consider these results as a pseudo-inverted body anomaly by means that we can delineate source bodies and get the density contrast place to place relatively using the results. The relative depth and geometry of the body anomaly is also can be infered and it ~~is~~ might be more useful to make it as an initial model for estimating anomalous density by gravity inversion techniques.

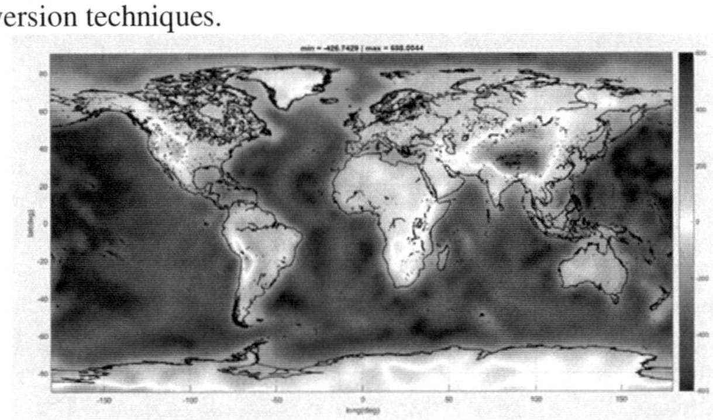

Figure 3. Crust-Regional Anomaly (CRA) extracts major informations that are easily observable e.g. lithospheric structures and plate boundaries. CRA is generated from BIMF5-BIMF9.

BEMD isolates and enhances gravity anomaly for helping the interpretation of the data. BEMD facilitates other method to isolate gravity anomalies by separating the gravity effects of known or hypothesized geological sources. From the results, for instance, it can be identified some anomalous bodies which have very wide wavelength but still associated with noticeable structure from surface (BIMF5-BIMF7). Those BIMFs reconstruction together with BIMF8 and BIMF9 resemble major Earth structures such as oceanic – continental crust boundaries, thicker regions of continental crust, some varieties in oceanic crust density, main lithospheric and mantle structures, and divergent plate boundary regions. In short, the resultant of BIMF5-BIMF9 extracts major informations that are observable from the original data. These groups of anomalies are suitable for deep and major Earth's structure interpretations. As we just notice, the term of regional anomaly is not always about the related depth of source body but it might be in the size of some form of density basins or containers where smaller bodies assemble within. It must be recognized that some anomalies that are broad and with relatively low curvature values may also be related to shallow geological features that are themselves broad and without abrupt changes of slope [19]. Generally we know two obvious density containers; continental and oceanic crust. We notify them as Crust Regional Anomaly (Figure 3).

Smaller bodies jostle inside the CRA. BIMF5-BIMF9 from which CRA was generated are analogue to the reference of any body smaller than them as in BIMF1-BIMF4. Smaller anomalies we usually consider as residual certainly have smaller body, restricted in dimension at some locals. We typically use this kind of anomalies in much of our practical works. The residuals have smaller ambiguity than those of regional anomalies. BIMF1-4 show most of the residual anomalies around the globe. IMF1 is the most detail and the nearest to the surface. Next come BIMF2, 3 and 4 bringing less detail compared to BIMF1. This kind of body anomalies collections surely can help us in interpreting and modeling earth subsurface in different scales. CRA can effectively be used as practical tool to quickly extract *surface-residual anomaly* (SRA) from any CBA data, providing higher resolution data, only by using formulae : SRA = CBA - CRA. Surface-residual anomaly (SRA) stands for CBA anomalies separated from CRA. After such separation between CRA and SRA, one can do simple and robust BEMD technique to the SRA so BIMF1-4 can be obtained.

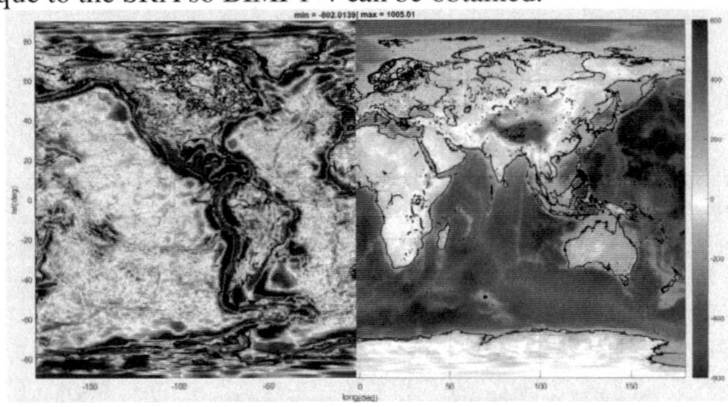

Figure 4. *Surface-Residual Anomaly* SRA (left) compared to CBA (right). SRA = CBA – CRA.

Dealing with data in the size of WGM is clearly not easy. Meanwhile, reducing the size of WGM data to get smaller data to process is unbearable as it will fail to give complete decomposition results. BIMF will come up with fewer amount. For the size of WGM data, it could take several weeks and great amount of memories to decompose the data completely, though it also depends on utilized computational framework. To make it efficient, we keep the CRA to all new measurement data so the further processing and modeling stage can be performed to give good interpretation of new data. Demonstrating the result of removing CRA from the new higher resolution data is given in Figure 4. Simple matrices subtraction could reveal the surface-residual anomaly. It gives the detail insight to our data with ease.

SRA has the preserving amplitude without any distortion, implying that the SRA can savely be used as modeling input. True residual of our target anomaly can be extracted from SRA using BEMD or other anomaly separation methods. SRA is not a type of residual anomaly and CRA not the regional one according to common terminology. Based on our experience, residual resulted from exploration scaled measurements using spectral analysis is comparable to the BIMF1, while the rest is considered as regional. Rarely we found the BIMF2 or above representable enough to our residual results. Nevertheless other author using polynomial separation method found residual as someway similar to the resultant of BIMF1 to BIMF3 [4]. It does not matter, the most important thing for interpreting the data is we have to be sure what we keep inside our data and which part we put away. That way, we can build and interpret better geological models. The comparison between SRA and BIMF1 (the most shallow and local residuals) is given in Figure 5.

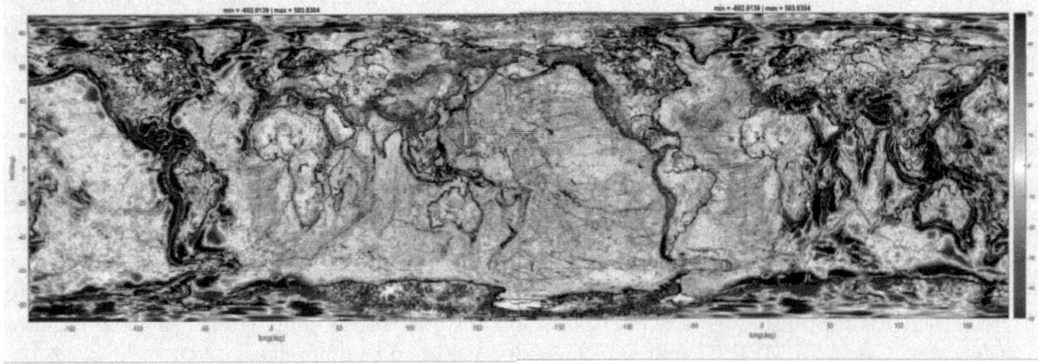

Figure 5. Comparison between SRA (left and right edges) with BIMF1 (centered)

5. Conclusions

BEMD resolve all the anomalies contained in WGM data in a few number of BIMFs with excellent resolution in separating one anomaly from others. BEMD reveals those physical structure and groups them together so we can make hypotheses of the geological bodies from only gravity anomaly. One of those earth physical structure is *crust-regional anomaly* (CRA). CRA due to the gravity anomaly drifts is caused by continental and oceanic masses. Removing CRA from any measurement data will reveal *surface-residual anomaly* (SRA). Therefore, the CRA can be used as reference to inspect shallow body anomaly from any new measurement data. Each BIMF or a resultant of a few selected of their combinations provides us a very good understanding to hypothesize geological structure and mass bodies distribution. They are better input for qualitative interpretation, quantitative analysis, inversion and modeling. That is because we have dealt with a great amount of ambiguity in gravity method caused by anomaly mixing which hides true physical structure of geological matters.

6. Acknowledgement

We would like to acknowledge the generosity of Institut Teknologi Sumatera (ITERA) in providing partial-funding for this research.

References

[1] Selley R C, Cocks L, Robin M, Plimer I R 2005 *Elsevier Academic Press*
[2] Hinze W J, von Frese R R Rb, Saad A H 2012 *United Kingdom: Cambridge University Press*
[3] Gupta H K 2011 *Springer*
[4] Rahim A M Al- 2015 *Arab J. Geosci.* DOI. 10.1007/s12517-015-2118-7
[5] Firdaus R, Abdurrahman D, Ekawati G M, Wahyudi E J 2012 *Proceeding PIT HAGI*
[6] Pei Yao, Wu Yan-Gang, Jia Da-Cheng 2012 *Res. J. of App. Sci., Eng. and Tech.* **4(21)** 4227-4236
[7] Bonvalot S, Balmino G, Briais A, Kuhn M, Peyrefitte A, Vales N 2012 *CGMW-bgi-cnes-ird Ed., Paris.*
[8] Huang N E, Shen Z, Long S R, Wu M C, Shih H H, Zheng Q, Yen N-C, Tung CC, Liu H H 1998 *Proc. R. Soc. Lond.* **A 454**:903-993
[9] Battista B M, Knapp C, McGee T, and Goebel V 2007 *Geophysics* **72** : H29-H37
[10] Han J and Van der Baan M 2013 *Geophysics* **78**: O9-O19
[11] Huang J W and Milkereit B 2009 *CSPG CSEG CWLS Convention* : 205-210
[12] Bekara M and Van der Baan M 2009 *Geophysics* **74** : V89-V98
[13] Linderhed A, 2002 *Proceedings of SPIE* **4738** 1-8
[14] Nunes J C, and Delechelle E 2003 *Image Analysis* 295-296
[15] Hasan H H, Pierce J W 2008 *CSEG RECORDER, Focus Article* **25** 26-30
[16] Huang J N, Zhao B B, Chen Y Q, Zhao P D 2010 *Comput. Geosci.* **36(7)** 987-995
[17] Morris C S, Tianyou L, Innocent Ndoh Mbue 2010 *J. Am. Sci.* **6(7)** 183-187
[18] Skeels D C 1947 *Geophysics* **12** 43-56
[19] Skeels D C 1967 *Geophysics* **32** 872-876

The Utility of Free Software for Gravity and Magnetic Advanced Data Processing

Hendra Grandis* **and Darharta Dahrin**

Applied and Exploration Geophysics Research Group, Faculty of Mining and Petroleum Engineering, Institut Teknologi Bandung, Jalan Ganesha 10 Bandung 40132, Indonesia

*corresponding author: grandis@itb.ac.id

Abstract. The lack of computational tools, i.e. software, often hinders the proper teaching and application of geophysical data processing in academic institutions in Indonesia. Although there are academic licensing options for commercial software, such options are still way beyond the financial capability of some academic institutions. Academic community members (both lecturers and students) are supposed to be creative and resourceful to overcome such situation. Therefore, capability for writing computer programs or codes is a necessity. However, there are also many computer programs and even software that are freely available on the internet. Generally, the utility of the freely distributed software is limited for demonstration only or for visualizing and exchanging data. The paper discusses the utility of Geosoft's Oasis Montaj Viewer along with USGS GX programs that are available for free. Useful gravity and magnetic advanced data processing (i.e. gradient calculation, spectral analysis etc.) can be performed "correctly" without any approximation that sometimes leads to dubious results and interpretation.

1. Introduction

Advances in potential field (gravity and magnetic) data acquisition have led to a vast amount of data with high resolution that need to be processed efficiently. Both gravity and magnetic methods share common basic concepts such that the data processing techniques, especially the advanced data processing for interpretation, can be applied for both methods. Advanced data processing is commonly applied to grid data, among others, to enhance anomalies present in the data. The filtering (enhancement and suppression) of anomalies relies on the spatial frequency content of the data. Therefore, the filtering technique is mainly based on the spectral analysis of the data and the process is performed in the wave-number or spatial frequency domain. However, a combination of processing in the space and wave-number domain is sometimes necessary to obtain better results [1,2].

The gravity advanced data processing and interpretation necessitates heavily on the availability of computational tools, more specifically computer program and software. Lack of computational tools and also awareness of the basic concepts of the methods may lead to doubtful practice (i.e. use of approximate techniques or incorrect methods) and results [3]. The paper aims at introducing the use of Geosoft's Oasis Montaj Viewer among many freely available tools to perform advanced gravity processing in both space and wave-number domain. Both synthetic and field data were used to illustrate the utility of the software. The paper is not intended to support or promote only one particular software. It is only a support for our need in addition to our laboratory developed codes and programs for teaching and research activities. Other alternatives are equally recommended.

2. OASIS Montaj Viewer and USGS GX Programs

The Oasis Montaj Viewer is a free software from Geosoft Inc. that allows to view and to share data in grid (.grd) and database (.gdb) file formats. The viewer can also convert grids and export maps as images in various supported formats. In addition, basic plotting profile of the database can also be performed [4]. In this paper, we present the most basic functions of the software in accordance with the need of advanced gravity data processing, i.e. contour shading, filtering etc. for teaching and instructional purposes.

2.1. Installation and Basic Menu

The Oasis Montaj Viewer is available for download from Geosoft's website, the latest being the version 8.5 (www.geosoft.com). With almost similar functionalities, we use the older version 6.4 for the sake of compatibility with our licensed version. Complementary modules called Geosoft's executables or GX's are provided by USGS (pubs.usgs.gov/of/2007/1355/). The exact sites for download might be changed, however, they can be easily searched from the internet.

After installation of both the Oasis Montaj Viewer and USGS GX's programs, it is necessary to move or to copy the resulting files with the usgs prefix (usgs*.*) into the same directories under the Oasis Montaj Viewer directory. The directories are designated as /bin, /gx, /omn and /ger, for binaries, GX's, additional menus and error messages respectively. When the Oasis Montaj Viewer is launched, we can create or open a project file. Then, only basic menus are shown. In the **GX** menu, the **Load Menu** is used to load an additional menu, e.g. to load USGS's programs menu by selecting the file usgs.omn in the dialog box. Once the **USGSV** is loaded (V for viewer, otherwise **USGS** is used for licensed Oasis Montaj) calculations using USGS's GX programs can be performed (see Figure 1). In principle, the installation process is straightforward and the use of the software is clearly explained in the manual [5].

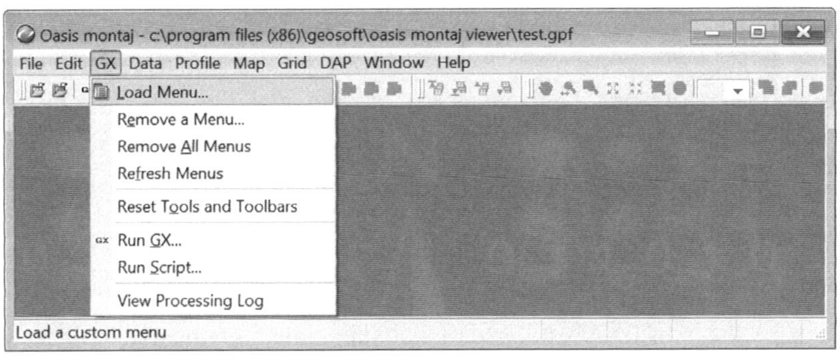

Figure 1. The Oasis Montaj Viewer's window with **GX** menu selected to load a custom menu, in this case **USGSV**, with V for viewer.

2.2. Working with Grids and Maps

Only data in the form of grid can be processed by using the Oasis Montaj Viewer. Therefore, we must convert our gravity (or magnetic) data usually distributed arbitrarily in the survey area into a uniform grid by a gridding process. The database (.gdb) and also the grid (.grd) files that can be loaded into the software are proprietary format of Geosoft and can only be created by using the licensed version of the Oasis Montaj. However, a grid file created with the Golden Software's Surfer can also be used as the input of the software, i.e. by specifying the Surfer's grid format in the input dialog.

The grid can be directly displayed from the **`Grid`** menu and **`Display Grid`** sub-menu as shown in Figure 2a. The grid can be first converted to the Geosoft's format by selecting **`Copy/Convert grid`** sub-menu (see Figure 2a) and then processed using available menu in **`USGSV`**. The grid conversion is usually preferred to avoid specifying the grid format each time we input the data. Once the processes are performed and several grid files are generated, then the grids can be displayed and exported to images with standard formats, i.e. jpg, png, emf etc. (Figure 2b). These images can be imported again into Surfer or other software for map annotations, since the free or viewer version of the Oasis Montaj does not allow for annotated map creation.

(a)

(b)

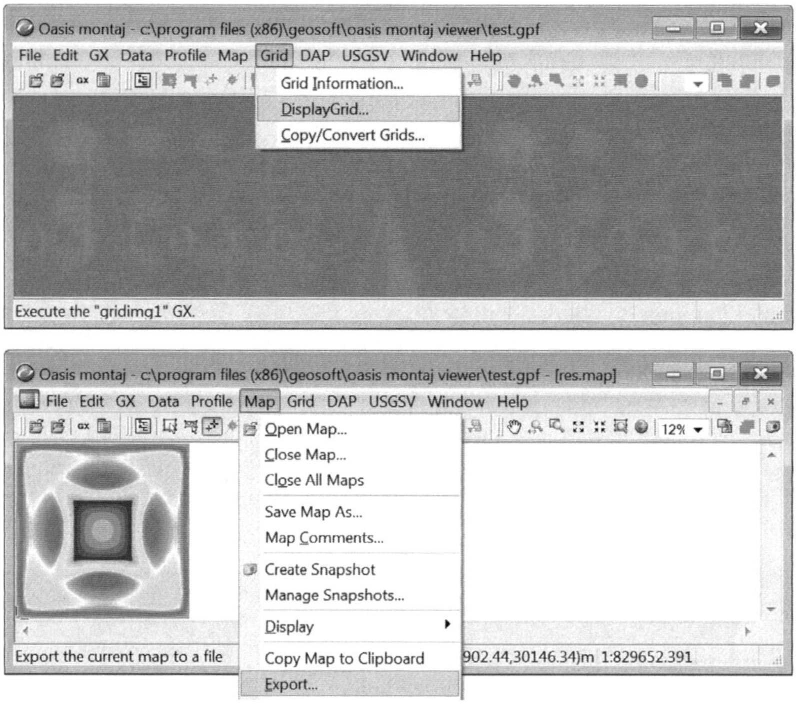

Figure 2. (a) **`Grid`** menu with sub-menu **`Display Grid`** to input and display a grid file and sub-menu **`Copy/Convert Grids`** for grid format conversion, (b). **`Map`** menu with sub-menu **`Export`** for converting the displayed grid as image.

3. Examples with Synthetic Data

We illustrate the utility of the software by calculating gravity gradients or derivatives of the gravity anomaly (g_z) of a 3D block model representing a basin with vertical slopes and different depths (2 km and 4 km) and with a density contrast of -0.5 gr/cm^3. We used the academic licensed software provided by UBC-GIF [6] for the 3D gravity forward modeling, while another alternative is to use the FORTRAN code from Blakely [7].

The total horizontal derivatives or Horizontal Gradient Magnitude (HGM) is commonly used to enhance the anomalous source's boundaries [1],

$$HGM = \left(\left(\frac{\partial g_z}{\partial x} \right)^2 + \left(\frac{\partial g_z}{\partial y} \right)^2 \right)^{1/2} \qquad (1)$$

The HGM can be obtained directly by using the **`Grid Utilities`** sub-menu of **`USGSV`** menu and select **`Total Gradient from derivatives`** (see Figure 3a). However, the horizontal gradient can be calculated component by component ($\partial g_z/\partial x$ and $\partial g_z/\partial y$) by selecting the sub-menu **`Grid Spatial Filtering`** and **`Horizontal Gradients`** as shown in Figure 3b. By using appropriate **`Algebraic operations`** in **`Grid Utilities`** sub-menu, then the HGM can be obtained.

(a)

(b)

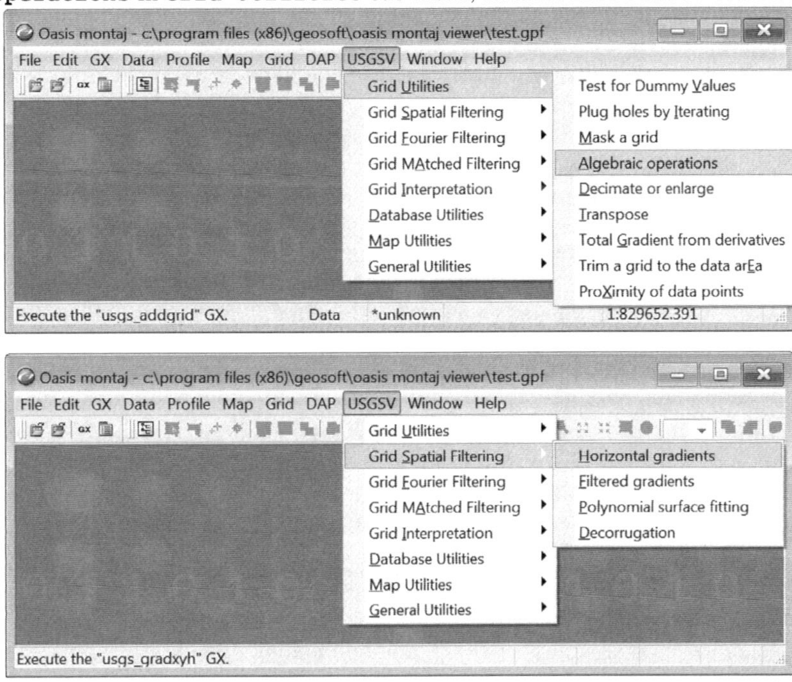

Figure 3. (a) **`USGSV`** menu is loaded with **`Grid Utilities`** and **`Algebraic operations`** are selected, (b). **`Grid Spatial Filtering`** and **`Horizontal gradients`** sub-menu for component by component calculation of the horizontal gradients.

The Singular Value Decomposition (SVD) in the space domain is expressed by [1,7],

$$\frac{\partial^2 g_z}{\partial z^2} = -\left(\frac{\partial^2 g_z}{\partial x^2} + \frac{\partial^2 g_z}{\partial y^2}\right) \tag{2}$$

By performing similar procedures as before (horizontal gradient calculation component by component twice), we obtain second horizontal derivatives, i.e. $\partial g_z^2/\partial x^2$ and $\partial g_z^2/\partial y^2$. The SVD can be obtained by **`Algebraic Operation`** of grids in the **`Grid Utilities`** sub-menu of **`USGSV`** (Figure 3a). As an alternative, the SVD can also be calculated in the wave-number domain based on the following Fourier transform pair [7],

$$\frac{\partial^2 g_z}{\partial z^2} \Leftrightarrow |\mathbf{k}|^2 G_z \quad \text{with} \quad |\mathbf{k}|^2 = k_x^2 + k_y^2 \tag{3}$$

where G_z is the Fourier transform of FFT of g_z, while k_x and k_y are wave-number in x- and y-direction respectively. It is obvious from the squared $|\mathbf{k}|$ that the SVD tends to enhance high frequency content of the data, including the noise. The First Vertical Derivative (FVD) can also be used to delineate the anomalous source's boundaries while limiting the high frequency amplification. The FVD is calculated in the frequency domain by using the similar Fourier transform pair to equation (3) with $n =$

1. Furthermore, the equation (3) can be generalized to calculate any integer or fractional (n) derivatives of the gravity anomaly [7], i.e.

$$\frac{\partial^n g_z}{\partial z^n} \Leftrightarrow |\mathbf{k}|^n G_z \quad \text{with} \quad |\mathbf{k}| = \sqrt{k_x^2 + k_y^2} \tag{4}$$

The **`Forward Fourier transform`** of the gravity anomaly, the **`Basic Fourier domain filtering`** and the **`Inverse Fourier transform`** sub-menus to perform the calculation of FVD and SVD in the wave-number domain is shown in Figure 4.

The results of the basic gradient calculation or operation on grids are presented in Figure 5. It is obvious that the shallower basin in the western (left) part of the area is associated with the smaller negative gravity anomaly and *vice-versa* (Figure 5a).This difference in amplitude is represented by a relatively distinguishable different values of the HGM (Figure 5b) and FVD (Figure 5c), while in the SVD (Figure 5d) the difference is less obvious. The latter may be resulted from the very small magnitude of the SVD. Therefore, it can be very dubious to draw any conclusion or interpretation, e.g. types of fault based only on the SVD value [3].

Figure 6 shows West-East profiles crossing the center of the anomalous zone for the gravity anomaly, HGM, FVD and SVD. The anomaly (associated with a basin formed by two adjacent blocks) and its gradients interfere such that the profiles are not so simple anymore, compared to those associated with a simple block model [3]. The situation can be encountered with real or field gravity data where anomalies are superposition of many smaller scale anomalies leading to difficulties in interpretation based on gradient or derivative values both on maps or profiles. Qualitative interpretation of gravity gradients should be performed cautiously and limited only for simple models.

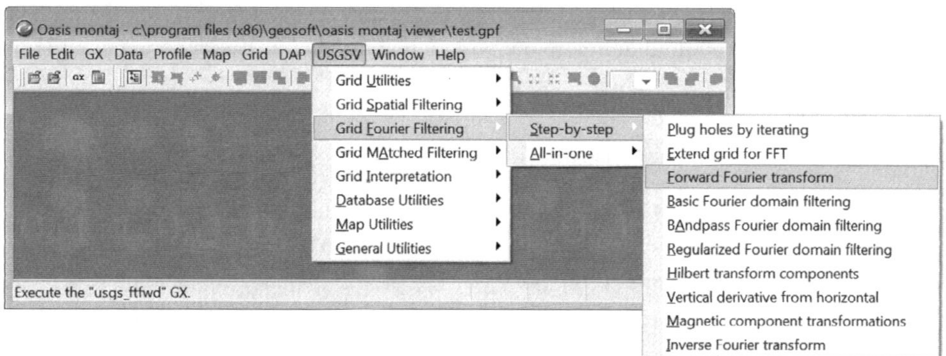

Figure 4. Menu and sub-menu for **`Grid Fourier filtering`**, e.g. FVD, SVD computations in the wave-number or Fourier domain.

4. Concluding Remarks

The paper demonstrates the utility of a freely available software, namely the Geosoft's Oasis Montaj Viewer complemented with the USGS GX's modules for advanced gravity data processing. Although the examples presented in this paper involve only gravity data, the software can also applicable for magnetic data with several typical applications relevant to magnetic data, e.g. Reduction to the Pole (RTP), Reduction to the Equator (RTE), etc. The use of the software may avoid approximate calculation of gradients, especially SVD, using only values on several profiles and consider the results valid to make gradient maps which is certainly dubious, as has been presented and discussed in [3].

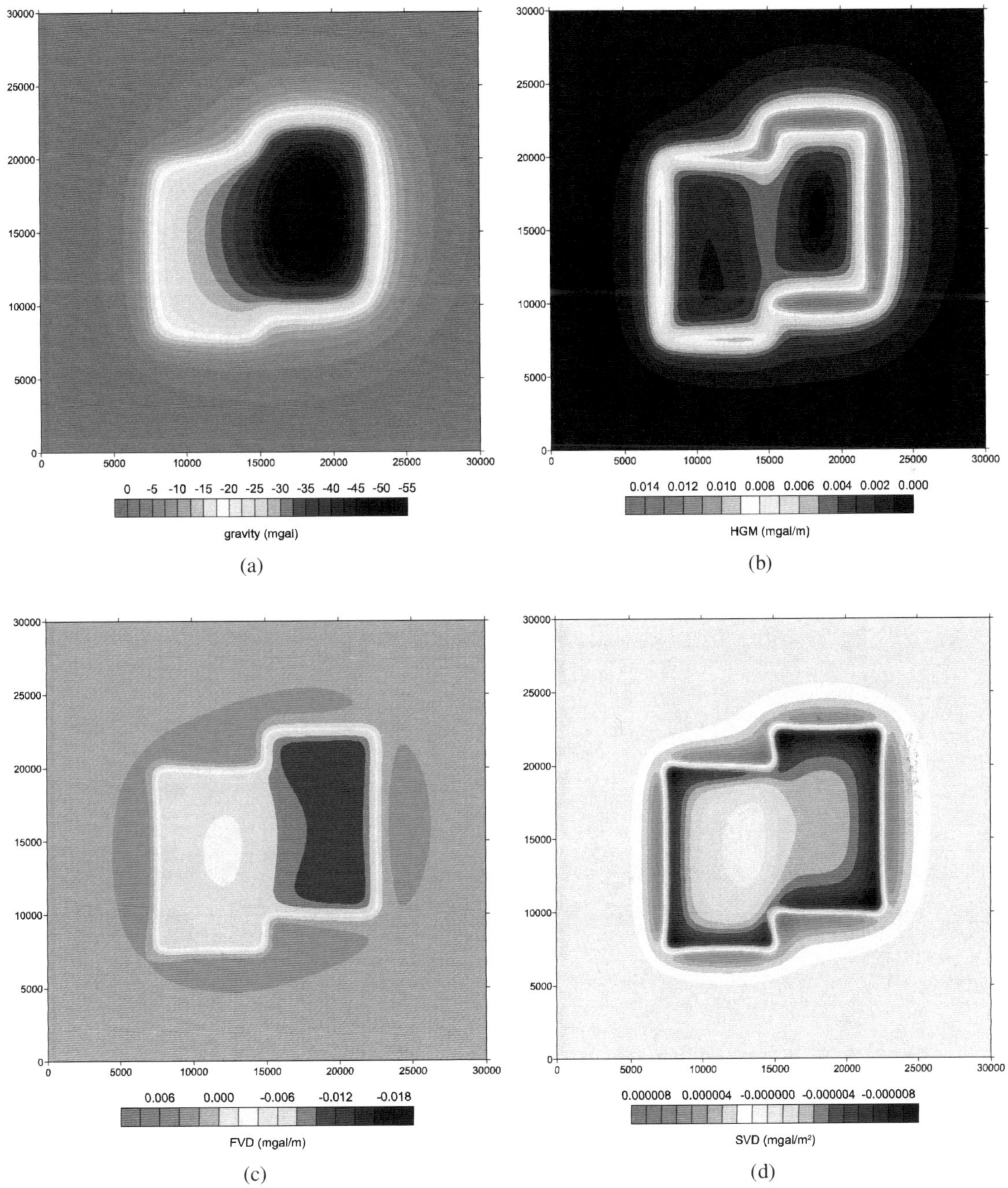

Figure 5. (a) Gravity anomaly of a block model representing a basin with vertical edges, different depth (2 an 4 km) and a density contrast of -0.5 gr/cm^3, (b) Horizontal Gradient Magnitude (HGM), (c) First Vertical Derivative (FVD), (d) Second Vertical Derivative (SVD).

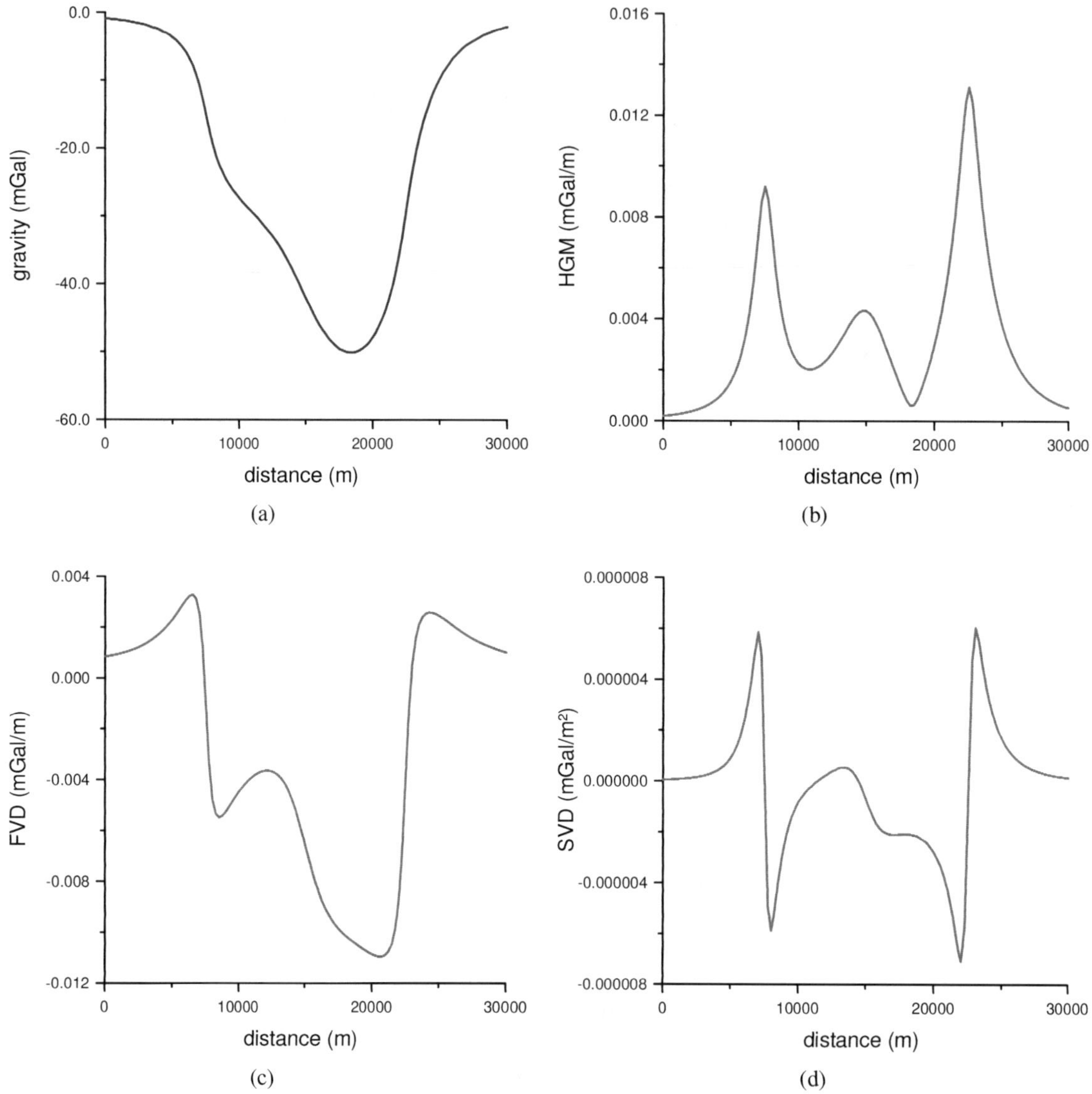

Figure 6. West-East profiles crossing the center of the anomalous zone of Figure 5. (a) Gravity anomaly, (b) Horizontal Gradient Magnitude (HGM), (c) First Vertical Derivative (FVD), (d) Second Vertical Derivative (SVD).

The utility of the software for gravity and magnetic data processing in the wave-number domain is yet to be explored further. There are common practices found in some works by Indonesian practitioners and academics that the spectral analysis is performed only along profiles, i.e. with a 1D FFT. The results in the form of frequency cut-off (for filtering) are averaged from several profiles arbitrarily distributed on an area. Then, the filtering is done by applying a 2D moving average technique to the data on map, using parameters from the spectral analysis transformed into window length or grid size. Although we cannot pinpoint particular references and proof that the results are erroneous, such process does not conform with the fact that the data are 2D spatially distributed. The widely accepted practice is to perform a 2D spectral analysis leading to the radially averaged spectrum based on which the filtering process is performed in the wave-number domain [8,9]. It is expected that the freely available software combined with additional tools can be used to avoid such dubious practice.

The paper is not intended to endorse only one software. There are numerous software and computer programs freely available for academic and research purposes. The academic community (lecturer and students) can use them for their experiment, teaching and even studies on real data. The possibility is almost limitless. It is the creativity of lecturers and students in using available resources that may advance geophysical applications in Indonesia, starting from education and academic institutions. Finally, there can be differences in the interpretation of the same geophysical data, but they have to be based on the correct and reliable data processing.

References

[1] LaFehr T R and Nabighian M N 2012 Fundamentals of Gravity Exploration *Society of Exploration Geophysicists (SEG)*
[2] Fairhead J D 2015 Advances in Gravity and Magnetic Processing and Interpretation *European Association of Geoscientists and Engineers (EAGE) Publications*
[3] Grandis H and Dahrin D 2016 The use and mis-suse of the Second Vertical Derivative (SVD) in gravity data with the emphasis of Indonesian cases (in prepraration)
[4] www.geosoft.com/support/downloads/viewers/oasis-montaj-viewer
[5] www.geo.mtu.edu/~jdiehl/Homework4550/ViewerTutorial.pdf
[6] Li Y and Oldenburg D W 1998 *Geophysics* **63** 109-119
[7] Blakely R J 1996 Potential Theory in Gravity and Magnetic Applications *Cambridge University Press*
[8] Likkason O K 2011 Spectral Analysis of Geophysical Data, in Chen, D.M. (ed.), Advances in Data, Methods, Models and Their Applications in Geoscience *Intech*
[9] Buttkus B 2000 Spectral Analysis and Filter Theory in Applied Geophysics *Springer*

Time-lapse microgravity data acquisition in baseline stage of CO_2 injection Gundih pilot project

Eko Januari Wahyudi[1], R Marthen[1], Y Fukuda[2], and Y Nurali[3]

[1]Geophysical Engineering, Institut Teknologi Bandung, Indonesia
[2]Kyoto University, Japan
[3]PERTAMINA EP, Indonesia

Abstract. Indonesia is conducting research on CO2 sequestration and monitoring in Gundih area. Geophysical methods are applied in the study area as monitoring technologies of CO2 storage in the subsurface. In this paper, we will describe data quality and baseline geophysical survey (2014 and 2016) as part of time-lapse microgravity (TLM) data acquisition. The same set of two relative gravimeters (Scintrex CG5) were used in 2014 and 2016 gravity data acquisition to provide TLM map. TLM can be useful to monitor the mass increase and decrease due to significant activities in the reservoir, but for this project TLM signal (caused by the CO2 injection in Gundih's reservoir) estimated very small. Considering current project status, no injection yet, the data acquisition in 2014 and 2016 will be analysed to help us understand non-target anomalies in near surface. Preliminary analysis of gravity differences between 2014 and 2016 shows correlation pattern of time-lapse microgravity with land use in the study area. Rice fields and villages in the Western part of the study area correlate with negative TLM anomalies and forest area in the Eastern part of the study area correlate with positive TLM anomalies.

1. Introduction

Indonesia is conducting research on CO_2 sequestration and monitoring near Gundih area. Carbon Capture Storage (CCS) Gundih pilot project started in 2012. CO_2 from Gundih area will be transported to the Jepon well close to the Blora regency in Central Java province, Indonesia. Jepon well as the candidate of injection well area already evaluated and the report completed in early 2015. CCS Gundih study area shown in figure 1.

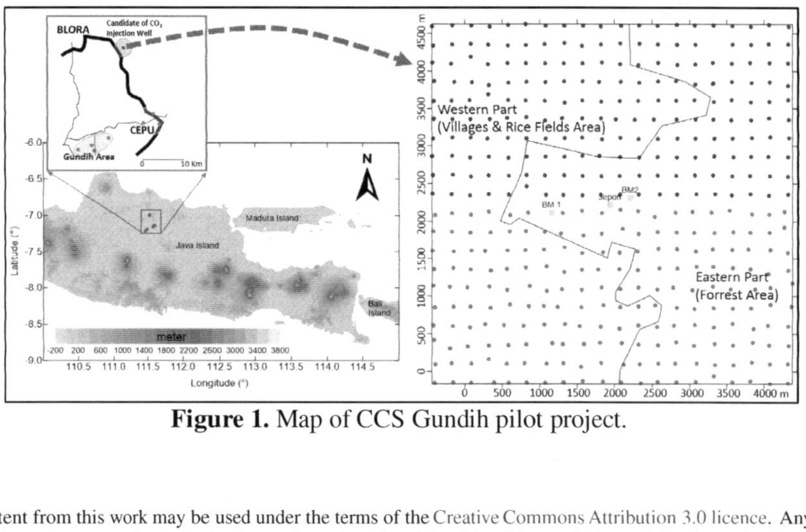

Figure 1. Map of CCS Gundih pilot project.

As we can see from project status illustration (figure 2), before injection we conduct baseline survey to monitor the future subsurface condition. Geophysical methods (seismic, TDEM, and gravity) are applied in the study area as monitoring technologies of CO_2 storage in the subsurface. In this project, we plan to carry out super hybrid gravity measurement using continuous gravimeter, absolute gravimeter, and relative gravimeter. Relative gravimeter used to observe gravity changes in grid stations that cover about 5x5 km^2 injection area. In this paper, we will report quality of relative gravimeter data from baseline survey (2014 and 2016).

Figure 2. CCS Gundih pilot project status since 2012 with illustration of surface condition changes in Jepon well.

Considering current project status, no injection yet, the data acquisition in 2014 and 2016 will be analysed to help us understand non-target anomalies in near surface. Land use in the surface is simplified into two types: villages with rice fields in the Western part of study area and forest in the Eastern part of study area. During time-lapse microgravity (TLM) data acquisition in Jepon well, it is possible we deal with different surface condition between two data acquisition. Observation data for each station in the surface need to consider the effect of near-surface that involving density changes. The surface condition of rice fields area can be very different due to farming activity (seeding, irrigation, cultivation, harvest, etc.), while the surface condition of tropical forest area can be very different due to weather condition that influence soil moisture and plantation growth.

There are only two stations that observe in benchmark and one station close to the well-pad. The other station location need to be revisited again to make sure we comeback in the same location. Most of the stations location was only marked by spray paint, hand-held GPS coordinate, photo documentation, and natural/artificial feature. During TLM data acquisition, it is possible we lost the stations after several months periods. Twice baseline survey of TLM in this study area can be useful to analyze the performance of land data acquisition in the study area.

2. Gravity Data Acquisition

In this project we adapt repeat gravity data acquisition methodology that detail explanation ca be found in several publications such as: [1], [2] and [3]. Grid survey that cover Jepon well area are referenced to a stable reference in some stations located in Cepu (at least 25 km). Regional stations are chosen far away from the injection well site in order to avoid gravity changes caused by future activity (well and surface facilities). Regional stations already observed using A-10 absolute gravimeter in 2013 and 2014. Our regional gravity data provide stable value within < 2 μGal different from A-10 absolute gravimeter observation.

Scintrex CG-5 options during 2014 and 2016 surveys were set as an automatic correction for auto-rejection, seismic filter, and tide correction. To compute tide correction, the coordinate were set with (111.6°E, 7.1°S), zone 49, and GMT difference -7. The other three Scintrex CG-5 options (raw data, terrain correction, and continuous tilt correction) was not set as auto calculation.

Gravity data are acquired using two Scintrex CG-5 at each station. Data are acquired in 30 seconds samples for 3-5 minutes, so we can get at least three relative gravity reading for each observation. For each observation, the operator will notice instrument level and gravity reading to maintain the quality of data reading. Each reading of gravity value is controled within ± 10 arcsec range in digital instrument level (tilt-X and tilt-Y). Stable station typically observed with the small value of error and standard deviation for each reading. For each observation, we decide to select gravity reading within ten μGal value range. Beside operator control, we also able to consider auto rejection for each gravity reading during on-site data processing. Auto-rejection with high percentage value provides a suggestion or second opinion that gravity reading is not good enough. Considering field condition, statistically our observation get < 3% bad leveling and > 80% gravity readings have small auto-rejection value (figure 3).

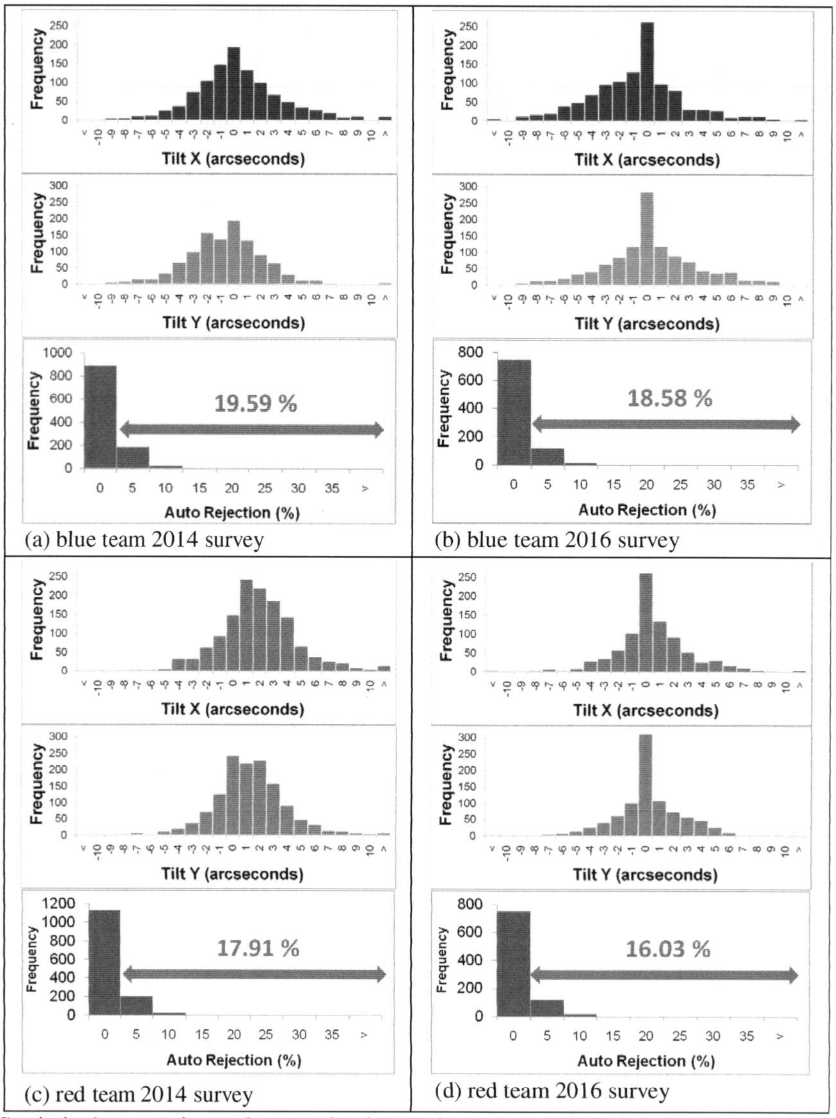

(a) blue team 2014 survey
(b) blue team 2016 survey
(c) red team 2014 survey
(d) red team 2016 survey

Figure 3. Statistical comparison of auto-rejection and gravimeter leveling: (a) blue team 2014 survey, (b) blue team 2016 survey, (c) red team 2014 survey, and (d) red team 2016 survey.

Comparison data acquisition in 2014 and 2016 shows in table 1. From first and second data acquisition we can get TLM within a period of 20 months (less than two years). First data acquisition conducted between 18-29 September 2014 and second data acquisition 18-28 May 2016. Based on

IOP Conf. Series: Earth and Environmental Science **62** (2017) 012047 doi:10.1088/1755-1315/62/1/012047

table 1, vertical gradient only conducted in first data acquisition, while in the second data acquisition we see effectivity improvement during data acquisition. We can increase observation and repeat sample during 10 days survey in 2016.

Grid stations that cover Jepon well area designed with the average of 250 meters interval. Day-by-day gravity data observation can be seen in figure 4. We have two teams to observe North and South grid. North grid stations were observed by the blue team using Scintrex CG-5 SN822, while South grid stations were observed by the red team using Scintrex CG-5 SN1092. Blue team and read team for each close loop data acquisition also observe one repeat station or in benchmark (BM-1 or BM-2). Repeat station observations will be use to provide repeatability during field data acquisition.

Table 1. Comparison grid gravity data survey in 2014 and 2016.

Survey	**2014**		**2016**	
Team	Blue	Red	Blue	Red
Date	19-29 Sep	18-29 Sep	18-28 May	18-28 May
Gravimeter	CG-5 822	CG-5 1092	CG-5 822	CG-5 1092
Tot. Time Duration	37560 min.	43020 min.	34380 min.	27090 min.
Good Leveling	98.02%	98.26%	97.92%	97.34%
Auto Rej. > 5%	19.59%	17.91%	18.58%	16.03%
Average daily drift	42 μGal	46 μGal	25 μGal	24 μGal
#Day(s)	10	11	10	10
#Reading(s)	1113	1379	903	1155
#Observation(s)	243	239	247	250
#Station(s)	206	202	203	202
#Repeat Obsv.(s)	37	37	44	48
#VG Station(s)	3	2	0	0

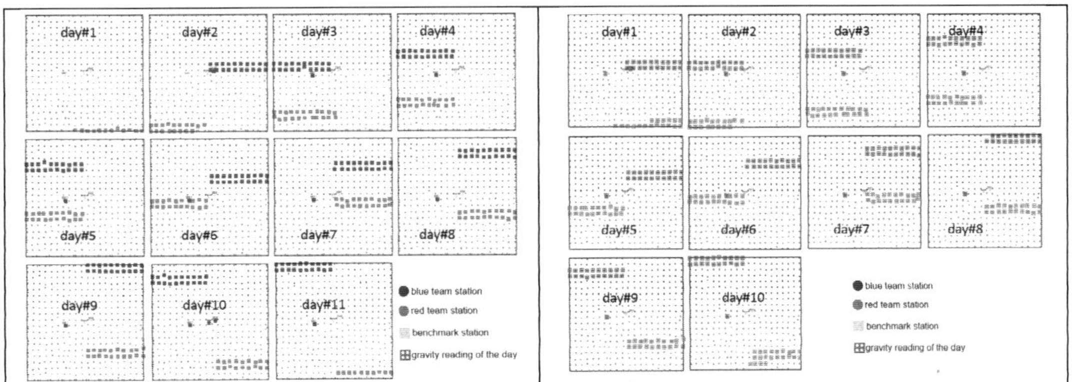

(a) 2014 survey (b) 2016 survey

Figure 4. Comparison of day-by-day gravity data observation: (a) 2014 survey and (b) 2016 survey.

3. On-site Data Processing

Before we conducted the second survey in 2016, we evaluate data quality from the first survey (2014). At the end of first survey we see higher drift correction and then we try to optimize the instrument performance before we go to the field in 2016. We conducted simple simulation (small loop data acquisition) in ITB campus, to set the best drift constant. We change Scintrex CG-5 setup parameter for drift constant as follow:

1. Blue team (CG-5 SN 822) → from 0.353 changed into 0.511
2. Red team (CG-5 SN 1092) → from 2.550 changed into 1.900

IOP Conf. Series: Earth and Environmental Science **62** (2017) 012047 doi:10.1088/1755-1315/62/1/012047

Figure 5 shows long-term drift comparison from the first and second survey. Gravimeters performance in 2014 and 2016 survey respectively estimated with linear drift approximation give rates in the range 313-350 μGal/day and 82-161 μGal/day. As we can see in figure 5, after both instrument drift constant changed we see significant improvement from gravimeters performance. Average daily drift as short-term drift (each loop) in 2014 and 2016 survey respectively are 42-46 μGal and 24-25 μGal (Table 1). Figure 6 shows the statistical comparison of drift correction value for gravity readings in 2014 and 2016 survey. Percentage gravity readings from 2014 and 2016 survey corrected with small drift value (< 35 μGal) respectively are 85.63% and 87.61%.

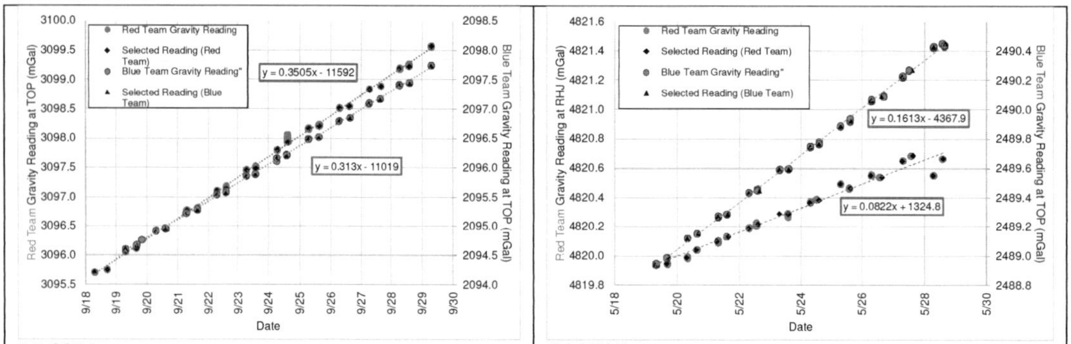

(a) 2014 survey (b) 2016 survey

Figure 5. Long-term drift comparison: (a) 2014 survey and (b) 2016 survey.

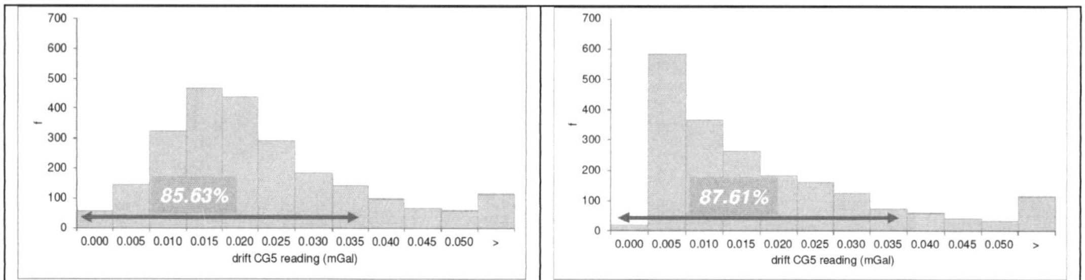

(a) 2014 survey (b) 2016 survey

Figure 6. Histogram comparison of drift correction using linear drift approximation applied in each gravity reading: (a) 2014 survey and (b) 2016 survey.

After drift analysis, we will describe instruments repeatability in field condition. We conducted repeatability measurement for some stations. The stations that measured more than once in different observation will be compared to understand gravimeters performance. Repeatability samples from 2014 and 2016 survey respectively are 74 and 92 (Table 1). Ideal gravity readings sample after corrected with linear drift will give the same value, but it's very common in gravity data acquisition using relative gravimeter will give deviation. Deviation from repeat stations shows in figure 7 describes repeatability test. We can notice from figure 7 that ± 10 μGal deviations from 2014 and 2016 survey respectively are 62.21% and 78.34%.

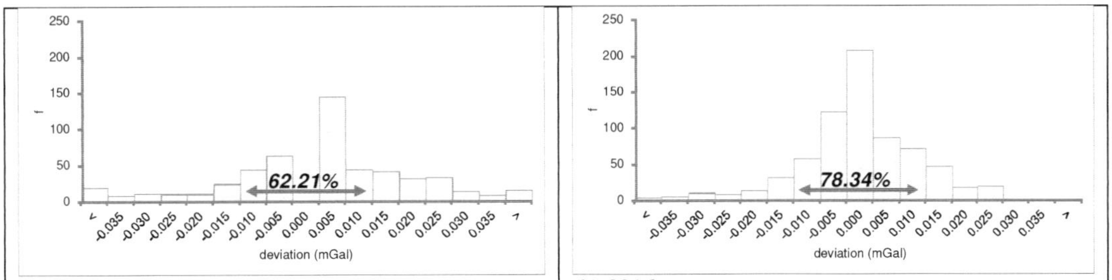

(a) 2014 survey (b) 2016 survey

Figure 7. Histogram comparison of repeatability test in repeat stations in (a) 2014 survey and (b) 2016 survey.

On-site data processing results provide us statistical proof of gravimeters performance in 2016 survey compare to the previous survey (2014). By conducting a simple simulation of small loop data acquisition before we go to the field can be significant preparation. Small loop data acquisition will provide information of consistency (repeatability) to distinguish small gravity variation.

4. Analysis and Discussion

Time-lapse microgravity (TLM) anomaly calculated by substracting gravity observation 2016 and 2014. After that TLM anomaly corrected with tripod height difference (that we measure in each datum in 2014 and 2016 survey). Histogram of TLM 2016-2014 in Jepon well area shown in figure 8a. From histogram figure 8a we can see that 87.53% data are in the range of -80 to 80 μGal. Outliers value that is considered very big from our TLM values are distributed in 51 stations (figure 8b). Some stations locations are very hard to find after 20 months, especially in forest area. Most of the stations location that produces big TLM value located in forest area (31 stations), while the other stations (20) are located in villages and rice fields area. Human activity in villages and rice fields may impact significant surface changes such as building construction, road construction, cultivation, and irrigation.

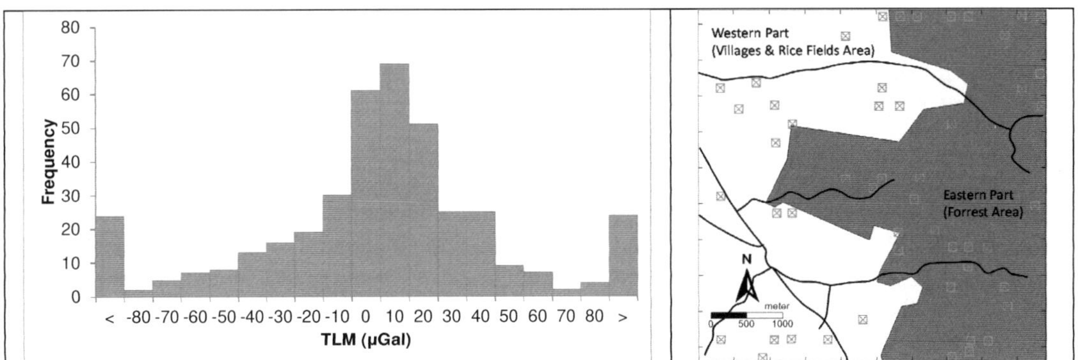

(a) (b)

Figure 8. (a) Histogram of TLM 2016-2014 in Jepon well area and (b) stations distribution that shows outlier TLM values (<-80 and >80 μGal).

TLM map that shows near surface and hydrology effect in Jepon well area shown in figure 9a. Figure 9a shows near surface effect because we can easily observe high-frequency signals from some stations that produce the bulls-eye pattern. There are interesting pattern from figure 9a that shows strong negative (marked with A and B) and positive anomalies (marked with C, D, and E). The strong anomalies (A, B, C, D, and E) close to the stations observation that have big different in linear drift correction (figure 9b). From our calculation, those station's corrected with linear drift correction that differs more than |±35| μGal between 2014 and 2016 survey.

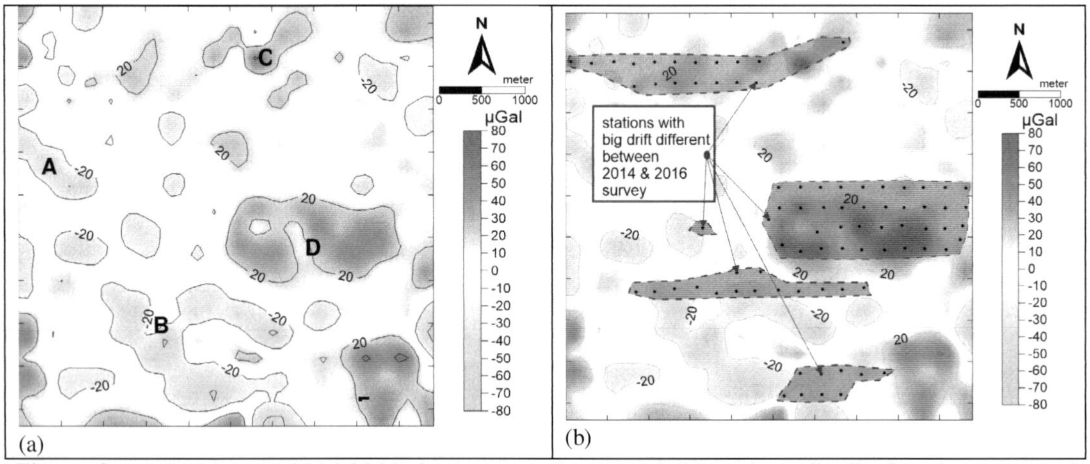

Figure 9. (a) TLM map (2016-2014) in Jepon well area and (b) stations distribution that corrected with big drift different between 2014 and 2016 survey.

Analysis of gravimeters performance from repeatability test suggests that stable value for our TLM anomalies should be ranged from -10 to +10 μGal. We also have uncertainties of tripod correction from 1 to 3.5 cm and translated to correction value in the range of 3 to 10 μGal. Considering that kind of argument and condition, we get near surface/hydrology effect in Jepon well area for 20 months in the negative and positive range value respectively are -20 to -80 μGal and +20 to +80 μGal. Stations distribution that gives TLM value that suspected as near surface/hydrology effect shows in figure 10.

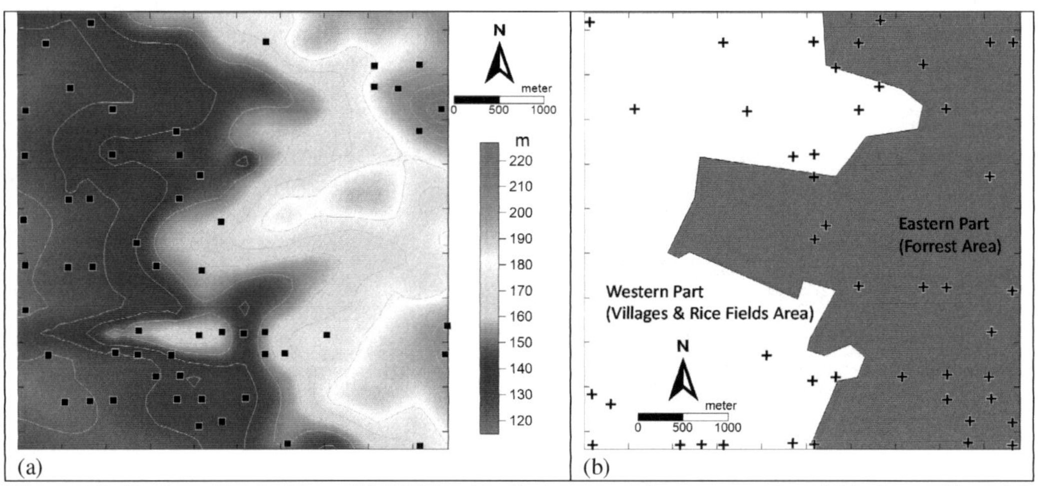

Figure 10. Stations distribution with TLM value that suspected as near surface/hidrology effect: (a) negative value (-20 to -80 μGal) and (b) positive value (20 to 80 μGal).

Figure 10a is topographic contour in Jepon well area with stations distribution of negative TLM value (-20 to -80 μGal). Stations distribution with negative TLM value dominantly located in the Western part of study area (local people villages and rice fields). Small number of stations distribution in the forest area (Eastern part) that provide negative anomalies located in the high topographic contour. Negative TLM value can be interpreted as the mass decrease in the near surface, such as lowering water table.

Figure 10b is land use map in Jepon well area with stations distribution of positive TLM value (20 to 80 μGal). Stations distribution with positive TLM value dominantly located in the Eastern part of the study area (forest). A small number of stations distribution in the Western part of the study area that provides positive anomalies can be caused by human activity such as cultivation, lowering ground for road construction, or subsidence. Positive TLM value can also be interpreted as the mass increase in the near surface, for example water run-off from higher topographic.

5. Conclusion

Preliminary analysis of gravity differences between 2014 and 2016 shows correlation pattern of time-lapse microgravity with land use in the study area. Rice fields and villages in the Western part of the study area correlate with negative TLM anomalies and forest area in the Eastern part of the study area correlate with positive TLM anomalies.

6. Acknowledgments

Authors acknowledge the support of the SATREPS project by JICA-JST, FTTM ITB, and other Gundih CCS pilot project that help gravity grid survey data acquisition.

References

[1] Allis R G, Gettings P, and Chapman D S 2000 *Proceedings Twenty-Fifth Workshop on Geothermal Reservoir Engineering Stanford University, Stanford, California* Precise Gravimetry and Geothermal Reservoir Management

[2] Debeglia N, and Dupont F 2002 Some critical factors for engineering and environmental microgravity investigations *Journal of Applied Geophysics* **50** 435

[3] Ferguson J F, Chen T, Brady J, Aiken C L V, and Seibert J 2007 The 4D microgravity method for waterflood surveillance II – Gravity measurement for the Prudhoe Bay reservoir, Alaska *Geophysics* **72** I33

Full Waveform Modelling for Subsurface Characterization with Converted-Wave Seismic Reflection

Wahyu Triyoso[1*], Madaniya Oktariena[2], Edycakra Sinaga[2], Firman Syaifuddin[3]

[1]Global Geophysics Research Group, Institut Teknologi Bandung
[2]Master Program of Geophysical Engineering, Institut Teknologi Bandung
[3]Seismic Exploration & Petrophysics Research Group, Institut Teknologi Sepuluh Nopember

*corresponding author's e-mail : wtriyoso@gmail.com

Abstract. While a large number of reservoirs have been explored using P-waves seismic data, P-wave seismic survey ceases to provide adequate result in seismically and geologically challenging areas, like gas cloud, shallow drilling hazards, strong multiples, highly fractured, anisotropy. Most of these reservoir problems can be addressed using P and PS seismic data combination. Multicomponent seismic survey records both P-wave and S-wave unlike conventional survey that only records compressional P-wave. Under certain conditions, conventional energy source can be used to record P and PS data using the fact that compressional wave energy partly converts into shear waves at the reflector. Shear component can be recorded using down going P-wave and upcoming S-wave by placing a horizontal component geophone on the ocean floor. A synthetic model is created based on real data to analyze the effect of gas cloud existence to PP and PS wave reflections which has a similar characteristic to Sub-Volcanic imaging. The challenge within the multicomponent seismic is the different travel time between P-wave and S-wave, therefore the converted-wave seismic data should be processed with different approach. This research will provide a method to determine an optimum converted point known as Common Conversion Point (CCP) that can solve the Asymmetrical Conversion Point of PS data. The value of γ (Vp/Vs) is essential to estimate the right CCP that will be used in converted-wave seismic processing. This research will also continue to the advanced processing method of converted-wave seismic by applying Joint Inversion to PP&PS seismic. Joint Inversion is a simultaneous model-based inversion that estimates the P&S-wave impedance which are consistent with the PP&PS amplitude data. The result reveals a more complex structure mirrored in PS data below the gas cloud area. Through estimated γ section resulted from Joint Inversion, we receive a better imaging improvement below gas cloud area tribute to the converted-wave seismic as additional constrain.

1. Introduction

The presence of rock saturant will change elastic properties of the rock depending on the type and quantity of the pore fluid. Gas cloud existence prone to contribute in bad imaging of the conventional P-wave seismic below the gas cloud area. Similar characteristic also found in Sub-Volcanic imaging, both gas cloud and Sub-Volcanic triggers an imaging quality degradation caused by energy scattering and attenuation in its high-contrast velocity zone. When conventional P-wave surveying fails to provide adequate seismic image, additional S-wave data might be just the solution. Shear wave is generally less sensitive to the rock saturant [1], so it is possible to use S-wave surveying to penetrate this gas cloud layer.

There are several ways to obtain shear wave information from seismic data. The easiest way and certainly cost less is by using multicomponent seismic survey. As the source was compressional, multicomponent seismic survey allows propagation of converted-wave or PS images that derives from a downgoing P-wave which converting to an upgoing S-wave at the reflector point. The PS data is the key to assist conventional PP seismic imaging in case such as gas cloud. By reducing ambiguities in PP seismic, multicomponent data assist in creating more accurate evaluation of imaging through gas cloud [2]. The PS seismic data provides highly improved subsurface imaging as compared to conventional P-wave data. Different rocks and its saturant could have a similar P-wave response, therefore create indistinguishable PP images. The joint usage of S-wave and P-wave seismic could help to characterize the rock structure, lithology type, and pore fluid [3]. Acknowledging this common practice, most of the unconventional reservoir problems are handled by P and PS seismic data. This combination is going to be a standard tool for subsurface imaging of challenging reservoirs.

However, there are several stumbling blocks that hinder a widely use of PS seismic data in assisting conventional PP datasets. The most prominent problem is related to the different event times between PP and PS seismic data. Although both of them derived from a same reflector, PP and PS seismic data show different reflectivity at different time. This condition leads to the necessity of applying different approach both in converted wave seismic processing and also in its further use in aiding conventional PP seismic data for reservoir characterization purpose. This paper will present a series of tricky processing flows for multicomponent seismic data, also the benefit of using converted-wave seismic in assisting conventional PP seismic data to solve seismic imaging below gas cloud area through Joint Inversion process.

2. Methodology

This research is intended to better understand the converted-wave seismic reflection using Full Waveform Modelling starts from the converted-wave seismic data processing to the application for reservoir characterization through Joint Inversion of PP-PS seismic data stack. To achieve this objective, the study is comprised into three main focus of workflow as described below.

2.1. Full Waveform Modelling

Full waveform simulation in seismic study is a common method to use in learning about seismic wave propagation through layered medium. For such a high risk reservoir, it is preferred to do modeling first before the real acquisition done in order to get the best design parameter for maximum result. Seismic data used in this research is a synthetic data derived from a real case of gas cloud recorded in its well log data. The purpose of modeling is to get better knowledge about wave propagation of seismic converted-wave with full wave seismic simulation. This modeling process used 2D approach to simulate converted-wave propagation and to produce synthetic data which consist of two-component receiver – vertical and horizontal component. The model builds using velocity model recreated from seismic velocity picking model of a real converted-wave seismic data processing. The velocity model consists of P-wave velocity model and S-wave velocity model of seismic line GOG004 crossing well ALPHA-01. The full model created by extracting the correlation between density and P-wave from the well log data and applying the correlation to create the low-frequency subsurface model (Figure 1). The presence of P-wave velocity decreasing is shown in the model relating to the purpose of this modeling that intend to study about the effect of seismic imaging below gas cloud zone.

The simulation shoots using elastic wave. The configuration of shooting geometry has 153 point shot numbers. The receiver uses 161 geophones with split-spread geometry. Each group interval ranges 25 m and 50 m for the shoot interval. The synthetic shot gather data of the full wave modelling is recorded in two components. They are vertical and horizontal component. Each component addresses a different purposes. The vertical component which common in conventional seismic reflection method is used for PP velocity analysis in the converted-wave seismic processing. The horizontal component is where the converted wave or PS seismic data well-recorded. After synthetic data is created, it is time to move on to next stage of this research, the converted-wave seismic data processing.

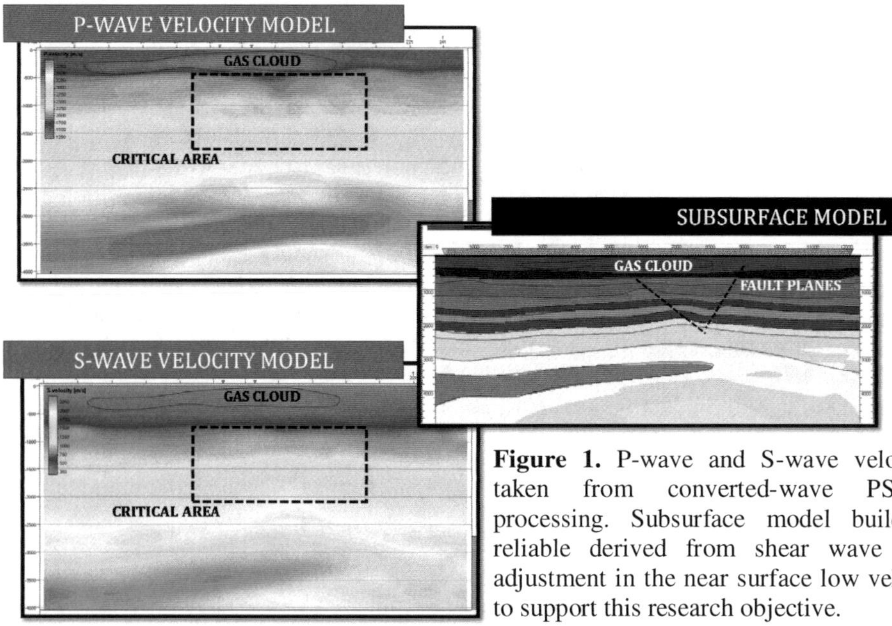

Figure 1. P-wave and S-wave velocity model is taken from converted-wave PSDM seismic processing. Subsurface model building is more reliable derived from shear wave velocity with adjustment in the near surface low velocity anomaly to support this research objective.

2.2. Converted-Wave Seismic Data Processing

A series of processing flows for both vertical component (PP) and radial component (PS) are applied to synthetic line GOG004. Among several purposes of this step, one of them is to eliminate the event time difference between PP and PS seismic data. Several assumptions are made in this 2D converted wave processing. First assumption, both PP and PS seismic data have the same geometry. Second assumption, in order to provide the final PP RMS Velocity, the vertical component (PP) should already fully-processed before converted-wave seismic processing begins.

The processing flow will be focused on providing more reliable estimation of Vp/Vs value through Asymptotic Conversion Point (ACP) Binning for each input of Vp/Vs. The next step is to register the PP RMS Velocity into the PS Data that are already processed through the ACP Binning. The general processing workflow are shown in Figure 2 below.

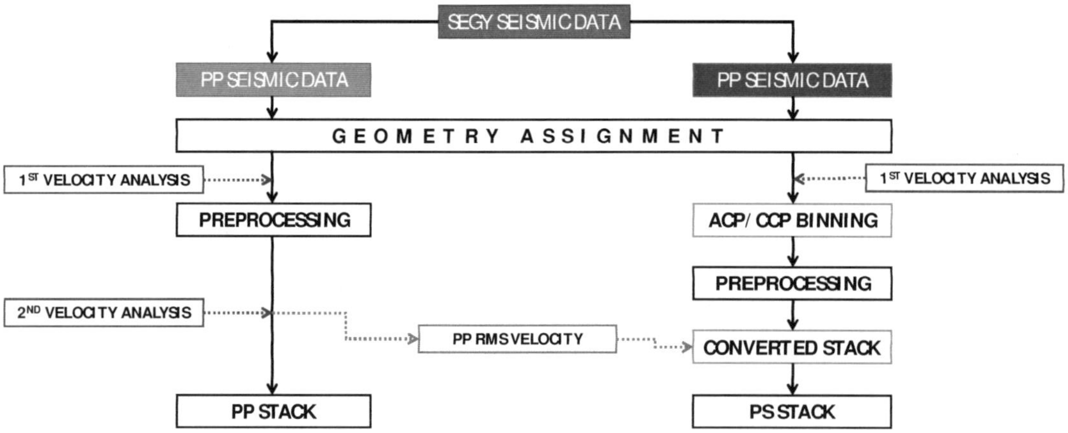

Figure 2. General processing workflow for PP and PS seimic data processing.

IOP Conf. Series: Earth and Environmental Science **62** (2017) 012048 doi:10.1088/1755-1315/62/1/012048

The processing flow between the PP and PS data are focusing on similar basic sequences, such as Geometry Assignment, Velocity Analysis, Pre-processing, Final Velocity Analysis, Stacking, and Post-Stack Time Migration. Additional sequences for PS processing are ACP binning and Converted-Wave Stacking. The reflection point of PS data is not at the Common Mid-Point (CMP) like in conventional PP seismic data. In order to predict the PS reflection point, converted-wave processing uses Asymptotic Conversion Point. In this step, an estimation of Vp/Vs is essential in order to investigate the Common Conversion Point (CCP) Fold. Arguably the optimum CCP Fold achieved at Vp/Vs value of 2.2 as shown in Figure 3 below. The later step is the extraction of RMS velocity from the PP Data which will be registered and applied into the PS data.

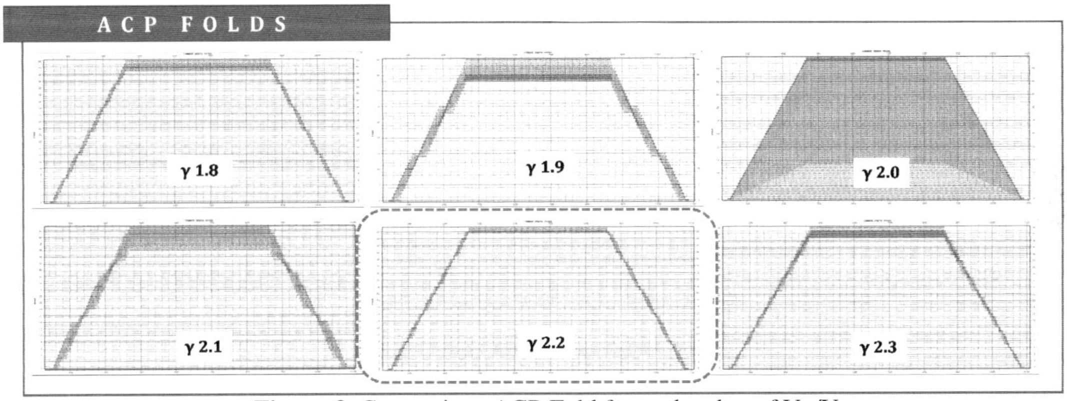

Figure 3. Comparison ACP Fold for each value of Vp/Vs

2.3. Joint Inversion of PP-PS Seismic Data Stack

There are several practical difficulties that should be addressed before applying Joint Inversion method to PP and PS seismic data. First and foremost is the need to ensure that the PP and PS reflections that will be combined in the inversion model coming from the same reflection point below the earth surface. This problem already addressed in converted-wave seismic data processing and would be furthermore enhanced the accuracy during Joint Inversion process.

Joint Inversion is an inversion that runs simultaneously. It is a model-based inversion that estimates the P- and S-wave impedance sections that are consistent with the PP and PS amplitude data [4]. The objective of simultaneous inversion is to invert directly from several seismic traces to multiple inverted results. The process also includes some form of coupling between the input variables [5]. Coupling of the PP and the PS data is based on the fact that the S-impedance and the density should be related to the P-impedance [6]. There are several conditions that should be fulfilled to ensure that the input data is suitable for the Joint Inversion process. First, the input data must have the same number of trace and geometry position. Second, the PP and PS seismic must be carefully processed preserving the amplitudes and using a valid velocity volume.

There are two step of the Joint Inversion process that need a special attention. Part one is Domain Conversion. The purpose of Domain Conversion is to ensure that both seismic sections are matching well before joining the two different domain data into one model in the next step. The Domain Conversion of the PS data to PP time is initially based on assumed Vp/Vs ratio of 2. Since this ratio does not represent the actual velocity structure, the converted PS data does not properly align with the PP data. Therefore, we correlate the P and PS well logs with the seismic data in the PP and PS domain and use the well velocities to correctly define the relationship between the two datasets to bring them into alignment at the well location. Part two is Joint Model Building. The purpose of Joint Model Building is to build a joint model from the two PP and PS domains and perform a Model Based Inversion that delivers results which are consistent with the PP - PS seismic amplitude volumes. The step by step procedure of Joint Inversion workflow is shown through Figure 4.

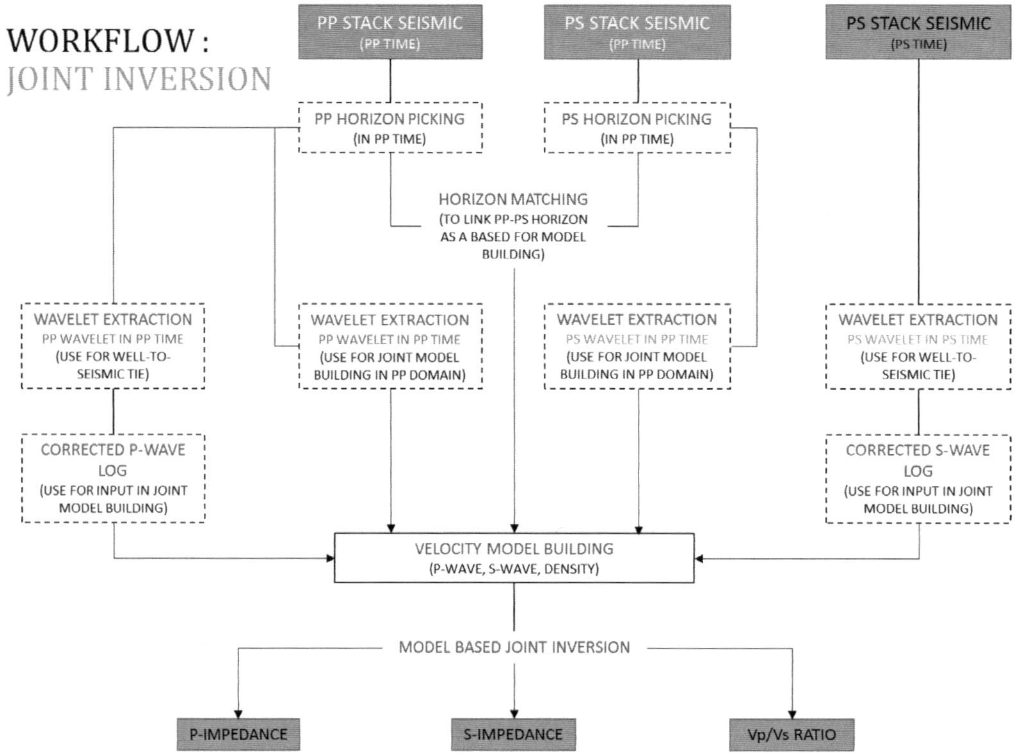

Figure 4. Flowchart shows the Joint Inversion PP-PS seismic workflow used in this research.

3. Results and Discussions

A comparison between the final stack of PP and PS seismic processing is needed to analyze whether these methods work in PS seismic processing. PP and PS Data is similar and suitable to be compared as shown in Figure 5 below. The effect of gas cloud is well-captured on synthetic PP seismic data. This phenomenon reducing the quality of the seismic image sub-layers below the accumulation, which is in term of structural image fault plane become blur and in term of amplitude level become shadowed as it is effected by presence of gas accumulation. Reflections from interfaces within and below the gas-charged channels are extremely poor on the stacked P-wave section, exhibiting reverberations and attenuation of high frequencies. As shown in Figure 5, amplitude level of the horizontal component show weaker than vertical component by comparison of the synthetic data.

The PS Seismic data could shows subsurface layering image more clearly. Notice the complex structure of seismic waves shown in each data circled with yellow line. The presence of the gas cloud does not affect the quality of seismic image. In term of structural image, PS seismic data could show discontinuity better than PP seismic data. The amplitude level of the PS seismic data shows more balance compare to PP seismic data in term of capturing the lithology contrast. The PS Data is better in the shallow time domain, and the PP Data is the opposite. This result is aligned with the fact that the PP seismic data has a higher energy level than the PS seismic data. The PP seismic stack appears more low frequency than the PS seismic stack which has a noticeably richer in number of event aspect as well as the ability to define structural edges despite the presence of gas cloud.

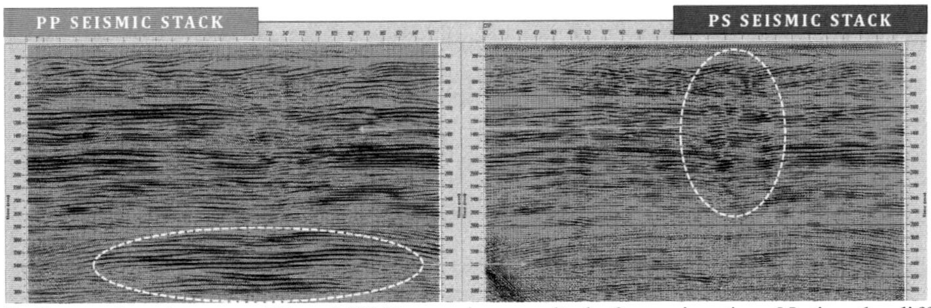

Figure 5. Comparison between PP and PS Processing results before migration. Notice the differences in structure details between PP and PS Data inside the dashed yellow circle.

The superiority of PS seismic data imaging is further proofed through the impedance result (Figure 6). While P-impedance section appears coherent as a flat layer cake around fault area below gas cloud zone, S-impedance section shows a sub-layer imaging that is quite similar to the synthetic model in term of discontinuity position. This is the main reason of the resulted Vp/Vs section more similar appearance to the S-impedance section rather than the P-impedance section. The S-impedance section achieves a better resolution and contains higher frequency content compared to the P-impedance section. This result shows that the S-impedance obtained through converted-wave datasets could be a useful alternative to assist P-impedance when it fails to map subsurface because of challenging phenomenon such as gas cloud or sub-volcanic. Vp/Vs ratio result from Joint Inversion also compared to Independent Inversion result (Figure 7) to showcase the benefit in simultaneously involved the converted-wave seismic through parameter estimation and calculation process. The main different between Independent Inversion and Joint Inversion located on the method of producing inverted PS Seismic Section. The Independent Inversion creates S-Impedance by secondary approach based on Acoustic Impedance inversion. The Joint Inversion creates S-Impedance by dealing Well-to-Seismic Tie directly in PS Domain and creating a model that involved the PS Seismic Stack on the building process. So, it is inevitable that the result from Joint Inversion, particularly the Vp/Vs section shows a better delineation and value stability tribute to the converted-wave seismic data complementing the conventional PP seismic data.

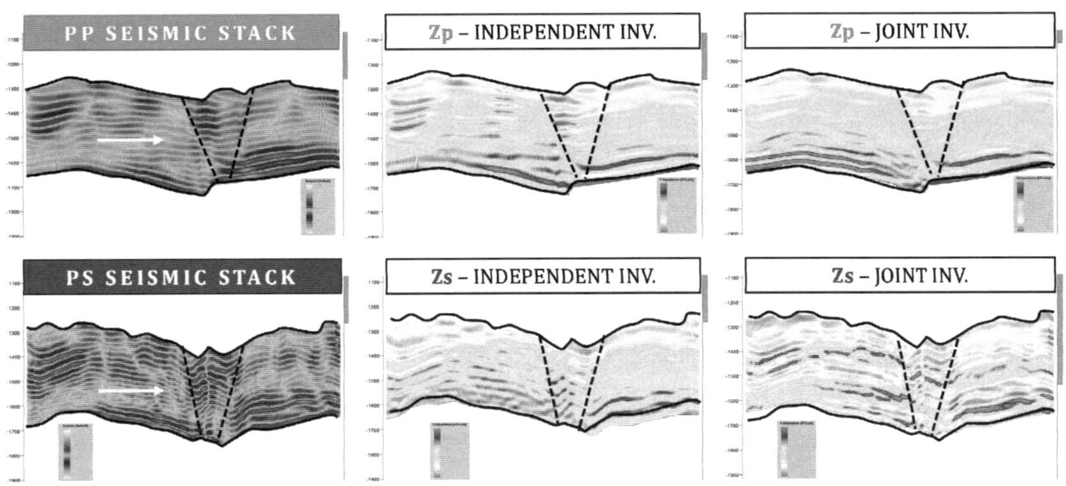

Figure 6. Comparison of impedance section from Joint and Independent Inversion. The critical area which is prone to imaging degradation triggered by gas cloud existence is pointed by the yellow arrow.

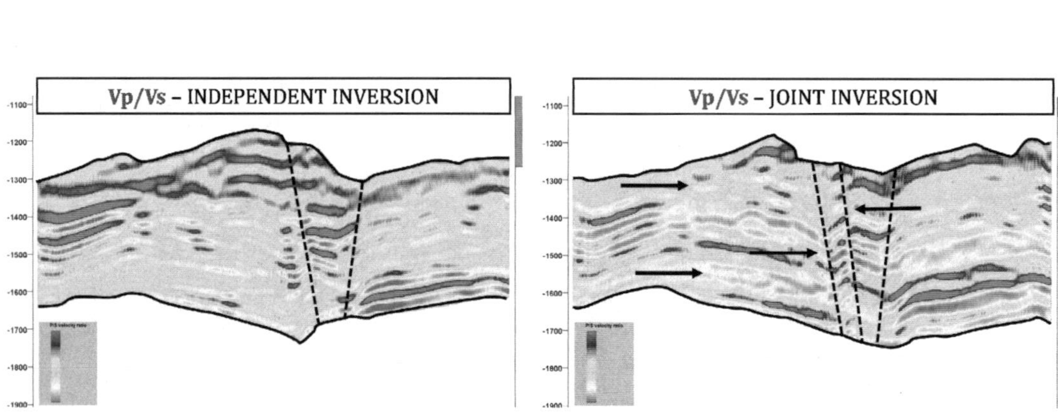

Figure 7. The variable coupling between Vp, Vs, and density which build the initial model of Joint Inversion results in a more stable Vp/Vs value, better delineation (pointed by black arrows), and the ability to define a clearer structural images compared to result from Independent Inversion. As can be seen through dashed black line, Joint Inversion could distinguish an additional fault planes sharply.

4. Conclusions

Answering the seismically and geology challenging area, the converted-wave seismic is just the right alternative in aiding the conventional PP seismic image which is sensitive to the rock characteristic and its saturant. The different in event time also reflectivity characteristic between the PP and PS seismic data sure is deemed to be a setback but not impossible to break through in order to enjoy the benefit of clearer seismic imaging. Through a full waveform modelling, the synthetic PP and PS seismic data shows that the PP seismic image is heavily affected by the presence of gas cloud while the PS seismic image shows a sharp fault boundary and more event recorded despite of having a same gas cloud problem as the PP seismic image. Although the converted-wave seismic data processing could be quite tricky because a different calculation of travel time between downgoing P-waves and upgoing S-waves, the Asymptotic Conversion Point Binning is helpful to accommodate the PS seismic data aligned with the PP seismic data. Joint Inversion method is applied in order to reap the superiority of PS seismic imaging by inverted it simultaneously with the PP seismic data. Using the fact that the PS seismic data related to PP seismic data, the Joint Inversion result creates a better imaging in delineation and stability value aspect compared to Independent Inversion.

References

[1] Danbom S H and Domenico S N 1986 Shear-wave exploration *Soc. Expl. Geophys.*
[2] Stewart R R, Gaiser J E, Brown R J, and Lawton D C 1999 Converted-wave seismic exploration : a tutorial *CREWES Research Report* **1999(1)**
[3] Margrave G F, Stewart R R, and Larsen J A 2001 Joint PP and PS seismic inversion *The Leading Edge* **20(9)** p 1048-1052.
[4] Stewart R R 1991 Joint P-P and P-SV Inversion *CREWES Research Report* **1990(2)**
[5] Stewart R R 1990 Joint P and P-S seismic inversion *CREWES Research Report* **1990(1)**
[6] Gardner G H F, Gardner L W, and Gregory A R 1980 *Geophysics* **39** p 770-780.
[7] Lu H and Hall K 2003 Converted Wave (2D PS) Processing *CREWES Research Report* **15**

Post-stack Deterministic Seismic Inversion using Sparse Optimisation Utilizing Gaussian Dense Layer Simulation

Ruhul Firdaus[1] and Gestin Mey Ekawati[1]

[1]Department of Geophysical Engineering, Faculty of Science, Institut Teknologi Sumatera, Jl. Terusan Ryacudu, Desa Way Hui, Kecamatan Jati Agung, Lampung Selatan 35365, Indonesia

E-mail : ruhulhajjaj@gmail.com

Abstract. The main objective in inverting seismic data is to estimate layer characteristics instead of layer boundaries. It is about getting as much as layer boundaries as seismically permitted with correct position and amplitude. However, the number of layers identified by seismic inversion methods is relatively in small number. Meaning, inversion methods tend to thicken the layers due to band-limited of seismic data and wavelet inaccuracy. It is well-accepted that sparse-layer inversion has the ability in recovering full RC series that are permittable seismically yet it tends to produce blocky impedance. Sparse-layer inversion is also computationally extensive because of large amount of competing relectivity model. In this paper, we test sparse optimisation inversion performance using gaussian dense-layer simulation instead of sparse-layer model.

1. Introduction

Principally, seismic inversion is a stratigraphic deconvolutional process[1]. Recorded seismic data are resulted from a sophisticated interaction phenomena between source wave, earth's transmitivity, acquisition setup, processing and receiver instrument. This type of data are than processed to meet a much more simple concept called the convolutional model[2]. In convolutional model, amplitude in a seismic trace is assumed as representation of earth reflectiviy but in much less resolutive version, filtered by the wavelet spectrum. The bandlimited nature of the source, the internal structure of the geophone which records the data, and also the complexity of the subsurface because of the attenuation and other properties are some of the reasons of losing most of the data bandwidth during the seismic survey. The seismic wave is greatly decreased in high frequency content mainly due to earth transmitivity, acquisition, the recording system, and processing effects whereas lack of lower band of frequency content mainly due to acquisition, receiver and the recording instruments [3]. This leads to bandlimit the seismic data and was believed to cause some difficulties getting reasonable result of inverted impedance [3].

The case is perhaps true, unless sparse-layer seismic inversion at the highest capacity can regenerate reflectivity series from band-limited seismic synthetic data with high resolution, correct position, accurate in amplitude and more importantly giving much better continuity [[5],[6]. In practice, by using real seismic data we never get close to such result particularly because of inappropriate wavelet estimation and noise existance [6]. Sparse layer seismic inversion uses sparse optimisation incorporating norm-l1 as a surrogate for norm-l0. Norm-l1 (sparsity) has advantages over norm-l2 (least-square) in several aspects e.g. more reliable with noisy data. Noisy data will still give proper result except the resolution is decreasing [7]. Norm-l1 was rarely implemented mainly because mathematically complex. All deterministic inversion method, espescially sparse spike including sparse layer inversion methods, tend to give the average version of reflectivities separated some distance as thickening layers. Averaging layers might cause wrong estimation of reflectivity coefficient by substituting low-varying reflectivity series into a single representable reflectivity which violates the facts observed in logs. Using those average layers also might restrict us to get full-bandwidth inverted impedance. Gaussian dense-layer simulation is proposed to overcome this problem. Using intensive reflectivity simulation, we present a demonstration of elaborating stochastic simulation to the deterministic inversion framework.

2. Methodology

Basis pursuit inversion method is a type of sparse layer inversion [1]. Using concept of basis pursuit decomposition [1], it attempt to evaluate all combinations of any pair reflectivity waveform that fit the real data [5,6,[8],[9]. The inversion use the convolutional theory to compute synthetic seismic from generated model so we can decide which model are most appropriate by evaluating minimum error between the data and the computed one. It optimizes the waveforms collection in *Ds* and find the coefficients which reflect the contribution of those waveforms on the real data by using equation:

$$\min \| s - \mathcal{D}_s \|_2 + \lambda \| x \|_1 \qquad (1)$$

After finding the coefficients *x*, then reflectivity series *r* can be obtained by simple matrix multiplication:

$$r = \mathcal{D}_r x \qquad (2)$$

where *Dr* is a reflectivity pair dictionary (even pairs and odd pairs) for those of the *Ds*. Using this kind of dictionary costs us many computation time and computer memory [6].

The main advantage of using basis pursuit decomposition is it permits us to add any other dictionary to be included within the over-complete dictionary [6]. We choose random gaussian reflectivity package to be added to the dictionary and reduce a great amount of reflectivity pairs. This scheme supposedly can provide the same quality but with less computational costs. It contains dense Gaussian reflectivity simulation which also represent the very thin layer series so it more likely to give us a broadband result. Comparative scheme of dictionary is shown in Figure 1.

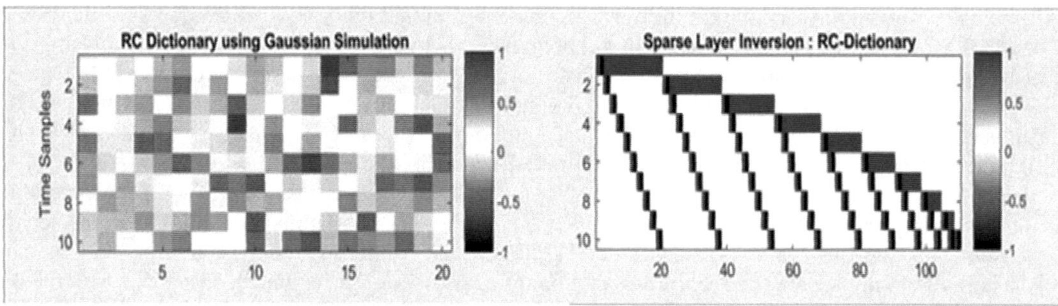

Figure 1. Scheme for reflectivity Dictionary.

3. Experiment on Synthetic Data

Using inversion technique described before, we put to the test two synthetic cases. Each case reflects real type of Earth's reflectivity series. First case represents relatively homogeneous layers and the other case represents slightly denser reflectivity (more heterogeneous layers). For both cases, as we notice in Figure 2, neither of them might be expected in synthetic seismic data after we convolve them with the Ormsby wavelet. In these cases, we made it noise free as we want to see how far this technique can invert reflectivity in ideal condition.

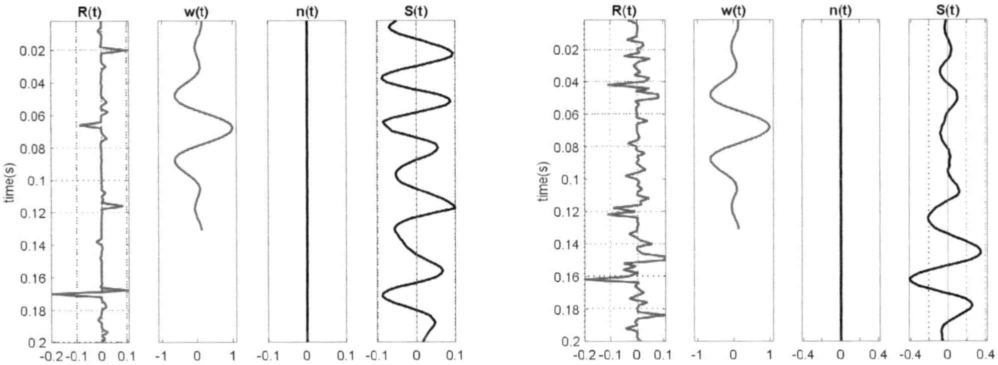

Figure 2. Two synthetic cases. First case is on the left. Second case is on the right.

First synthetic case shows spiky reflectivity separated in some distance while the seismic synthetic surprisingly shows more wiggle than we expected due to wavelet sidelobes. The contradictory phenomenon is shown in the second case. Less wiggle we can see on the seismic synthetic trace while the reflectivities are so intense. Using our method, we do the inversion to get the reflectivity back by assuming we know the wavelet exactly.

As we can see in Figure 3, reflectivity series can be obtained very accurately (Correlation R=0.98). On the left part, we can see the dictionary used to invert the seismic data. Notice that the data more likely represented by Gaussian simulation than to reflectivity pairs. Experiment on the second case also gives the result somewhat in kind, Figure 4. Here we notice that our proposed method can also give high resolution of inverted reflectivity series for both of the case but with less Dictionary thus more efficient in computation process.

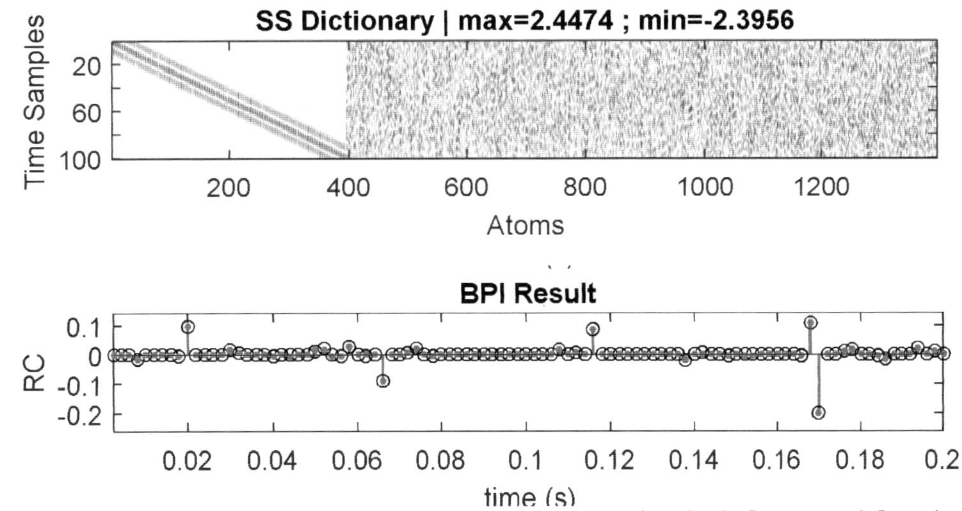

Figure 3. The first case result. Upper panel: Dictionary used to invert the seismic. Lower panel: Inversion QC. Inverted reflectivities (Red) fit accurately to the original reflectivities (Black).

4. Real Data Implementation

In this section, the real data of P-impedance log were used. The seismic trace was computed by doing forward modeling reflectivity log using a certain wavelet. In this case, we also assume noise free condition. So the seismic trace is assumed in perfect condition (see Figure 5). Different from the synthetic case, in the real data, we can get low frequency part of the impedance so the absolute broadband impedance can be demonstrated. From Figure 5, we also can see the spectrum of reflectivity log, P-impedance log and the synthetic trace. Wavelet used in this implementation is the same kind of in synthetic data before but with 90° rotation. It is surely band limited wavelet. If we consider the convolutional theory, it seems there are no chance to get all the reflectivity back as they are filtered by the spectrum of wavelet. The result in Figure 6 is astonishing because using this method we can recall them back with high level of accuracy (R=0.99).

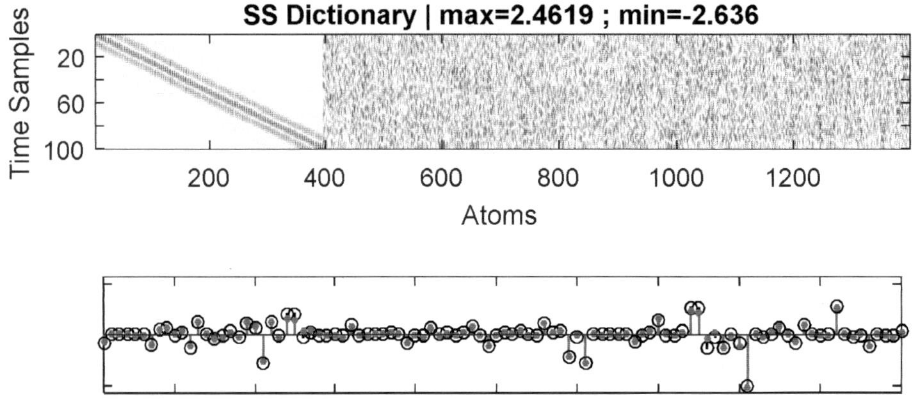

Figure 4. The second case result. Upper panel: Dictionary used to invert the seismic. Lower panel: Inversion QC. Inverted reflectivities (Red) fit accurately to the original reflectivities (Black).

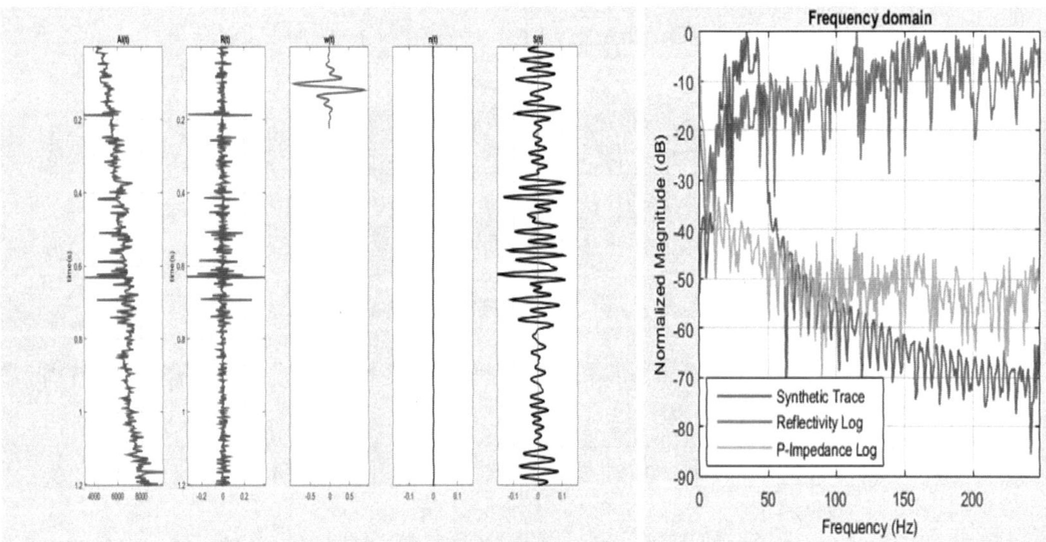

Figure 5. Real data from P-impedance log and its spectrum

Next, we compute the absolute broadband impedance by adding low frequency impedance from log data to the recursive process. The result (Figure 7), as expected, shows very good match between computed impedance and log impedance. Every thin layer can be resolved.

5. Conclusions

We have given a demonstration how full reflectivity series obtained from band-limited data due to wavelet filtering. This method can completely remove wavelet overprint from the data thus every reflectivity appear as a single spike. This make our intention to compute broadband impedance become possible with less time consuming. Here, we test our method in ideal condition where wavelet is exactly known and noise is completely zero. But in the real data, they always exist at some amount. As a consequence, predicted reflectivity can be poor in resolution or even worse giving wrong value. However, with good prediction of wavelet and careful processing steps, we at least have a chance to get broadband inverted AI. In the near future, we will implement this methodology in real seismic data.

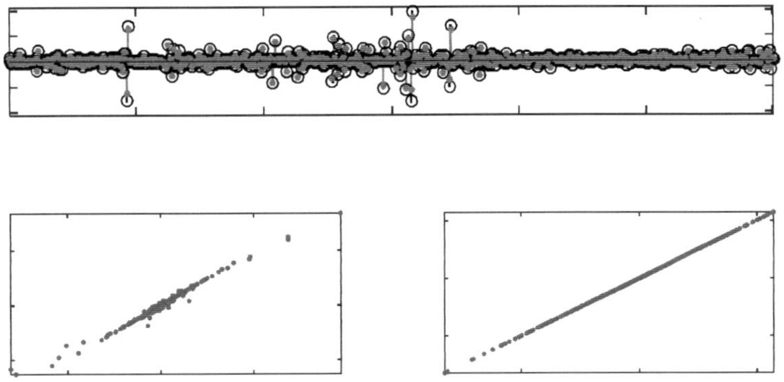

Figure 6. Computed reflectivity shows high correlation to the true reflectivity. Down left of the right panel shows crossplot between the two (Inverted RC v.s. Original RC)

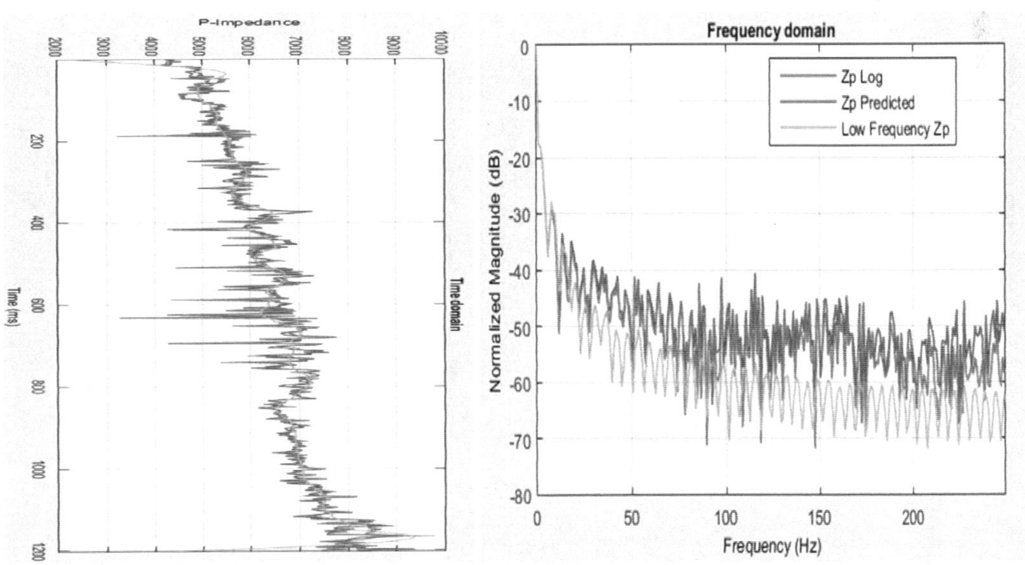

Figure 7. Comparison of inverted impedance and true impedance in time (left) domain and frequency (right) domain.

6. Acknowledgement
We would like to acknowledge the generosity of Institut Teknologi Sumatera in providing partial-funding for this research.

References
[1] Veeken P C H and Da Silva M 2004 *First Break* **22** 47-70
[2] Oldenburg D W, Scheuer T, and Levy S 1983 *Geophysics* **48** 1318-1337
[3] Aki K and Richards P G 2002 *Lamont-Doherty Earth Observatory of Columbia University*
[4] Sams M and Saussus D 2015 *Extended Abstract 3rd Int. Workshop on Rock Physics, Perth, Western Australia 13th-17th April 2015*
[5] Zhang R and Castagna J P 2011 *Geophysics* **76**(6) R147-R158
[6] Firdaus R dan Winardhi S 2014 *Proceedings PIT HAGI 39*
[7] Firdaus R 2014 *Master Thesis, Institut Teknologi Bandung, Indonesia*
[8] Bork J and Wood L 2001 71st *Annual International Meeting, SEG, Expanded Abstracts,* 510-513
[9] Puryear C I and Castagna J P 2008 *Geophysics* **73** (2) R37-R48
[10] Chen S S, Donoho D L, and Saunders M A 2001 *SIAM Review* **43** 129–159

The use of Graphic User Interface for development of a user-friendly CRS-Stack software

Rachmat Sule[1], Dythia Prayudhatama[1], Muhammad D. Perkasa[(1)], Andri Hendriyana[1], Fatkhan[1], Sardjito[2], Adriansyah[2]

[1] Institut Teknologi Bandung, Jalan Ganeca No. 10, Bandung, Indonesia
[2] (formerly) Upstream Technology Center, PT Pertamina

E-mail: rachmat@gf.itb.ac.id

Abstract. The development of a user-friendly Common Reflection Surface (CRS) Stack software that has been built by implementing Graphical User Interface (GUI) is described in this paper. The original CRS-Stack software developed by WIT Consortium is compiled in the unix/linux environment, which is not a user-friendly software, so that a user must write the commands and parameters manually in a script file. Due to this limitation, the CRS-Stack become a non popular method, although applying this method is actually a promising way in order to obtain better seismic sections, which have better reflector continuity and S/N ratio. After obtaining successful results that have been tested by using several seismic data belong to oil companies in Indonesia, it comes to an idea to develop a user-friendly software in our own laboratory. Graphical User Interface (GUI) is a type of user interface that allows people to interact with computer programs in a better way. Rather than typing commands and module parameters, GUI allows the users to use computer programs in much simple and easy. Thus, GUI can transform the text-based interface into graphical icons and visual indicators. The use of complicated seismic unix shell script can be avoided. The Java Swing GUI library is used to develop this CRS-Stack GUI. Every shell script that represents each seismic process is invoked from Java environment. Besides developing interactive GUI to perform CRS-Stack processing, this CRS-Stack GUI is design to help geophysicists to manage a project with complex seismic processing procedures. The CRS-Stack GUI software is composed by input directory, operators, and output directory, which are defined as a seismic data processing workflow. The CRS-Stack processing workflow involves four steps; i.e. automatic CMP stack, initial CRS-Stack, optimized CRS-Stack, and CRS-Stack Supergather. Those operations are visualized in an informative flowchart with self explanatory system to guide the user inputting the parameter values for each operation. The knowledge of CRS-Stack processing procedure is still preserved in the software, which is easy and efficient to be learned. The software will still be developed in the future. Any new innovative seismic processing workflow will also be added into this GUI software.

1. Introduction

Common Reflection Surface (CRS) Stack method is a new reflection seismic processing method that replaces the sequence processing steps of velocity analysis – NMO/DMO correction – stacking steps, which are always used in conventional stacking method. The Zero Offset (ZO) CRS-Stack method is introduced for the first time in 1997 by the Wave Inversion Technology (WIT) Consortium in Germany [3]. Since that time, its application to several seismic data that contains complex geological structures is widely used in all over the world, including in Indonesia. The development of CRS-Stack technology still continues until now. Even, a new version of CRS-Stack technique is appeared in 2009, which is then called as Partial CRS-Stack [1].

The original source codes of ZO and Partial CRS-Stack methods can be installed and run in standard computer hardware under Linux OS environment. So far, these versions are widely used for research purposes. For industries needs, it seems that they will be difficult to be accepted, since they are not user-friendly softwares. In October 2007, Institut Teknologi Bandung (ITB) and Pertamina EPTC agreed to conduct a research cooperation in order to develop several geosciences softwares. One of them is to develop a user-friendly reflection seismic data processing software that implement several new processing techniques, including ZO and Partial CRS-Stack methods. Since that time, several faculty staffs and students of Geophysical Engineering Study Program - ITB involve in this research project. This paper will describe the current situation of the developed CRS-Stack based software.

2. The Principle of ZO and Partial CRS-Stack Methods

The successful of CRS-Stack method has been proved by several conducted case studies abroad [5] and in Indonesia [6]. Based on their studies, ZO CRS-Stack method is proved as an alternative method to the existing conventional stack method, which could provide better seismic section that has higher S/N ratio and better continuity of reflections. The ZO CRS-Stack operator is a second-order traveltime approximation of paraxial ray, which is given in terms of central ray coordinate of midpoint-half offset (x_m, h) for segment curved reflector. It can be expressed as follow:

$$t^2_{(x_m,h)} = \left[t_0 + \frac{2\sin\alpha}{V_0}(x_m - x_0)\right]^2 + \frac{2t_0\cos^2\alpha}{V_0}\left[\frac{(x_m - x_0)^2}{R_N} + \frac{h^2}{R_{NIP}}\right] \quad (1)$$

where h is half source-receiver offset, m is the midpoint displacement with respect to the considered CMP position, and t_o corresponds to the zero offset two-way-traveltime (TWT). In this method, stacking is conducted throughout several midpoints and offsets, which is described as green surface (this surface is then called as CRS-Stack surface operator) in Fig. 1 (left), in order to get a stacked trace that is situated in x_o. One needs only to input the near-surface velocity (V_0), and there is no explicit macro-velocity model is needed. Therefore, this method is sometimes called as a technique that is independent from macro velocity model.

Based on Eq. (1), CRS-Stack depends on three kinematic wavefield attributes, i.e. emergence angle (α) and radii of curvature of two hypothetical normal incidence point wave and normal wave (R_{NIP} and R_N). The main issue in CRS-Stack is how to determine those three CRS attributes for each point (x_o, t_o) in a ZO section. This could be done by executing 3 (three) steps, as proposed by [4] by using the scheme of pragmatic search or by [3] by implementing extended search scheme. Those three steps are:

a) Automatic CMP Stack, i.e. an effort in order to obtain initial ZO stacking section and initial stacking velocity.
b) Linear and hyperbolic CRS-Stack attributes search, i.e. an effort to extract α, R_N, and R_{NIP} from the initial ZO stack section, which has been obtained from the first step.
c) Local optimization step, i.e. an effort that should be done in order to refine the initial CRS-Stack attributes by using an algorithm that can find the fittest operator to reflection travel-time data. In this step, one must test all possible attributes and choose one that lead to highest coherency value.

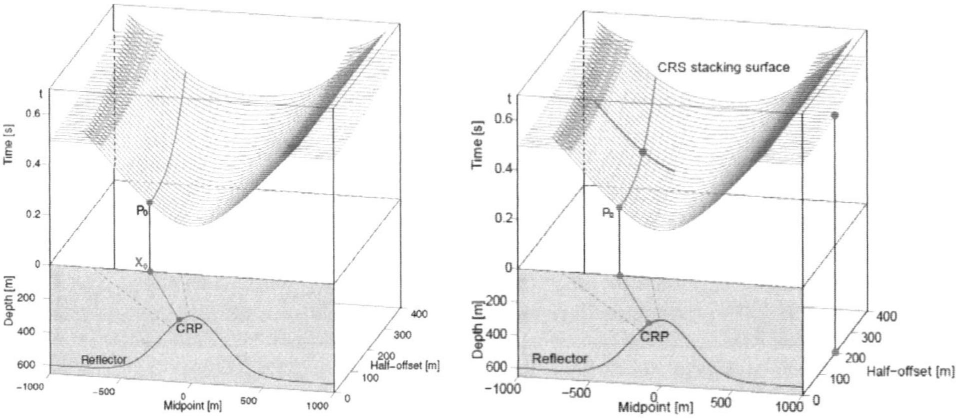

Figure 1. Left - ZO CRS-Stack surface (after Mann, 2002), right - Partial CRS-Stack surface. In Partial CRS-Stack method, the stacking process is conducted only along the red line and stored the result in the red dot [1].

The Partial CRS-Stack has similar formulation and tricks to the ZO CRS-Stack. The main difference between both methods is seen on the stacking procedure. In Partial CRS-Stack, the stacking process is carried out only along each half-offset (red line in Fig. 1 right) and the result is a stacked trace that is placed on a particular half-offset (red point in Fig. 1 right). This procedure is repeated for other offsets. As result, one will get a gather with higher fold than that can be provided by the conventional stacking method. This gather is then known as a supergather, which is displayed in Fig. 1 (right) as a purple line. Since the final output of Partial CRS-Stack is a supergather, one should still conduct velocity analysis, NMO/DMO correction and stacking, in order to obtain a stack seismic trace at x_0, which is part of a stack section. Detail procedures of Partial CRS-Stack method is described in [2].

By applying this method, the amount of data (traces) in a supergather can be increased. This will make velocity analysis will be easier to do, in which the velocity pattern will be more focused. The application of this method is very useful if seismic dataset consists irregular geometry, e.g. because the seismic lines crossed rivers, villages, complex topography areas, etc. This method could provide additional seismic traces for the gap areas.

3. The Development of CRS Stack GUI

The research was begun with learning process of CRS-Stack method. This could be done by doing literature study of several published papers, Master thesis and Ph.D dissertations, especially originated from Universität Karlsruhe, e.g. [2]; [3]; and [4]. Several B.Sc and M.Sc students of ITB have chosen the topic of CRS-Stack as their research project to pursue degrees from ITB. Since the original source codes from WIT consortium is obtained in March 2008, several ITB students that have good knowledge in information technology (IT) are involved in this research group. Their initial responsibility is to install the original source codes in the available laboratory's hardware.

To ensure the reliability of this CRS-Stack software, the original source codes are tested in several existing hardware platforms. Several synthetic dataset, which are resulted from elastic waveform modeling by using finite difference method, are produced in order to make synthetic studies and to validate the original programs. The students are asked to make several subsurface models, from a simple model until the complex one, which will be inputted later to the elastic waveform modeling software. Besides that, several real dataset are also provided from several oil companies in Indonesia and abroad for the same purposes. By using those datasets, we reproduce the stack sections by using CRS-Stack method, for different hardware and different operating system. Those results are then compared to the best results obtained from the conventional stacking method.

Up to this stage, several errors have been encountered for different systems. Next important step is to develop a software, since all errors have been fixed. On the same time, optimization of CRS-Stack program for several hardware and operating system must be done, this includes the implementation of CRS-Stack program on 64-bit machine with 64-bit operating system. Besides that, optimization on multi-core CPU to reduce the processing time is also an important issue that must be solved urgently.

The original CRS-Stack code developed by WIT-consortium was written in C++ programming language. Since there is a backward incompatibility in the code to newer C++ syntax, we solved this problem by installing GNU backward-compatibility software library, i.e. an additional library system, which may be different to be implemented for several operating systems. The recent status is that the CRS-Stack code can be compiled and run in 32-bit and 64-bit Linux operating system. Several compatibility tests have been done for several Linux distributors, such as Red Hat Enterprise Linux 5, Fedora Core 8, and open SUSE.

The second phase of the CRS-Stack software development is the development of the Graphical User Interface (GUI), including how automatic installation program can be implemented. The CRS-Stack programs are originally developed in the Seismic Unix (SU) environment, so it requires several SU programs involved in the CRS-Stack processing work-flow. Besides that, the possibility of other software (e.g. SEPlib) to be included must be concerned. Because of facing those challenges, it has been chosen the Java Programming to address those interoperability problems. Besides that, Java Programming consists less efforts to address backward compatibility problem. In the process of CRS-Stack software development, we apply the Agile Development Techniques. These techniques use iterative development as a basis, but advocate a lighter and more people-centric viewpoint than traditional approaches. Agile processes use feedback, rather than planning, as their primary control mechanism. The feedback is driven by regular tests and releases of the evolving software.

The research activities in the laboratory can be divided into 2 groups. There are research members that focus on the understanding and implementing the techniques of CRS-Stack, whereas other group members focus on computation and programming. The first group is composed mostly by the B.Sc and M.Sc students of geophysical engineering study program, who work with CRS-Stack programs for their final projects. As a student finalized his/her research and study at ITB, he/she has to provide a documentation related to his/her methods, programs and scripts, so that his/her experiences and knowledge can be transferred to other laboratory members. This effort is the way of laboratory's policy in order to preserve knowledge after a student is graduated.

In laboratory, the students are encouraged to use several open sources and commercial seismic processing software, instead of restricting them to use only one commercial seismic processing software. In the development of a new seismic data processing software, the source code of CRS-Stack is integrated with SU programs. The SU data format is adapted in order to perform CRS-Stack procedures. SU, which is developed by Colorado School of Mines (CSM), is a standard available seismic data processing programs and programming library in the laboratory. Moreover, the CRS-Stack procedures provides more possibilities to develop innovative processing procedures incorporating with other seismic imaging tools, in which other programs or seismic programming library may be included, e.g. SEPlib and Madagascar. In our cases, the NIP-wave tomography and PSDM by means of CRS attributes are also parts of interest. Both programs are also developed by WIT-Consortium and incorporating SEPlib inside the processing flow. Thus, the possibility of including other software without major revising in the GUI software is concerned.

During three years of development, three major changes of this software have been occurred, in which the main reason is in order to adapt and fulfill industrial needs. By experiencing other seismic software, such as PROMAX, Vista, and BotoSeis, several concepts are adapted and compiled from their design into our CRS-Stack GUI software. The users' opinions related to several available seismic processing software are accommodated in order to gain knowledge on how the CRS-Stack software should be designed.

4. Design of CRS-Stack GUI

The developed software is then named as ASPS (Advance Seismic Data Processing Software). The latest version of software interface can be seen in Fig. 2. Several existing features are as follow:

- A: The program lists, which are automatically generated. This CRS-Stack GUI Software search a particular folder path ($ASPSROOT), in order to obtains program headers information. To add a program into this existing GUI software, the developer has only to provide the program header and its corresponding shell script in order to perform program execution. This could be done without recoding and recompiling this GUI software. A program header is a file that describes inputs and program parameters to be displayed on the features E and F. We can add a program to a work-flow window D, by right-clicking the program icon in this display.
- B & C: The Area and Line tabs, which are designed to manage a project. This idea is adopted from PROMAX concept, in which each line has its own datasets. The data name in this line are managed and viewed by panel G. Every area tab can have several line tabs, and every line tab can have several work-flow panels. A work-flow panel consists of a combination of panel D, panel E, and panel F, which can be named by a user.

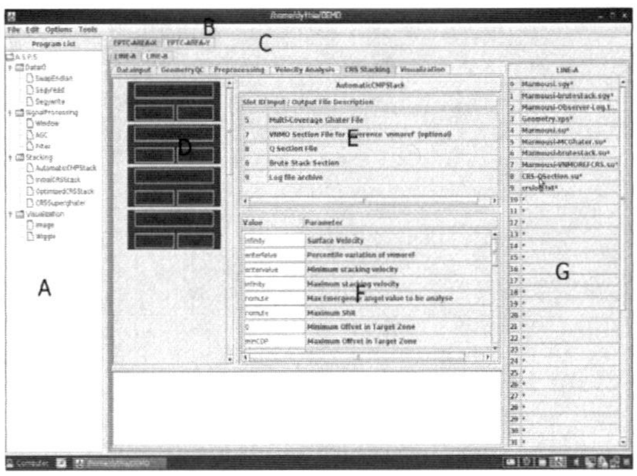

Figure 2. The latest Graphical User Interfaces

- D: Programs flow. A blue colored box represents a program, which can be called by right-clicking program icon in the panel A. A user must set input/output files in the panel E, and sets its program parameters in the panel F. There are two buttons in this icon, i.e. the delete and the run button. The "delete" button is used to remove a program from the work-flow. By right-clicking the "run" button, the user will trigger the shell script program to be executed.
- E: Input Output Panel. A user can set input/output file in this panel. This panel consists of two columns, which provide description on input/output file, and a slots ID (number) that corresponds to a Slots number in the datasets Viewer G. Thus, the user can only set input/output files, which have only been described in the panel G.
- F: Parameters panel. In this panel, a user can set the values of their parameters for a set of programs in the workflow.
- G: Datasets, This facility is adopted from PROMAX's concept of datasets, in which a user must insert a file into this datasets before the program is processed. By right-clicking this panel, a user can also view file properties and header.

All of those software features are so far working well. This software has been tested to process several datasets belong to some oil companies in Indonesia. A commercial course related to CRS-

Stack has also been conducted by our group in 2009, in which this software is used in the tutorial. So far the feed-back from the course participants is very good.

Figure 4 shows an advertisement poster that has been prepared for the launching of ASPS in 2010. Two stack sections are displayed in this poster, the upper stack section is the best section obtained from conventional stacking method, whereas the bottom section is the result of CRS-Stack method.

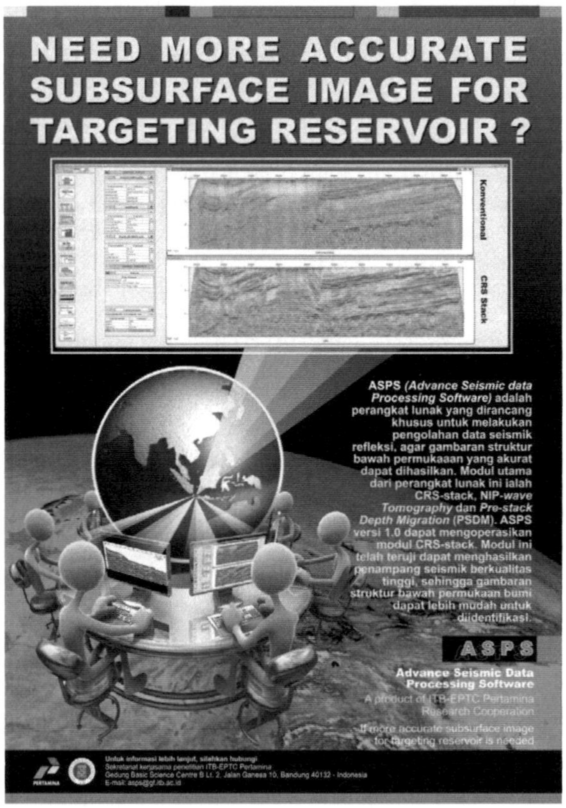

Figure 3. The advertisement poster of ASPS software. This software is ready to be launched in 2010.

5. Conclusions

The recent software is the initial achievement that can be given by the team members of ITB – Pertamina EPTC research cooperation, which its members are mostly Indonesian young scientists (ITB students and fresh graduated). The software is already tested and ready to use. Since this software is an Indonesian product and includes CRS-Stack method, we hope that this product can be accepted at least in Indonesian markets for the first stage.

The development of this software will still be continued. In a near future, the new version of this software will include CRS-Stack based of NIP-Wave Tomography and Pre-Stack Depth Migration (PSDM) modules. Any other new application derived from CRS-Stack methodology can always be added, so that this software will be one of the available software in the markets that can include several non-standard processing sequences, in which other commercial software will be difficult to fulfill those facilities. In order to make this plan feasible, special design requirement must be followed; Firstly: There are sustainable research and development activities in order to produce innovative seismic processing sequence; Secondly: The software must also provide high reliability to be implemented in the industrial business. Those two design requirements are very important to ensure the linkage between research and industry.

References

[1] Baykulov M 2009 Seismic imaging in complex media with the Common Reflection Surface stack *PhD thesis, University of Hamburg, Germany*

[2] Jäger R 1999 The Common-Reflection-Surface Stack- Theory and Application *Master's Thesis, Universität Karlsruhe*

[3] Mann J 2002 Extensions and Applications of the Common-Reflection-Surface Stack Method *Logos Verlag, Berlin*

[4] Müller T 1999 The Common Reflection Surface Stack Method: Seismic Imaging without explicit knowledge of the velocity model *PhD thesis, University of Karlsruhe*

[5] Pruessmann J, Coman R, Endres H, and Trappe H 2004 Improved imaging and AVO analysis of a shallow gas reservoir by CRS *The Leading Edge* **23** 915–918.

[6] Sule R, Hendriyana A, Dinanto E, Asikin A R and Adriansyah 2010 Improvement of seismic sections by implementing common reflection surface stack method *Proceeding of 122nd SEGJ Conference, Tokyo*

IOP Conf. Series: Earth and Environmental Science **62** (2017) 012051 doi:10.1088/1755-1315/62/1/012051

Enhancement of Seismic Stacked Section Obtained from a Complex Geological Structure by using Partial CRS-Stack Method

Ariesty R. Asikin[1], Andri Hendriyana[1], and Rachmat Sule[1]

[1] Geophysical Engineering Department, Institut Teknologi Bandung, Bandung, Jawa Barat, Indonesia

E-mail: ariestyasikin@gmail.com

Abstract. Partial CRS-Stack method is proved as an alternative method that can produce excellent subsurface image, especially if this method is applied to a seismic data that is acquired from complex subsurface structure areas. The application of this method will give a new hope in determining subsurface structures in a better way, especially if it is implemented to seismic data obtained from Indonesian region, which is dominated by complex geological structures. The Partial CRS-stack method is tested by using a seismic dataset, which is acquired from eastern part of Indonesia. Here, the continuity of reflectors cannot be seen clearly. To prove the ability of Partial CRS-stack method, its result will be compared with the result obtained from the conventional sequences. The stacked section resulted from Partial CRS-stack is much better than the result of conventional one. This could be understood, since the Partial CRS-stack method uses the information of reflectors along fresnel zone, instead conventional method that only uses information in a CDP. During its processing sequence, CRS kinematic wavefield attributes, e.g. emergence angle (α), radius curvature of normal ray (R_N) and radius curvature of normal incident point ray (R_{NIP}) must be determined previously, which indicates the location and behaviour of reflectors. As a conclusion, the Partial CRS-stack method is proved as a good alternative method to give better seismic sections. Because of this, the interpretation of unclear events that are seen in the conventional stack section can be avoided.

1. Introduction

The most common used geophysical method for oil and gas prospecting is reflection seismic method. This method utilizes the principle of seismic wave propagation inside the earth. Since sedimentary basin is composed by sedimentary layers, some of seismic energy will be reflected back to the surface if the waves reach subsurface interfaces that have contrast acoustic impedances. The arrivals of seismic waves will be registered by receivers that are deployed near earth surface. Those registered data will be processed in order to obtain seismic sections, which illustrate subsurface condition. However, if the subsurface structures are really complex, it would be difficult to obtain good seismic sections, although modern acquisition instrumentations and parameters are used. The success of obtaining good results is depend on what kind of processing technique is implemented to that data.

IOP Conf. Series: Earth and Environmental Science **62** (2017) 012051 doi:10.1088/1755-1315/62/1/012051

Conventional stacking method, which embraces the concept of common depth point (CDP), will normally not overcome the classical problem, i.e. how reliable seismic sections can be obtained from an area with complex subsurface geological structures. This conventional method assumes that a reflector composed by many reflector points. Each reflector point, which is part of a reflector plain, is then called as CDP. The concept of CDP can still be used as long as the reflectors are flat or have a uniform dip. If the reflectors are undulated, this concept can actually not be used anymore, since several sources-receivers pairs that develop a CDP gather could not represent a desired CDP point. Thus, special seismic processing technique must be found out in order to substitute the concept of conventional stacking method.

One of advanced processing methods that could overcome those above problems is the CRS-Stack method. This method is based on an assumption that a reflector is composed by many reflection surface segments, which is identical to the width Fresnel zone at a depth. All seismic traces that correspond to this Fresnel zone (which are developed by several source-receiver pairs on a surface) involve in the determination of a stacked seismic trace. Thus, CRS-Stack method is implemented by involvement of more seismic data in order to sum up or stack them into a stacked trace. Seismic data processing by using CRS-Stack method is also known as a multi-focusing method.

The Partial CRS-Stack method is introduced for the first time in 2008 [1]. This method is the generalization of previous existing Zero Offset (ZO) CRS-Stack method, which is introduced in 1998 by the Wave Inversion Technology (WIT) consortium ([4], [3]). Several previous applications of CRS-Stack method are found in published papers, e.g. [6] and [2]. Those previous studies showed that the ZO CRS-Stack is successful in handling seismic data taken from an area with complex subsurface geological structures. Besides better S/N ratio is achieved, the continuity of reflectors is obviously seen on the results of ZO CRS-Stack method. This paper will expose some achievements if Partial CRS-stack method is implemented. In order to fulfill this purpose, a real seismic dataset obtained from complex subsurface conditions is used, and this result is compared to the result obtained from conventional stacking method.

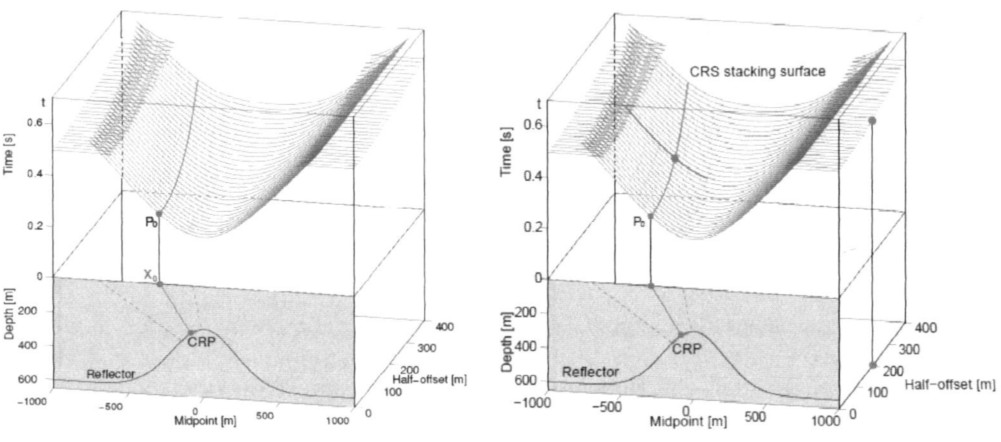

Figure 1. Left: ZO CRS-Stack surface [4] and Right: Partial CRS-Stack surface. The stacking process is conducted only along the red line and stored the result in the red dot [1].

2. Partial CRS-Stack Method

The principle of ZO CRS-Stack and Partial CRS-Stack method is actually quite similar. The significant difference between both methods is seen on the representation of their operators. In the ZO CRS-Stack, traces summation (or stacking process) is carried out along the CRS-Stack surface that is drawn as green surface in Fig. 1 (left). This summation/stacking is conducted in order to obtain a stacked trace at x_0. This procedure is repeated for other positions. As final output, a stacked section is obtained.

In the Partial CRS-Stack method, the traces summation is carried out in every half-offset. The curvature red line displayed in Fig. 1 (right) illustrates all seismic data that contribute to develop a stacked trace (red point in Fig. 1 right). This procedure is repeated for other half-offset, in order to obtain other stacked traces that are later situated along curvature purple line. This curvature purple line is actually identical to the CMP gather embraced in the conventional stacking method. However, this curvature purple line consists more stacked traces, in which the information related to reflector's dip for each trace is already considered. This curvature purple line is then known as a supergather, which is the final product of Partial CRS-Stack. The supergathers will be more regular than the original CMP gathers.

Before conducting this method, same procedures that are valid in ZO CRS-Stack method must be done previously, i.e. all efforts in order to obtain CRS attributes of emergence angle (α), and the two radii curvature (R_{NIP} and R_N). Those CRS attributes are needed in order to perform stacking process, as it is formulated in Eq. (1). Basically, those CRS attributes can be determined either after performing Automatic CMP-Stack or after CRS-Stack Attributes Optimization [4].

$$t^2_{(x_m,h)} = \left(t_0 + \frac{2\sin\alpha}{v_0}(x_m - x_0)\right)^2 + \frac{2t_0\cos^2\alpha}{v_0}\left[\frac{(x_m - x_0)^2}{R_N} + \frac{h^2}{R_{NIP}}\right] \quad (1)$$

The calculation of partial CRS stacking surface is conducted in a chosen CMP location for every specified sample $A(t_A,h_A)$, as can be seen in Fig. 2. An accurate zero offset time (t_0) and corresponding CRS attributes must be found first. One must search a traveltime hyperbola that is close to the traveltime curve, in which the sample A is situated. The traveltime hyperbola search is conducted within the range $[0;t_A]$ for the corresponding CRS parameters (α, R_N and R_{NIP}). The best fit curve that is chosen is the hyperbola curve that gives minimum deviation between the computed and the observed traveltime for sample A. Thus, the best zero offset traveltimes (t_0') is now obtained, including the corresponding CRS attributes. Since the t_0' time can have only discrete values, the computed t_0 time may deviate slightly from the t_0'.

The t_0 can now be calculated by using α and R_{NIP} attributes that corresponds to t_0', as it is described in Eq. 2.

$$t_0 = -\frac{h_A^2\cos^2\alpha}{V_0 R_{NIP}} + \sqrt{\left(\frac{h_A^2\cos^2\alpha}{V_0 R_{NIP}}\right)^2 + t_A^2} \quad (2)$$

R_{NIP} is assumed to be positive because it is a reflector depth. The negative solution from Eq. 2 is ignored. The obtained t_0 from Eq. 2 is then used in the equation 1 to obtain Partial CRS-stacking surface, as it is written in Eq. 3. This stacking surface is used to sum up the data coherently.

$$t^2[x_m,h] = \left(-\frac{h_A^2\cos^2\alpha}{V_0 R_{NIP}} + \sqrt{\left(\frac{h_A^2\cos^2\alpha}{V_0 R_{NIP}}\right)^2 + t_A^2} + \frac{2\sin\alpha}{V_0}(x_m - x_0)\right)^2 + \frac{2\cos^2\alpha}{V_0}\left(-\frac{h_A^2\cos^2\alpha}{V_0 R_{NIP}} + \sqrt{\left(\frac{h_A^2\cos^2\alpha}{V_0 R_{NIP}}\right)^2 + t_A^2}\left(\frac{(x_m - x_0)^2}{R_N} + \frac{h_A^2}{R_{NIP}}\right)\right) \quad (3)$$

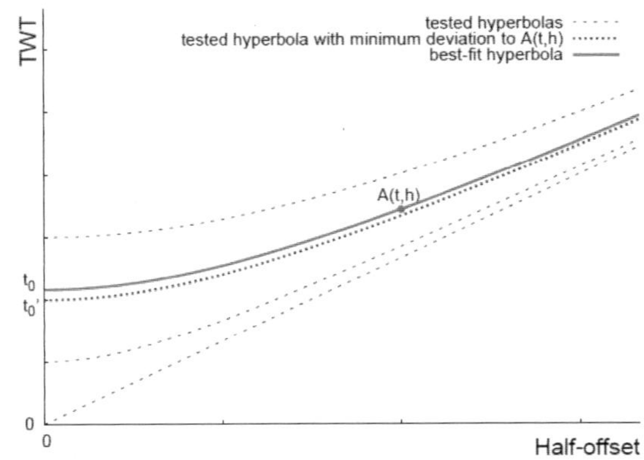

Figure 2: Hyperbola travel time curves that are used to get the most suitable for sample $A(t_A,h_A)$ in the CRS supergather [1].

3. Implemented Case Study

Implementation of Partial CRS-Stack method is conducted by using an onshore seismic dataset. The obtained result will be then compared with the best result, which is obtained from implementing conventional stacking method. The seismic data processing by using conventional stacking method has been done by a third party. The data is acquired from a complex geological structure in eastern part of Indonesia, in which intensive tectonic processes are occurred there. Besides that, the topographic variation is very strong, so that static correction must be applied carefully during pre-processing step. Some acquisition parameters are listed as follow: shot interval = 25 m, receiver interval = 25 m, near offset = 6.25 m, far offset = 4500 m, and maximum fold coverage = 60.

One of the most important parameters in the CRS-Stack processing flows is the aperture size. This value can be determined by estimating the size of Fresnel zone, which can be obtained from the projection surface of the first Fresnel zone. In its calculation, three CRS kinematic wavefield attributes (α, RN and RNIP), which indicate the location and behavior of reflectors, will influence the obtained value of Fresnel zone estimation.

The output of the CRS Stack method is 'supergathers', which consist more traces than the conventional gathers have. It means that the fold number will increase if Partial CRS-Stack is applied. This is one of the positive implications given by Partial CRS-Stack, because more data involvement occurred here, i.e. throughout the Fresnel zone. Figure 3 shows an example, which shows the difference between a conventional gather (left figure) and its corresponded supergather (right figure). Although 2.4 seconds is set as maximum aperture, the supergather still shows significant improvement. The reflectors are seen more clearly and the content of random noise is minimized.

The velocity spectra of both gathers are seen in Fig. 4. The velocity pattern in the velocity spectra of supergather (right figure) is described in a better way in. The interpreter will be easier to pick the velocity values in that velocity spectra, especially in the area that is restricted by tallow circle. As comparison, the velocity spectra obtained from conventional method is seen in Fig. 4 (left).

The best stack section obtained from conventional stack method is seen in Fig. 5 (top), whereas the stacked section resulted from Partial CRS-Stack method is displayed in Fig. 5 (bottom). If both sections are compared, it can be seen very clearly that the result from Partial CRS-Stack is much better than the result from conventional stack method, in which the reflectors look more continuous than the conventional result. Some deeper reflections, which are not so clear in conventional stacking section, are appeared in the Partial CRS stack results.

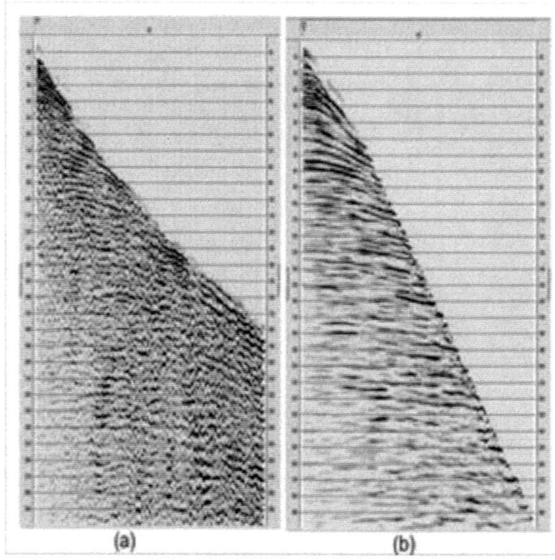

Figure 3. The difference between seismic gather obtained from conventional stacking method (left) and Partial CRS Stack method (right). Maximum aperture of 2.4 s is set for this example, which is not the maximum time.

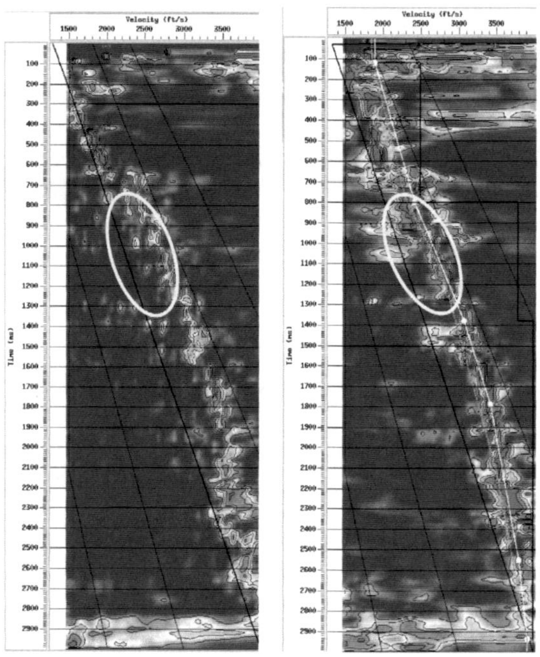

Figure 4. Velocity spectra obtained from conventional stacking method (left) and Partial CRS-Stack method (right).

If both sections in Fig. 5 are compared, it can be seen several differences, some of them are marked with black circles. The reflections that are unclear in the conventional stack section are now become more clearly in the stack section produced by Partial CRS-Stack. The shape of synclinal structure in the bottom figure can be better recognized, especially the base of that synclinal structure. The random noises that usually appeared between reflectors in the conventional stack section are now seen very clearly in the stack section produced by Partial CRS-Stack method. It makes some internal characters between reflectors can be better interpreted for seismic stratigraphy analysis purpose.

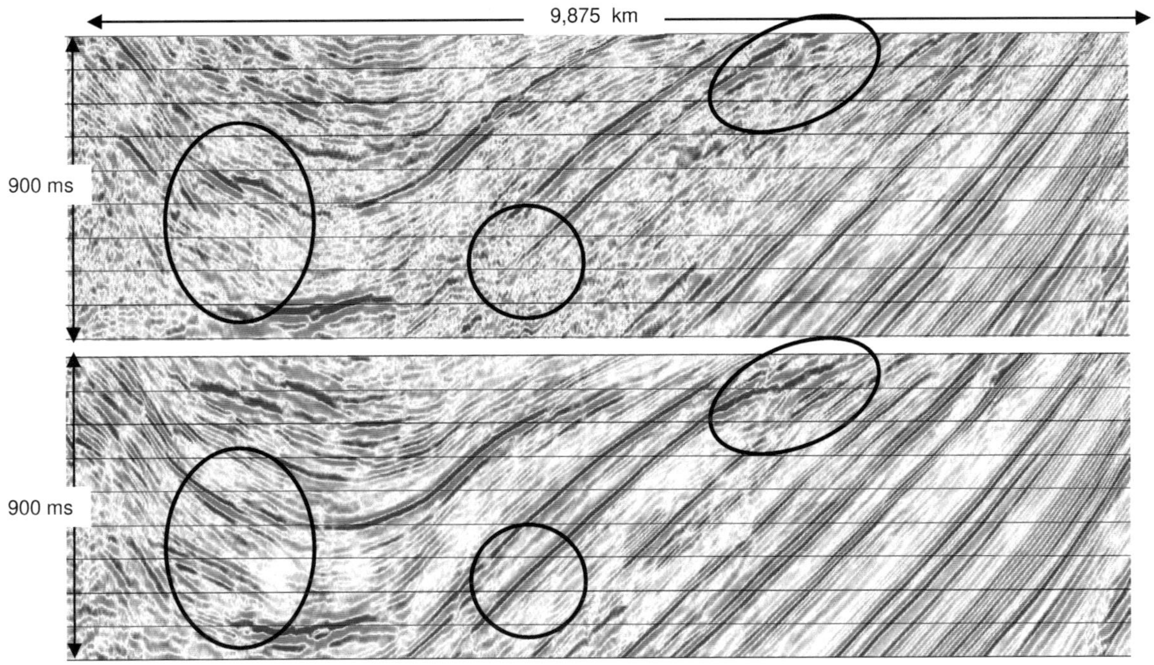

Figure 5. The difference between stack section obtained from conventional stacking method (top) and from Partial CRS-Stack method (bottom). The reflection continuities and S/N ratio are enhanced in the result of Partial CRS-Stack method.

4. Conclusion

The example showed in this paper proved the significant ability of Partial CRS-Stack method in enhancing the quality of seismic data. The significant improvement could be seen, if this method is implemented to the data that is acquired from an area with complex subsurface geological structures. If one faces irregular seismic data, this method can provide supergathers as output data, which contain denser traces. The fold number can be increased by implementing Partial CRS-Stack. Finally, this method could be suitable to be implemented for exploration purposes in Indonesian region, since most of Indonesia's geological conditions are very complex due to active tectonic processes.

References

[1] Baykulov M 2009 Seismic imaging in complex media with the common reflection surface stack *Ph.D thesis, University of Hamburg*

[2] Garabito G, Cruz J C, Eiras J, and Queiroz N P 2003 Application of CRS stack to seismik data of Amazon Paleozoic Basin *8th Proceeding of International Congress of Brazilian Geophysical Society* Rio de Janeiro - Brazil.

[3] Jäger R, Mann J, Hoecht G, and Hubral P 2001 *Geophysics* **66**(1) 97–109.

[4] Mann J 2002 Extensions and applications of The Common-Reflection-Surface Stack Method *Logos Verlag, Berlin*

[5] Müller T, Jäger R, and Hoecht G 1998 Common reflection surface stacking method – imaging with an unknown velocity model *68th Annual Internat. Mtg., Soc. Expl. Geophys.* p 1764–1767

[6] Trappe H, Gierse G, and Pruessmann J 2001 Case studies show potential of common reflection surface stack-structural resolution in the time domain beyond the conventional NMO/DMO stack *First Break* **19**

IOP Conf. Series: Earth and Environmental Science **62** (2017) 012052 doi:10.1088/1755-1315/62/1/012052

Hypocenter Determination Using a Non-Linear Method for Events in West Java, Indonesia: A Preliminary Result

Shindy Rosalia[1], Sri Widiyantoro[2], Andri Dian Nugraha[2], Hasbi Ash Shiddiqi[3], Pepen Supendi[3, 4], and Wandono[4]

[1]Master Program of Geophysical Engineering, Faculty of Mining and Petroleum Engineering, Institut Teknologi Bandung
[2]Global Geophysics Research Group, Faculty of Mining and Petroleum Engineering, Institut Teknologi Bandung
[3]Earth Sciences Graduate Program, Faculty of Earth Sciences and Technology, Institut Teknologi Bandung
[4]Meteorological, Climatological, and Geophysical Agency of Indonesia, Jakarta

E-mail: shindy.rosalia31@gmail.com

Abstract. West Java, part of the Sunda Arc, has relatively high seismicity due to subduction activity and faulting. The first step of tomography study in order to infer the geometry of the structure beneath West Java is to conduct precise earthquake hypocenter determination. In this study, we used earthquake waveform data taken from the regional Meteorological, Climatological, Geophysical Agency (BMKG) network from South Sumatra to central Java. We have repicked P and S arrival times from about 800 events in the period from April 2009 to December 2015. We selected the events which have azimuthal gap < 210° and phase more than 8. The non-linear method employed in this study used the oct-tree sampling algorithm from NonLinLoc program to determine the earthquake hypocenters. The hypocenter location results give better clustering earthquakes which are correlated well with geological structure in the study region. We also compared our results with BMKG catalog data and found that the average hypocenter location difference is about 12 km in latitude direction, 9.5 km in longitude direction, and the average focal depth difference is about 19.5 km. For future studies, we will conduct tomographic imaging to invert 3-D seismic velocity structure beneath the western part of Java.

1. Introduction

West Java is located in the central Sunda Arc, i.e. an active plate boundary zone between oblique subduction of the Indo-Australian plate beneath Sumatra and perpendicular subduction along Java [1]. This tectonic activity has formed fault structure in the western part of Java, such as Cimandiri Fault [1]. Historical seismicity of West Java before 1921 has been studied and revealed that West Java has 2 major earthquake events with magnitude greater than 7.5, i.e. in 1903 and 1921 [2]. Meanwhile for the last decade, there were also two major earthquakes occurring in this area. On 17 July 2006, a thrust faulting earthquake with Mw 7.7 occurred on the boundary between the Indo-Australian and Sunda plates producing tsunami in Pangandaran [3]; and on 2 September 2009 a reverse faulting earthquake with Mw 7 hit south of West Java and caused a landslide in Cikangkareng [4]. Due to its high seismicity level, it is important to have a better knowledge of the geometrical structure beneath West

IOP Conf. Series: Earth and Environmental Science **62** (2017) 012052 doi:10.1088/1755-1315/62/1/012052

Java, for example by employing a tomographic method, which requires accurate earthquake hypocenters as input.

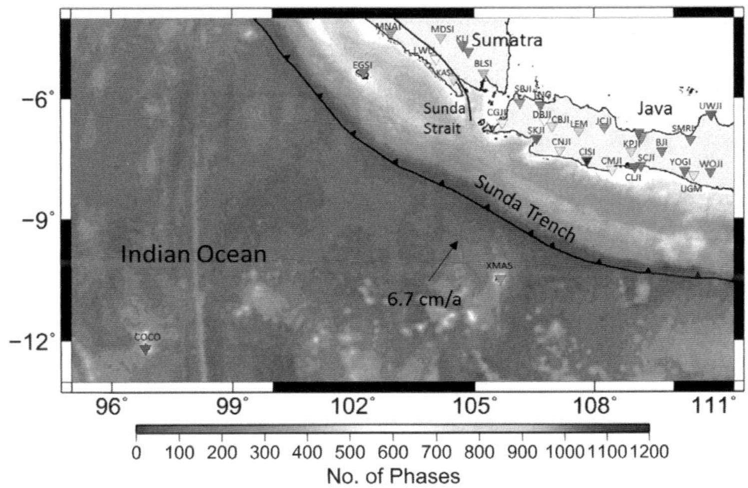

Figure 1. Map showing the location of seismographic stations (inverted triangles) used in this study. Colours depict the number of phases picked for each station.

Previous studies related to seismicity and tomography in West Java and its surroundings were conducted by using a global data set beneath the Sunda Arc [5], regional and local earthquake data catalogue [6, 7, 8], volcano-tectonic earthquake [9, 10, 11], employing non-linear tomography for eastern Sunda Arc [12], and relocating hypocenters with a 3D velocity structure of West Java [13, 14]. In this study, we used different data set and method to determine earthquake hypocenters.

2. Data and Method

We repicked P and S-wave arrival time data taken from BMKG waveforms for 891 events in West Java and its surrounding area from April 2009 to December 2015. We used 33 BMKG stations from South Sumatra to central Java (Figure 1). We selected the events which have azimuthal gap < 210° and phase more than 8 in order to have a good constraint and quality of location result. We also conducted a quality control of the repicking process by plotting Wadati diagram.

After the repicking process, we applied a hypocenter determination technique using the non-linear NLLoc program [14, 15]. This program contains a probabilistic, direct search method which employs the algorithm developed by Tarantola and Valette [16]. Unlike a linearized approach that can give convergent problems, e.g. the solution can be trapped in a local minimum, the non-linear approach can overcome such a convergent problem and can identify irregular volumetric, probability density function better. This method has been applied successfully in Yellowstone [17] and Southwest Iberia [18]. For this study, we used the global package from NLLoc with AK135 spherical velocity model [19] and oct-tree importance sampling algorithm [20] to determine the hypocenter.

3. Result and Discussion

Hypocenter determination result consists of 829 events located using ~12,000 P and ~4,900 S arrival times. The NLLoc result shows a significant improvement in epicentre locations (Figure 2a) by refining more clustered earthquakes related to geological features. The epicentre shifting vectors and rose diagram show that the direction of the shifting is trending to Northeast-Southwest, perpendicular to the trench (Figure 2b). Histogram of NLLoc residual travel times also shows a better data fit compared to those of BMKG (Figure 2c). The average hypocenter location difference between NLLoc and BMKG is 12 km in latitude direction with maximum shifting 136 km, 9.5 km in longitude direction with maximum shifting 142 km, and 19.5 km in depth with maximum shifting 157 km. The improvement may be due to the repicking process so that arrival times are more accurate and/or because of the velocity model used and station distribution, in which most of the stations are located in the Java Island.

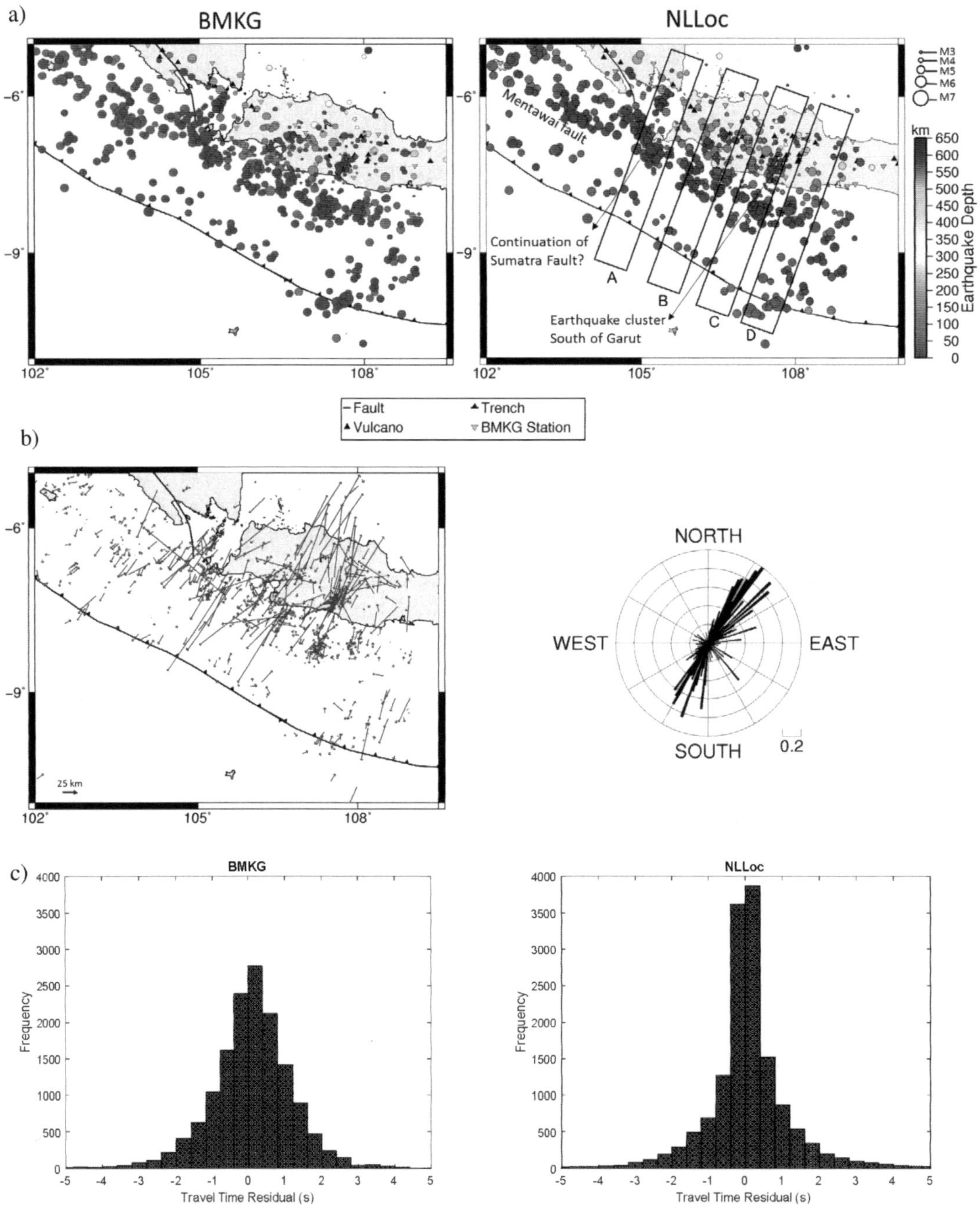

Figure 2. a) Comparison of West Java seismicity from BMKG catalog (left) and NLLoc result (right). Colours indicate earthquake depths and the size of dots depicts event magnitudes. Boxes represent

areas used to plot the cross sections shown in Figure 3. The NLLoc result gives a better event clustering compared to BMKG. b) Map representing epicenter shifting from BMKG location to NLLoc location (left). Rose diagram showing the major direction of the shifting that is perpendicular to the trench. c) Histograms of residual times of BMKG (left) and NLLoc result (right).

Seismic activity in West Java is dominantly distributed to the south of the island, i.e. near the trench, and in the fore-arc owing to the subduction of the Indo-Australian plate. Shallow clustered earthquakes on land are probably related to faulting, e.g. Cimandiri Fault. We also observed some interesting earthquake clusters (Figure 2a). For instance: an earthquake cluster in the south of Sumatra which is related to Mentawai Fault; an earthquake cluster that shows a lineation in the south of the Sunda Strait that is probably the continuation of Sumatra Fault; and an earthquake cluster in the south of Garut. However, we still need further studies in order to confirm these intriguing event clusters.

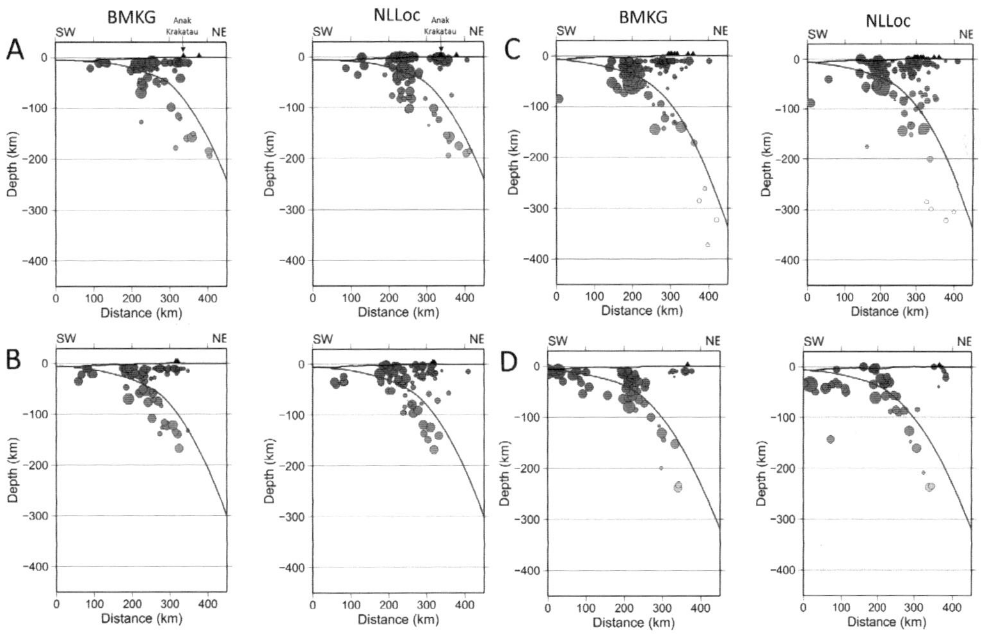

Figure 3. Cross sections of West Java: (A) through the Sunda Strait, (B) Pelabuhan Ratu and Cimandiri Fault, (C) earthquake cluster in the south of Garut, and (D) earthquake cluster near the trench. Colours represent earthquake depths and blue line depicts the slab 1.0 model [15]. In general the NLLoc result gives stepper earthquake distributions compared to those inferred from BMKG hypocenters and slab 1.0.

Based on the cross sections in Figure 3, the NLLoc result shows a better earthquake clustering that is in good agreement with geological features, especially in shallow part. One of the problems if we use the BMKG catalog as input to a tomographic study is the 10 km fixed depth (see Figure 3a, b, c). We can see that below the mountain and its surrounding area, there are lineations of events of 10 km depth in the BMKG location. Our NLLoc result improves the fixed depth and produces earthquake clusters below the mountain, e.g. beneath Mt. Anak Krakatau (Figure 3a). We also observed that there is an earthquake sequence trending from thethe slab surface upwards, close to the 2 September 2009 Mw 7 earthquake, indicating a back-thrust fault in the area of the 2009 event. From the NLLoc result, we infer a slab in the eastern part of the study area (Figure 3d), which is steeper than those inferred from the BMKG location and the slab model 1.0 [21].

4. Conclusions

We have conducted hypocenter determination using a non-linear method for events in West Java. The new hypocenter locations can improve the location of 10 km fixed depth from the BMKG catalog. Furthermore, the new locations also give better clustered earthquakes which are correlated with geological structures from previous studies. This hypocenter location data set will be used as input in our near future works which will include an application of a double-difference tomography to obtain a 3D velocity structure beneath West Java.

Acknowledgments

We are grateful to the Meteorological, Climatological, and Geophysical Agency of Indonesia (BMKG) for the waveform and catalog data; and Prof. Anthony Lomax for his software NLLoc and Seisgram2Kv.7.0 used in this study. We are also grateful to PMDSU 2015 scholarship from the Ministry of Research, Technology, and Higher Education of the Republic of Indonesia awarded to SR. Most figures were produced using the GMT Software [22]. Topography and bathymetry data were taken from SRTM 30 Plus [23], while plate boundary was adopted from Bird [24].

References

[1] J A Malod, Komar Karta, M O Beslier, and M T Zen Jr. 1995 *J. Southeast Asian Earth Sci.* **12** 85-93
[2] K R Newcomb and W R Mccann 1987 *J. Geophys. Res.* **92** 421–439
[3] J. Mori, W. D. Mooney, Afnimar, S. Kurniawan, A. I. Anaya, and S. Widiyantoro 2007 *Seismol. Res. Lett.* **78** 201–207
[4] Earthquake Engineering Research Institute 2009 *EERI Spec. Earthq. Rep.* 7–8
[5] S Widiyantoro and R D van der Hilst 1996 *Science* **271** 1566-70.
[6] P Supendi and A D Nugraha 2016 *AIP. Conf. Proc.* **1730** 020002 DOI: 10.1063/1.4947370
[7] A D Nugraha, H A Shiddiqi, S Widiyantoro, Sutiyono, T Handayani 2016 *AIP. Conf. Proc.* **1730** 020002 DOI: 10.1063/1.4947371
[8] A D Nugraha, H A Shiddiqi, S Widiyantoro, M Ramdhan, Wandono, Sutiyono, T Handayani, H Nugroho 2015 *AIP. Conf. Proc.* **1658** 030002 DOI: 10.1063/1.4915010.
[9] R V Ry, A Priyono, A D Nugraha, A Basuki 2016 *AIP. Conf. Proc.* **1730** 020004 DOI: 10.1063/1.4947372
[10] A D Nugraha, S Widiyantoro, A Gunawan, G Suamtika 2013 *J. Math. Fund. Sci* **45(1)** 17-28 DOI: 10.5614/ j.math.fund.sci.2013.45.1.2
[11] A Priyono, G Suantika, S Widiyantoro, A D Nugraha 2011 *International Journal of Tomography and Statistics* **17 (S11)** 17-28
[12] S Widiyantoro, J D Pesicek, C H Thurber, and M Wi 2011 *SE Asian Gatew. Hist. Tectonics Aust. Collis.* 139–155
[13] A P Sakti and A D Nugraha 2012 *Thesis, Institut Teknologi Bandung (Unpublished)*
[14] A Lomax and Jean Virieux 2000 *Advances in Seismic Event Location,* ed Clifford H Thurber and Nitzan Rabinowitz (Amsterdam: Kluwer) pp 101-134
[15] A Lomax, A Michelini, and A Curtis 2014 *Encyclopedia of Complexity and System Science,* ed R A Meyers (New York: Springer) pp 2449-73
[16] A Tarantola and B Valette 1982 *J. Geophys.* **50** 159-170
[17] S Husen and R B Smith 2004 *Bull. Seismol. Soc. Am.* **94** 880–896
[18] G Pinho 2008 Probabilistic non-linear earthquake location in a 3 – D velocity model
[19] B L N Kennett, E R Engdahl, and R Buland 1995 *Geophys. J. Int.* **122** 108–124
[20] A Lomax and A Curtis 2001 *Geophys Res Abstr* [Online]. Available: http://alomax.free.fr/nlloc/octtree/OctTree.html. [Accessed: 25-Sep-2016].
[21] G P Hayes, D J Wald, and R L Johnson 2012 *J. Geophys Res.* **117** 1–15

[22] P Wessel and W H F Smith 1998 *Eos, Trans. Am. Geophys. Union* **79** 579–579
[23] J J Becker, D T Sandwell, W H F Smith, J Braud, B Binder, J Depner, D Fabre, J Factor, S Ingalls, S H Kim, R Ladner, K Marks, S Nelson, A Pharaoh, R Trimmer, J Von Rosenberg, G Wallace, and P Weatherall 2009 *Mar. Geod.* **32** 355–371
[24] P Bird 2003 *Geochemistry, Geophys. Geosystems*, vol. **4**

IOP Conf. Series: Earth and Environmental Science **62** (2017) 012053 doi:10.1088/1755-1315/62/1/012053

Hypocenter Relocation of Earthquake Swarm in West Halmahera, North Molucca Region, Indonesia by using Double-Difference Method and 3D Seismic Velocity Structure

Andri Dian Nugraha[1,*], Hasbi Ash Shiddiqi[2], Sri Widiyantoro[1], Nanang T Puspito[1], Wahyu Triyoso[1], Samsul Wiyono[3], Daryono[3], Wandono[3], Shindy Rosalia[4]

[1]Global Geophysics Research Group, Faculty of Mining and Petroleum Engineering, Institut Teknologi Bandung, Ganesha 10, Bandung 40132, Indonesia
[2]Earth Sciences Graduate Program, Faculty of Earth Sciences and Technology, Institut Teknologi Bandung, Ganesha 10, Bandung 40132, Indonesia
[3]Geophysical Engineerinng, Faculty of Mining and Petroleum Engineering, Institut Teknologi Bandung, Ganesha 10, Bandung 40132, Indonesia
[4]Meteorological, Climatological, and Geophysical Agency (BMKG) of Indonesia, Jakarta, Indonesia

*corresponding author's email: nugraha@gf.itb.ac.id

Abstract. The earthquake swarm events sequence occurred in west Halmahera, north Molucca, Indonesia for the period of October 2015 to February 2016 as reported by Meteorological, Climatological, and Geophysical Agency (BMKG) of Indonesia. There were tenths swarm events with Magnitude larger than four in the region during the period. In this study, we used the earthquake catalog data compiled by BMKG to improve the location of swarms event in west Halmahera, north Molucca, Indonesia. We relocated 86 swarm events by applying teleseismic double-difference method and 3D seismic velocity model. The focus depth of swarm events mainly concentrated at depth of 5 to 12 km at south-east of Jailolo volcano. Our preliminary interpretation the earthquake swarms may be related to the stress change around the deep magma region of the volcano.

1. Introduction

Earthquake swarms are generally defined as a sequence of events closely clustered in time and space without mainshock and these events can be used for forecasting eruptions [1]. It means swarms are sequence of the quakes of similar magnitude to the mainshock. Earthquake swarms are usually common in volcanic regions [2]. The previous study on AD 2000 Izu island earthquake swarm by Toda et al. [3] concluded that gradual and sustained increase in stressing rate caused by magma intrusions and eruptions, extrusion or creep event will lead to producing earthquake swarms. The phenomenon of earthquake swarms were also detected by BMKG network in west Halmahera, north Molucca, Indonesia in October 2015 to February 2016. During this time period, there were tenths swarm events with Magnitude larger than four in the region. The destructive swarm event has been reported occured on November 20, 2015 with Magnitude of 4.9 which caused houses and public facilities damaged in the region. The Molucca region has complex and unique tectonic and geological setting, high seismicity [4,

5, 6, 7] and volcano activities as shown in Figure 1. The previous tomographic inversion study by [8] showed the two opposing subducted slabs of the Molluca Sea plate beneath arc-arc collision zone between Sangihe and Halmahera. In this study, we attempted to determine an accurate earthquake swarms hypocenter by applying double-difference relocation method [9] and regional 3D seismic velocity model [8].

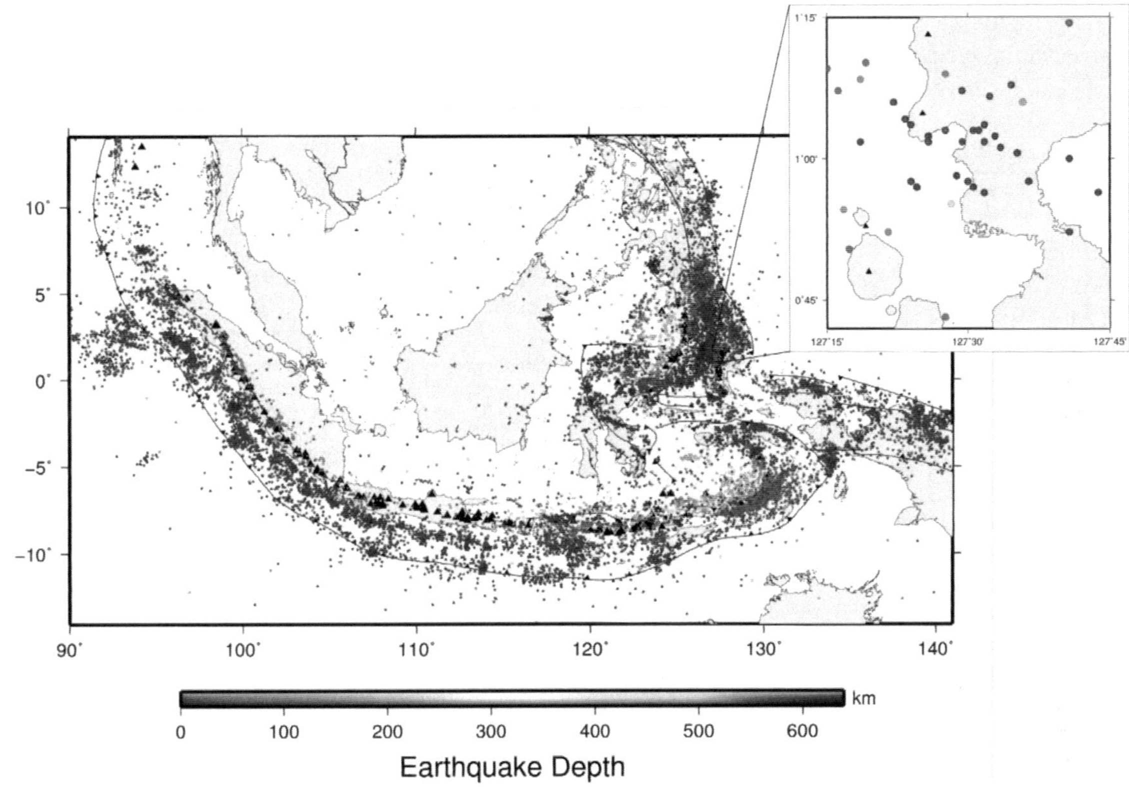

Figure 1. Distribution of earthquake epicenter with Magnitude > 4 (dot circle) around Indonesia region compiled by BMKG for the time period of April 2009 to January 2017. Top right is zoom in of the earthquake swarms location in west Halmahera, north Molucca, Indonesia.

2. Data & Method

We relocated 86 swarm events in west Halmahera region which occurred on November 2015. These events have magnitude range from 2.5 to 4.8. We used teleseismic double-difference (DD) relocation method [9] by using 3D velocity model from Widiyantoro and van der Hilst [8]. The arrival time data were obtained from Indonesian Meteorology, Climatology, and Geophysics (BMKG). We used around 30 stations at local and regional distances to relocate the swarm events.

3. Results and Discussions

We have been successfully determined 86 earthquake swarms event in west Halmahera, north Molucca, Indonesia by applying DD method. Our preliminary results show the relocated swarm events were distributed more clustered at south east of Jailolo volcano (Figure 2). The focus depth of swarm events and the fixed depth at depth of 10 km from BMKG data catalogue were improved significantly. The relocated swarm events have a focus depth from about 5 km to 30 km and commonly focused at depths of 5 to 12 km (Figure 3). The location of these events in region of active tectonic and volcano are a key to interpret the source mechanism of events, whether its related to stress changed in magma region or creep processes around the fault. Our preliminary results show the relocated swarm events are located close to Jailolo volcano. So our interpretation these events may be related to stress change of deep magma activity around the volcano. However, our results need to compare with others geological, geodetic and geophysical methods. The dense local networks are also needed to determine more precise hypocenter determination and advance seismological study. Statistically, the travel time residual of phase pairs for relocated swarms event using DD method in this study show much better values than the initial swarms event location (Figure 4).

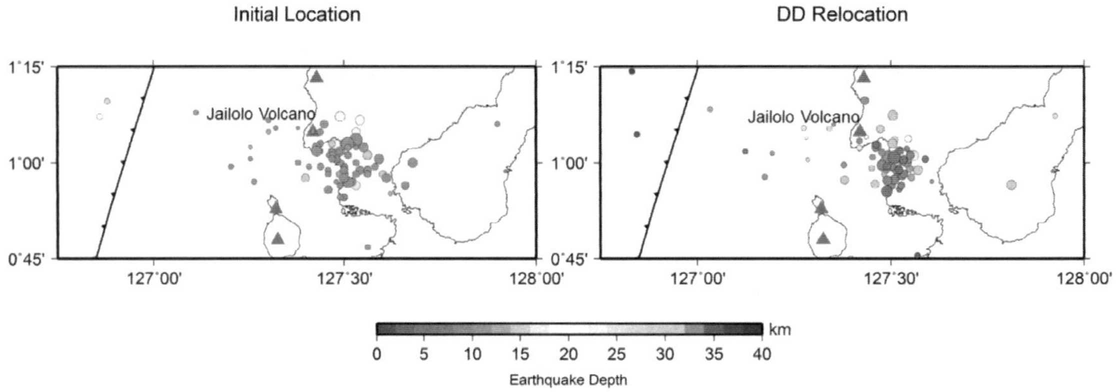

Figure 2. Earthquake swarms epicenter location before relocation (left panel) and after relocation using DD method and 3D seismic velocity model in this study (right panel).

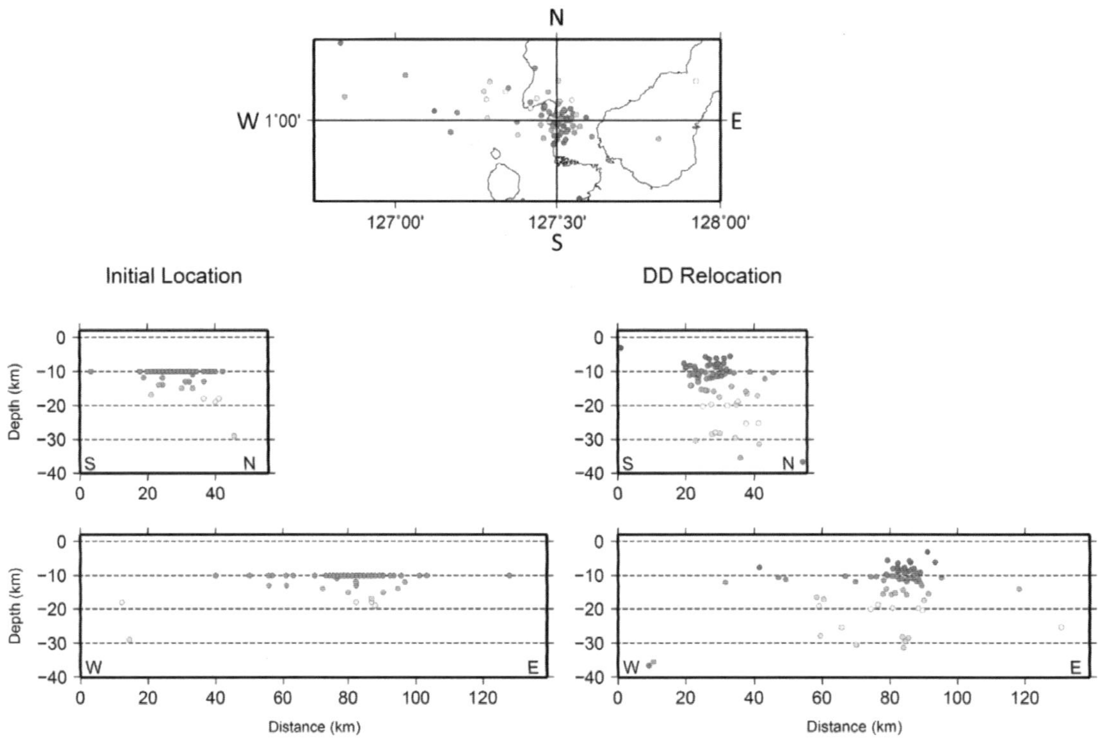

Figure 3. Vertical section of N-S and W-E before (left panel) and after relocation of earthquake swarms hypocenter distribution in this study (right panel), respectively.

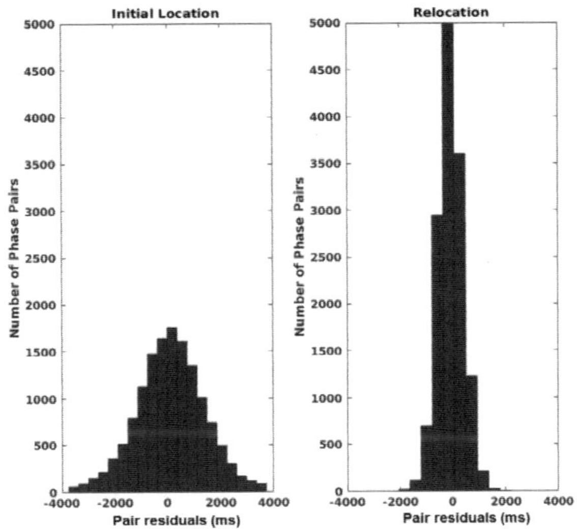

Figure 4. Histogram travel time residuals of phase pairs for initial locations (left) and relocations (right) of swarms events in this study.

4. Conclusions

We used the earthquake catalog data compiled by BMKG to improve the location of swarm events in west Halmahera, north Molucca region, Indonesia. We determined 86 swarm events by applying DD method and 3D seismic velocity model. The swarm events mainly distributed at depth of 5 to 12 km at south-east of Jailolo volcano. Our preliminary interpretation these swarm events may be related to stress change around the deep magma region of the volcano.

Acknowledgment

We gratefully thank BMKG for catalog data used in this study. This study supported by "Hibah Penelitian Unggulan Perguruan Tinggi 2015-2016 dari KEMRISTEKDIKTI, Indonesia"

References

[1] Mogi K 1963 *Bulletin of the Earthquake Research Institute* **41** 615-658.
[2] Benoit J P, Stephen R M 1996 *U.S. Department of the Interior, U. S. Geological Survey, Open-File Report*, **96-69**
[3] Toda S, Stein R S, Sagiya T 2002 *Nature* **419** 58-61
[4] Shiddiqi H A, Widiyantoro S, Nugraha A D, Ramdhan M, Wiyono S H, Wandono W 2016 *AIP. Conf. Proc.* **1730** 020010 doi:10.1063/1.4947378.
[5] Utama M R J, Nugraha A D, Puspito N T 2015 *AIP. Conf. Proc.* **1658** 030022 doi:10.1063/1.4915030
[6] Shiddiqi H A, Widiyantoro S, Nugraha A D, Ramdhan M, Wandono, Sutiyono, Handayani, T, Nugroho H 2015 *AIP. Conf. Proc.* **1658** 030011 doi:10.1063/1.4915019
[7] Nugraha A D, Shiddiqi H A, Widiyantoro S, Ramdhan M, Wandono, Sutiyono, Handayani, T, Nugroho H 2015 *AIP. Conf. Proc.* **1658** 030002 doi:10.1063/1.4915010
[8] Widiyantoro S, Van der Hilst R 1997 *Geophys. J. Int.* **130** 167-182
[9] Pesicek J D, Thurber C H, Zhang H, DeShon H R, Engdahl E R, Widiyantoro S 2010 *J. Geophys. Res.* **115** B10303 doi:10.1029/2010JB007443

Study on 2-D Crustal Shear Wave Splitting Tomography along The Sunda-Banda Arc Transition Zone

Syuhada Syuhada[1,2],Nugroho D. Hananto[3], Chalid I. Abdullah[4], Nanang T. Puspito[5], Tedi Yudistira[5]and Titi Anggono[1]

[1]Research Centre for Physics - Indonesian Institute of Sciences (LIPI), Tangerang, Indonesia
[2]Graduate Research on Earthquake and Active Tectonics (GREAT), Institut Teknologi Bandung, Indonesia
[3]Research Centre for Geotechnology-LIPI, Bandung, Indonesia
[4]Faculty of Earth Sciences and Technology, Institut Teknologi Bandung, Indonesia
[5]Global Geophysics Research Group, Faculty of Mining and Petroleum Engineering, Institut Teknologi Bandung, Indonesia

E-mail: hadda9@gmail.com

Abstract. The Sunda-Banda Arc transition zone is an active region characterized by a change in tectonic regime from subduction of Indo-Australia oceanic lithosphere along the eastern part of Sunda Arc to collision of the Australian continental crust with islands arc in the western part of the Banda Arc. This complicated tectonic setting causes this area is an ideal place to study the crustal deformation along the plate boundary. The density contrast between the Australian continental crust and Indo-Australia oceanic crust in the transition zone may cause large stresses around the boundary between them. These plate boundary forces may control the distribution pattern of the deformation in the subduction to collision transition zone. The geometry of this deformation can be investigated using shear wave splitting (seismic anisotropy) study. We conduct shear wave splitting measurements from local earthquakes recorded at 17 broadband seismic stations around the Sunda-Banda arc transition zone. The 2-D delay time tomography is then applied to determine the first order approximation of lateral varying anisotropic layers due to the local effect of geological structures. We observe strong anisotropy regions which coincide with the geological features as possible causes of anisotropy in the Sunda-Banda Arc transition zone. For instance, the high anisotropy zone found in Timor Island can be related to the alignment of metamorphic and igneous rocks, whereas the high anisotropy area around Sumba Island might correspond to the interaction of Sumba basement with the Australian margin increasing the frictional strength at the plate boundary.

1. Introduction

The evidence of seismic anisotropy in the crust has been investigated through shear wave splitting analysis providing information on the deformation related to the regional geodynamic processes. The shear wave splitting analysis determines the fast direction of the split S waves (ϕ) and delay time between the split S waves (δt). There are two mechanisms responsible for crustal anisotropy: stress induced anisotropy and structural anisotropy [1, 2]. The previous study of shear wave splitting from local event has been conducted in the collision domain of Sunda-Banda transition zone [3]. This study observed that the crustal anisotropy in this area is not only governed by stress induced anisotropy, but also by structure-controlled anisotropy. In this report we extend our data set using 17 GEOFONE-IA broadband three component seismometers covering two different tectonic domains, subduction and collision. We use ISC database to search shallow events (< 30 km) in the study area recorded between 2008 and 2015. We also utilize the automated shear wave splitting (MFAST) developed by Savage *et al.* [4] to estimate anisotropy parameters. We employ two-dimensional tomography method of Johnson *et al.* [5] to analyze spatial anisotropy distribution in the region. The result of this research will be used to map and to identify regions with high and low degree anisotropy that might be correlated to the possible causes of anisotropy.

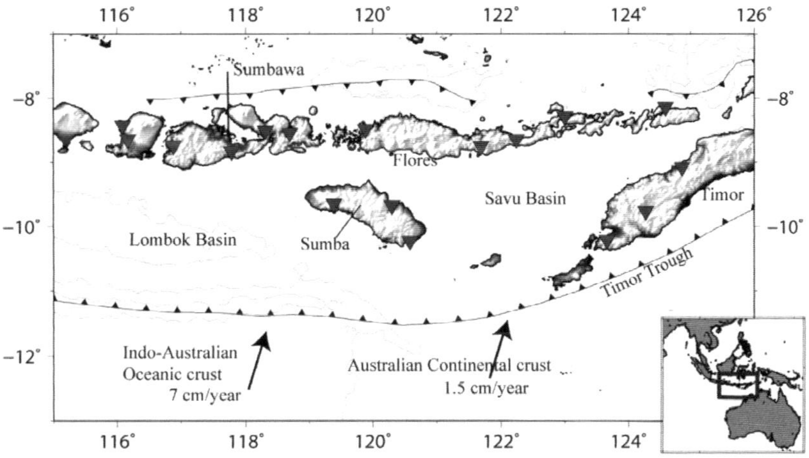

Figure 1. Map showing the tectonic setting of Sunda-Banda arc transition zone with the inverted blue triangles representing the seismic station used in this study.

Sunda-Banda arc transition zone (figure 1) is characterized by a change in tectonic setting from subduction of the Indo-Australian plate in west of Sumba to continent-island arc collision in east of Sumba [6]. The boundary of this transition zone is marked by the presence of Sumba Island, which is believed to be an isolated micro-continent block trapped in the forearc [7, 8]. However, the origin and evolution of Sumba Island is still debated [7]. The convergence rate at the west of Sumba is around 70 mm/yr [9] and decreases to ~15 mm/yr near the Timor Trough [10]. A previous seismic reflection study conducted in the west of Sumba observed that this region heavily faulted and fractured into 5-10 km wide blocks [11]. In the basin located east of Sumba, major structures are dominated by normal faults [12] and controlled by insertion of the promontory Australian continental crust beneath Banda arc. Furthermore, active volcanic islands mark the forearc boundary in the north of Sumba. A major structural discontinuity separating Sunda and Banda arc might exist between Sumbawa and Flores Islands as observed and proposed by geological and GPS studies [6, 13]. Timor Island marks the south boundary of the collision zone. This island is formed as a result of the collision between Australian plate and Banda arc [14]. The structure of the island is composed of complex group of rocks involving variety of deformed rocks including mafic and ultramafic rocks which mix together [15, 16].

2. Method

2.1. Shear-wave splitting

We implement splitting measurement using the automatic shear wave splitting algorithm, MFAST code [4]. This method is based on a grid search inversion of Silver and Chan [17] algorithm to compute splitting parameters (the fast polarization direction ϕ and delay time δt) for a given time window (figure 2). The method also incorporates cluster analysis introduced by Teanby *et al.* [18] to search the most stable measurement over a range of time windows. Multiple filters are used, and the best filter is chosen based on the product of the signal-to-noise ratio and bandwidth. Uncertainties of the measurements are computed using F test for the selected time window by seeking the 95% confidence level for the optimum values of ϕ and δt. The selected splitting results are then graded from A to D reflecting good and bad results, based on the stability of the results obtained from the cluster analysis. Only the results with grades of at least B are selected for further processing. It is important to note that we only use the rays with incidence angles less than 35° to avoid contamination from the S-P conversion at the surface. The incidence angles are calculated using the TauP Toolkit [19] with the 1-D velocity models derived from receiver function study at each seismic station [20]. We observe that the models contain the low velocity layer at the near surface bending seismic rays into near-vertical propagation.

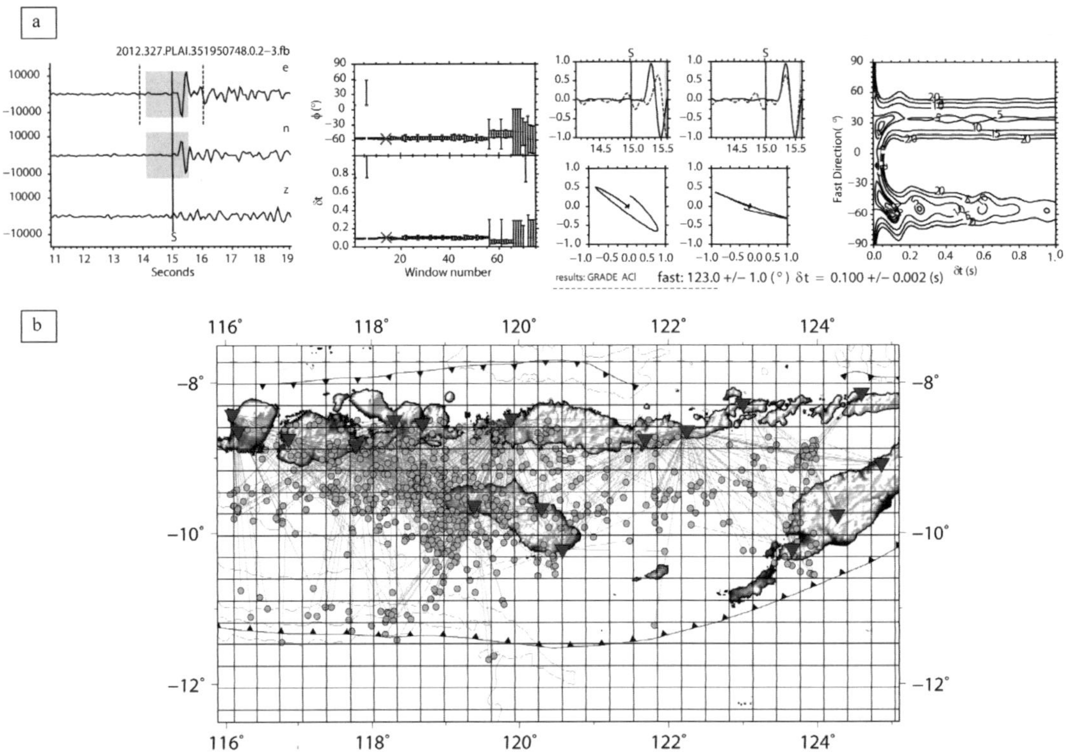

Figure 2. (a) An example of the splitting measurement obtained from MFAST. The left panel shows the three component seismogram with the S wave manual picking. The middle panels depict the cluster analysis to provide the best solution of splitting parameters, the waveforms and the particle motions of shear wave before and after correction for the splitting measurement, respectively. The right panel shows the contour diagram for the best solution denoted by the blue cross. (b) The earthquake epicenters (grey dots), raypaths (red lines) and grids used for the splitting analysis and inversion 2-D delay time tomography.

2.2. *2-D delay time (δt) tomography*

To constrain the regions with low and high anisotropy, we employ 2-D delay time (δt) tomography using TESSA package developed by Johnson *et al.* [5]. The method measures the strength anisotropy in each of grid blocks assuming that the delay time (δt) is accumulated along the raypath. Thus, the relation between δt and path length L for each grid block can simply be written as:

$$\delta t_r = \sum_{b=1}^{n} S_b \times L_{rb}$$

where δt_r represents the estimated delay time of the ray r, S_b is the strength anisotropy for each grid block and L_{rb} is the path length of the ray r travelling within each of the n blocks. This equation can be solved using least square inversion to estimate the strength anisotropy. The solution satisfies the boundary condition in which the minimum value of strength anisotropy in each block is always greater than 0 km/s and the maximum values cannot exceed the maximum δt observed for a single block. Although this linear relationship reflects contrasts in the observed apparent δt rather than accurate values, the results still provide the first order approximation of heterogeneous anisotropic structure [5]. In this study, we use regular gridding with 32 km square blocks (figure 2).

3. Result and discussion

A large spatial variation in the splitting results obtained from the crustal events indicates complex anisotropic sources beneath the region [3] causing the interpretation difficult. Thus, to determine the average structure in the regions with low and high anisotropy and the most obvious trend of fast polarizations, we utilize 2-D delay time tomography and spatial averages of fast directions. Figure 3 depicts the δt tomography results using high-quality splitting parameters computed from 930 event-station pairs. We apply the ray coverage within the block, the checkerboard test results and the resolution maps above 5 (i.e. this value is chosen by considering that some blocks contain 8 passing rays) to define the limit of statistical significance illustrated by the white boundary lines in figure 3.

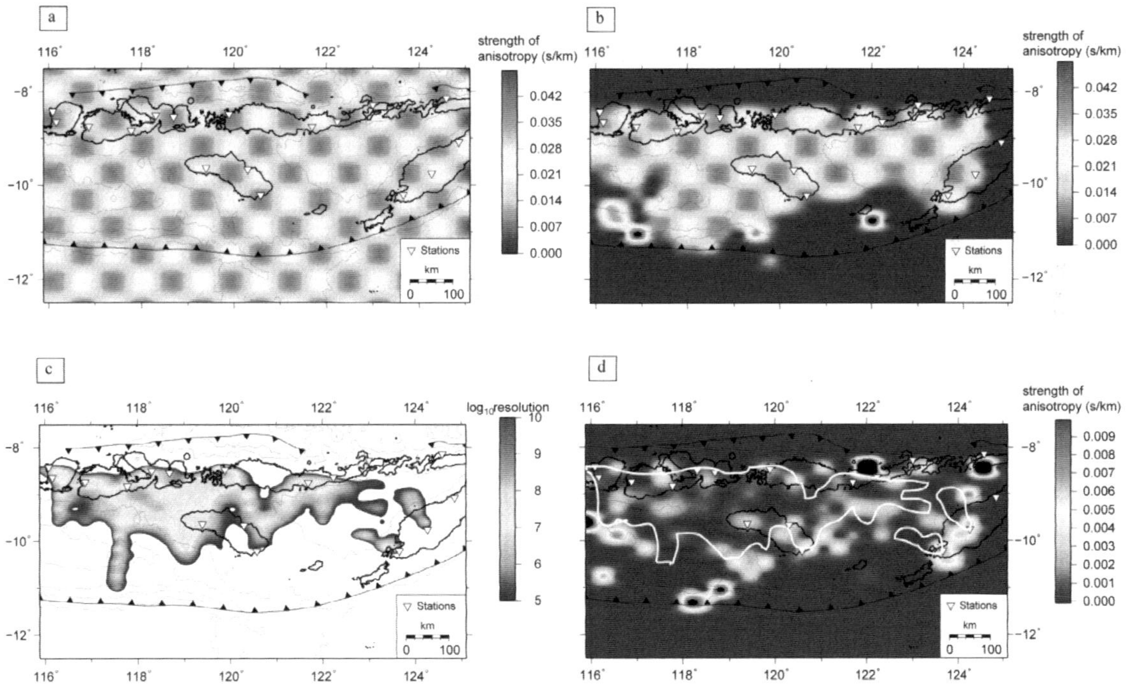

Figure 3. (a) Input for checkerboard test for 2-D delay time tomography. (b) Output for checkerboard test for 2-D delay time tomography using synthetic data created from the true rays passing through the checkerboard. (c) Resolution data inversion showing the limit of statistical significance computed from the model variance matrix [5]. (d) 2-D δt tomography using high-quality splitting parameters computed from 930 event-station pairs around the Sunda-Banda arc transition zone. The region with higher strength anisotropy is marked by warm colour (red to yellowish green), the weak anisotropic zone is denoted by blue colour. The white boundary line marks the region satisfying the resolution limit.

The inversion results indicate that the collision domain exhibits higher anisotropy (> 0.003 s/km), which may reflect the influence of the continent-arc collision. The high anisotropy found in this region coincides quite well with the region that experience crustal thickening (e.g. Savu Basin and Sumba) due to the insertion of Australian continental crust underneath southern part of the Banda arc. We observe possible high anisotropy beneath Lombok Basin. For the shallow depth, the area of Lombok Basin is characterized by more seismically active compared to the area in the east of Sumba Island [e.g. 21]. Previous seismic reflection studies show that that the crustal rocks beneath Lombok Basin are heavily fractured [e.g. 11, 22], suggesting that this area is actively deformed. The strong anisotropy is also observed in Timor Island. However, the island has very little earthquake activity, indicating stress induced anisotropy is not the only main source of anisotropy in this island. Thus, we suggest that the strong anisotropy in this island might correspond to the crustal fault zone or alignment of high anisotropic mineral such as gneisses, schist, and amphibolites as observed also by geophysical and geological studies [23, 24].

4. Conclusions

We apply 2-D shear wave splitting tomography to map lateral distribution of the possible sources of anisotropy in the Sunda-Banda transition zone. The result is in good agreement with structural features observed by previous geological and geophysical studies. For example, high anisotropy on Lombok Basin, Sumba, and Timor might correspond to the heavily fractured rocks associated to the active seismicity, fluid-filled crack related to the low-velocity zone due to insertion of continental crust, and the alignment of metamorphic and igneous rocks formed during later stages of continental-arc collision, respectively.

5. Acknowledgments

We acknowledge Indonesian Agency for Meteorology, Climatology, and Geophysics (BMKG) and GFZ Potsdam for their seismogram data. We are also grateful to Prof. Martha Savage and Jessica Johnson for providing softwares used in this study.

References

[1] Boness N L and Zoback M D 2004 *Geophys. Res. Lett.* **31** L15S17
[2] Boness N L and Zoback M D 2006 *Geology* **34** (10) 825-828
[3] Syuhada, Nugroho D H, Nanang T Puspito, Titi Anggono and Tedi Yudistira 2016 AIP Conference Proceedings **1719** 030042
[4] Savage M K, Wessel A, Teanby N A and Hurst A W 2010 *J. Geophys. Res.* **115** B12321
[5] Johnson J H, Savage M K and Townend J 2011 *J. geophys. Res.* **116** B12303
[6] Audley-Charles M G 1975 *Tectonophys* **26** 213-228
[7] Rutherford E, Burke K and Lytwyn J 2001 *J. Asian Earth Sci.* **19** 453– 479
[8] Abdullah C I, Rampnoux J P, Bellon H, Maury R C and Soeria-Atmadja R 2000 *J. Asian Earth Sci.* **18** 533-546
[9] Curray J R 1989 *Neth. J. Sea. Res.* **24** 131– 140
[10] Bock Y, Prawirodirdjo L, Genrich J F, Stevens C W, McCaffrey R, Subarya C, Puntodewo S S O and Calais E 2003 *J. geophys. Res.* **108** (B8) 2367
[11] Lueschen E, Mueller C, Kopp H, Engels M, Lutz R, Planert L, Shulgin A and Djajadihardja Y 2011 *Tectonophysics* **508** 1-4
[12] Rigg J W D and Hall R 2012 *Mar. Pet. Geol.* **32** 76-94
[13] Nugroho H, Harris R A, Amin W L and Bilal M 2009 *Tectonophys* **479** 52-65
[14] Harris R A and Audley Charles M G 1987 *Memoir Geological Society of China* **9** 45-61

[15] Barber A J, Audley-Charles M G and Carter D J 1977 *J. Geol. Soc. Australia* **24** 51–62
[16] Charlton T R 1989 *Aust. J. Earth. Sci.* **36** 263-274
[17] Silver P and Chan G 1991 *J. Geophys. Res.* **96** 16.429-16.454
[18] Teanby N A, Kendall J M and van der Baan M 2004 *Bull. Seismol. Soc. Am.* **94** 453–463
[19] Crotwell H P, Owens T J and Ristema J 1998 *Seismol. Res. Lett.* **70** 154-160
[20] Syuhada S, Hananto N D, Abdullah C I, Puspito N T, Anggono T and Yudistira T 2016 *Accepted Acta Geophysica*
[21] Ely K S and Sandiford M 2010 *Tectonophys.* **483** 112–124
[22] Van derWerff W 1996 *J. Southeast Asian Earth Sci.* **14** 331–349
[23] Karig D E, Barber A J, Charlton T R, Klemperer S E and Hussong D M 1987 *Geol. Soc. Am. Bull.* **93** 18–32
[24] Helmers J, Sopaheluwaken J, Tjokrosapoetro S, and Nila E S 1989 *Neth J Sea Res* **24** 357-371

Comparative Analysis of Peak Ground Acceleration Before and After Padang Earthquake 2009 Using Mc. Guirre Method

Diah Ayu Rahmalia[1*] and Hesti Nilamprasasti[1]

[1]Geophysics Subdepartment, Department of Physics, Faculty of Science and Mathematics, Gadjah Mada University, Sekip Utara, Yogyakarta 55281, Indonesia.

*Corresponding author's e-mail : diah.ayu.rahmalia@mail.ugm.ac.id

Abstract. We have analyzed the earthquakes data in West Sumatra province to determine peak ground acceleration value. The peak ground acceleration is a parameter that describes the strength of the tremor that ever happened. This paper aims to compare the value of the peak ground acceleration by considering the b-value before and after the Padang earthquake 2009. This research was carried out in stages, starting by taking the earthquake data in West Sumatra province with boundary coordinates 0.923° LU - 2.811° LS and 97.075° - 102.261° BT, before and after the 2009 Padang earthquake with a magnitude ≥ 3 and depth of ≤ 300 km, calculation of the b-value, and ended by creating peak ground acceleration map based on Mc. Guirre empirical formula with Excel and Surfer software.

Based on earthquake data from 2002 until before Padang earthquake 2009, the b-value is 0.874 while the b-value after the Padang earthquake in 2009 to 2016 is 0.891. Considering b value, it can be known that peak ground acceleration before and after the 2009 Padang earthquake might be different. Based on the seismic data before 2009, the peak ground acceleration value of West Sumatra province is ranged from 7,002 to 308.875 gal. This value will be compared by the value of the peak ground acceleration after the Padang earthquake in 2009 which ranged from 7,946 to 372,736 gal.

Keywords: Padang Earthquake, b-value, Peak ground acceleration, Mc. Guirre method

1. Introduction

Indonesia is the area of triple junction plate convergence such as Indo-Australian Plate, Eurasian Plate, and Pacific Plate. It moves and interacts each other like Indo-Australian plate plunging to Eurasian plate which known as subduction [1]. Several islands in Indonesia are on the border of subduction zones such as Sumatra Island. West Sumatra especially Padang as the capital of West Sumatra province have a high risk of an earthquake because the location is close to the subduction zones. An Earthquake is an event of ground movement due to wave propagation generated by tectonic activity [2]. Based on [3] data an earthquake have moment magnitude 7.6 had hit West Sumatra on 30th September 2009 at 17:16:09 GMT with its epicenter located at coordinates LS 0.84 - 99.65 BT about 60 km WNW of the city of Padang with depth 71 km. The highest number of deaths and damage is in the Padang and surrounding.

Earthquakes are caused by the stress that accumulates continuously and release suddenly [4]. The accumulation of stress before the earthquake in 2009 known from the b value by Gutenberg-Richter equation [5]. Padang earthquake in 2009 has about 809 deaths and 2498 injuries. Most of the victims died due to crushed by building [3]. Which one cause the destruction by the earthquake is the peak ground acceleration. The peak ground acceleration is the acceleration value due to land in a place of earthquakes in a given period where the value is calculated at the point of research at the Earth's surface based on the data of earthquakes with the greatest value calculation [6]. Using earthquake data from USGS, the peak ground acceleration is calculated by Mc. Guirre method. The purpose of this paper is comparing and to map the value of peak ground acceleration before and after Padang earthquake 2009 to reduce seismic risk, from 7 years before and after Padang earthquake 2009.

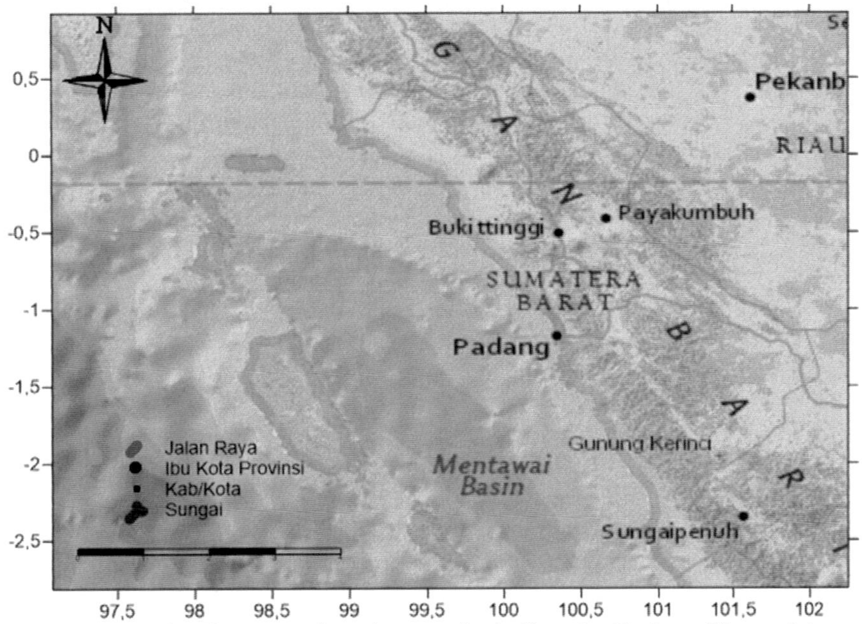

Figure 1. The map of study area, including the Padang City and its surrounding with coordinates 0.923° LU - 2.811° LS and 97.075° - 102.261° BT.

Figure 1 Shows that the earthquake data was taken from USGS with coordinates 0.923° - 2.811°LU and 97.075° - 102.261° BT as the area of West Sumatra, Padang city, and its surrounding, from 2002 until to the day before the mainshock in 2009 and at that day until 2016. West Sumatera, Padang city and surrounding had chosen an observed area because it has high seismicity and high population.

2. B-Value

Kirbani [7] represents the relation between magnitude and number of magnitude as follows:

$$\log N = a - bM \quad (1)$$

Where N is the number of magnitude M or larger which a cumulative curve, a is productivity, and b is called b value as a gradient of the graph LogN vs. M, and it has served of the tectonic parameter. B value depends on the stress level of a region where the stress value inversly towads b value [8][9][10].

3. Peak Ground Acceleration

Peak Ground Acceleration (PGA) is the value of the maximum ground acceleration due to earthquake shaking at a location. The definition PGA, in general is the accelerations of maximum ground vibrations that occurs at a point of a particular position in an area that is calculated by the result of all the earthquakes that happened at a definite time which concern to the magnitude and distance of its hypocenter.

PGA calculation can be done by several methods, one of the calculation methods is the empirical formula of Mc Guirre [11], as follows:

$$E[v] = a10^{bM}(R + 25)^{-c} \qquad (2)$$

where a = 472,3; b = 0,278; c = 1,301 and E[v] is the magnitude of the acceleration estimation in units of gal, M is the magnitude of the surface wave in the Richter scale; R is the distance between the hypocenter to the receiver (receiver station waves) in units of kilometers. Surface seismic acceleration can be expressed with "g", due to the similarity to Earth's gravity, either in decimal or percentage which can be expressed in m / s^2, where 1 g = 9.81 m / s2 or when converted into "gal", then 1 Gal = 0.01 m / s^2 with 1 g = 981 Gal.

The calculation of magnitude in Indonesia, (especially BMKG) use the calculation of body wave magnitude (Mb) and local magnitude (ML), it is necessary to convert the magnitude become surface magnitude (Ms). Gutenberg has redefined the relationship from the third magnitude as follows:

$$Mb = 0{,}56\ Ms + 2{,}9 \text{ and } Mb = 1{,}7 + 0{,}8\ ML - 0{,}01\ ML^2 \qquad (3)$$

So, Ms can be derived from the relationship above as follows :

$$Ms = 1.8Mb - 5.2 \qquad (4)$$

4. The Influence of Ground Motion toward Vulnerability

The Tectonic earthquake caused by the alignment process due to large deformation are derived by movements in the earth [12]. The Earthquake which has high magnitude and shallow hypocenter can damage the area near of epicenter and its surrounding.

Earthquake cause a wave of energy is release and propagation [2]. The wave will be constructive if the frequency of seismic wave is equal to the natural frequency of an area [13]. The natural frequency of an area depends on the lithology [14]. Based on geoengineering, soft, and unconsolidated rocks have a higher risk when shaken by a wave of earthquakes, because of a greater amplification compared with the most compact rock (Central Geological Survey, 2007). The high risk of damage can be caused by strong ground motion in soft lithology [15]. Therefore, in the earthquake-resistant infrastructure planning, analysis and the selection of ground motion parameters are absolutely necessary to obtain seismic load plan.(Irsyam et.al 2010) [15].

Table 4.1. Risk Level of Earthquake (BMKG, 2016)

Skala SIG BMKG	Warna	Deskripsi Sederhana	Deskripsi Rinci	Skala MMI	PGA (gal)
I	Putih	TIDAK DIRASAKAN (*Not Felt*)	Tidak dirasakan atau dirasakan hanya oleh beberapa orang tetapi terekam oleh alat.	I-II	<2,9
II	Hijau	DIRASAKAN (*Felt*)	Dirasakan oleh orang banyak tetapi tidak menimbulkan kerusakan. Benda-benda ringan yang digantung bergoyang dan jendela kaca bergetar.	III-V	2,9-88
III	Kuning	KERUSAKAN RINGAN (*Slight Damage*)	Bagian non struktur bangunan mengalami kerusakan ringan, seperti retak rambut pada dinding, genteng bergeser ke bawah dan sebagian berjatuhan.	VI	89-167
IV	Jingga	KERUSAKAN SEDANG (*Moderate Damage*)	Banyak retakan terjadi pada dinding bangunan sederhana, sebagian roboh, kaca pecah. Sebagian plester dinding lepas. Hampir sebagian besar genteng bergeser ke bawah atau jatuh. Struktur bangunan mengalami kerusakan ringan sampai sedang.	VI-VIII	168-564
V	Merah	KERUSAKAN BERAT (*Heavy Damage*)	Sebagian besar dinding bangunan permanen roboh. Struktur bangunan mengalami kerusakan berat. Rel kereta api melengkung.	IX-XII	>564

5. Methodology

To get the peak ground acceleration values using Mc. Guirre method is done in several steps as follows.

- The first stage consists of:
  a. The data of earthquake parameters is collected from West Sumatra and its surrounding areas derived from the USGS. The data that used are seismic data which recorded from 2002 to 2009 and data from 2009 until 2016.
  b. The data is being selected at a magnitude ≥ 4 SR and depth of ≤ 300 km
  c. The body magnitude (Mb) or other magnitude quantities are converted into the surface magnitude (Ms) as in equation (4).
  d. The seismicity maps are created based on seismic data acquired using Python software, which consists of seismicity maps before and after the Sumatra earthquake in 2009.
- The second step, The seismicity map of Padang and surrounding with the coordinate 0.923 ° LU – 2.811 ° LS and 97.075° - 102.261° BT is gridded into 0.5 x 0.5 degree.
- The third step consists of:
  a. The distance of the epicenter is calculated by the equation as follows:

$$I = \sqrt{x^2 + y^2} \qquad (5)$$

  Where I as epicenter, x as latitude and y as longitude. The distance of epicenter which are obtained then made in units of kilometers.

  b. The distance of hypocenter is calculated by this following equation.

$$R = \sqrt{I^2 + h^2} \qquad (6)$$

  Where R as hypocenter, I as epicenter and h as the depth of an earthquake.

- The fourth step consists of:
  a. The value of peak ground acceleration in every point is calculated either before or after Padang earthquake 2009 using Mc. Guirre equation (2) and then, the highest PGA in every grid is chosen
  b. Based on the highest PGA that has been chosen in every grid, PGA that have same value will be drawn in contour map using Surfer Software.

6. Result and Discussion

Catalog earthquake data is obtained from the United States Geological Survey (USGS) either before or after Padang earthquake 2009 from 2002 to the day before Padang mainshock 2009 and after the mainshock until 2016. Based on the earthquakes catalog, obtained the relationship between seismicity and b-value for seven years before and after the earthquake.

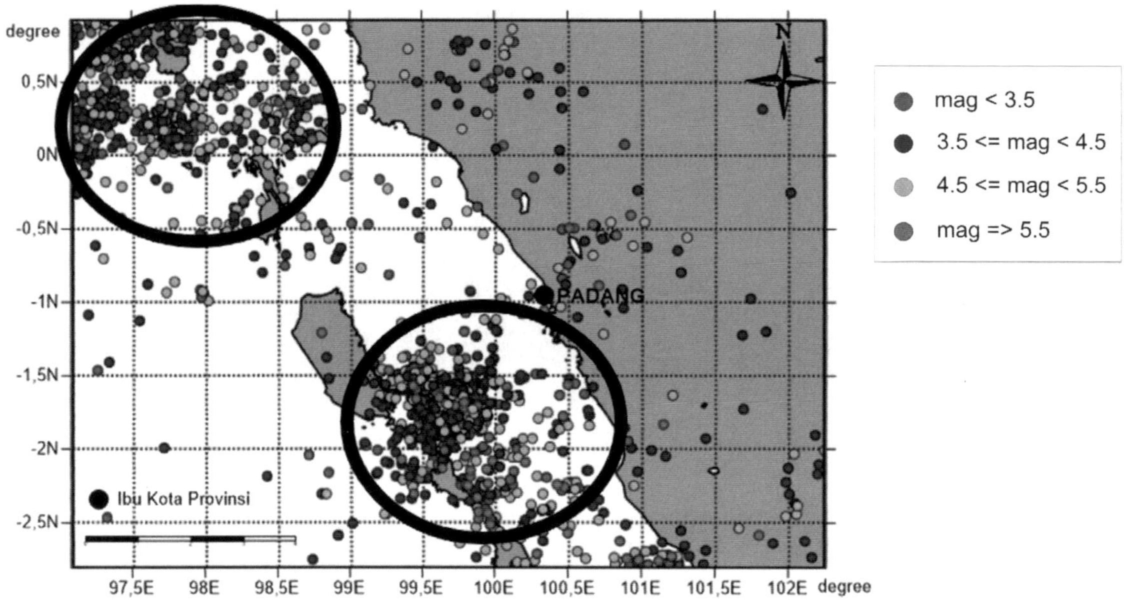

Figure 2. Seismicity Map of Padang and its Surrounding Area from 2002 to 2009

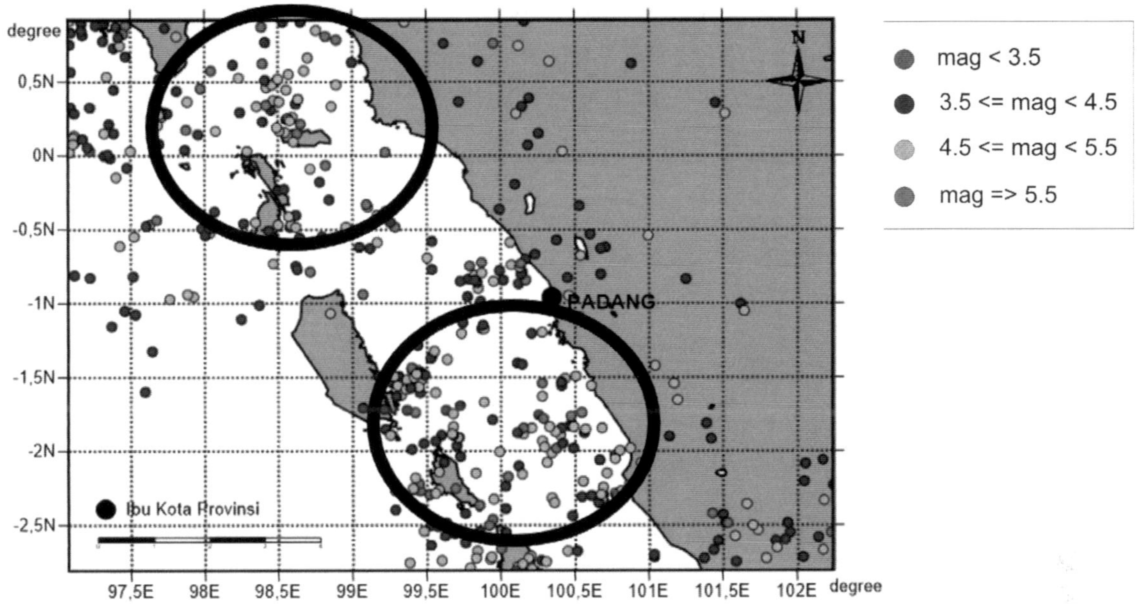

Figure 3. Seismicity Map of Padang and its Surrounding Area from 2009 to 2016

Figure 2 and figure 3 show the number of the earthquakes with the magnitude range 4.0 to 7.7 Ms. The distribution color from blue towards red indicates the higher magnitude of an earthquake. Figure 2 explains the distribution of seismicity as many as 1518 events in Padang and surrounding from 2002 until the day before the earthquake of 2009. The allocation of earthquakes centered in Mentawai Strait and Hindia Ocean. The seismicity for post-seismic shown by figure 3 in this period, USGS recorded 442 events from 2009 to 2016. Seven years before Padang earthquake, there is no seismicity gap like after the Padang earthquake that have a space in every earthquake event. Based on the comparison of seismicity before and after the earthquake, the seismic intensity for before Padang earthquake is higher than after Padang earthquake.

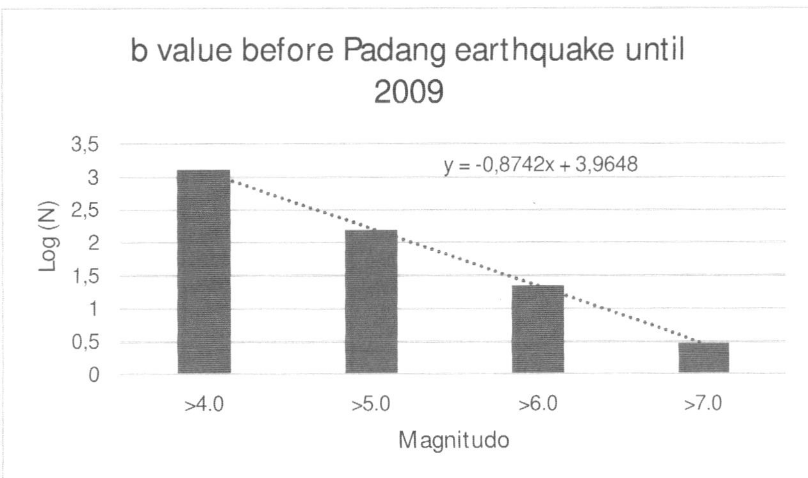

Figure 4. Graph of log N vs Magnitude Before Padang Earthquake 2009

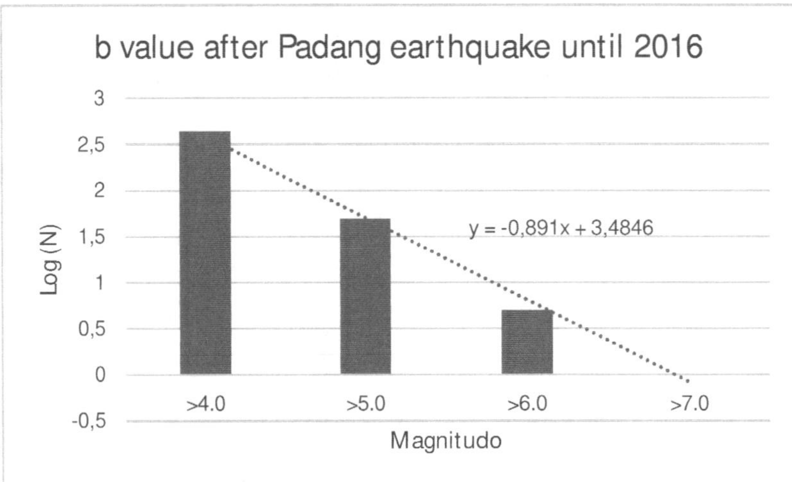

Figure 5. Graph of log N vs Magnitude After Padang Earthquake 2009

B value is also obtained from earthquakes catalog, b-value itself is the ratio between the number of the event with magnitude [7]. B value for pre-seismic and post-seismic is 0.874 and 0.891, where b value graph is shown in figure 4 and figure 5 as the gradient of the line.

B value of pre-seismic and post-seismic increase for 0.0168, the increasing value indicate releasing stress towards lower magnitude after the earthquake happens. It is characterized by the b value of pre-seismic lesser than post-seismic b-value [8]. It can also be seen from seismic intensity and distribution of higher magnitude after the earthquake 2009 is lower than before the earthquake 2009. Releasing stress over the ground characterized by minor earthquakes that occurred intensive few days after the mainshock which is the release of energy from the remnants of the previous large earthquake [17]

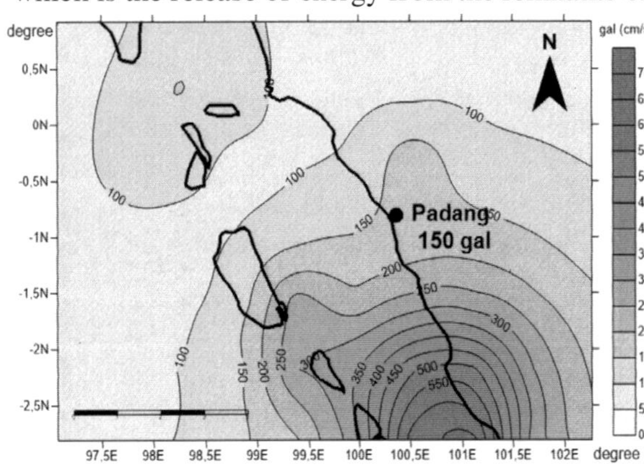

Figure 6. Maximum PGA values Map Before Padang Earthquake (2002-2009)

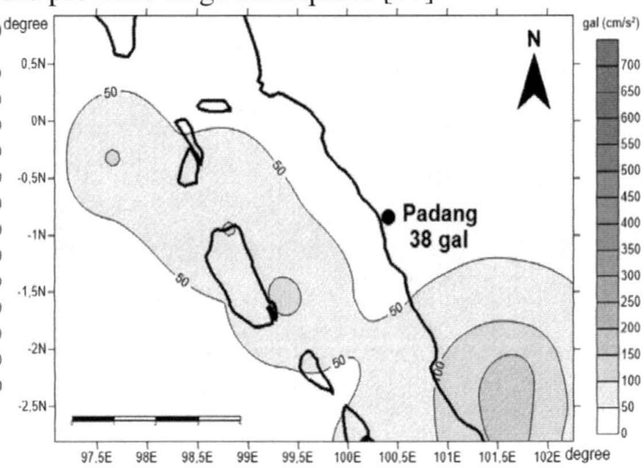

Figure 7. Maximum PGA values Map After Padang Earthquake (2009-2016)

PGA value obtained by calculation using mc. Guirre method with seismicity data from USGS either pre-seismic and post-seismic 2009. The maximum PGA value between seven years before and after Padang earthquake 2009 is different. The maximum PGA value in Padang and its surrounding before the mainshock is about 150 gals, while the maximum PGA value after the mainshock is about 38 gals in Padang and its surrounding. Increasing PGA value is caused by the difference a number of event for before and after Padang earthquake. The number of the phenomenon after the mainshock is less than before the mainshock. Also, distribution of magnitude with its depth also affects to PGA value as well.

According to National Disaster Management Agency (BNPB), an earthquake on September 30th 2009 caused many casualties and destroyed many facilities. Based on BNPB and Implementing Disaster Management Coordination Unit (Satkorlak PB) West Sumatra, the largest casualties and damage were in Padang city, Padang Pariaman city, and Agam regency. The earthquake killed more than 800 people, and more than 2000 facilities were destroyed.

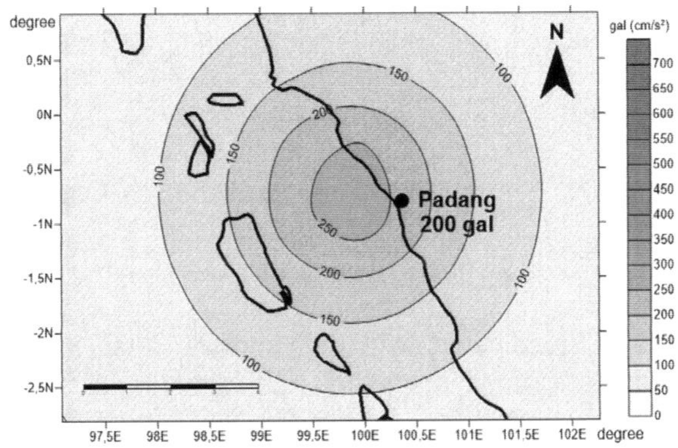

Figure 8. Peak Ground Acceleration Map After Padang Earthquake (2009-2016)

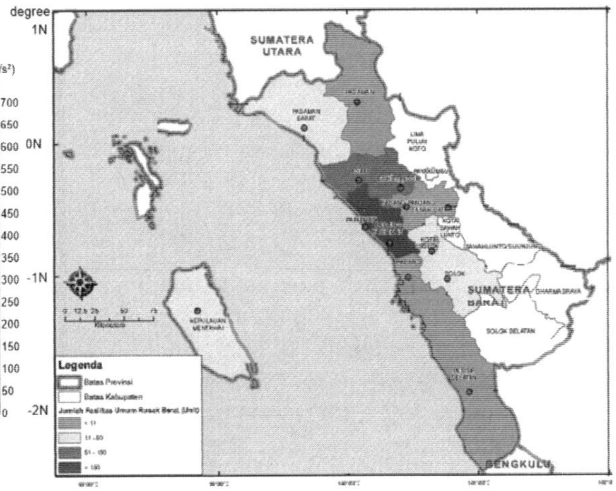

Figure 9. Casualties Map of Padang Earthquake 2009

Based on the colateral damage map by BNPB and the maximum PGA value at the day the mainshock has similarity area for high impact of the mainshock that indicates by red color in figure 9 and highest PGA value in figure 9 is in Padang and surrounding include Padang, Padang Pariaman, and Agam. So, we can conclude that when the PGA value is below to 200 gal such as maximum PGA value before and after the earthquake in 150 gals and 38 gals, it doesn't cause the significant damage or casualties, but when the PGA is about 200 gal or more like PGA value at the day the mainshock, it can cause significant effect like damaging many facilities even kill peoples. So, PGA value had relation with casualties and damaged by considering magnitude, the distance of epicenter and site effect [15]. According to Tabel 1 (BMKG, 2016) maximum PGA value at the day the mainshock in Padang and surrounding is 200 gals has risk level 'Moderate Damage' when the earthquake happens.

7. Conclusion

Based on seismicity data from USGS, had known the comparison of the b value for seven years before and after Padang earthquake by Gutenberg-Richter equation is 0.874 and 0.894. Increasing b-value indicates releasing stress towards lower magnitude.

Mc. Guirre method applied accordingly to calculate the peak ground acceleration (PGA) in Padang city and its surrounding. The maximum PGA value before the mainshock is about 150 gals, while after the mainshock is about 38 gals, and at the day the mainshock is about 200 gal, even more were located in Padang and its surrounding.

Most deaths people and damage were in Padang, Padang Pariaman, and Agam. The casualties data by BNPB and the maximum PGA value at the day 2009 have the similarity area for high impact of the mainshock and highest PGA value. The maximum PGA value of Padang is about 200 gals, even more were located about in Padang, Padang Pariaman, and Agam too. The PGA value had relation with casualties and damaged by considering magnitude, the distance of epicenter and site effect [15]. According to Tabel 1 (BMKG, 2016), the maximum PGA value at the day the mainshock in Padang and surrounding indicate 'Moderate Damage' when the earthquake happens.

8. Acknowledgments

We would like to thank Ikatan Alumni Geofisika (IAGF), Direktorat Kemahasiswaan (DIRMAWA) Gadjah Mada University for the financial support, Prof. Dr. Kirbani Sri Brotopuspito, Dr.rer.nat. Wiwit Suryanto M.Si,Theodosius Marwan Irnaka, S.Si., M.Sc as the Lecturers at Geophysics subdepartment of Gadjah Mada University, Ekrar Winarta as the staff of Study Energy Center UGM, Hastin C. Diantari as the staff Geoseismal Laboratory, and Anang Sahroni as the assistant of Computational Method.

References

[1] H Th Verstappen 2010 Indonesian Landforms and Plate Tectonics *Jurnal Geologi Indonesia*, Vol. 5 No. 3 September 2010: 197-207

[2] Paul Denthon 2000-2006 British Geological Survey, Natural Environment Research Council *www.bgs.ac.uk*

[3] Badan Nasional Penanggulangan Bencana (BNPB) 2009 Laporan Harian PUSDALOPS BNPB *www.bnbp.go.id* Jakarta, Indonesia

[4] Suleyman S. Nalbant, John McCloskey, Sandy Steacy and Aykut A Barka 2002 Stress Accumulation and Increased Seismic Risk in Eastern Turkey *Earth and Planetary Science Letters* 195 (2002) 291^298

[5] Kirbani Sri Brotopuspito and Wahyudi 2011 Eruption of The Kelud Volcano and b-Value of Its Surrounding Earthquakes, *Geophysics Subdepartment Gadjah Mada University* Yogyakarta, Indonesia

[6] E Irwansyah, E Winarko, Z E Rasjid, and R D Bekti 2013 Earthquake Hazard Zonation Using Peak Ground Acceleration (PGA) Approach *IOP Publishing Journal of Physics* : Conference Series 423 (2013) 012067, doi:10.1088/1742-6596/423/1/012067

[7] Guternberg, B and Richter, C. F. 1954 Seismicity of the Earth *Princeton University Press*, p 440

[8] Scholz C H 1968 The frequency-magnitude relation of microfracturing in rock and its relation to earthquakes *Bull. Seismol. Soc. Am* 58, 399-415

[9] C A Serino, K F Tiampo, and W Klein 2011 New Approach to Gutenberg-Richter Scaling *Physical Review Letters* 106, 108501

[10] B. Enescu, D. Enescu, and K ITO 2009 Values of b and p: Their Variations and Relation to Physical Process For Earthquakes in Japan *Rom. Journ. Phys* Vol 56

[11] C Lomnitz and E Rosenblueth 1976 Seismic Risk *Elsevier Scientific Publishing Company*, ISBN 0-444-41494-0

[12] Waluyo 2012 Seismology *Geophysics Subdepartment of Gadjah Mada University* Yogyakarta, Indonesia

[13] Borman P 2003 New Manual of Seismological Observasy Practise *Geoforschung Zentrum Potsdam* Vol. 1 Netherland

[14] Marjiono, Soehaimi, and Setyawan J H 2007 Mikrozonasi Daerah Kendari dan Sekitarnya Berdasarkan Respon Tanah Setempat *Pusat Survey Geologi* Bandung, Indonesia

[15] Mashyur Irsyam *et al* 2010 Ringkasan Hasil Studi Tim Revisi Pata Gempa Indonesia *BNPB* Indonesia

[16] Kirbani S B, Tiar Prasetya, F M Widigdo 2006 Percepatan Getaran Tanah Maksimum Daerah Istimewa Yogyakarta 1943-2006 *Jurnal HAGI* Vol. 7 pp 19-22

[17] Kirbani S B 2013 Study on Time Series of Energy Released by Global and Indonesian Tectonic Earthquakes in 1973-2013 *Presented at HFI SFN XXVI and ICTAP III* October 2013, Malang, Indonesia

[18] Mark Petersen, Stephen Harmsen, Charles Mueller, Kathleen Haller, James Dewey, Nicolas Luco, Anthony Crone, David Lidke, and Kenneth Rukstales 2007 Documentation for the Southeast Asia Seismic HazardMaps *U.S. Department of the Interior U.S. Geological Survey*

[19] USGS 2016 Catalog Earthquake *http.usgs.gov*

[20] M. B. Sørensen, K. Atakan, and N. Pulido 2007 Simulated Strong Ground Motions for the Great *M* 9.3 Sumatra–Andaman Earthquake of 26 December 2004 *Bulletin of the Seismological Society of America*, Vol. 97, No. 1A, pp. S139–S151, doi: 10.1785/0120050608

Precise Hypocenter Determination around Palu Koro Fault: a Preliminary Results

Muhammad Fawzy Ismullah M.[1], Andri Dian Nugraha[2], Mohamad Ramdhan[3,4], Wandono[4]

[1]Master Program of Geophysical Engineering, Faculty of Mining and Petroleum Engineering, Institut Teknologi Bandung, Ganesha 10, Bandung 40132, Indonesia
[2]Global Geophysics Research Group, Faculty of Mining and Petroleum Engineering, Institut Teknologi Bandung, Ganesha 10, Bandung 40132, Indonesia
[3]Doctorate Program of Earth Science, Faculty of Earth Sciences and Technology, Institut Teknologi Bandung, Ganesha 10, Bandung 40132, Indonesia
[4]Agency for Meteorology, Climatology, and Geophysics (BMKG), Jakarta, Indonesia

E-mail: mallaniung@gmail.com

Abstract. Sulawesi area is located in complex tectonic pattern. High seismicity activity in the middle of Sulawesi is related to Palu Koro fault (PKF). In this study, we determined precise hypocenter around PKF by applying double-difference method. We attempt to investigate of the seismicity rate, geometry of the fault and distribution of focus depth around PKF. We first re-pick P-and S-wave arrival time of the PKF events to determine the initial hypocenter location using Hypoellipse method through updated 1-D seismic velocity. Later on, we relocated the earthquake event using double-difference method. Our preliminary results show the distribution of relocated events are located around PKF and have smaller residual time than the initial location. We will enhance the hypocenter location through updating of arrival time by applying waveform cross correlation method as input for double-difference relocation.

1. Introduction

Sulawesi area is located in plate tectonic interaction like Eurasia, Indo-Australia, Pacific, Philippine Sea and Sundaland [1]. That condition made complex tectonic pattern which signed by seismicity and volcanoes activity. High seismicity activity in the middle of Sulawesi is related to Palu Koro fault (PKF).

The first step to study about seismicity and his relation to fault is hypocenter determination. Hypocenter required to study structure, tectonic regime and processes that trigger seismic activity. Hypocenter accuracy depend on factors, one of them is seismic wave velocity structure.

Many techniques had developed to determine hypocenter become precisely. One of them is double difference method [2]. This method based on fact that if the hypocenter separation between two hypocenter is small enough recorded at the same station compared to event – station distance then the ray path can be considered identical along their length. The double-difference methods have been successfully aplied to determine precise hypocenter around active fault and subduction zone in Indonesia region [3,4,5,6,7,8,9]

2. Methodology

In this research we used 226 PKF events waveform from BMKG's network during January 2011 – December 2016. We re-picked P- and S-phase arrival time from PKF events waveform to determine initial hypocenter. Initial hypocenter determined by Hypoellipse method [10]. Output from Hypoellipse consist of origin time, hypocenter, *RMS*, observed arrival time, calculated arrival time and residual time.

1-D velocity model used as initial model is iasp91 1-D velocity model [11]. That 1-D velocity model updated by coupled velocity – hypocenter method [12] on Velest [13]. Generally, updated 1-D velocity model higher than initial 1-D velocity model, see **Table 1**.

Table 1. Comparison between Initial and Updated 1-D Velocity Model.

Depth (Km)	Initial Velocity Model (Km/s)		Updated Velocity Model (Km/s)	
	P wave	S wave	P wave	S wave
20	5.8	3.36	5.99	3.46
30	6.5	3.75	6.17	3.58
35	8.04	4.47	8.08	4.55
71	8.044	4.48	8.08	4.55

Initial hypocenter and updated 1-D velocity model were used to relocate PKF event. RKF event relocated using double different method. Double different method based on fact that if the hypocenter separation between two hypocenter is small enough recorded at the same station compared to event – station distance then the ray path can be considered identical along their length. It means that the difference of travel time between two events recorded at the same station may be attributed to the differences in spatial separation. Double different applied on HypoDD software [14]. HypoDD required catalog travel time data, waveform cross correlation data and both of them. We used catalog travel time data to our preliminary results.

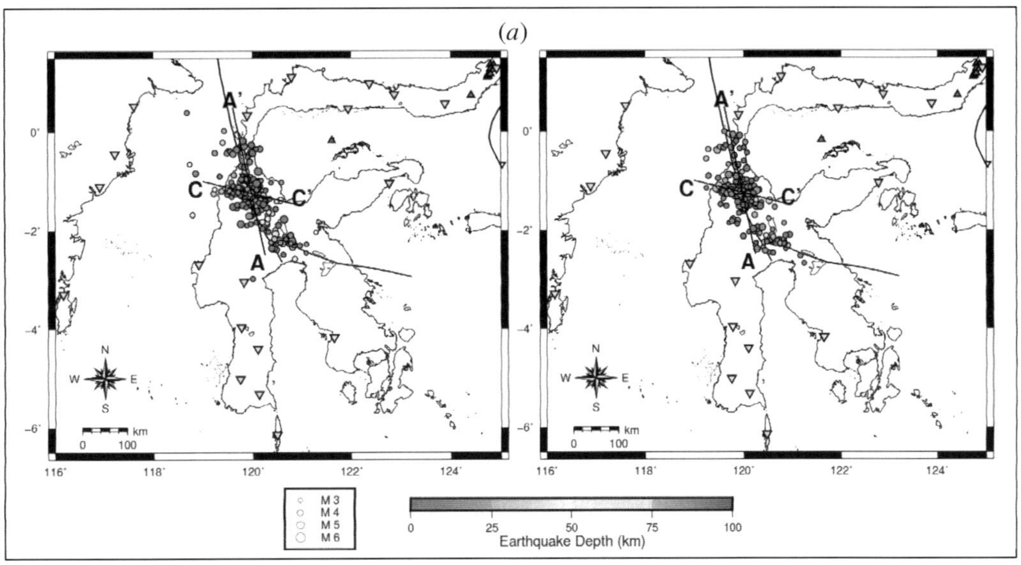

Figure 1. (be continued in next page)

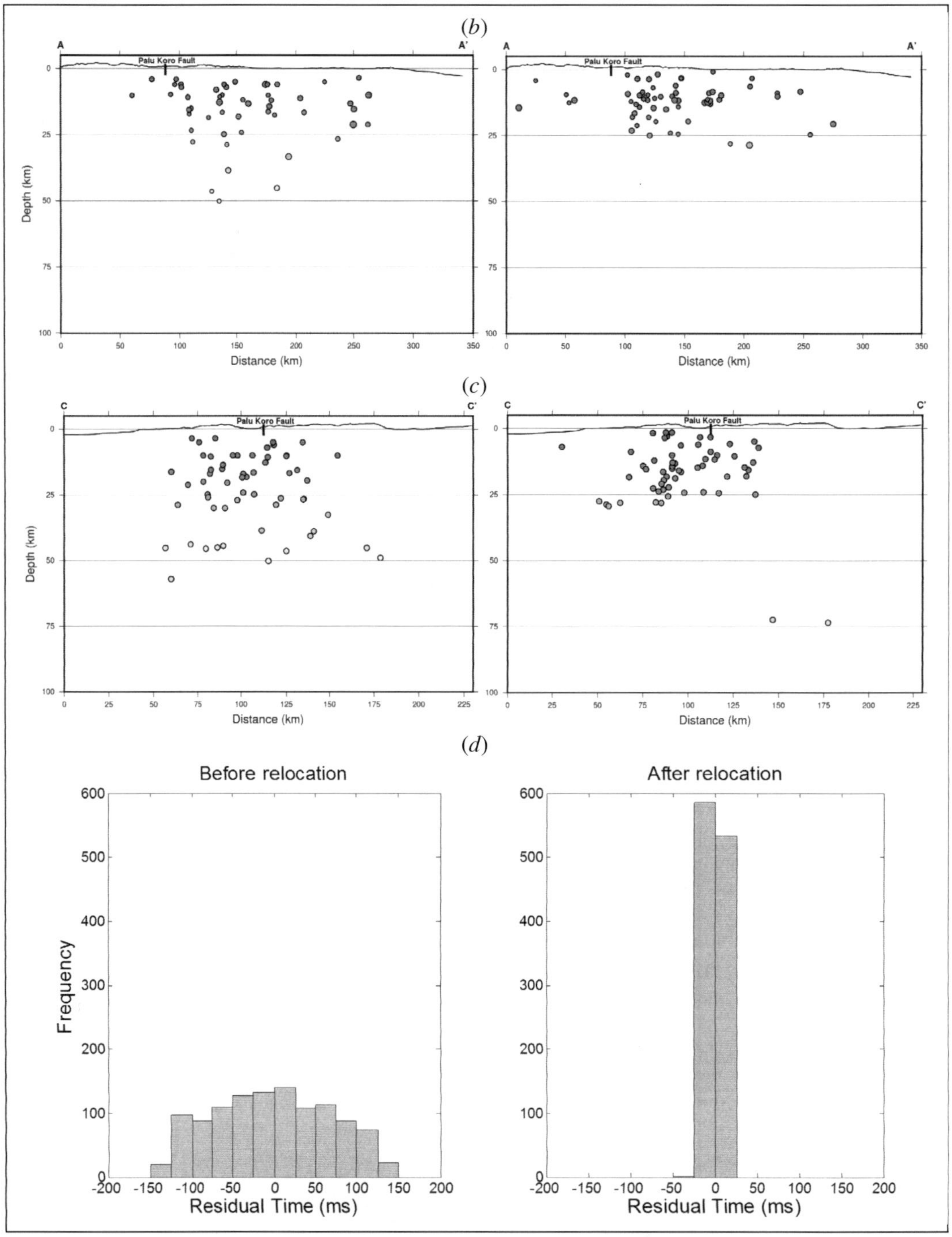

Figure 1. Comparison results from Hypoellipse method (left) and double difference method (right). Part (*a*) show epicenter location. Circle dot is epicenter, blue inverted triangle is stations, red triangle is volcanoes, black line is fault and black line with triangle is subduction. Part (*b*) and (*c*) shows hypocenter location on A-A' and C-C' section from part (*a*). Part (*d*) show comparison double difference residual time.

3. Results

We relocated 226 PKF events from initial hypocenter by HypoDD. We evaluated the relocated hypocenter according to visual and statistic. Visually based on hypocenter distribution and statistically based on comparison their double difference residual time.

In Fig. 1, we can look that the relocated epicenter is closer and agree with PKF than the initial epicenter. The outliers on the relocated epicenter is less than the initial epicenter. Average movement from the initial epicenter is 88.714 Km. In section, average movement of hypocenter is up to 3.79 Km. Average depth from relocated hypocenter is 14.37 Km, more shallow than 18.17 Km from initial hypocenter. Generally, relocated hypocenter is more shallow than initial hypocenter, but not for all event.

Double difference residual time comparison between before and after relocation showed also in Fig. 1. Residual time is a difference between observed and calculated travel time. The quality of model can determine by residual time. The good quality signed by residual time value collected around 0 (zero) and on the contrary. The double difference residual time form initial hypocenter's around -134.249 until 134.491 ms with his average is -2.3568 ms. The maximum frequency of residual time around 0 (zero). The double difference residual time from relocated hypocenter is smaller than the initial hypocenter. The average double difference residual time of relocated hypocenter is 0.1929 ms and around -25.91 until 22.568 ms.

4. Conclusions

Our preliminary result that the hypocenter determination used double difference method showed hypocenter location more precisely. That is signed by the distribution of relocated hypocenter are located around PKF and have smaller residual time than the initial relocation. We don't interpret our preliminary result about seismicity and tectonic regime because there are next step to get hypocenter precisely. After used the catalog travel time data, we are going to use waveform cross correlation data and both of them to improve hypocenter determination. We consider, the relocation hypocenter obtained in this study will be usefull as input for advance seismological, seismic hazard mitigation in PKF and surround it.

5. Acknowledgements

The authors gratefully thanks the use of earthquake waveform and catalog data form BMKG, Indonesia. Near surface geophysics lab., Faculty of Mining and Petroleum Engineering, Institut Teknologi Bandung for supporting this study. *Beasiswa Unggulan 2014* from Ministry of Research, Technology and Higher Education Indonesia which have funded master scholarship program at Geophysical Engineering, Faculty of Mining and Petroleum Engineering, Institut Teknologi Bandung.

References

[1] DeMets C, Gordon R G, Argus D F 2010 *Geophys. J. Int.* **181** 1–80
[2] Waldhauser F and Ellsworth W L 2000 *Bull. Seismol. Soc. Am.* **90** 1353 – 1368.
[3] Shiddiqi H A, Widiyantoro S, Nugraha A D, Ramdhan M, Wiyono S H, Wandono W 2016 *AIP. Conf. Proc.* **1730** 020010 DOI:10.1063/1.4947378
[4] Sabtaji A, Nugraha A D 2015 *AIP. Conf. Proc.***1658** 030005 DOI: 10.1063/1.4915013
[5] Aprilia Puspita C, Nugraha A D, Puspito N T 2015 *AIP. Conf. Proc.* **1658** 030021 DOI: 10.1063/1.4915029
[6] Utama M R J, Nugraha A D, Puspito N T 2015 *AIP. Conf. Proc.* **1658** 030022 DOI: 10.1063/1.4915030.
[7] Shiddiqi H A, Widiyantoro S, Nugraha A D, Ramdhan M, Wandono, Sutiyono, Handayani T, Nugroho H 2015 *AIP. Conf. Proc.***1658** 030011 DOI: 10.1063/1.4915019.
[8] Nugraha A D, Shiddiqi H A, Widiyantoro S, Ramdhan M, Wandono, Sutiyono, Handayani T, Nugroho H 2015 *AIP. Conf. Proc.* **1658** 030002 DOI: 10.1063/1.4915010.
[9] Ramdhan M, Nugraha A D 2013 *AIP. Conf. Proc.* **1554** 242-244 DOI: 10.1063/1.4820330.

[10] Lahr J C 1989 HYPOELLIPSE/Version 2.0: A computer program for determining local earthquakes hypocentral parameters, magnitude, and first-motion pattern: U.S. Geological Survey *Open-File Report* 89-116. http://pubs.usgs.gov/of/1989/0116/report.pdf

[11] Kennett B L N and Engdahl E R 1991 *Geophys. J. Int.* **105** 429-465

[12] Kissling E, Ellsworth W L, Eberhart-Phillips D, Kradolfer U 1994 *J. Geophys. Res* **99** 19635-19646

[13] Kissling E 1995 Program VELEST USER'S GUIDE – *Short Introduction* Institute of Geophysics ETH Zurich

[14] Waldhauser F 2001 HypoDD A Program to Compute Double-Difference Hypocenter Locations *U.S Geol. Surv. Open File Rep.* 01-113

Preliminary Result of The Influence of Earthquake Stress Change and The Implication for Soputan Volcano and Lokon Volcano

I G K S Bunaga[1,2] and M F Nugraha[2]

[1] Geophysical Station of Mataram, Jl. Adi Sucipto No. 10, Mataram City, Indonesia
[2] Meteorology, Climatology, and Geophysics Agency (MCGA), Jl. Angkasa 1 No. 2, Jakarta, Indonesia

E-mail: satriabunaga.bmkg@gmail.com

Abstract. On February 6, 2016, an eruption occurred on the Northern Sulawesi arm, particularly Soputan volcano. One day earlier, Lokon volcano located close to Soputan volcano was decreased its status from standby to alert level by the Center for Volcanology and Geological Hazard Mitigation (CVGHM). The different reactions of two volcanoes proposed the question why the increment activity just happened in Soputan volcano. This uniqueness made us to suggest that static stress of earthquake may control the magmatic systems. We investigate here the earthquake-volcanism interaction through static stress changes by using Coulomb failure stress associated with an earthquake occurred in the Northern Molluca Sea on 25 November 2015. We slice the same dip for the each region in vertical cross sections. Therefore, the Coulomb failure stress pattern can be investigated beneath the study area. Our results suggest that Coulomb failure stress was increased by 0.3 $\times$ 10^{-3} to 0.4 $\times$ 10^{-3} bar below the Soputan's region. Lokon's region, the stress was reduced by -0.1 $\times$ 10^{-3} to -0.4 $\times$ 10^{-3} bar. The positive change may perturb magma overpressure leading to eruption and promoted volcanic earthquakes. The situation was very different that Lokon volcano ran into reduction activity and volcanic earthquakes were discourage due to stress shadow. We show that the difference volcanic response were likely controlled by static stress of the earthquake.

1. Introduction

The Sulawesi island lies near the triple junction between of the Sunda, Australia, and Philippines plates [1,2,3,4]. The collision caused the subduction of the Molluca Sea Plate which has an inverted U shape subduction. The plate dive to the east of Halmahera arc and to the west of Sangihe arc [5]. One of the consequence of the activity is volcanic arcs in Northern Sulawesi Arm, such as Lokon volcano and Soputan volcano.

An unusual phenomenon between two volcanoes were occurred in Northern Sulawesi arm. According to national news, Soputan volcano and Lokon volcano have a difference response at almost the same time. The day before the eruption of Soputan volcano on 6 February 2016, Lokon volcano was decreased its status set by Center for Volcanology and Geological Hazard Mitigation (CVGHM) due to reduction activity.

Numerous studies were conducted to examine the link of some processes to volcanic eruptions. The activity are generally preceded by measurable indications, one of which is the influence of earthquake occurrence [6,7,8,9]. In this study, we investigate how this interaction works, between earthquake and volcano. We analyze the stress change caused by an earthquake which occurred close to the volcanoes before Soputan volcano's erupted. Based on these result, we point out that the earthquake may promote the difference responses of two volcanoes by analyzing Coulomb failure stress distribution.

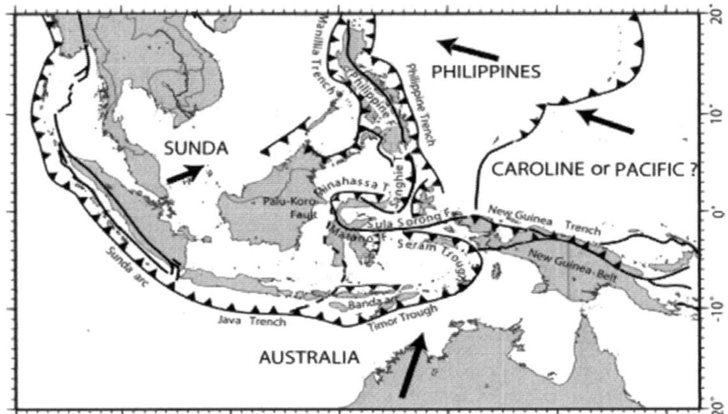

Figure 1. Regional tectonic features of Sunda-Australia-Philippine plates. The dark line indicate the fault line [1]

2. Data

To obtain stress change pattern on the study area, we utilized the Coulomb failure stress distribution. We used an earthquake data archived from the Global Harvard Centroid Moment Tensor (CMT Global Harvard) (figure 2 and table 1). Nodal plane II was used which is consistent with the tendency of the 25 November earthquake's aftershocks based on the mainshock stress change in which the determination already applied with the same method by previous workers [10].

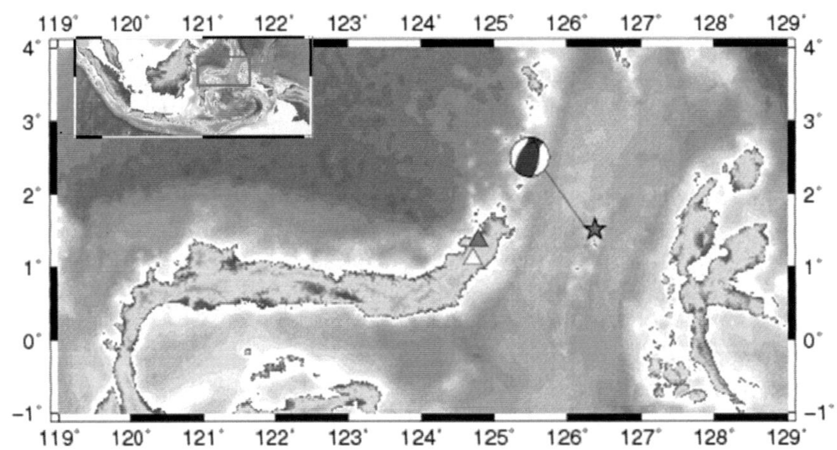

Figure 2. An earthquake in Northern Molluca Sea reported by Global CMT. Red star shows the epicenter of the 25 November 2016 (Mw 5.7) and focal mechanism (blue and white ball) with the fault plane (strike = 1°, dip = 52 °, rake = 64 °). Triangles indicate Soputan volcano (△) and Lokon volcano (▲). Map was created using the Generic Mapping Tool (GMT) [11].

Table 1 Focal mechanism of the 25 November 2015 earthquake

Mw	Nodal plane I (°)			Nodal plane II (°)			Source
	Strike	Dip	Rake	Strike	Dip	Rake	
5.7	219	45	119	1	52	64	Global CMT

We also used 3102 events which were obtained from Meteorology Climatology and Geophysics Agency (MCGA) from 2009 to 2016 to conduct tomographic inversion in Northern Sulawesi Arm area and its vicinity (figure 3). The events were recorded by 13 stations and AK 135 was used velocity model.

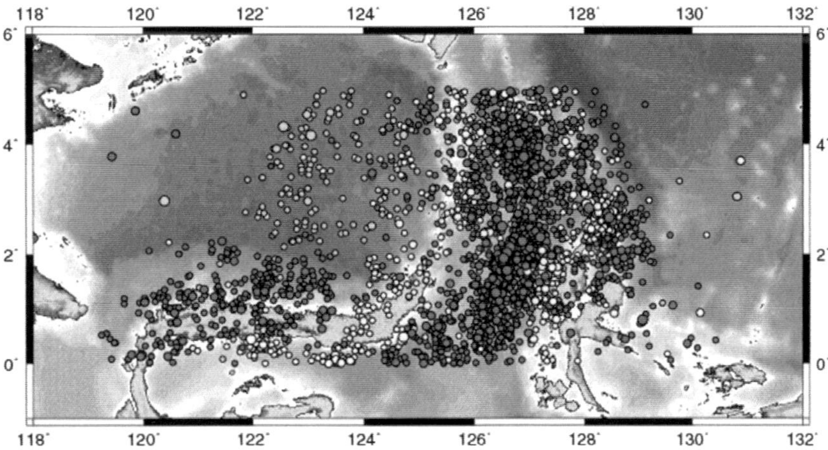

Figure 3. The seismicity between 2009 to 2016 of Northern Sulawesi Arm area and its vicinity. The colors of circle indicates the depth of event (red = 0-100 km, yellow = 100-300 km, and green > 300 km). Map was created using the Generic Mapping Tool (GMT) [11]

3. Theory and Methodology

3.1 Coulomb Failure Stress

We examined whether coulomb failure stress change of the earthquake related to the Soputan eruption and the Lokon volcano. In this equation, geometry information, slip direction, and coefficient of friction are utilized into stress change calculation using Coulomb 3.3 software [12,13]. The equation can be represented as:

$$\Delta CFF = \Delta\tau + \mu'\Delta\sigma \quad (1)$$

where ΔCFF is Coulomb stress increment, $\Delta\tau$ is the change in the shear related to the slip distribution (positive in slip direction), μ' is the apparent coefficient of friction correspond to the change of pore pressure,and $\Delta\sigma$ is the changes in the normal stress (positive when the fault is unclamped) [12] [14,15]. If $\Delta CFF > 0$, the stress are loaded into medium or fault plane and brought closer to failure. If $\Delta CFF < 0$, the fault plane is relaxed or stress shadow. It means the medium lost the stress after the first earthquake occurred [15]. The spatial pattern of stress changes is displayed into map with value of ΔCFF using the software.

3.2 Tomographic inversion

To reconstruct the medium or to determine the tectonic pattern beneath Northern Sulawesi Arm whether the existence of volcanoes are influenced by tectonic activity, here we applied delay time tomography algorithm. This method able to minimalize the difference between calculation time (Tcal) using ray tracing pseudo-bending method [16] and observation time (Tobs). The method is applied in a packet program which is simulps 12 [17].

4. Results and Discussion

To confirm the volcanoes existence are influenced by the Molluca Sea Plate, here we made seismic tomography (figure 4).

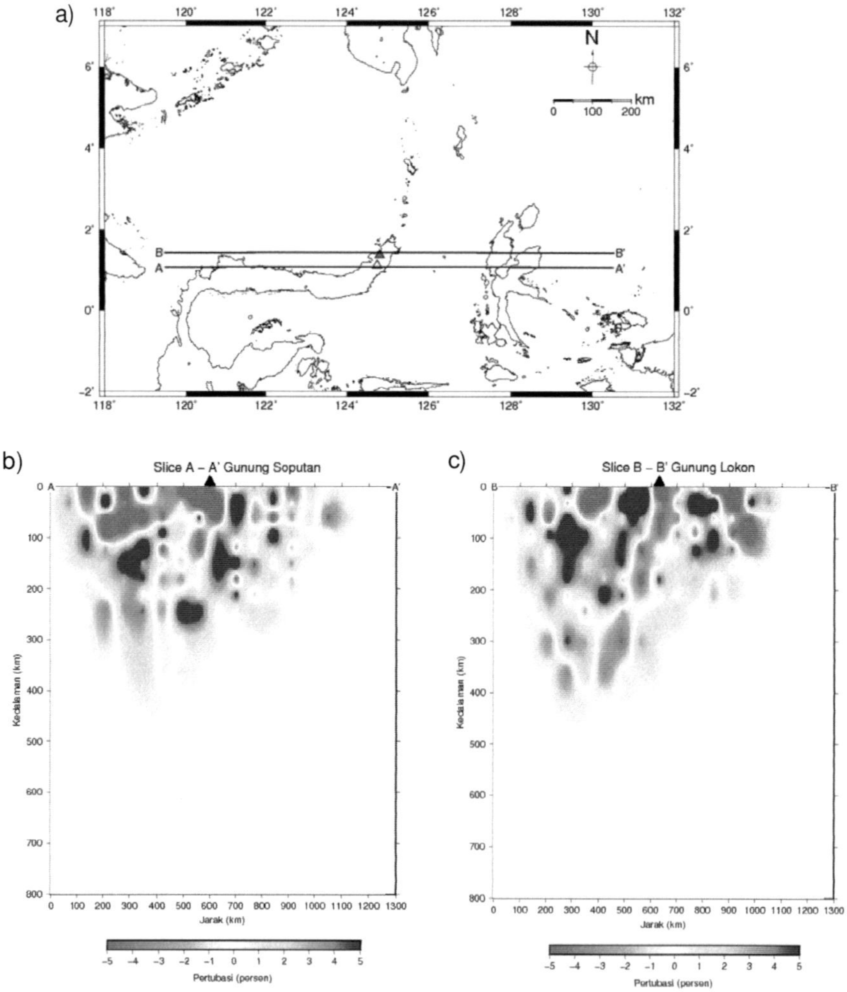

Figure 4. (a) The location of Soputan volcano and Lokon volcano, solid lines and triangles depict cross section line and the volcanoes, Soputan volcano (△) and Lokon volcano (▲). (b) and (c) The results of tomographic inversions below the volcanoes and its vicinity with contour scale of perturbation or anomaly value of seismic wave velocity in percentage between -5% to +5% correspond to AK 135 velocity model.

The tomograms for the structure below the volcanoes are shown in figure 4. According to figure 4, we consider that the opposing subduction of the Molluca Sea plate was defined by positive perturbation (blue color). On the other hand, negative perturbation dominated beneath volcanoes which indicated partial melting or magma chamber. So we infer that both volcanoes are located on subduction zone and the existence of the volcanoes are close ties of the tectonic process. Moreover, we presume that the earthquake occurrence which is one of effect of tectonic activity may influence the activity of those volcanoes.

To understand how the earthquake may affect the volcanoes condition, we calculated the stress change into maps. The map of stress distribution of 25 November 2015 earthquake shown in figure 3. Positive change ($\Delta CFF > 0$) and negative change ($\Delta CFF < 0$) in maps are represented with red and blue colors, respectively. The pattern of the Coulomb stress change increased in the SW-NE, while a lobe of stress reduction or stress shadow dominated in the SE-NW. The stress pattern were imaged laterally and vertically at a depth of 50 km (figure 5 and figure 6).

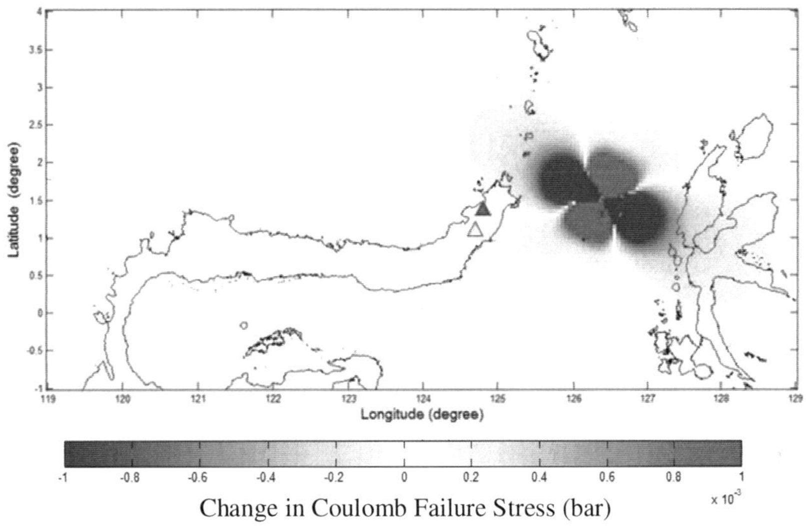

Figure 5. A map of Coulomb stress changes of the 25 November 2015 earthquake. Triangles indicate Soputan volcano (△) and Lokon volcano (▲).

Figure 6 shows that the positive stress (warm colors, 0.1×10^{-3} - 0.3×10^{-3} bar) slightly impacted the Soputan volvano's zone, but another was influenced slightly by the negative stress (blue colors, from -0.1×10^{-3} to -0.4×10^{-3} bar). The models show that the zone of yellow-red and blue was distributed through the cross-section. The positive stress which was distributed beneath Soputan Volcano at a depth $\geq$ 15 km may affect the magma chamber, changed the magma overpressure and triggered volcanic earthquakes where the magma chamber was approximately at > 8 km [18] and the possibility to impress the magma overpressure was close enough.

On the other hand, the reduction stress was dominant beneath Lokon volcano at a depth 0 km - 30 km which influenced the volcano's medium to be relax. Moreover, the magma chamber was detected at a depth 2 km – 3 km [19]. So, the magma chamber of Lokon may be undisturbed and the volcanic earthquakes occurrence may be discourage. These analyses resolve this unusual phenomenon and we consider that the enhancement and reduction in stress change able to control the volcanoes circumstance.

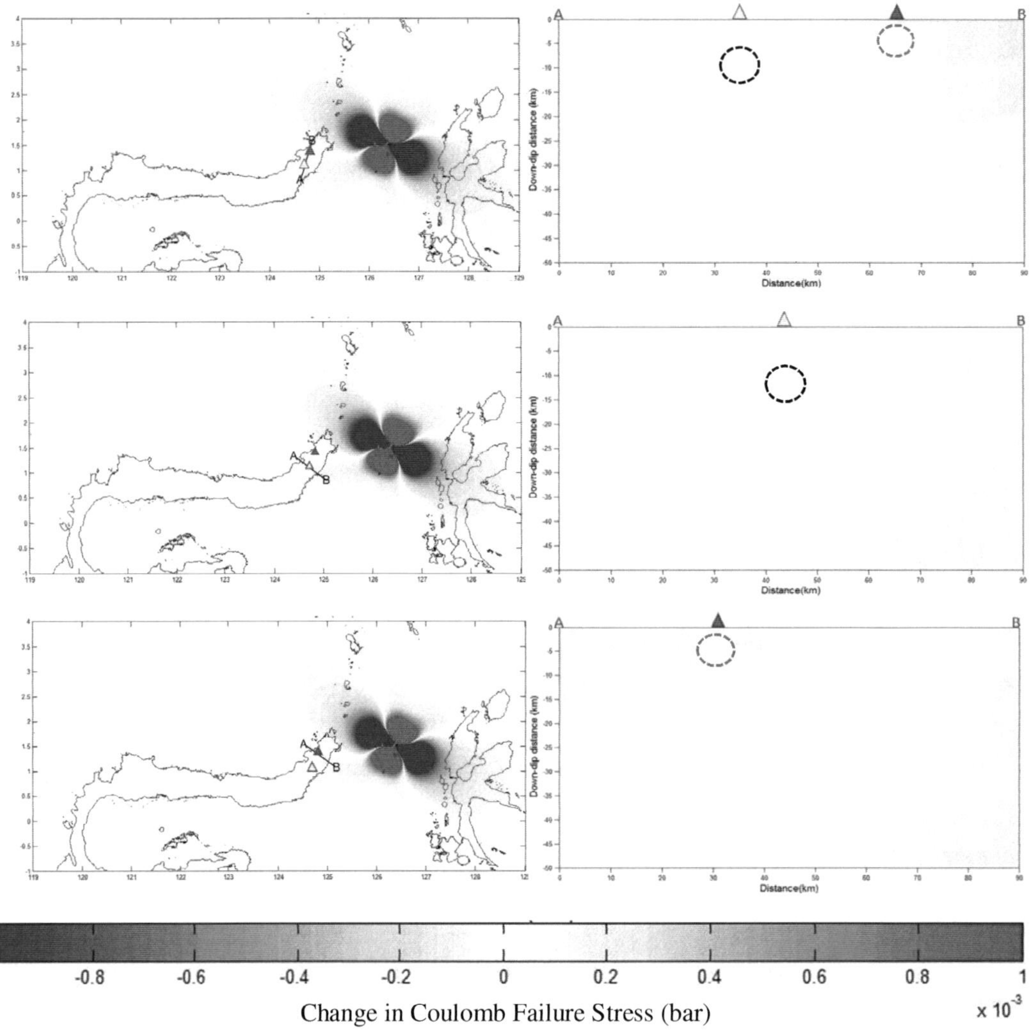

Figure 6. Change in Coulomb stress at profiles along the Soputan volcano and Lokon volcano triggered by Mw 5.7 earthquake. Triangles indicate Soputan volcano (△) and Lokon volcano (▲). The left shows the location of the volcanoes and slice lines (solid line). The right shows the cross-section along volcanoes lines at a depth of 50 km and the location magma chamber each volcano which is represented by dashes circle (black and red are the magma chamber of Soputan volcano and Lokon volcano, respectively).

5. Conclusions

The activity between Soputan volcano and Lokon volcano are coupled with distribution of Coulomb failure stress of the earthquake. Moreover, our study perhaps provide new information for controlling volcanic activity from this unusual phenomenon and this study will be clearly seen if several studies are combined to figure it out how this phenomenon works.

IOP Conf. Series: Earth and Environmental Science **62** (2017) 012057 doi:10.1088/1755-1315/62/1/012057

References

[1] Socquet A, Simons W, Vigny C, McCaffrey R, Subarya C, Sarsito D, Ambrosius B and Spakman W 2006 *J. Geophys. Res.* **111** B08409
[2] Silver E A, McCaffrey R and Smith R B 1983 *J. Geophys. Res.* **88** 9407-9418
[3] Hall R and Wilson M E J 2000 *J. Asian Earth Sci.* **18** 781-808
[4] Gómez J M, Madariaga R, Walpersdorf A and Chalard E 2000 *Bull. Seis. Soc. Am.* **90** 739-751
[5] McCaffery R, Silver E A and Raitt R W 1980 *Am. Geophys. Union.* **23** 161-177
[6] Walter T R and Amelung F 2006 *J. Geophys. Res.* **111** B05204
[7] Feuillet N, Cocco M, Musumeci C and Nostro C 2006 *Geophys. J. Int.* **164** 697-718
[8] Manga M and Brodsky E 2006 *Annu. Rev. Earth Planet Sci.* **34** 263-291
[9] Watt S F L, Pyle D M and Mather T A 2009 *Earth and Planetary Sci. Let.* **277** 399-407
[10] Bunaga I G K S, Suardi I and Deviyanti N E 2016 *IOP Conf. Series: Earth and Env. Sci.* **29** 012017
[11] Wessel P and Smith W H F 1998 *Eos Trans. Am. Geophys. Union* **79** 579
[12] Lin J and Stein R S 2004 *J. Geophys. Res.* **109** B02303
[13] Toda S, Stein R S, Dinger K R and Bozkurt B 2005 *J. Geophys. Res.* **110** B05S16
[14] King G C P, Stein R S and Lin J 1994 *B. Seis. Soc. Am.* **84** 935-953
[15] Harris R A 1998 *J. Geophys. Res.* **103** 24347-24358
[16] Um J and Thurber C 1987 *Bull. Seis. Soc. Am.* **77** 972-986
[17] Evans J R, Eberhart P D and Thurber C H 1994 *User's Manual For SIMULPS12 For Imaging* $_{vp}$ *and* $_{vp/vs}$*: A Derivative Of The "Thurber" Tomographic Inversion SIMUL3 For Local Earthquakes and Explosions* (America: United State Geological Survey) 94-431
[18] Dawid S, Ferdy and Pasau G 2015 *Jurnal Ilmiah Sains* **15** 88-93
[19] Anthe S, Pasau G and Tanaunama A 2015 *Jurnal Ilmiah Sains* **15** 27-32

Institute of Physics
Dirac House, Temple Back
Bristol BS1 6BE UK

ISSN: 1755-1307
ISBN 978-1-5108-3958-8

CURRAN ASSOCIATES INC.
proceedings.com